DRILL HALL LIBRARY
MEDWAY

Landslide Risk Assessment

Landslide Risk Assessment

Second edition

E. Mark Lee and David K.C. Jones

Published by ICE Publishing, One Great George Street, Westminster, London SW1P 3AA.

Full details of ICE Publishing sales representatives and distributors can be found at: www.icevirtuallibrary.com/info/printbooksales

First published 2004
This second edition 2014

Other titles by ICE Publishing:

Shock Transmission Units in Construction.
D.J. Patel. ISBN 978-0-7277-5713-5
Environmental Geotechnics, Second edition.
R.W. Sarsby. ISBN 978-0-7277-4187-5
Effective Site Investigation. Site Investigation in Construction Series Second edition.
Site Investigation Steering Group. ISBN 978-0-7277-3505-8

www.icevirtuallibrary.com

A catalogue record for this book is available from the British Library

ISBN 978-0-7277-5801-9

© Thomas Telford Limited 2014

ICE Publishing is a division of Thomas Telford Ltd, a wholly-owned subsidiary of the Institution of Civil Engineers (ICE).

All rights, including translation, reserved. Except as permitted by the Copyright, Designs and Patents Act 1988, no part of this publication may be reproduced, stored in a retrieval system or transmitted in any form or by any means, electronic, mechanical, photocopying or otherwise, without the prior written permission of the Publisher, ICE Publishing, One Great George Street, Westminster, London SW1P 3AA.

This book is published on the understanding that the author is solely responsible for the statements made and opinions expressed in it and that its publication does not necessarily imply that such statements and/or opinions are or reflect the views or opinions of the publishers. Whilst every effort has been made to ensure that the statements made and the opinions expressed in this publication provide a safe and accurate guide, no liability or responsibility can be accepted in this respect by the author or publishers.

Whilst every reasonable effort has been undertaken by the author and the publisher to acknowledge copyright on material reproduced, if there has been an oversight please contact the publisher and we will endeavour to correct this upon a reprint.

Associate Commissioning Editor: Jennifer Saines
Production Editors: Imran Mirza and Vikarn Chowdhary
Market Development Executive: Catherine de Gatacre

Typeset by Academic + Technical, Bristol
Index created by Pauline Davies
Printed and bound by CPI Group (UK) Ltd, Croydon CR0 4YY

Contents

	Preface to the first edition	ix
	Preface to the second edition	xi
	Scientific notation	xiii

01 — Background to landslide risk assessment — 1

1.1.	Introduction	1
1.2.	What is risk?	1
1.3.	A view of risk	4
1.4.	Black Swans and Perfect Storms	6
1.5.	What is landslide risk?	7
1.6.	Hazard and vulnerability	8
1.7.	From hazard to risk	12
1.8.	Categories of risk	13
1.9.	Why do landslide risk assessment?	14
	References	17

02 — The basic elements of landslide risk assessment — 21

2.1.	Introduction: the risk assessment process	21
2.2.	Description of intention	23
2.3.	Hazard assessment	23
2.4.	Consequence assessment	27
2.5.	Risk estimation	31
2.6.	Risk assessment and its relationship to risk evaluation and risk management	33
2.7.	Risk perception and risk communication	35
2.8.	Landslide risk assessment procedures	40
2.9.	Risk assessment as a decision-making tool	41
2.10.	Structure of the book	42
	References	44

03 — Landslide hazard — 47

3.1.	Introduction	47
3.2.	Landslide mechanisms and type	49
3.3.	Landslide behaviour	52
3.4.	Potential for landsliding	56
3.5.	Landslide velocity	58
3.6.	Landslide travel distance	60
3.7.	Nature of landslide hazards	63
3.8.	Landslide intensity	74
3.9.	Landslide susceptibility and hazard zoning	79
3.10.	Hazard models	83
3.11.	Uncertainty, assurance and defensibility	84
	References	88

04 — Qualitative and semi-quantitative risk assessment — 97

4.1.	Introduction	97
4.2.	Risk registers	100
4.3.	Relative risk scoring	102
4.4.	Risk ranking matrices	109
4.5.	The FMECA approach	115

		4.6.	Qualitative risk assessment: an easy option?	124
		References		126
05		**Introduction to probability and quantitative assessment**		**129**
		5.1.	Uncertainty and probability	129
		5.2.	Probability distributions	143
		5.3.	Judgement and the use of experts	171
		5.4.	Probability assessment and reliability methods	181
		References		197
06		**Estimating the probability of landsliding**		**205**
		6.1.	Introduction	205
		6.2.	Approaches to estimating landslide probability	207
		6.3.	Statistical methods: the use of incident databases	209
		6.4.	Statistical methods: historical frequency assessment	212
		6.5.	Statistical methods: the use of landslide magnitude–frequency curves	216
		6.6.	Statistical methods: estimating landslide exceedence probability	225
		6.7.	Statistical methods: estimating probability of cliff recession through simulation models	229
		6.8.	Conceptual models: normalisation of base-rate frequency	236
		6.9.	Conceptual models: estimating probability from landslide-triggering events	243
		6.10.	Conceptual models: estimating probability through expert judgement	252
		6.11.	Reliability methods: estimating probability through use of probabilistic stability analysis	267
		6.12.	Estimating probability: precision or pragmatism?	272
		References		273
07		**Exposure**		**281**
		7.1.	Introduction	281
		7.2.	Spatial probability: stationary assets	282
		7.3.	Spatial and temporal probability: non-stationary assets	286
		7.4.	Occupancy and population models	292
		7.5.	Reducing exposure through landslide detection	294
		References		296
08		**Vulnerability**		**297**
		8.1.	Introduction: multiple meanings of vulnerability	297
		8.2.	Physical vulnerability	299
		8.3.	Physical vulnerability of buildings and infrastructure	300
		8.4.	Human vulnerability	307
		8.5.	Societal and social vulnerability	311
		References		316
09		**Estimating the consequences**		**321**
		9.1.	Introduction	321
		9.2.	Elements at risk	324
		9.3.	Using the historical record	324
		9.4.	Categories of adverse consequences	329

		9.5.	Loss of life and injury	330
		9.6.	Direct impact on buildings, structures and infra-structure (direct economic costs/losses)	334
		9.7.	Other economic costs that may be estimated in advance	336
		9.8.	Intangible losses	337
		9.9.	Uncertain consequences	339
		9.10.	Consequence models	339
		9.11.	Multiple-outcome consequence models	359
		9.12.	Complex outcomes and uncertain futures	367
			References	369
10		**Quantifying risk**		**375**
		10.1.	Introduction	375
		10.2.	Current annual risk	376
		10.3.	Cliff recession risk	381
		10.4.	Comparing the risks associated with different management options	388
		10.5.	Individual risk	396
		10.6.	Societal risk	409
		10.7.	Statistics are signs from God?	416
			References	418
11		**From risk estimation to landslide management strategy**		**421**
		11.1.	Introduction to landslide risk management	421
		11.2.	Assessment criteria	425
		11.3.	Risk acceptance criteria: legal frameworks	426
		11.4.	Acceptable or tolerable risks: the ALARP principle in the UK	427
		11.5.	Individual risk criteria	429
		11.6.	Societal risk criteria	431
		11.7.	Corporate risk management: major accident risk criteria	435
		11.8.	Applying the ALARP principle: loss of life	437
		11.9.	Applying the ALARP principle: economic risk	439
		11.10.	Environmental protection	443
		11.11.	Environmental acceptability	447
		11.12.	Corporate risk management: the risk matrix approach	447
		11.13.	Future uncertainty: implications for landslide management	449
		11.14.	Risk assessment, decision-making and consultation	452
			References	453
12		**Future challenges**		**459**
		12.1.	Introduction	459
		12.2.	Quantitative risk zonation: lessons from Ventnor, Isle of Wight	459
		12.3.	Landslide risks and major project schedules: lessons from the Camisea pipeline, Peru	461
		12.4.	Risk assessment and management of extreme events: lessons from Fukushima, Japan	464
		12.5.	Landslide management and the environment: lessons from Easton Bavents	466
		12.6.	Future uncertainty: valuation of environmental resources	468
		12.7.	Future uncertainty: global change	470

	12.8.	Evidence for climate change	471
	12.9.	Future uncertainty: climate model predictions	472
	12.10.	Increased rainfall	474
	12.11.	Sea-level change	476
	12.12.	Uncertainty and risk assessment	484
		References	485
13		**Glossary of terms**	**491**
		Index	**201**

Preface to the first edition

In recent years it has become fashionable to talk about risk in a variety of contexts and to actively undertake risk assessments. Closer inspection reveals, however, that in many instances the notions of risk that are employed differ greatly or are unclear, and that the risk assessments themselves are often vague and of limited scope. This situation is not helped by the fact that until recently the literature has itself been characterised by diverse views on the nature of risk and the ideal sequence of steps required in the undertaking of a risk assessment. Hence the idea of producing a book to clarify the situation with reference to landslide risk assessment, a subject of considerable interest to both authors.

Originally envisaged as a book of examples to actively assist field scientists in developing landslide risk assessments, the final version has evolved to include more detailed consideration of the nature of landslide hazard and risk, as well as their relevance in a global context. Nevertheless, the worked examples gleaned from personal experience and the literature are considered a key feature of the book and it is hoped that the blend of consultancy experience and academic background has produced a text that all readers will find both stimulating and useful. As Confucius said: '*Man has three ways of learning: firstly, by meditation, this is the noblest; secondly, by imitation, this is the easiest; and thirdly, by experience, this is the most bitter.*'

It is important to stress that this is not a book about landsliding but about how landsliding can affect human society. Whilst an understanding of landsliding is crucial in the development of a landslide risk assessment, it is only the first step in the process. Landslide risk assessment is concerned with establishing the likelihood and extent to which future slope failures could adversely impact society. Two landslides can be identical in all physical respects yet pose very different levels of risk, emphasising that the focus should not be on the *what* but on the *so what*

Other important underlying issues that are worth stressing at the outset are:

- *Uncertainty* is an inevitable part of the risk assessment process because of incomplete knowledge of both the probability of future events and their consequences. All risk assessments need to be supported by a clear statement of the uncertainties in order to inform all the parties of what is known and unknown, and the weight of evidence for what is only partly understood.
- *Precision*. Statements as to the probability of landsliding and the value of adverse consequences can only be *estimates*. The temptation for increasing precision in the risk assessment process needs to be tempered by a degree of pragmatism that reflects the reality of the situation and the limitations of available information. Numbers expressed to many decimal places can provide a false impression of detailed consideration, accuracy and precision. The use of numbers also conceals the fact that the potential for error is great because of the assumptions made and the computations involved. Estimates will generally need to be 'fit for purpose' rather than the product of a lengthy academic research programme. The quality of a landslide risk assessment is related to the extent to which the hazards are recognised, understood and explained – aspects not necessarily related to the extent to which they are quantified.

- *Expert Judgement.* The limited availability of information dictates that many risk assessments will rely on expert judgement. Indeed, Fookes (1997) noted that the art of geological or geotechnical assessment is 'the ability to make rational decisions in the face of imperfect knowledge'. Because of the reliance on judgement, it is important that effort be directed towards ensuring that judgements can be justified through adequate documentation, allowing any reviewer to trace the reasoning behind particular estimates, scores or rankings. Ideally the risk assessment process should involve a group of experts, rather than single individuals, as this facilitates the pooling of knowledge and experience, as well as limiting bias.
- *Defensibility.* The world is becoming less tolerant of the losses caused by landslides, especially those associated with the failure of man-made or man-modified slopes. Increasingly, engineers get blamed for their actions or inactions. In an increasingly litigious world there will be a need for practitioners to demonstrate that they have acted in a professional manner appropriate to the circumstances.
- *Trust.* It is important for practitioners to appreciate that their judgements about risk may provoke considerable disagreement and controversy, especially if the judgements have implications for property values or the development potential of a site. Acceptance of the risk decisions by affected parties is critical to the successful implementation of landslide risk management strategies. Such acceptance is dependent on the establishment of trust and this, in turn, is dependent on openness, involvement and good communications.

The completion of the book has brought sharply into focus the debt of gratitude owed to many others. Both authors would like to acknowledge the huge influence of Professor Peter Fookes in nurturing the involvement of geomorphologists in the 'real world' of consultancy. They would also like to thank all those colleagues who have, over the years, provided them with stimulation when working on landslides, most especially Professor Denys Brunsden, Dr John Doornkamp, Dr Jim Griffiths, Professor John Hutchinson, Dr Jim Hall, Dr Roger Moore, Dr Fred Baynes (remember Bakuriani), Rick Guthrie, Mike Sweeney (*Shoa*), David Shilston, Saul Pollos, John Charman, Dr Alan Clark and Maggie Sellwood (forever frozen in Whitby). They are also grateful to Jane Pugh and Mina Moskeri of the London School of Economics for producing 34 of the diagrams (EML did the easy ones), and to their wives Claire and Judith for putting up with the seemingly endless disruption to normal life.

Mark Lee (marklee626@btinternet.com)

David Jones (d.k.jones@lse.ac.uk)

May 2004

Preface to the second edition

Ten years have passed since we produced the first edition of this book. During this time, the application of landslide risk assessment has increased dramatically, from a relatively few pioneering studies to a burgeoning array of case histories and research projects. There is now an enormous diversity of studies available in the public domain that fall under the broad heading of 'landslide risk management'. These include the following.

- Research and development studies, in which the focus is on specific aspects of the landslide risk assessment process, such as the statistical analysis of landslide inventories, modelling of aspects of landslide behaviour (e.g. run-out and travel distance) or the vulnerability of particular types of structure to particular landslide events.
- Technical-based assessments, in which landslide risk is an extension of a detailed investigation of the landslide hazard within an area or at a particular site. The focus tends to be on understanding the nature and extent of the landslide threat, rather than identifying and valuing the full range of potential consequences.
- Management-based assessments, in which landsliding may be just one of a range of possible threats to a particular asset. The focus tends to be on providing sufficient information on the landslide threat and consequences to support decision-making.

The literature is dominated by the first two types of studies. Unfortunately, this gives the impression that landslide risk assessment and management is a technical subject that can be studied in isolation from the broader social, political and business contexts. In our experience, this is definitely not the case. Effective landslide risk studies should be based on collaboration between specialists from a variety of backgrounds, including planners, engineers, project managers, health and safety officers, and risk analysts, as well as geoscientists. This has been perhaps the most important lesson to emerge from the last 10 years of working on a wide range of consultancy projects, discussions with colleagues and exposure to the emerging 'risk cultures' in many organisations.

In order to accommodate these and other developments it proved necessary that the book was not merely updated but largely remodelled and expanded, growing from seven to 12 chapters. There is now greater consideration of the nature of risk in Chapter 1, all the main statistical techniques used to develop probabilities are now gathered together in one chapter (Chapter 5), and there is greater consideration of the nature of future uncertainties and challenges (Chapter 12). There is also greater discussion of landslide risk assessments as applied to pipeline developments and submarine locations, as well as of the new frontier of 'Black Swans', 'Perfect Storms' and 'Dragon Kings'.

Following on from the acknowledgements made in the first edition, the debts to our colleagues just keep mounting. Special mentions are needed this time for Andrew Hart and John Perry (Atkins), Jim Clarke, Trevor Evans, Andy Hill and David Waring (BP), Dr Andy Mills (Halcrow), Chris Arnold and Lizzie Bardsley (Motts), Kevin Styles and Bob Sas (Fugro Hong Kong), Steve Perry and Jon Hart (GeoRisk Solutions Hong Kong), Wayne Savigny and Mike Porter (BGC, Vancouver), Nick Jackson, and Walter Gil and Bezhan Azanidze (WREP, Georgia), along with Doug Nyman, Jean Audibert and Jim Hengesh (*remember 12.6.2006*).

All that remains is to thank our long-suffering wives Claire and Dame Judith for allowing us to take time off from household chores.

Mark Lee (marklee626@btinternet.com)

David Jones (d.k.jones@lse.ac.uk)

Scientific notation

Numbers are presented in a variety of ways in this book. In addition to the conventional forms, scientific notation is used. Numbers in scientific notation always consist of an integer to the left of the decimal point and the remainder of the number after the decimal point, with the entire number expressed as a multiple of 10; alternatively, the power of 10 can be expressed as the number of decimal places to the right (positive exponent) or left (negative exponent) of the first significant digit:

Number	Scientific notation	
0.0001	1.0×10^{-4}	1.0 E-4
0.001	1.0×10^{-3}	1.0 E-3
0.01	1.0×10^{-2}	1.0 E-2
0.1	1.0×10^{-1}	1.0 E-1
1	1.0×10^{0}	1.0 E0
10	1.0×10^{1}	1.0 E1
100	1.0×10^{2}	1.0 E2

Beware of the liberal use of decimal places, as they can convey a sense of precision that may not be warranted by the judgements and assumptions made in assessment processes. However, we have generally avoided rounding figures up or down in order to help the reader see where the answers have come from.

Landslide Risk Assessment
ISBN 978-0-7277-5801-9

ICE Publishing: All rights reserved
http://dx.doi.org/10.1680/lra.58019.001

Chapter 1
Background to landslide risk assessment

1.1. Introduction

Doubtless many readers will expect a book on landslide risk assessment to begin with a detailed consideration of landslides, their nature, form, generation, distribution and potential to cause damage. However, there are good grounds for arguing that this would be a mistake, because the main purpose of this book is to explore how the rapidly expanding field of risk assessment can be applied in the context of landsliding. Thus, although the nature of landsliding will have to be examined at some point because of its role as the cause of harm or loss, the most logical place to start is rather with the nature of risk. This is important, as risk is now recognised to be 'a universal concept, inherent in every aspect of life' (Chicken and Posner, 1998), an element of which increasingly pervades all decision-making.

Emphasis on risk and its assessment has grown dramatically in recent decades, raising the subject from relative obscurity as the preserve of the financial sector and certain hazardous industries and high-tech activities, to achieve such prominence that some claim it to be one of the most powerful concepts in modern society – a transformation that is sometimes called the *risk revolution*. Such dramatic growth has produced a huge and confusing – some would say confused – literature that is highly compartmentalised because contributions focus on a wide variety of different subjects, are written from a range of disciplinary backgrounds or perspectives, and often employ differing terminologies, so that the results can appear contradictory. One of the main purposes of this introductory chapter is to clarify terminology and provide a specific framework for *landslide* risk assessment.

As regards landsliding itself, the relevant aspects are discussed in Chapter 3. For the moment, the following statement from Jones (1995a) provides a sufficient basis to proceed (also see Chapter 9):

> It has to be recognised that the terms 'landslides', 'mass movements' or 'slope movements' are convenient umbrella terms which cover a multitude of gravity-dominated processes that displace relatively dry earth materials downslope to lower ground by one or more of the three main mechanisms: falling, flowing (turbulent motion of material with a water content of less than 21%) and sliding (material displaced as a coherent body over a basal discontinuity or shear plane/surface). In reality these three basic mechanisms combine with geologic and topographic conditions to produce a bewildering spectrum of slope failure phenomena, many of which do not conform to stereotype views of landsliding, thereby resulting in serious underestimation of the extent and significance of landslide hazard.

1.2. What is risk?

The word 'risk' is used frequently and in many different contexts. To most readers of this book, risk is the term used to describe the likely scale and magnitude of future harms or adverse consequences arising from the impact of hazards such as landslides. This is consistent with the traditional scientific and technical view of risk. For example, the Royal Society Study Group (Royal Society, 1992) defined risk as

1

a combination of the probability, or frequency, of occurrence of a defined hazard and the magnitude of the consequences of occurrence.

However, the concept is often used in other ways by different academic disciplines and practitioners. This has led to the development of diverse views regarding its precise nature, as was testified by the disagreements between the scientists and social scientists involved in the preparation of the Royal Society Study Group Report *Risk: Analysis, Perception and Management* (Royal Society, 1992). Because of this diversity of views, it is useful to consider some of the main variations of relevance to landslide investigations.

1.2.1 Financial risk

Financial risk is often defined as the uncertainty associated with any investment; that is, risk is the possibility that the actual return on an investment will be different from (i.e. lower than) its expected return. A key concept in finance is the notion that an investment that carries a higher risk also has the potential of a higher return.

The coupling of risk, reward and uncertainty is consistent with the earliest notions of risk. The word 'risk' derives from the early Italian *risicare*, which means 'to dare'. In the Middle Ages, the term *risicum* was used specifically in the context of sea trade and its ensuing legal problems in cases of loss or damage (Luhmann, 1996). The costs incurred owing to the possibility of ships sinking or being waylaid by pirates were offset by the rewards to be gained from ships that made it to their destinations with cargo. It was not uncommon to lose half or more of the cargo along the way, but the high prices that the goods commanded at their final destinations still made this a lucrative endeavour for both the owners of the ships and the sailors who survived. The unpredictability of events and unforeseeable outcomes were simply 'Acts of God', attributable to the goddess Fortuna. 'Prudence' was the human quality to make balanced judgements between good and bad outcomes.

In his classic book, *Risk, Uncertainty and Profit*, the economist Frank Knight (1921) considered risk to be related to uncertainty. He distinguished between

- *classical (a priori) probabilities*, as derived from a game of chance (e.g. rolling dice), where there are a finite number of different possible outcomes, all of which are assumed to be equally likely
- *statistical (frequency) probabilities* that can be obtained through the analysis of historical data
- *estimated (degree of belief) probabilities or opinions* that people make when 'there is no valid basis of any kind for classifying instances'.

Knight (1921) suggested that the a priori and statistical probability categories reflect 'measurable uncertainty', whereas estimates represent 'unmeasurable uncertainty', and he defined risk as follows:

> To preserve the distinction ... between measurable uncertainty and an unmeasurable one we may use the term 'risk' to designate the former and the term 'uncertainty' for the latter.

In Knight's opinion, an ever-changing world brings new opportunities for businesses to make profits, but also means that there is imperfect knowledge of future events. Therefore, risk applies to situations where the outcome is unknown but the odds can be accurately measured. Uncertainty, on the other hand, applies to situations where it is not possible to know all the information that is needed in order to set accurate odds. Knight's definition remains relevant for economists, where measurable uncertainty relates to insurance and unmeasurable uncertainty is the reality for entrepreneurs and speculators (Holton, 2004). However, for other disciplines, his distinctions are less relevant, and, in the case of landslide studies, *degree of belief* is an important basis of risk assessment.

1.2.2 Risk and safety

Risk can also be about safety (or danger) and minimising the threat to individuals: for example, 'hospital budget cuts place babies at risk'. The creation of regulatory bodies such as the Health and Safety Executive in the UK (e.g. HSE, 2001) is an example of society coping with risks encountered in the workplace. Although legislation varies around the world, the general principle is that those who create risks from work activity are responsible for protecting workers and the public from their adverse consequences. This typically places a duty on employers to assess risks and base their control measures on the results of these assessments.

In the UK, the Courts have ruled that, in the context of the Health and Safety at Work etc. Act, 1974, risk means the exposure to the 'possibility of danger' rather than 'actual danger' (Court of Appeal in *Regina* v. *Board of Trustees of the Science Museum*, 1993; see HSE, 2001). The focus is directed towards the uncertain dangers that people are exposed to in the workplace and, implicitly, the harm that may be caused. This definition incorporates three main elements: exposure to danger, harm and uncertainty.

1.2.3 Risk, exposure and health

The US Environmental Protection Agency (EPA) considers risk to be the chance (i.e. probability) of harmful effects to human health or to ecological systems resulting from exposure to an environmental 'stressor'. A stressor is any physical, chemical or biological entity that can induce an adverse response. Stressors may adversely affect specific natural resources or entire ecosystems, including plants and animals, as well as the environment with which they interact. For example, exposure to toxic air pollutants can increase the chance (the 'risk') of an individual developing cancer.

The EPA uses risk assessment to characterise the nature and magnitude of health risks to humans (e.g. residents, workers and recreational visitors) and ecological receptors (e.g. birds, fish and wildlife) from chemical contaminants and other stressors that may be present in the environment. The process involves the evaluation of the frequency and magnitude of human and ecological exposures that may occur as a consequence of contact with a contaminated medium. The evaluation of exposure is then combined with information on the inherent toxicity of the chemical (i.e. the expected response to a given level of exposure or dose) to predict the probability, nature and magnitude of the adverse health effects that may occur. The dose–response relationship for a specific pollutant describes the association between exposure and the health effect – that is, it estimates how different levels of exposure to a pollutant change the likelihood and severity of health effects.

This concept of health risk incorporates exposure to a threat (i.e. the stressor), changes in vulnerability depending on the severity of the threat (i.e. the dose–response relationship), harm and uncertainty. Note that this is a slightly different use of the term 'exposure' than is adopted in this book (see later in this chapter).

1.2.4 Risk as uncertain outcomes

The recently published ISO 31000 standard from the International Standardisation Organisation (ISO, 2009) provides a broader definition of risk as the 'effect of uncertainty on objectives'. In this definition, uncertainties include events (which may or may not happen) and uncertainties caused by ambiguity or a lack of information. The focus is on unexpected outcomes arising from an organisation's action, either positive or negative. The definition differs from the prevailing perception that risks are to do with 'hazards' or about 'things that go wrong'. In this instance, risk is considered to be a neutral event, it can be positive or negative, and the main elements of risk are exposure to a situation (not necessarily danger) and uncertainty (Holton, 2004).

Implicit in the ISO 31000 definition is the notion that risk is also about decision-making, choice and management rather than simply being a measurable condition of the environment. Managing risk is a process of optimisation that makes the achievement of objectives more likely. It is concerned with changing the magnitude and likelihood of consequences, both positive and negative, so as to achieve a net increase in benefit (Purdy, 2010).

1.2.5   Risk and damaging events

Risk is frequently used in the context of the probability of failure in technological systems (e.g. a nuclear plant meltdown or an aircraft accident), failure of engineering structures (e.g. a dam failure) or the occurrence of extreme natural hazard events (e.g. a meteor or comet impact or a tsunami generated by the collapse of an island volcano cone), where significant levels of loss are assumed but not specified. Even the probabilities of hurricanes, floods, volcanic eruptions or large earthquakes impacting on particular areas are referred to as risk (Cutter, 1993). The focus is on incidents that initiate consequence scenarios; that is, 'top events' (see Chapter 2).

This is considered to be a weaker representation of risk, where it is assumed that adverse consequences will result but no attempt is made to calculate them. This line of approach was rejected by Warner (Royal Society, 1992) because of the clearly developed distinction between 'hazard' and 'risk'. However, it is a pragmatic approach to a basic level of risk assessment, especially where any form of loss can be considered to be highly undesirable. For example, Lee and Charman (2005) describe how landslide risk assessment for oil and gas pipelines is often focused on the probability of pipeline rupture, rather than a full costing of the consequences of rupture. In this context, risk is expressed as the probability of a damaging event that leads to adverse consequences.

These examples illustrate that, owing to the flexibility of the English language, there has been some blurring of the distinction between hazard and risk. However, references to 'increased risk of rain' or 'growing debris flow risk' are misleading, since they refer to the increased likelihood of the occurrence of phenomena rather than an increase in the likelihood and scale of adverse consequences.

1.3.   A view of risk

The concept of 'risk' probably emerged with the developments in science and technology during the eighteenth and nineteenth centuries. Sjöberg (1987) wrote that 'risk has become an urgent social concern since the 1960s', especially in Western societies, because 'across a wide spectrum of risks, many people are no longer content to accept the inevitability of adverse effects as being natural or to be expected'. In a similar vein, Beck (1992) wrote that 'risk may be defined as a systematic way of dealing with hazards and insecurities and introduced by modernization itself'.

Bernstein (1996) argued that the mastery of risk, as a result of the evolution of systematic methodologies to handle future events that may have adverse effects, defines the boundary between modern times and the past. Growth in knowledge dictates that the threats are many and ever increasing in variety and scale, mainly as a result of developments in science and technology. Similarly, the consequences can be exceptionally diverse, and are increasing owing to valuation of former intangibles such as environmental conditions.

Risk is a *human concept*; the product of the human mind. It is also a human-centred concept. It provides opportunities, while exposing people to outcomes that may be undesirable. This coupling of risk and reward lies at the core of the risk concept. As the opportunities and undesirable outcomes tend not to be valued similarly by different individuals, groups or even societies, it is necessary to view risk as not merely a human concept but also a *cultural construct*. In other words, every individual has a slightly different construction of the risk they face in terms of the likelihood and scale of threat and the potential for harm or loss, based on background, past experience, beliefs and so on (see Chapter 2). Thus, what may constitute acceptable (*tolerable* is the preferred term) levels of risk vary across the world, and also change with time; a good example of the latter is the way in which views as to what constitute tolerable levels of health risk have changed dramatically in Europe over the last hundred years as expectations of health and safety have risen.

Despite the contrasting definitions of risk introduced earlier, it is clear that the focus is on exposure to threats and uncertain future outcomes. It is concerned with yet to be realised consequences (predominantly adverse consequences) or, to quote the sociologist Giddens (1999), 'risk resides in the future'. Herein lies a fundamental problem, since, despite advances in science, humans still have limited knowledge and thus imperfect appreciation of recognised or known risks, while remaining unaware of yet to be identified risks that may emerge in the future.

Clearly, if all possible future outcomes of an event, action or circumstance were known, along with the likelihood of each, then the computation of risk would appear to be relatively straightforward. However, it is more normal for there to be a lack of certainty regarding the full range of possible outcomes or the likelihood of each. This is uncertainty in the true sense, and probability theory represents the scientific analysis of uncertainty with reference to *likelihood*. Calculations of risk can still be undertaken and will be of value, but the results will become more and more vague as the levels of uncertainty increase. Indeed, there will be a gradual gradation from quantitative estimations to qualitative estimations of risk. But in some cases there will be so little information regarding possible outcomes and likelihoods that no meaningful guesses can be made. This is ignorance.

The then US Secretary of Defence Donald Rumsfeld's much ridiculed statement made on 12 February 2002 is absolutely correct in the context of risk (US Department of Defence, 2002):

> There are known knowns; there are things we know that we know. There are known unknowns: that is to say, there are things that we know we don't know. But there are also unknown unknowns – there are things we do not know we don't know.

It is now generally recognised that risk is ubiquitous and that no human action or activity can be considered to be risk-free. Risk can be increased, decreased and transferred, but rarely eradicated. For example, the eradication of the smallpox virus has not entirely removed the risk, as has been highlighted by concerns that stocks of the virus held in laboratories could be exploited by terrorists. Prevailing approaches to risk management, therefore, seek to keep risk within tolerable levels and to minimise risk where it is economically feasible, technologically practical, commercially necessary or politically desirable.

It is, however, possible to talk of the removal of risk in specific contexts through the process of risk avoidance. Thus, in the case of flood hazard, the development of flood control schemes (reduction of hazard) and relocation of assets away from hazardous areas to safe locations (avoidance of risk) may be viable options, but flooding will still pose a risk in other contexts and any relocation may well increase exposure to other risks. Thus, the once fashionable aspirations for a 'zero-risk society' are now seen to be pure fantasy.

1.4. Black Swans and Perfect Storms

In recent years, the risk agenda has been forced to focus on major crises such as the 2007 financial market crisis and the ongoing global recession, the 2010 *Deepwater Horizon* explosion and oil spill in the Gulf of Mexico and the 2011 Tohoku earthquake and tsunami in Japan. These, and similar rare events, have been labelled as 'Black Swans' or 'Perfect Storms' by many in the risk management industry to convey a sense of unpredictability or extreme unlikelihood (Paté-Cornell, 2012). It has been suggested that labelling the events in this way might be an attempt to deflect responsibility away from those parties whose behaviour contributed to the crisis (Catanach and Ragatz, 2010). If the event could not have been predicted or its causes were so unique, then how could anyone be held accountable for contributing to its occurrence?

Black Swan theory describes events that are a surprise (to the observer), have a major impact, and after the fact are often inappropriately rationalised with the benefit of hindsight (Taleb, 2010). The expression 'Black Swan' derives from a Latin saying, *'rara avis in terris nigroque simillima cygno'* ('a rare bird in the lands, very much like a black swan'). It was a common expression in 16th-century London as a statement of impossibility, as the black swan was presumed not to exist. After the Dutch explorer Willem de Vlamingh discovered black swans in Western Australia in 1697, the term changed to mean a perceived impossibility that might later be disproven.

Taleb (2010) considered the 11 September 2001 terrorist attacks on the World Trade Centre and the Pentagon in the USA to be an example of a Black Swan event. It was a shock to all observers and the ramifications continue to be felt in many ways: increased levels of security and 'preventive' strikes or wars waged by Western governments. The attack, using commercial airliners as weapons, was virtually unthinkable at the time. However, with the benefit of hindsight, it has come to be seen as a predictable incident in the context of the evolutionary changes in terrorist tactics. Thus, the towers had been designed to withstand the impact of small aircraft but not fully fuelled medium-sized airliners.

Black Swan events may be the 'unknown unknowns'. However, the ability to rationalise a Black Swan event in hindsight separates it from being pure chance. After the event, the cause can be seen or calculated and the precursor evidence seen to have been there all along. For example, the 2011 Tohoku earthquake and tsunami are thought by some to have been Black Swan events (e.g. Stein *et al.*, 2012). A magnitude 9 ($M_w = 9$) earthquake offshore generated a huge tsunami that overtopped the tsunami defences, causing over 19 000 deaths and at least $200 billion of damage (Normile, 2012), including the crippling of the Fukushima nuclear power plant.

The earthquake released about 150 times the energy of the $M_w = 7.5$ earthquake that had been expected in the region (e.g. Stein *et al.*, 2012). No $M_w = 9$ events had been recorded in the area, and this seemed consistent with the prevailing hypothesis that the subduction dynamics of the Japan Trench precluded the generation of giant earthquakes. However, subsequent research has revealed that instead of only some subduction zones being able to generate $M_w = 9$ earthquakes, it is now believed that many or all can. If a $M_w = 9$ event had been expected, then the areas at risk from the resulting 15 m high tsunami would have been predictable. The Fukushima nuclear reactors had been designed for a maximum wave height of 5.7 m and not the 15 m-high wave that arrived. If a full risk assessment had been carried out looking at geological, seismic, reactor safety and emergency services factors, it may have concluded that the chain of events that occurred on 11 March were entirely possible.

A 'Perfect Storm', on the other hand, requires a combination of factors that together magnify the impact of each factor considered independently. The term was made popular by Sebastian Junger (1991) in his book about the 1991 Halloween Nor'easter storm that sank the swordfishing boat *Andrea Gail*. The storm was the product of combining the influences of three different weather patterns: warm air from a low-pressure system coming from one direction, a flow of cool and dry air from another direction generated by a high-pressure area, and tropical moisture provided by Hurricane Grace. The Perfect Storm premise is that none of these factors was individually powerful enough to create the resulting storm; it was the product of synergism.

The 2007 financial crisis has been called a Perfect Storm in order to explain how seemingly ordinary events could combine to cause havoc on the global markets: a combination of questionable economic policies, weak bank strategies, inappropriate derivative usage, mismanaged deregulation, lax oversight and new accounting (Catanach and Ragatz, 2010). Ali Velshi, CNN's chief business correspondent, stated that 'it was a perfect storm... it was a lack of regulation, it was greed and creativity in the financial industry, and it was an American dream that got off track'. Even Russian Prime Minister Vladimir Putin, in a keynote speech at the 2009 World Economic Forum, told delegates that the crisis constituted a 'Perfect Storm' (Catanach and Ragatz, 2010).

Black Swans and Perfect Storms describe different aspects of uncertainty. Black Swan events represent ignorance or lack of fundamental knowledge, where not only the magnitude and frequency of an event are unknown but also the very possibility of the event itself (*epistemic uncertainty*; see Chapter 5). This is essentially a problem of relying on statistical datasets that cover limited time spans and hence do not include extremely rare events. On the other hand, Perfect Storms represent the joint occurrence of random known events (*aleatory uncertainty*). The probability of each individual event might be determined from statistics. If the events are independent, the probability of a conjunction is simply the product of the probabilities of each event (see Chapter 5). However, if they are dependent, the probability of their conjunctions may be severely underestimated by this product.

Increasing awareness of Black Swans and Perfect Storms emphasises how the risk landscape has become more complex over the last decade. In many cases, it is no longer acceptable to confine landslide risk assessments to those events that are represented in the historical record or can be anticipated at some point in the future (e.g. the failure of a seawall). The potential for extreme events with catastrophic consequences should not be ignored. The lesson from past Black Swan events is that they may have been unforeseen, but they were not unforeseeable.

1.5. What is landslide risk?

In many areas of the world, landsliding is a hazard that poses significant danger to people, property, economic activity and the environment. It is a danger that many geoscientists believe can be measured and quantified through analysis of historical records, modelling or expert judgement (see Chapter 5).

The International Society of Soil Mechanics and Ground Engineering (ISSMGE) Technical Committee on Risk Assessment and Management defines landslide risk as a measure of the probability and severity of an adverse effect to life, health, property or the environment (ISSMGE, 2004). This measure can be estimated as the probability of an adverse event times the consequences if the event occurs. This is basically the same as the Royal Society definition of risk (Royal Society, 1992).

Landsliding is almost exclusively an undesirable event for much of society. As a result, it is felt that the ISO 31000 definition of risk as the 'effect of uncertainty on objectives' is less appropriate than the traditional technical view of risk. For this reason, we define landslide risk in this book as

> the potential for adverse consequences, loss, harm or detriment as a result of landsliding, as viewed from a human perspective, within a stated period and area.

1.6. Hazard and vulnerability

Hazard is another human/cultural construct. It is the label applied by humans to objects, organisms, phenomena, events and situations that have the attribute of

> potentially affecting adversely humans and the things that humans value, as viewed from a human perspective.

Hazard is a perceived danger, peril, threat or possible source of harm or loss. Other useful definitions of hazard are

- 'a situation that in particular circumstances could lead to harm' (Royal Society, 1992)
- 'the potential for adverse consequences of some primary event, sequence of events or combination of circumstances' (BSI, 1991)
- 'threats to humans and what they value: life, well-being, material goods and environment' (Perry, 1981)
- 'the probability that a particular danger (threat) occurs within a given period of time' (ISSMGE, 2004).

Unfortunately, the term 'hazard' has been somewhat eclipsed in recent years owing to widespread infatuation with the term 'risk', stimulated by the 'risk revolution'. Regrettably, the wholly respectable field of hazard assessment and prediction has also become swamped by, and largely subsumed within, what some authors (e.g. Sapolsky, 1990) view as a 'collective mania with risk'. In reality, 'hazard' and 'risk' are complementary but distinct, with hazard focusing on the 'causes' of harm or loss and risk focusing on the 'consequences' – a distinction that is crucial to the understanding of risk assessment. Hazard assessments continue to have a valuable and distinct contribution in the establishment of zoning policies (hazard zonation) and in the establishment and activation of emergency procedures, including warnings and evacuations.

For hazard to exist, situations have to arise or circumstances occur where humans and the things that humans value can be adversely affected. Thus, landslides on remote uninhabited mountains or islands are not hazards. Even if humans and the things that humans value occur in the same area as the landslides, the landslides must be capable of causing costs from a human perspective to be classified as a hazard. Clearly, size (magnitude) and nature of hazard (landslide) are important determinants, but they are not the sole determinants, since of at least equal importance is the notion of vulnerability.

Vulnerability is yet another human construct, and can be defined simply as 'the potential to suffer harm', or more fully as

> the potential for humans and the things that humans value to suffer harm or adverse impact from a human perspective.

It is the reverse of robustness, resilience or durability, and in some contexts other words may be used, such as fragility.

The recognition of vulnerability results in a simple generalised relationship:

hazard event × vulnerability = adverse consequences

This reveals the crucial role of vulnerability in imparting the attribute of 'hazard' to phenomena or circumstances, since without vulnerability there cannot be hazard; that is, phenomena are merely phenomena!

It also has to be recognised that vulnerability is a very complex subject, because it can be applied to people, communities, societies, industries, economies, structures, infrastructure, communications, ecosystems and environments.

Developing this notion further means that the simple equation above can be expressed as:

$$H \times (E \times V) = C$$

where H is a specific hazard event; E is the total value of all threatened items valued by humans, known as *elements at risk* and including population, artefacts, infrastructure, economic activity, services, amenity, etc.; V is the vulnerability or the proportion of E adversely impacted by the hazard event; and C represents the adverse consequences of the hazard event.

However, if the impact of a particular type of hazard (i.e. landsliding) is considered in a generic sense, then it immediately becomes obvious that both *elements at risk* and *vulnerability* vary over time. Slight changes in the timing of a landslide can result in very different levels of detriment depending on exactly what is present to be impacted upon. As a result, some consider that vulnerability can be regarded as consisting of two distinct aspects:

1. The level of potential damage, disruption or degree of loss experienced by a particular asset or activity subjected to a hazard event (i.e. a landslide) of a given intensity: for example, the different vulnerability of a timber-framed building to slow ground movement, compared with a rigid concrete structure.
2. The proportion of time that an asset or person is exposed to the hazard (i.e. being in the 'wrong place' at the 'wrong time'): for example, the contrasting exposure of a person walking underneath an overhanging rock, compared with the occupancy of a stationary beach hut. If, say, a vehicle takes 10 minutes to travel through a road cutting prone to rockfalls, then the driver is exposed to the hazard for $10/(60 \times 24) = 0.007$ of a day. If this journey is repeated 300 times during a year, then the driver's exposure becomes $(10 \times 300)/(60 \times 24 \times 365) = 0.006$ of a year.

In the second case, however, it is not vulnerability that has actually changed in the short term but the value of elements at risk. A person is equally vulnerable to a falling rock irrespective of whether or not he or she is actually present when the rock falls. What is of crucial importance is *presence* or *absence* at the time when the event occurs. If the person is present within a threatened area, then the potential for loss is greater, because the value of elements at risk has been raised by the addition of a highly valued and vulnerable element (i.e. temporal vulnerability). This variation is known as *exposure*.

Temporal vulnerability was cruelly displayed by two relatively small landslides that occurred in Dorset, UK in July 2012, following three months of above-average rainfall. The first incident (7 July 2012) involved two people who were killed when their car was crushed by falling debris from a landslide bringing down part of the Beaminster Tunnel portal. The second (on a warm, sunny 24 July) involved a cliff fall at Burton Bradstock that killed a young woman and narrowly missed two others while her mother lay sunbathing 30 m away, unaware of the tragedy. In both cases, the timing of the events was of crucial importance.

Theoretically, exposure E^* is a measure of the ever-changing value E and vulnerability V of elements at risk within an area, so that

$$E \times V = E^*$$

and the general relationship becomes

$$H \times E^* = C$$

or

adverse consequences = hazard × exposure

When it comes to developing risk assessments, the problem of temporal vulnerability has to be tackled in a different way. The values and vulnerabilities of every different category of element at risk are retained, but each is multiplied by a figure representing the proportion likely to be present at the time of impact by the hazard. This different measure of exposure E_x results in the simple equation becoming

$$H \times (E \times V \times E_x) = C$$

where H is the specific hazard, E is the maximum potential value of all elements at risk, V is the vulnerability or the proportion of E that could be reduced by the hazard event, and E_x is the proportion of each category of elements at risk present at the time of the event.

As a consequence, the generic relationship becomes

$$H \times \sum (E \times V \times E_x)$$

where E, V and E_x have to be computed for each and every relevant category of element at risk.

1.6.1 Geohazards, environmental hazards and disasters

Hazards, such as landslides, are the agents that can cause loss and thereby contribute to the generation of risk. As a result, their magnitude–frequency characteristics are important in helping to determine risk, but they are not in themselves risk. Thus, it is incorrect to consider the probabilities of differing magnitudes of hazard as estimations of risk – they are merely measures of hazardousness. The 100-year flood, the recurrence interval of a magnitude 8 earthquake in a particular region or the probability of a major cliff fall are all essential ingredients in estimating risk, but are not actual measures of risk. Size and intensity are clearly of relevance, but 'when', 'where' and 'what is there to be impacted upon' are of even greater importance.

A re-emphasis of the significance of the term 'hazard' also requires some clarification of associated terms. Hazards have traditionally been subdivided into *natural* and *human-made* categories, later

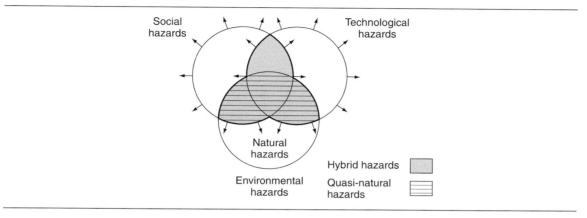

Figure 1.1 The hazard spectrum. The arrows indicate the expansion of the different spheres and the resulting increased overlap. (Based on Jones, 1993)

subdivided into *natural*, *technological* and *societal* categories, with *natural hazard* defined as 'those elements of the physical environment harmful to humans and caused by forces extraneous to human society' (after Burton and Kates, 1964).

However, it is now recognised that the growing extent and complexity of human activity has had a dramatic influence on the operation of environmental systems, thereby blurring this simple division, with an increasing proportion of hazards now interpreted as *hybrid hazards* (Figure 1.1), reflecting their complex origins, including the well-known *na-tech* (natural-technological) and *quasi-natural* groups. It is also true that the term 'natural hazard' has increasingly fallen out of favour to be replaced by 'biogeophysical, biophysical or geophysical hazard' (*geohazard* for short).

As landsliding can be a natural phenomenon as well as human-accentuated and human-induced, it has to be included within both the geohazard and hybrid hazard categories (quasi-natural and na-tech) under the general label of environmental hazard. The term *environmental hazard* remains highly contentious, with some authors continuing to argue that it refers to hazards that degrade the quality of the environment (Royal Society, 1992), while others follow the definition of Kates (1978) as 'the threat potential posed to humans or nature by events originating in, or transmitted by, the natural or built environment'. The latter is to be preferred.

Disaster is yet another human/cultural construct and a word much beloved by the media. There is an extensive literature on the various aspects of disaster, but the term essentially means a level of impact or adverse consequences of sufficient severity that either outside assistance is required to facilitate the recovery process or the detrimental effects are long-lasting and debilitating, – for example, the United Nations Office for Disaster Risk Reduction (UNISDR, 2009) states that a disaster occurs when losses 'exceed the ability of the affected community or society to cope using its own resources'. Disaster is thus a term that refers to the relative intensity of the impact (e.g. 40% of inhabitants killed or 60% of buildings destroyed) rather than the absolute level of impact as measured by deaths or number of buildings destroyed. It has to be used in context, which means that there is no absolute scale of disaster, despite the fact that the term is usually associated with sheer numbers and concentrations of casualties and damage. The size of an event (e.g. a landslide) is not a basis for using the term. Indeed, many studies state that disasters are more to do with people, societies and economies than with hazard events. It is the significance of vulnerability and exposure in facilitating disaster that has resulted in growing criticism of the term 'natural disaster' and led to recommendations that its use should be discontinued. The relationship between disaster, calamity and catastrophe is also open to debate, and it is suggested here that use of all three terms should be avoided if at all possible.

1.7. From hazard to risk

Risk is concerned with the likelihood and scale of future adverse consequences that are the product of the interaction of hazard, vulnerability and exposure. Thus, both the magnitude–frequency characteristics of hazard and variations in vulnerability and exposure over space and time contribute to risk (Figure 1.2).

The simple relationships described earlier can therefore be turned into basic statements of risk, so long as the probabilities of specific magnitudes of events are known. In the case of landsliding, the following relationship can be produced, based on Carrara (1983), Varnes (1984) and Jones (1995b)

$$R_s = P(H_i) \times (E \times V \times E_x)$$

where R_s is the specific risk, or the expected degree of loss due to a particular magnitude of landslide H_i occurring within a specified area over a given period of time; $P(H_i)$ is the hazard, or the probability of a particular magnitude of landslide H_i occurring within the specified area and time frame; E is the total value of the elements at risk threatened by the landslide hazard; V is the vulnerability, or the proportion of E likely to be affected detrimentally by the given magnitude of landslide H_i expressed as either a percentage of E or on a scale of 0 to 1; and E_x is the exposure, or the proportion of total value likely to be present and thereby susceptible to being adversely affected by the landslide, expressed on a scale of 0 to 1.

In reality, the varied components of E have to be disaggregated and each considered separately, which is one reason why full risk assessments are so complex. It also has to be recognised that exposure is both complex and variable because it consists of two distinct components, namely

Figure 1.2 Diagrammatic representation to show how risk is the product of hazard and vulnerability. (Based on Coburn and Spence, 1992)

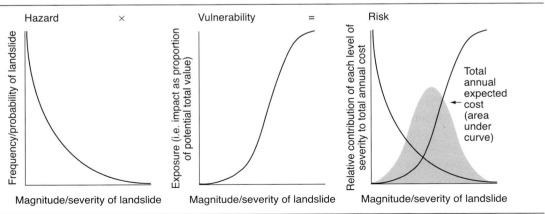

- *fixed or static assets*, such as buildings, which change in number, construction, size and value over time
- *mobile assets*, such as humans or modes of transport, whose presence and concentration display marked short-term fluctuations, as well as long-term trends (see Chapter 7).

As a consequence, the above relationship should be rewritten as

$$R_s = P(H_i) \times \sum (E \times V \times E_x)$$

Note that E, V and E_x have to be computed for every relevant category of element at risk.

Finally, it should be noted that the *total risk* R posed by landsliding is the sum of the calculations of specific risk for the full range of potential magnitudes of landslides:

$$R = \sum_{i=1}^{n} R_{s,i}$$

1.8. Categories of risk

Just as there are many types of hazard, so too there are numerous categories of risk. An often-quoted fundamental division is between natural and human-made risks, otherwise known as external and manufactured risks (Giddens, 1999). This clear-cut distinction does not survive close scrutiny. Perhaps a better division is into societal, technological and environmental risks (see Figure 1.1 for a basis), although there is much disagreement as to what should be included within each of these groups.

The term 'environmental risk' is especially problematic, with some arguing that it is concerned with adverse consequences to the environment (Royal Society, 1992), whereas others take a broader view by extending the definition of environmental hazard proposed by Kates (1978) into the following definition of environmental risk:

> An amalgam of the probability and scale of exposure to loss arising from hazards originating in, or transmitted by, the physical and built environments.

Such a broad definition clearly includes a wide range of sources of risk, including those emanating from geohazards.

It is normal to attribute risk to each particular hazard agent or cause of loss, so that landslide risk (i.e. the risk emanating from landsliding) is merely one component of environmental risk, irrespective of whether it is *natural* or *manufactured*. But landslide risk can also fall within, and be subsumed by, other categories of risk, such as earthquake risk and volcanic risk.

Viewed from a different perspective, each broad area of human activity is also affected by particular types of risk, which may be termed *risk domains*. Thus, there is economic risk, political risk, engineering risk, financial risk, reputational risk and so on. But, to many, the most important risks are those faced by humans themselves, in terms of death, injury, disability, incapacitation, etc., which can be expressed in a number of different ways.

- *Societal risk* is the likelihood of death or injury within a society (usually a nation state or large administrative unit) due to a specified event (e.g. a landslide of particular magnitude) or a

particular category of events (e.g. landsliding). It is usually defined as the product of the frequency of occurrence of a specified hazard and the number of people in a given population suffering (or likely to suffer from) a specified level of harm, and is normally restricted to events potentially capable of causing large-scale loss of life, injury, etc. (see Chapter 10).
- *Individual risk* is the likelihood of death or injury to an individual, and can be calculated by dividing the societal risk by the number of individuals exposed to the hazard (although there are other measures of individual risk; see Chapter 10).
- *Group risk* is the risk faced by particular groups within society, based on activity (e.g. climbers), occupation (e.g. farmers) or other relevant divisions (e.g. males).

1.9. Why do landslide risk assessment?

Detailed slope investigation and analysis has long been the domain of the geotechnical engineer, and no doubt the question that such specialists will ask is 'why do risk assessment?' when standard approaches to stability analysis appear to have served the engineering profession well over the years. There are, in fact, a number of good reasons, many of which have been detailed by Ho *et al.* (2000), and essentially concern the need to consider broader issues and to build upon and extend what they call the current *pragmatic* or deterministic approach. Indeed, they argue that the traditional approach of obtaining indices from standard tests, the pragmatic discounting of minor factors as unimportant and extreme factors as unlikely, and the use of the Factor of Safety (see Chapter 5), which they describe as 'an experienced-based index, intended to aid judgement and decision-making', have together only been adequate for routine situations and problems. To further their argument, they go on to point out the following.

- Geotechnical failures are not a rarity and are sometimes disastrous.
- Conventional stability analysis with traditional Factors of Safety are not always capable of averting undesirable performance (Morgenstern, 1991).
- Over-designing costs money and often works against achieving an elegant solution.
- Excessive use of codes of practice can result in unsatisfactory performance if done by the inexperienced. For example, over-conservatism in the assessment of natural hillsides could prohibit new development, result in reductions in land values and raise fears regarding the integrity of existing developments, while lack of recognition of potential failures could result in dire consequences.

Over the last two decades, risk assessment methods have begun to be applied to quantify the risk of slope failures as a result of growing pressure on the geotechnical community from the following sources.

- Clients who want to know their exposure to risk and assign priorities (e.g. Lee and Charman, 2005; Evans *et al.*, 2007).
- Regulatory requirements by governments (e.g. AGS, 2007a; Besson *et al.*, 1999; Cave, 1992; Couture, 2011a; DRM, 1990; Garry and Graszk, 1997; Graszk and Toulemont, 1996; Scottish Executive, 2006).
- Public bodies who are increasingly concerned about the adequacy of safety systems or measures, especially after disasters.

In Hong Kong, there has been a significant move towards the development of a risk-based approach to supplement the conventional *geotechnical* approach for particular types of slope problems. Ho *et al.* (2000) suggest that the increasing use of risk assessment in Hong Kong was due to a number of factors, including

- There is growing realisation that considerable uncertainties are associated with ground and groundwater conditions, especially given the inherent variability of weathered profiles and tropical rainstorm characteristics; even slopes that have previously been assessed as being up to the required standards can have a fairly high failure rate (e.g. Ho and Lau, 2010; Morgenstern, 2000; Whitman, 1984, 1997; Wong and Ho, 2000).
- A risk-based approach assists in prioritisation of the retro-fitting of stabilisation measures to smaller-sized slopes with less serious failure consequences and the development of a rational strategy to deal with this category of slopes.
- A risk-based approach facilitates the communication of the realities of landslide risk to the public.

Landslide risk in areas of natural terrain, for example, is evaluated using a qualitative risk assessment and design framework known as the Design Event approach (Ng et al., 2002). Mitigation measures to protect a site are determined with reference to a landslide magnitude associated with a notional return period estimated from site conditions and the historical record (i.e. a design event). Depending on the susceptibility of the hillside to failure and the potential consequence at the site, the design event may be

- a 'conservative event' representing a reasonably safe, but not overly cautious, estimate of the hazard that may affect the site, with a notional return period of the order of 100 years, and is generally based on the largest historical landslide over the past 50–100 years
- a 'worst credible event', which is a very conservative estimate such that the occurrence of a more severe event is highly unlikely, with a notional return period of the order of 1000 years, and generally corresponding to the largest credible landslide determined by the interpretation of historical landslide data, geomorphological evidence in the catchment and any other relevant evidence from similar terrain in Hong Kong.

In Australia, landslide risk assessment has become an important component of the land development process, stimulated by a landslide in July 1997 at the New South Wales alpine resort of Thredbo, which destroyed two ski lodges and killed 18 occupants. At the subsequent Coronial Inquiry, the coroner recommended that the Building Code of Australia be amended to include geotechnical considerations when assessing and planning community developments in hillside environments, and that the guidelines for landslide risk management prepared by the Australian Geomechanics Society (AGS, 2000) should be taken into account in this exercise (Hand, 2000). In response, the Australian Building Codes Board engaged the AGS to develop guidelines for landslide risk management (Leventhal and Walker, 2005a, b). In 2007, the AGS published Practice Note Guidelines for Landslide Risk Management, one of a series of three guidelines related to LRM that have been prepared by AGS with funding under the National Disaster Mitigation Programme (AGS 2007a–d). The practice note provides advice both to practitioners as to the performance of project-specific landslide risk assessment and management and to government officers on the interpretation of the reports they receive. Relatively detailed procedures are provided for practitioners to follow regarding landslide likelihood/probability of occurrence, consequence analysis, risk estimation, and risk assessment and management. The commentary document (AGS, 2007b) provides an example of a risk evaluation case history and is a useful step-by-step illustration of how to apply the procedures.

In Canada, under the Federal Emergency Management Act (Canada, 2007) and as part of its current departmental planning responsibilities for emergency preparedness, Natural Resources Canada (NRCan) is responsible for developing and maintaining a civil emergency plan for the provision of technical information and advice related to several geophysical hazards, including landslides. As part of this responsibility, NRCan's Earth Science Sector is developing, through its Geoscience for Public Safety Program, national technical guidelines and best practices related to landslides. A series of Geological Survey of Canada Open File reports were published in 2012 that form the basis of the Canadian Technical Guidelines and Best Practices on Landslides. These reports are intended to provide Canadian engineers, geoscientists and other landslide practitioners with a framework for landslide risk assessment and management in Canada (e.g. Couture, 2011a, b; Porter and Morgenstern, 2012; VanDine, 2012).

In the last decade, guidelines for disaster risk management have also been published in Germany (Kohler et al., 2004). The guidelines are directed to emergency aid, reconstruction and food security programmes in regions threatened by geophysical hazards. The document describes general procedures for risk analysis, with three examples illustrating applications of the procedures to floods, droughts and erosion. Although the guidelines are not specific to landslides, the general procedures can be applied to landslides.

Guidelines for landslide likelihood/probability of occurrence mapping and land use planning have been published in Switzerland (OFAT/OFEE/OFEFP, 1997). The objectives are

- Guide geotechnical specialists in their practice for the evaluation of landslide risks
- Assist politicians in their decision process on land use planning
- Inform land owners of the potential hazards that can affect their properties.

The document provides basic tools and a standard methodology and terminology, with examples for developing hazard maps and land use planning throughout Switzerland (e.g. Lateltin, 2002).

Notable advances in landslide risk assessment have also occurred in France with the preparation of Risk Prevention Plans (e.g. Leroi et al., 2005) and in Italy, especially following the legislation after the 1998 Sarno disaster, when over 160 people were killed by mudflows (e.g. Aleotti et al., 2000; Cardinali et al., 2002; Eusebio et al., 1996; Sorriso-Valvo 2004). The Sarno decree (enforced by Law 267) required Basin Authorities in Italy to carry out the following tasks:

1. complete the reconnaissance of geological–hydraulic phenomena threatening property and lives in Italy
2. evaluate the hazard levels
3. assess the risk
4. urgently enforce rules for land use and management, including urban plans.

Over the last decade or so, risk-based methods have been developed and applied in a wide range of situations. These include the economic evaluation of coastal landslide and cliff recession problems in the UK (e.g. Hall et al., 2000; Lee et al., 2000, 2001), forestry practices in British Columbia (e.g. Fannin et al., 2005; VanDine et al., 2002; Wise et al., 2004), the impact of peat slides on wind farm development in Scotland (e.g. Scottish Executive, 2006), safety along the Canadian National Railway and Canadian Pacific Railway networks (e.g. Evans et al., 2005), threats to the integrity of onshore oil and gas pipelines (e.g. Lee and Charman, 2005) and the vulnerability of the rural road network in Nepal (e.g. Petley et al., 2007; Sunuwar et al., 2005).

Landslide hazard and risk studies are also becoming of increasing importance in supporting the planning and development of offshore oil and gas fields, such as in the Gulf of Mexico, the Caspian Sea and the West Nile delta (e.g. Evans et al., 2007; Jeanjean et al., 2003), as well as in more hostile environments. For example, the Ormen Lange field is located in the Norwegian Sea, 120 km from the coastline, in water depths of about 800–1100 m. The field lies within the Storegga landslide, one of the world's largest known submarine slides, with an estimated slide volume in excess of 3000 km^3. The landslide occurred around 8000 years ago, triggering a major tsunami for which evidence has been found along the coasts of Norway, Scotland and the Faroe Islands (e.g. Solheim *et al.*, 2005). Given the potentially catastrophic consequences of a similar event today, it was deemed essential to clarify and quantify the risks associated with submarine slides in the area in order to obtain approval for field development from the authorities. A major effort was therefore undertaken to evaluate the stability of the slopes in the Ormen Lange area and to quantify the risks (e.g. Nadim *et al.*, 2005b).

Risk assessment is not, therefore, merely a new fad or fashion, but a broader framework for considering the threat and costs produced by landsliding and for examining how best to manage both landslides and the risk posed by landslides.

REFERENCES

Aleotti P, Baldelli P and Polloni G (2000) Hydrogeological risk assessment of the Po River Basin (Italy). In *Landslides: In Research, Theory and Practice* (Bromhead EN, Dixon N and Ibsen ML (eds)). Thomas Telford, London, pp. 13–18.

AGS (Australian Geomechanics Society) (2000) Landslide risk management concepts and guidelines. *Australian Geomechanics* **35(1)**: 49–52.

AGS (2007a) Practice note guidelines for landslide risk management 2007. *Australian Geomechanics* **42(1)**: 63–114.

AGS (2007b) Commentary on the Practice note guidelines for landslide risk management. *Australian Geomechanics* **42(1)**: 115–158.

AGS (2007c) Guideline for landslide susceptibility, hazard and risk zoning for land use planning. *Australian Geomechanics* **42(1)**: 13–36.

AGS (2007d) Australian GeoGuides for 'Slope management and maintenance'. *Australian Geomechanics* **42(1)**.

Beck U (1992) *Risk Society: Towards a New Modernity*. SAGE Publications, London.

Bernstein PL (1996) *Against the Gods: The Remarkable Story of Risk*. John Wiley and Sons.

Besson L, Durville JL, Garry G *et al.* (1999) *Plans de prevention des risques naturels (PPR) – Risques de mouvements de terrain*. Guide méthodologique. La Documentation Française, Paris (in French).

BSI (1991) BS 4778: Quality vocabulary. BSI, London.

Burton I and Kates RW (1964) The perception of natural hazards in resource management. *Natural Resources Journal* **3**: 412–441.

Canada (2007) *Emergency Management Act*, Bill C-12, 22 June 2007.

Cardinali M, Reichenbach P, Guzzetti F, Ardizzone F, Antonini G, Galli M, Cacciano M, Castellani M and Salvati P. (2002) A geomorphological approach to the estimation of landslide hazards and risks in Umbria, Central Italy. *Natural Hazards and Earth System Sciences* **2**: 57–72.

Carrara A (1983) Multivariate models for landslide hazard evaluation. *Mathematical Geology* **15**: 403–427.

Catanach AH Jr and Ragatz JA (2010) 2008 Market Crisis: Black Swan, Perfect Storm or Tipping Point? *Bank Accounting and Finance*, April–May: pp. 20–26.

Cave PW (1992) Natural hazards, risk assessment and land use planning in British Columbia: progress and problems. *Proceedings of First Canadian Symposium on Geotechnique and Natural Hazards, Vancouver, Canada*. BC Geological Survey Branch Open File 1992-15, pp. 1–12.

Chicken JC and Posner T (1998) *The Philosophy of Risk*. Thomas Telford, London.

Coburn A and Spence R (1992) *Earthquake Protection*. Wiley, Chichester.

Couture R (2011a) *Introduction – National Technical Guidelines and Best Practices on Landslides*. Geological Survey of Canada, Open File 6765.

Couture R (2011b) *Terminology – National Technical Guidelines and Best Practices on Landslides*. Geological Survey of Canada, Open File 6824.

Cutter SL (1993) *Living with Risk*. Edward Arnold, London.

DRM (Délégation aux risques majeurs) (1990) *Les études préliminaires à la cartographie réglementaire des risques naturels majeurs*. La Documentation Française, Paris (in French).

Eusebio A, Grasso P, Mahtab A and Morino A (1996) Assessment of risk and prevention of landslides in urban areas of the Italian Alps. In *Landslides* (Senneset K (ed.)). Balkema, Rotterdam, pp. 189–194.

Evans SG, Cruden DM, Bobrowsky PT *et al.* (2005) Landslide risk assessment in Canada: a review of recent developments. In *Landslide Risk Management* (Hungr O, Fell R, Couture R and Eberhardt E (eds)). Balkema, Rotterdam, pp. 351–363.

Evans TG, Usher N and Moore R (2007) Management of geotechnical and geohazard risks in the West Nile Delta. *Proceedings of the 6th International Offshore Site Investigation and Geotechnics Conference: Confronting New Challenges and Sharing Knowledge*. Society for Underwater Technology (SUT), London.

Fannin RJ, Moore GD, Schwab JW and VanDine DF (2005) Landslide risk management in forest practices. In *Landslide Risk Management* (Hungr O, Fell R, Couture R and Eberhardt E (eds)). Balkema, Rotterdam, pp. 299–320.

Garry G and Graszk E (1997) *Plans de prévention des risques naturels prévisibles (PPR) – Guide général*. La Documentation Française, Paris (in French).

Giddens A (1999) *Runaway World*. Profile Books, London.

Graszk E and Toulemont M (1996) Plans de prévention des risques naturels et expropriation pour risques majeurs. Les mesures de prévention des risques naturels de la loi du 2 février 1995. *Bulletin des Laboratoires Ponts et Chaussées* **206**: 85–94 (in French).

Hall JW, Lee EM and Meadowcroft IC (2000) Risk-based assessment of coastal cliff recession. *Proceedings of the ICE: Water and Maritime Engineering* **142**: 127–139.

Hand D (2000) *Report on the Inquest into the Deaths Arising from the Thredbo Landslide*. State of New South Wales, Australia, Attorney General's Department – Office of the NSW Coroner.

Ho KKS and Lau JWC (2010) Learning from slope failures to enhance landslide risk management. *Quarterly Journal of Engineering Geology and Hydrogeology* **3**: 33–68.

Ho K, Leroi E and Roberds B (2000) Quantitative risk assessment: application, myths and future direction. *Proceedings of the Geo-Eng Conference, Melbourne, Australia,* Publication 1, pp. 269–312.

HSE (Health and Safety Executive) (2001) *Reducing Risks, Protecting People*. HSE Books/HMSO, Norwich.

Holton GA (2004) Defining risks. *Financial Analysts Journal* **60**: 19–25.

ISO (International Standardization Organization) (2009) ISO 31000: Risk management – principles and guidelines. ISO, Geneva.

ISSMGE (International Society of Soil Mechanics and Ground Engineering) (2004) Risk Assessment – Glossary of Terms. TC32, Technical Committee on Risk Assessment and Management Glossary of Risk Assessment Terms – Version 1, July 2004.

Jeanjean P, Hill A and Taylor S (2003) The challenges of confidently siting facilities along the Sigsbee Escarpment in the Southern Green Canyon area of the Gulf of Mexico. Framework of integrated studies. *Proceedings of the Offshore Technology Conference, Houston, TX, USA*, Paper 15156.

Jones DKC (1993) Environmental hazards in the 1990s: problems, paradigms and prospects. *Geography* **78**: 161–165.

Jones DKC (1995a) The relevance of landslide hazard to the International Decade for Natural Disaster Reduction. In *Landslides Hazard Mitigation*. The Royal Academy of Engineering, London, pp. 19–33.

Jones DKC (1995b) Landslide hazard assessment. In *Landslides Hazard Mitigation*. The Royal Academy of Engineering, London, pp. 96–113.

Junger S (1997) *The Perfect Storm: A True Story of Men Against the Sea*. WW Norton, New York, NY.

Kates RW (1978) *Risk Assessment of Environmental Hazard*. ICSU/SCOPE Report 8. Wiley, Chichester.

Knight FH (1921) *Risk, Uncertainty and Profit*. Hart, Schaffner & Marx, Boston, MA (republished 1964 by Century Press, New York, NY).

Kohler A, Julich S and Bloemertz L (2004) *Guidelines: Risk analysis – A Basis for Disaster Risk Management* (Danaher P (transl.)). German Society for Technical Cooperation (GTZ), Federal Ministry for Economic Cooperation and Development, Eschborn, Germany.

Lateltin OJ (2002) Landslides, land-use planning and risk management: Switzerland as a case-study. In *Instability: Planning and Management* (McInnes RG and Jakeways J (eds)). Thomas Telford, London, pp. 89–96.

Lee EM and Charman JH (2005) Geohazards and risk assessment for pipeline route selection. In *Terrain and Geohazard Challenges Facing Onshore Oil and Gas Pipelines* (Sweeney M (ed.)). Thomas Telford, London, pp. 95–116.

Lee EM, Brunsden D and Sellwood M (2000) Quantitative risk assessment of coastal landslide problems, Lyme Regis. In *Landslides: In Research, Theory and Practice* (Bromhead EN, Dixon N and Ibsen ML (eds)). Thomas Telford, London, pp. 899–904.

Lee EM, Hall JW and Meadowcroft IC (2001) Coastal cliff recession: the use of probabilistic prediction methods. *Geomorphology* **40**: 253–269.

Leroi E, Bonnard C, Fell R and McInnes R (2005) Risk assessment and management. In *Landslide Risk Management* (Hungr O, Fell R, Couture R and Eberhardt E (eds)). Balkema, Rotterdam, pp. 159–198.

Leventhal AR and Walker BF (2005a) Risky business – development and implementation of a national landslide risk management system. In *Landslide Risk Management* (Hungr O, Fell R, Couture R and Eberhardt E (eds)). Balkema, Rotterdam, pp. 401–409.

Leventhal AR and Walker BF (2005b) Risky business – the development and implementation of a national landslide risk management system for Australia, the state of play, September 2005. *Australian Geomechanics* **40(4)**: 1–14.

Luhmann N (1996) *Modern Society Shocked by its Risks*. Social Sciences Research Centre Occasional Paper 17, Department of Sociology, University of Hong Kong.

Morgenstern NR (1991) Limitations of stability analysis in geotechnical practice. *Geotechnia* **61**: 5–19.

Morgenstern NR (2000) Performance in Geotechnical Practice. Inaugural Lumb Lecture. *Transactions of the Hong Kong Institution of Engineers* **7**: 2–15.

Nadim F, Kvalstad TJ and Guttormsen T (2005) Quantification of risks associated with seabed instability at Ormen Lange. *Marine and Petroleum Geology* **22**: 311–318.

Ng KC, Parry S, King JP, Franks CAM and Shaw R (2002) *Guidelines for Natural Terrain Hazard Studies*. GEO Report 138.

Normile D (2012) One year after the devastation, Tohoku designs its renewal. *Science* **335**: 1164–1166.

OFAT/OFEE/OFEFP (Office fédéral de l'aménagement du territoire/Office fédéral de l'économie des eaux/Office fédéral de l'environnement, des forêts et du paysage) (1997) *Prise en compte des dangers dus aux mouvements de terrain dans le cadre de l'aménagement du territoire. Recommandations*. OCFIM 310.023f, Berne (in French).

Paté-Cornell E (2012) On 'Black Swans' and 'Perfect Storms': risk analysis and management when statistics are not enough. *Risk Analysis* **32**: 1823–1833.

Perry AH (1981) *Environmental Hazards in the British Isles*. George Allen & Unwin, London.

Petley DN, Hearn GJ, Hart A, Rosser NJ, Dunning SA, Oven KJ and Mitchell WA (2007) Trends in landslide occurrence in Nepal. *Natural Hazards* **43**: 23–34.

Porter M and Morgenstern N (2012) *Landslide Risk Evaluation – Canadian Technical Guidelines and Best Practices on Landslides*. Geological Survey of Canada, Open File report.

Purdy G (2010) ISO 31000 – setting a new standard for risk management. *Risk Analysis* **30**: 881–886.

Royal Society (1992) *Risk: Analysis, Perception and Management. Report of a Royal Society Study Group*. Royal Society, London.

Sapolsky H (1990) The politics of risk. *Daedalus (Proceedings of the American Academy of Arts and Sciences)* **119(4)**: 83–96.

Scottish Executive (2006) *Peat Landslide Hazard and Risk Assessments: Best Practice Guide for Proposed Electricity Generation Developments*. Scottish Executive, Edinburgh.

Sjöberg L (ed.) (1987) *Risk and Society: Studies in Risk Generation and Reactions to Risk*. Allen and Unwin, London.

Solheim A, Bryn P, Berg K, Sejrup HP and Mienert J (2005) Ormen Lange – an integrated study for the safe development of a deepwater gas field within the Storegga Slide Complex, NE Atlantic continental margin; executive summary. *Marine and Petroleum Geology* **22(1–2)**: 1–9.

Sorriso-Valvo M (2004) Landslide risk assessment in Italy. In *Landslide Hazard and Risk* (Glade T, Anderson M and Crozier MJ (eds)). Wiley, Chichester, pp. 697–732.

Stein S, Geller RJ and Liu M (2012) Why earthquake hazard maps often fail and what to do about it. *Tectonophysics* **562–563**: 1–25.

Sunuwar L, Karkee MB and Shrestha D (2005) A preliminary landslide risk assessment of road network in mountainous region of Nepal. In *Landslide Risk Management* (Hungr O, Fell R, Couture R and Eberhardt E (eds)). Balkema, Rotterdam, pp. 411–422.

Taleb NN (2010) *The Black Swan: The Impact of the Highly Improbable*. Penguin, London.

UNISDR (United Nations Office for Disaster Risk Reduction) (2009) UNISDR Terminology on Disaster Risk Reduction. See http:/www.unisdr.org/we/inform/terminology#letter-d (accessed 28/06/2013).

United States Department of Defence (2002) DoD News Briefing – Secretary Rumsfeld and General Myers. Feb. 12 2002. See http:/www.defence.gov/Transcripts/Transcripts.aspx?TranscriptID=2636.

VanDine DF (2012) *Risk Management – Canadian Technical Guidelines and Best Practices on Landslides*. Geological Survey of Canada, Open File 6996.

VanDine DF, Jordan P and Boyer DC (2002) An example of risk assessment from British Columbia, Canada. In *Instability: Planning and Management* (McInnes RG and Jakeways J (eds)). Thomas Telford, London, pp. 399–406.

Varnes DJ (1984) *Landslide Hazard Zonation: A Review of Principles and Practice*. Engineering Geology Commission on Landslides and other Mass Movements on Slopes, UNESCO, Paris.

Whitman RV (1984) Evaluating calculated risk in geotechnical engineering. *Journal of the American Society of Civil Engineers* **110**: 143–188.

Whitman RV (1997) Acceptable risk and decision-making criteria. *Proceedings of the International Workshop on Risk-Based Dam Safety Evaluation, Trondheim, Norway*.

Wise M, Moore G and VanDine D (2004) *Landslide Risk Case Studies in Forest Development Planning and Operations*. British Columbia Ministry of Forests, Forest Science Program.

Wong HN and Ho K (2000) Learning from slope failures in Hong Kong. Keynote Paper. *International Symposium of Landslides 2000, Cardiff*. CD-ROM. Available from e.bromhead@kingston.ac.uk.

Landslide Risk Assessment
ISBN 978-0-7277-5801-9

ICE Publishing: All rights reserved
http://dx.doi.org/10.1680/lra.58019.021

Chapter 2
The basic elements of landslide risk assessment

2.1. Introduction: the risk assessment process

Risk assessment has been defined as 'the structured gathering of the information available about risks and the forming of judgements about them' (DoE, 1995). It is widely claimed that its main function is to produce objective advisory information that effectively links science with decision-making, thereby providing the basis for better-informed decisions. However, the extent to which any risk assessment process can actually be considered to achieve objective results is debatable, and will be considered later.

At the simplest level, risk assessment can be envisaged to address four main questions.

1. What can go wrong to cause adverse consequences?
2. What is the range of possible adverse consequences and their patterns of severity?
3. What is the probability or frequency of occurrence of different severities of adverse consequences?

The answers to these three questions then allow the fourth question to be addressed.

4. What can be done, at what cost, to manage and reduce unacceptable levels of adverse consequences?

Many models of risk assessment exist, and there is a varied, and sometimes confusing, set of terminology to accompany the various approaches. The range of models is a reflection of the differing contexts within which forms of risk assessment are applied. The sequential scheme presented below is recommended for general use (Figure 2.1).

1. Description of intention.
2. Hazard identification.
3. Estimation of magnitude and frequency/probability of hazards.
4. Identification of the likely range of adverse consequences that identified hazards will generate.
5. Estimation of magnitude range of adverse consequences.
6. Estimation of frequency/probability of differing magnitude of adverse consequences.

Combining the results of Stages 5 and 6 results in the next stage:

7. *Risk estimation*, which may be described as a combination of the likely adverse outcomes/ adverse consequences of an event and the probability/probabilities of occurrence.

21

Landslide Risk Assessment

Figure 2.1 Flow chart of the stages in the risk assessment process

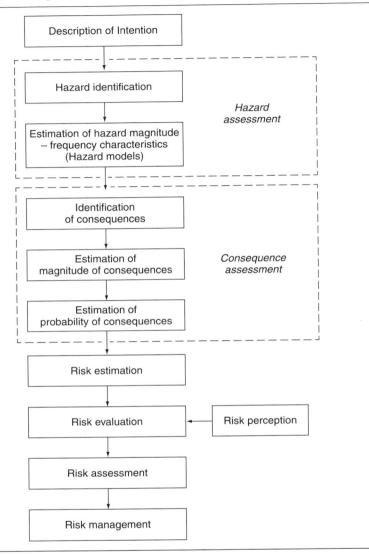

This has often been considered to be the conclusion of the risk assessment process. However, posing the question 'so what?' results in the addition of three further stages:

8 *Risk evaluation*, which is concerned with determining the significance of estimated risks for those affected. This late stage in the process requires the gathering of information about how affected, or potentially affected, people feel about the potential threats to valued objects, processes, amenities, environments and so on, an area of study generally referred to as *risk perception*. This aspect of risk assessment is becoming of increasing importance, since in many contexts decisions about risk issues are no longer seen to be the sole preserve of the scientific

elite (experts) but a matter of legitimate concern for all those people who are, or could be, affected (stakeholders).
9 Risk estimation and risk evaluation together inform *risk assessment*, where the results should be combined with some form of cost–benefit analysis (or benefit–cost analysis) to facilitate decisions as to (i) whether or not to proceed with the proposed strategy/project, (ii) whether to modify the proposed strategy/project so as to avoid high levels of risk or (iii) whether to accept, reduce or minimise some or all of the remaining identified risks by the implementation of management strategies.
10 *Risk management*, which involves the actual implementation of the identified risk reduction strategies.

However, it has to be recognised that the form of risk assessments actually undertaken with reference to landsliding will differ greatly in terms of emphasis and details owing to variations in scale, purpose, time constraints and availability of data. A risk assessment undertaken to provide a general indication of the threat across an area or region (somewhat misleadingly termed 'global' by Wong *et al.*, 1997) will be different to that undertaken for a development at a particular site.

2.2. Description of intention

This crucial initial stage of the risk assessment process involves two activities.

- *Screening* is the process by which it is decided whether or not a risk assessment is required, or whether it is required for a particular element, or elements, within a project.
- *Scoping* defines the focus of enquiry and the spatial and temporal limits of the resulting risk assessment. These may be determined on purely practical grounds such as budget constraints, time constraints, staff availability or data availability. On the other hand, there may be strong arguments for limiting the spatial and temporal dimensions of a risk assessment, especially in view of the uncertainties of predicting and valuing longer-term consequences.

Scoping and screening should culminate in the production of a clear and detailed *statement of intent/ description of intention* – an important document in an increasingly litigious society.

2.3. Hazard assessment

Hazard assessment is the first active stage of a risk assessment and is sometimes referred to as *hazard auditing* or *hazard accounting*. This stage focuses upon the development of hazard models, together with estimation of the nature, size (magnitude) and frequency characteristics of hazardous events (i.e. landsliding and landslide-generated hazards) within the parameters established in the scoping process (see Chapter 3). Should this involve, or even focus upon, the possibility of a major, potentially highly destructive, event (i.e. catastrophic failure), then the hazard assessment must concentrate on establishing the likely magnitude, character, *time to onset* and *speed of onset* of the envisaged event, or sequence of events. Speed of onset (i.e. suddenness) and violence of activity are both important in determining the extent to which risk can be reduced through the implementation of monitoring programmes, the establishment of emergency action plans, the development of warning systems and the adoption of evacuation procedures. Any activities that seek to reduce the element of surprise associated with hazardous events will reduce risk.

Geophysical events (geohazards), including landslides, simply do not just occur – there has to be preparatory activity that leads to the build-up of energy, followed by an imbalance between forces and constraints that leads to the release of energy, which, in turn, is followed by dynamic activity (i.e. the hazard event). This can be described in terms of a three-phase model of hazard:

Incubation → trigger → event

However, this model is an over-simplification in many situations, especially with reference to the post-trigger sequence of events. First, the trigger leads to the *initial event*, which may not necessarily be the most important event in terms of size or impact; indeed, it may not necessarily have any great hazard potential. However, this initial event may, in turn, directly cause further events (e.g. unloading produced by a small failure leading to the formation of further landslides, which may be far larger, more violent and have much greater impact potential). Thus, the sequence has to be extended to include these *primary events* that flow directly from, and are intimately associated with, the initial event. Thus, a more realistic four-phase model of hazard is

incubation → trigger → initial event → primary events

This sequence can be envisaged to result in three groups of outcomes (Figure 2.2): *further hazards*, *adverse consequences* and *benefits*. Focusing for the moment on further hazards, three main groups can be distinguished.

Figure 2.2 The hazard event cascade. The diagram shows how incubation in both the 'physical' and 'human' systems interact to produce a hazard event, resulting in a range of outcomes, including further hazards, adverse consequences and benefits

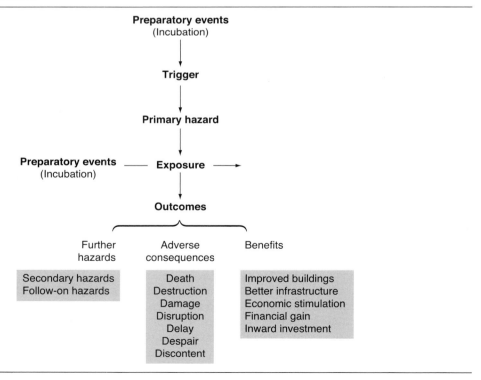

- *Post-event hazards* occur after the main hazard event sequence has concluded and are a product of the specific geophysical system returning towards stability (i.e. the system is relaxing), as is the case with the aftershocks following major earthquakes and the landslides that continue to occur after major slope failure events.
- *Secondary hazards* are different geohazards generated by the main hazard event sequence. Examples of these are the destructive tsunamis generated by earthquakes and major landslides, and the often-unexpected river floods caused by the failure of landslide-generated dams, which can, under certain circumstances, generate further instability elsewhere owing to slope undercutting. The crucial point about secondary hazards is that they can have an impact upon areas that are sometimes far removed from the primary hazard location, often after a considerable time lag (e.g. floods on the River Indus generated by landslides in headwater areas in the Karakoram Mountains).
- *Follow-on hazards* are rather different forms of hazard, generated by the primary events but occurring after variable lengths of time. Fires caused by overturned stoves, electrical short-circuits and broken gas pipes, localised flooding caused by broken supply mains or sewers, disease, looting, and famine can all be grouped under this heading.

Hazard assessment can be a surprisingly difficult task, especially where complex or compound hazards are involved, and it is necessary to analyse carefully the phenomena that produce *hazard*. As a result, the use of simple labels such as 'fall', 'flow' and 'slide' may be insufficient, because the purpose of the study is to identify the extent and scale of the processes and ground deformation mechanisms that have the potential to cause damage and destruction (Figure 2.3). So, once the main types of hazard have been identified, the focus should then be on the detail.

This stage should not be restricted to ascertaining the magnitude and frequency characteristics of the principal damaging event (e.g. the main landslide), although there does need to be an assessment of the likely maximum size and extent of landslides and any envisaged *secondary hazards* and *follow-on hazards*, so that the largest and most complex event can be envisaged. Equally important is the identification of possible sequences, or chains, of hazards that could develop from an initial event. Establishing models for such *hazard sequences* can be difficult for medium- to large-scale events that are intermediate between the simple, small-scale, clearly defined and spatially confined event on the one hand and the high-magnitude low-frequency catastrophic event resulting in a zone of total destruction.

Scenarios have to be developed regarding *what could happen* using a combination of hindsight reviews of relevant past impacts (analogues) and expert judgement. The possibilities that an initial hazard event could lead to a range of further possible hazardous processes that in turn could result in further possible hazards can be portrayed by a branching network of events, or hazard sequences, known as an *event tree* (see Chapter 5).

There are various approaches for expressing the likelihood of hazard events, with the following four being those most commonly encountered.

- Identified but unspecified likelihood for areas and locations where hazard and vulnerability obviously co-exist but for which no data exist, such as the case of a new house built at the base of a cliff or on an ancient landslide (with landslides then presenting a 'threat' to the house). Such subjective and vague statements are generally only of value at the reconnaissance level.

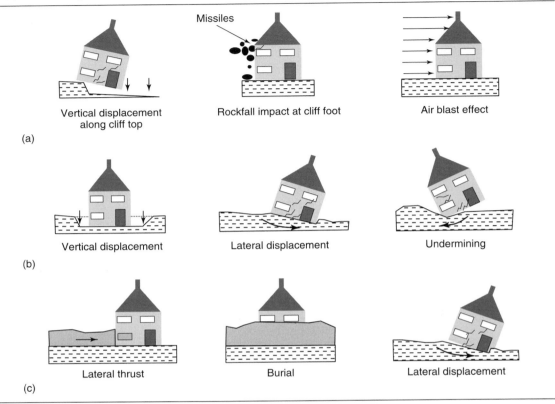

Figure 2.3 Types of structural damage associated with different types of landslide movements: (a) falls, (b) slides and (c) flows. (Adapted from Léone *et al.*, 1996)

- Qualitative expressions of likelihood using categories with word labels. MAFF (2000) advocates the following five-category scheme:
  frequent likely to occur many times during the period of concern
  probable likely to occur several times during the period of concern
  occasional likely to occur sometime during the period of concern
  remote unlikely, but possible
  improbable can be assumed, for most purposes, that it will not occur.
  Qualitative expressions of likelihood using numbered categories, such as from 1 to 5 or from 1 to 10, where 1 represents very rare (exceedingly low probability) and the uppermost value of 5 or 10 represents very high likelihood/certainty.
- Quantitative estimations of frequency and probability. The former are usually expressed in terms of average number of occurrences per unit time, or return periods/recurrence intervals. Probabilities are also determined with reference to time periods and expressed either as percentage probability or a decimal value between 0 and 1. These estimations can be described in terms of 'likelihood' along the lines adopted by the Inter-Governmental Panel on Climate Change (IPCC, 2012):
  extremely likely/virtually certain   >95%
  very likely   90–95%

likely	>66%
more likely than not	>50%
unlikely	<33%
very unlikely	<10%
extremely unlikely	<5%
exceptionally unlikely	<1%.

These estimates should be accompanied by a statement of the levels of uncertainty or confidence.

In reality, studies of the potential threats posed by landsliding can result in a large number of relevant hazard scenarios, all of which should be explored individually as to the nature and magnitude of processes likely to be generated (e.g. falling, thrusting, surface deformation or lateral pressure) that combine to produce the possible range of credible future threats. The resulting event trees merely represent the first stage of complexity in the risk assessment process.

It is important to appreciate that landslide hazard can be dynamic over time and space. Thus, three aspects of change need to be considered when contemplating how *present threat* may be converted into *future threat*, within the time frame established by the scoping process.

- There are changes to the biophysical environment that may increase or decrease the magnitude, frequency and character of landsliding over time (see Chapter 12).
- There are development processes that may increase or decrease the hazard potential of future landsliding owing to changes in the patterns of exposure (see Chapter 1).
- There are actions that may be taken to limit the potential for landsliding to pose costs to society, such as stabilisation measures. Such measures are increasingly being termed 'barriers', and might include a programme of inspection and monitoring, the construction of rock fences or debris flow culverts and bridges along a road (e.g. Bromhead, 2005), or stress-relieving of a buried pipeline to ensure that the strains imposed by very slow ground movement do not build up and cause rupture (e.g. BP, 2008).

2.4. Consequence assessment

The next major stage in the risk assessment process, and one that is crucial to the determination of risk, is *consequence assessment*, which, as its name suggests, attempts to identify and quantify the full range of adverse consequences arising from the identified patterns and sequences of hazard. But if predicting hazard sequences appears difficult, then identifying and quantifying consequences introduces new realms of uncertainty.

In most situations, it is possible to envisage a cascade of hazards and consequences flowing from an initiating event (see Figure 2.2). In the case of major geophysical events, losses are usually attributed to *primary*, *secondary* and *tertiary* processes. For example, earthquake impacts can be grouped into those resulting directly from ground shaking (including landslides), those produced by the subsequent fires, tsunamis, floods, etc., and those associated with human suffering and economic effects. Even in the case of simple situations, such as a cliff fall, it is possible for the ramifications to spread far and wide owing to chains of events, known as *accident sequences* (Figure 2.4), which illustrates how chance has a major role in determining consequences. However, where slope failures are large or extensive and affect concentrations of human activity and wealth, such as urban areas or industrial complexes, the numbers of identifiable potential accident sequences are huge and produce complex webs of both causes and consequences (event trees) that are exceptionally difficult

Landslide Risk Assessment

Figure 2.4 Diagram showing how a 'hazard event sequence' (the horizontal series of boxes) can interact with human activity to result in an 'accident sequence' (the vertical sequence of boxes)

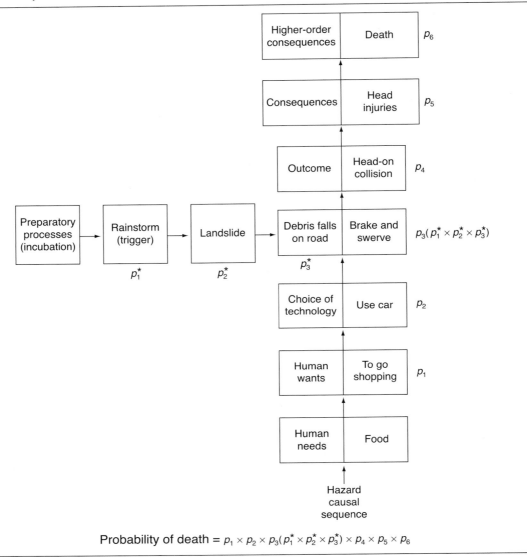

Probability of death = $p_1 \times p_2 \times p_3(p_1^* \times p_2^* \times p_3^*) \times p_4 \times p_5 \times p_6$

to unravel and virtually impossible to predict with any degree of accuracy. This highlights the applicability of the classic statement

There is only one past, but unlimited possible futures

Consequence assessment will, except in the cases of small and clearly defined problems and projects, inevitably have to be broad-brushed and based on a combination of

- analogies with known patterns of consequences from similar situations elsewhere (analogues)

- a range of scenarios produced by panels of experts
- computer modelling.

This is because exposure and vulnerability are dynamic over time (see Chapter 1). As 'risk resides in the future', it is necessary to recognise that future landslide events will probably have an impact upon patterns of land use, socio-economic activity and wealth that have changed from those that exist at the time of analysis. So it is important to envisage that the incubation of hazard and the incubation of exposure evolve on separate paths until they intersect when a slope failure materialises (see Figure 2.4). The further into the future an assessment seeks to go, the greater the potential scale of variation from the prevailing conditions at the time of the assessment – hence the obvious reaction to limit the temporal reach.

The problems associated with consequence assessments can be constrained to some extent by the careful use of scoping and clearly specified descriptions of intention. However, care needs to be taken not to make both the scoping and scenario development processes too restrictive. It has to be recognised that even the apparently most unlikely events or combinations of circumstances can, and indeed do, occur (e.g. Black Swans and Perfect Storms; see Chapter 1). The media has a tendency to refer to such events as *freak* occurrences, somehow implying that they are outside rational explanation and are therefore unpredictable. Good scenario development should nevertheless be able to envisage such possible occurrences within the category of *worst-case scenarios*, although computation of likelihood and actual levels of consequence may prove very difficult.

There is usually a very poor correlation between the magnitude of the primary event and the magnitude of adverse consequences. Very minor events can, through chance sequences of interaction that serve to amplify consequences, result in huge impacts. For example, a minor slump derails a freight train carrying toxic and inflammable chemicals so that it plunges into a river upstream of a town, and so on.

The possibility of huge impacts is referred to as *disaster potential* or *catastrophe potential*, supposedly depending on the magnitude and spatial extent of the possible event, but there is no agreed demarcation, so choice is usually down to personal preference. However, disaster potential need not reflect the magnitude of the hazard event but rather the vulnerability of people, structures and infrastructure at particular sites. As societies are increasingly dominated by blame cultures that seek redress and retribution when *things go wrong*, it is important that risk assessments do not ignore the possibilities of disastrous or catastrophic outcomes, even though the probabilities of occurrence may appear incredibly small.

The end product of consequence assessment should be an expression of the likelihood of differing magnitudes of adverse outcomes within a specified area resulting from the previously identified range of hazards. Likelihood of adverse consequences can be expressed in the same ways as likelihood of hazard.

- Where time and/or data are extremely limited, magnitude can be expressed as exceeding a certain threshold, but with no detail. Thus, the resulting estimations will be concerned with the likelihood of occurrence of a damaging landslide capable of creating adverse consequences or serious adverse consequences. Such simple 'will/will not' alternatives are highly subjective and only of value at a reconnaissance level.
- Magnitude can be expressed qualitatively using categories described by words to represent differing levels of adverse consequences. It is recommended that no more than five categories be

used, as larger numbers tend to result in confusion. Four categories are advocated in DoE (1995) including:
- severe
- moderate
- mild
- negligible.

Other terms that might be employed include *extremely severe, very severe* and *minor*. Such terms have a wide spectrum of applicability, including consequences for human populations, physical structures, infrastructure, environmental quality, ecological status and socio-economic activity, but need to be defined clearly and carefully.

- It is also possible to express qualitative estimations in terms of numerical gradings or numerical scoring scales, usually from 0 to 10, with 0 meaning no observable effects and 10 signifying total destruction. Alternatively, the principal adverse consequences can be weighted according to nature or level of harm posed by differing magnitudes of events; for example, ranging from 1 (no significant damage) via 50 (minor structural damage) and 500 (major structural damage or injury) to 2000 (loss of life) (DETR, 2000).

The desired objective of risk estimation is for magnitudes of adverse consequences to be expressed quantitatively in terms of numbers of deaths, severe injuries, minor injuries and people displaced or otherwise affected, together with the costs in terms of buildings, infrastructure, economic activity, environment, ecology and so on. The ultimate aim here is to get all losses expressed in comparable units, and preferably the same units – that is, monetary terms – but this remains extremely problematic because many adverse consequences are difficult to value. As a result, many assessments result in a mixture of quantitative and qualitative data, and it is sometimes the case that one outcome is highlighted, usually death, although this is a far from satisfactory measure of impact.

Consequences can also be probabilistic, since the losses incurred in a particular event can be dependent on a unique combination of circumstances. The inherent randomness in the main factors involved in accidents (e.g. the build-up of people in a particular location or the time of day or night) dictates that future landslides of the same magnitude cannot be expected to cause the same levels of adverse consequences. 'Barriers' may also be present that constrain or influence the realisation of particular consequence scenarios, including early-warning systems and evacuation procedures or the construction of oil-spill containment ponds.

It should be apparent that consequence assessment can prove extremely complex and time-consuming, since it can involve the creation of event trees of hazard, which then yield further branching accident sequences of consequences. Unsurprisingly, a more focused approach has been developed for use with valuable structures and infrastructure, which has been called 'the risk bow-tie'. The approach appears to have been developed by Imperial Chemical Industries (ICI) in the late 1970s (Lewis and Hurst, 2005). In the early 1990s, the Royal Dutch Shell Group adopted the bow-tie method as the company standard for analysing and managing risks. Over the last decade, the method has spread outside the oil and gas industry to include the aviation, mining, maritime, chemical and healthcare industries.

Bow-tie analysis focuses on identifying the 'incident' or 'state' that initiates the various consequence scenarios that might develop. These so-called 'top events' can also be referred to as 'initiating events' by risk specialists (e.g. Vinnem, 2007). An example of a top event would be the rupture of a pipeline following the occurrence of a hazard such as landslide movement. This will result in a loss of containment and release of product, initiating a range of possible consequence scenarios (e.g. Lee

Figure 2.5 The risk assessment 'bow-tie'

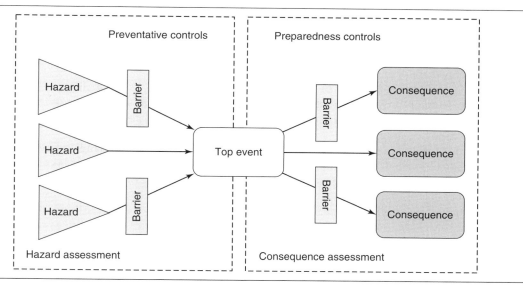

and Charman, 2005). In other situations, the top event may be landslide run-out reaching a building or transport infrastructure.

This approach requires that the nature of the 'top events' be established first, with the conditions that would cause them (i.e. the hazards) then being evaluated, together with the resulting consequences. In this instance, starting with one of a series of 'top events', the task is to establish the sequence, or sequences, of hazard events that would create the various conditions, and then determine the consequences that would result. In the case of the 'causes', the results are a series of branching networks that contract down to one at the top event; that is, all the combinations of circumstances that could result in one specific category of top event, akin to the 'fault trees' developed to analyse the failures of complex systems such as aircraft (see Chapter 5). In the case of the consequences, the results are branching networks of accident sequences (event trees) that broaden out from the 'top event'. The resultant portrayal of the risk concept takes the form of a 'bow-tie' (Figure 2.5; e.g. HSE, 2006), with the two wings representing the hazard and consequence sequences, and the central knot representing the 'top event' or primary hazard.

Bow-tie analysis is widely used in industry because it is an effective means of communicating risk issues to non-specialists. It can be used to demonstrate the linkages between hazards, consequences, risk controls (barriers and precautionary measures) and the overall risk management system (e.g. Hartford, 2009).

2.5. Risk estimation

Risk estimation represents an amalgam of the magnitude and probability of the hazard(s) and the magnitude and probability of the identified potential adverse consequences. The estimations have to be done in such a way that the results are understandable and comprehensible by others (i.e. non-specialists) and be capable of comparison with other risks. This implies the need for some standardisation of units and nomenclature.

Achieving this goal is problematic because risk assessments can be undertaken for a great variety of reasons and at very different scales. For example, landslide risk assessments may be undertaken in the contexts of

- specific small-scale engineering/construction projects such as road cuttings, individual buildings or other structures
- clearly defined small sites of known instability
- large-scale engineering projects such as dam sites
- linear construction projects through landslide-prone areas, such as roads, railways and pipelines
- large areas of known or potential instability, such as stretches of coastal cliffs, escarpments or major landslide complexes
- investigations of the possible spatially extensive effects of major engineering projects, such as the widespread response of slopes to new road-lines or the creation of a reservoir lake
- a really extensive assessment concerned with evaluating landslide risk due to potentially high-impact geohazard events such as earthquakes, widespread changes in land use (e.g. urbanisation) or as a consequence of climate change.

Combining the different approaches to estimating the likelihood of the hazard(s) and the likelihood and magnitude of adverse consequences produces a spectrum of risk estimations, extending from the exceptionally vague at one extreme to the relatively precise at the other. Vague risk estimations are often heavily based on hazard assessments that are extended to indicate risk. Thus, maps of landslide hazard may be interpreted in the context of risk. Similarly, predictions of major hazard events can also form basic risk estimations if specific levels of adverse consequences are assumed to follow (as with the various published hazard intensity scales). However, it has to be recognised that this use of hazard assessments as a proxy for risk assessment is imperfect, since although 'hazard' generates 'risk', the relationship is not a linear one; indeed, it is this practice that has contributed to the blurring of the important distinction between hazard and risk.

Disregarding the primitive forms of risk identification mentioned above results in the recognition of three broad groups of risk estimation:

- *Qualitative risk estimations* are those where both likelihood and adverse consequences are expressed in qualitative terms. They are therefore highly subjective estimations. One of the most widely used approaches is to combine the qualitative scoring of adverse consequences and the qualitative scoring of likelihood/probability in order to produce a matrix (Figure 2.6). The cells of the matrix can be labelled *high*, *medium*, *low* and *near-zero/negligible* to indicate the level of risk (as in Figure 2.6) or, if a numerical scheme is adopted, the numbered coordinates of each cell can be summed to yield the same end product, even though the number of cells may be greatly increased. Such processes facilitate the comparison of different risks and provide an indication of relative levels of risk, but it must be recognised that the results represent a gross oversimplification of reality. Indeed, this whole approach is capable of producing serious errors unless used critically and subjected to expert review.
- *Semi-quantitative risk estimations* are, as the name implies, combinations of qualitative and quantitative measures of likelihood and consequence. More usually, it is probabilities of frequency that are known, or assumed, while levels of consequence remain elusive.
- *Quantitative risk estimations*, combine values of detriment with probabilities of occurrence. It must be noted that such an approach frequently does not produce a single answer, since there can be probability distributions for both magnitude and frequency of the hazard event, as well as for a range of possible losses.

Figure 2.6 An example of a risk matrix

	Consequences			
Probability	Severe	Moderate	Mild	Negligible
High	high	high	medium/low	Near zero
Medium	high	medium	low	Near zero
Low	high/medium	medium/low	low	Near zero
Negligible	high/medium/low	medium/low	low	Near zero

The relative merits of *quantification* versus *qualitativism* is one of the seven major debates within risk management discussed in Royal Society (1992) and expanded upon in Hood and Jones (1996). It is quite natural for scientists and engineers to seek to employ numerical values and mathematical relationships in their work on risk, and therefore to extol the virtues of quantification. Numbers also provide a clear impression of detailed consideration, accuracy and precision – attributes much appreciated by decision-makers. However, it has to be recognised that quantitative risk estimation is much more demanding than qualitative estimation, in that it can require enormous amounts of data and considerable computational effort. The use of numbers also conceals the fact that the potential for error is great because of the assumptions made and the computations involved. Thus, while the value of the quantitative approach is widely accepted, it should not be seen as the only valid form of risk estimation. In many situations, constraints of time, resources and lack of data will make it impossible to produce anything other than a qualitative estimation. Such estimations are useful in their own right, and can often be further improved by inputs of expert judgement (see Chapter 5).

The extent to which risk estimation can be viewed as an objective process has also been a matter of much debate. Many scientists still hold to the view that there are two main categories of risk assessment – *objective risk* and *subjective risk* – with the former produced by scientific investigation and the latter resulting from human evaluations or what is sometimes called *the human factor*. As a result, quantitative risk estimation has long been seen to be the result of objective investigation. However, the very fact that the scoping process excludes certain risks and emphasises others, the fact that assumptions are made that reflect investigators' backgrounds, experience or emphasis of the project, together with the recognition that the value of *elements at risk* is often disputed (e.g. what is the value of an ancient monument?), indicate that all risk estimations are, to varying degrees, subjective (e.g. Stern and Fineberg, 1996). Thus, a better distinction is between *statistical* and *non-statistical* portrayals of risk.

2.6. Risk assessment and its relationship to risk evaluation and risk management

The distinction between *risk evaluation* and *risk assessment* is not always clear in practice, so it is sensible to consider them together. As further problems exist in terms of the distinction with *risk estimation*, the position of *risk evaluation* as a distinct stage appears somewhat precarious.

Risk evaluation is, quite simply, a judgemental process designed to ascertain just how significant the estimated risks are and thereby to inform the subsequent risk assessment process as to the best course of future action, including the nature of risk management required. However, the distinction between risk estimation and risk evaluation has become blurred and subject to different interpretations. Four distinct, but interrelated, elements of risk evaluation (RE)/risk assessment (RA) can be distinguished.

- Placing values on all aspects of detriment so that they can be compared and combined (RE).
- Comparing the risks and benefits of proposed activities and developments (RE/RA).
- Ascertaining how people who either are, or could be, affected by risk view the risks they face (RE).
- Deciding on what constitutes tolerable levels of risk (RE) and acceptable courses of future action (RA).

Many of these aspects will fall beyond the competence of the scientists or engineers responsible for developing the risk estimation, but it is nevertheless important to recognise their existence and to understand that they form crucial elements of the risk assessment process.

The ultimate aim of risk assessment is to get all losses expressed in the same units, for example, monetary units. This is extremely problematic, because many aspects of adverse consequences cannot be expressed in monetary terms as there is no market. For example, it is difficult to place a value on *a scenic view* or *the environment* or *an amenity,* despite the efforts of environmental economists (e.g. Pearce *et al.*, 2006). In addition to these so-called intangibles, there are artefacts, monuments, buildings and elements of cultural history that have values to individuals and groups that differ greatly from their replacement costs (e.g. a Saxon church or Stonehenge). Economists use a number of techniques, including contingent valuation, in an attempt to provide values for these items (see DETR, 2000), but problems remain. This is especially true in the case of human life, where techniques employed to establish what is known as the *value of a statistical* life (VOSL; see Chapter 9) have provoked outrage on the grounds of being morally offensive and leading to discriminatory results. Nevertheless, such approaches are necessary if risk estimations are to be realistic, despite the fact that there is huge scope for disagreement. It is this disagreement that has led to some authors placing the process within risk evaluation, while others see it as part of risk estimation.

Comparing risks and benefits of subsequent developments is another important aspect of risk evaluation and suffers from the problem that benefits, as is the case with risk, include both tangible and intangible elements. Therefore, while it might be relatively easy to compare the risks and benefits associated with the construction of a simple structure, for example, a retaining wall, such comparisons are normally more difficult and involve the risk perceptions of those affected and the resulting trade-offs between perceived risks and perceived benefits. As stated in DoE (1995):

> perceptions of risk and benefit, and of the values of intangibles such as quality of life will lead to different views on where to strike the balance between risks, costs and benefits which will vary from group to group.

Although this statement was made in the context of more general environmental considerations, it has some relevance in the broader aspects of landslide risk assessment.

The risk assessment stage is essentially the point at which the nature and scale of the threats are known, the likely costs are established and the possible risk reduction strategies are available for consideration. Decisions will have to be taken, generally after public consultation, as to exactly which risk reduction approach or approaches are to be adopted, if any: engineering, land cover change, development control. The actual risk management process is beyond the scope of this book, but essentially involves the interplay of technical, organisational and financial considerations to achieve the outcome(s) specified at the risk assessment stage.

2.7. Risk perception and risk communication

While risk evaluation is essentially a science-based activity largely focusing on what are called *expected utility methods* based on statistical estimations of risk, it does also involve consultations with the wider public. This is an aspect of the risk assessment process that many scientists and engineers consider problematic because it involves the risk perception of non-specialists. At the simplest level, perception is how people's knowledge, beliefs and attitudes (i.e. their socio-cultural make-up) lead them to interpret the stimuli and information that they receive. As a consequence, how people view potential risks and benefits is frequently both different and varied from that of scientists and engineers, and can result in frustration due to problems of communication.

Risk perception, according to the Royal Society (1992)

> involves people's beliefs, attitudes, judgements and feelings, as well as the wider social or cultural values and dispositions that people adopt, towards hazards and their benefits.

The result is *perceived risk*, which is (Royal Society, 1983)

> the combined evaluation that is made by an individual of the likelihood of an adverse event occurring in the future and its likely consequences.

Perceived risk can therefore be likened to a risk estimation made by an individual within a bounded rationality framework – in other words, a subjective assessment based on an imperfect view of probable outcomes, biased by belief, experience and personal disposition towards risk. That said, it has to be noted that because risk is a cultural construct, it follows that 'if risk is perceived then it is real'. As a consequence, people's unscientific perceptions of risk cannot simply be considered as irrelevant and dismissed, particularly where future losses may be incurred by individuals. There has been much work on establishing what are the main influences on human perceptions of risk and attitudes towards risk, some of which will be discussed below, but any reader interested in developing further understanding of risk perception is directed to Royal Society (1983, 1992), Lofstedt and Frewer (1998) and Slovic (2000).

Starr (1969) developed three 'laws of acceptability' of risk, which have survived in modified form up to the present:

1. Risks from an activity are acceptable if they are roughly proportioned to the third power of benefits for that activity.
2. The public will accept risks from voluntary activities, or, if chosen voluntarily, risks that are roughly 1000 times as great as they would tolerate from involuntary hazards, or from hazards imposed upon them, even though the same level of benefits is provided.
3. Risk acceptability is inversely proportional to the number at risk.

In the context of point 2, it has to be noted that natural hazards (geohazards) have traditionally been viewed as events outside *normal life* (hence the term *Acts of God*) and therefore as not voluntarily chosen. Despite scientific studies that have produced explanations for geohazards and thereby led to their increasing internalisation within normal life (see Hood and Jones, 1996), this perception of unacceptability persists for most geohazards, including landslides.

Subsequently, much questionnaire work in the 1970s and 1980s on individuals' perceptions of risk and riskiness ('expressed preference'), revealed how lay-people's views of risk often differ from the statistical assessments of scientists. For example, in the USA, it was found that women tended to over-estimate the risks from such things as nuclear power, flying, police work, fire fighting, hunting, mountaineering, skiing and spray cans, while underestimating the risks from surgery, X-rays, electricity, food preservatives and swimming (Table 2.1). At roughly the same time, Lichtenstein *et al.* (1978) showed that people tend to over-estimate the risk of rare adversities and to underestimate the risk of more common ones; in other words, people underestimate the risks associated with familiar events or behaviour (e.g. the risks posed within kitchens are widely underestimated).

By the late 1980s and based largely on the work of Slovic *et al.* (1980), it was generally considered that there were three main influences on risk perception, which were, in rank order

1. the horror of the hazard and its outcomes, the feeling of lack of control, possibility of fatal consequences, catastrophe potential (*dread*)
2. the unknown nature of the hazard, unobservable, unknown, new, delayed in terms of its manifestation (*unknown*)
3. the number of people exposed to the risk (*exposure*).

The opposite of *dread* is *controllability* or *preventability*, the opposite of *unknown* is *familiar*, and exposure is simply measured from high to low.

Landsliding normally scores low on all three measures, except in those areas where (i) there is a history of landslide activity, (ii) there is widespread landslide reactivation, (iii) there is progressive coastal cliff retreat or (iv) a landslide disaster has recently occurred nearby. It therefore appears that the statement 'out of sight, out of mind' is genuinely applicable in the case of landsliding.

As with most lines of research, later studies have introduced greater complexity, and the risk perception of individuals is currently considered to be strongly influenced by

- experience, especially of the activities and hazards involved
- environmental philosophy
- world view
- race, gender and socio-economic status
- catastrophe potential of the threat (dread)
- voluntariness
- equity and nature of threat to human generations.

The significance of *world view* arises from the work of anthropologists in exploring how risk is culturally constructed. It essentially began with the work of Douglas and Wildavsky (1983), who addressed the question as to why some cultures select certain dangers to worry about, while other cultures see no cause for concern, and has been refined in later studies by Schwarz and Thompson (1990) and Thompson *et al.* (1990) (for a general discussion, see Adams, 1995). Of crucial importance were the studies by the ecologist Holling (1979, 1986), who noted that managers of managed ecosystems in the developing world, when confronted with similar problems, often adopted different management environmental risks, such as those posed by slope instability, the views of lay-people may be significantly influenced by the media, pressure groups, past experience or the adverse impact of similar sounding, but actually dissimilar, events elsewhere. The resulting *heightening* or *dampening* of notions of threat, are known as the *social amplification of risk* and the *social attenuation of risk* (Kasperson and Kasperson, 1996).

Lay-people's perceptions of risk are both complex and varied. As a consequence, their reactions to risk often differ considerably, both within a population and, more importantly, with the judgements of environmental managers and decision-makers based on estimations of statistical risk involving probabilities produced by scientific investigations. This has, understandably, led to remarks that lay-people's *subjective assessments* have no role in the risk assessment process. However, as virtually all risk assessments are now seen to be subjective to a greater or lesser degree, and the debate about *narrow participation* versus *broad participation* (Royal Society, 1992; Hood and Jones, 1996) has led to an increased tendency for inclusivity in discussions on risk issues, so public participation has come to be seen to be both necessary and useful. As clearly stated in DoE (1995), the view that risk perception has nothing to do with risk evaluation because it is subjective while the scientific approach is objective 'is an idealised view which does not correspond to the world as it is and how decisions are taken'.

This view is reinforced in DETR (2000): 'while risk management decisions should be based on the best scientific information available . . . an important step is the creation of a constructive dialogue between stakeholders affected by or interested in risk problems' – to which can be added the caveat that all the parties involved must study the relevant data so that outcomes will not be swayed on the basis of beliefs, dogmas and misinformation.

The need for public involvement and the establishment of good dialogues is brought sharply into focus when it is appreciated that *risk* and *trust* are intimately associated (e.g. Cvetkovich and Lofstedt, 1999). If there is no trust between the public and those undertaking the risk assessment, then the results may not be believed and the proposed outcomes could well be opposed.

Consideration of risk perceptions and dialogues leads on, quite naturally, to a brief examination of *risk communication*. Risk communication evolved from risk perception to gain pre-eminence and has been defined in Kasperson and Stallen, 1991 as:

> Any purposeful exchange of information about health or environmental risks between interested parties. More specifically, risk communication is the act of conveying or transmitting information between parties about:
> (*a*) levels of health or environmental risks;
> (*b*) the significance or meaning of health or environmental risks;
> (*c*) decisions, actions or policies aimed at managing or controlling health or environmental risks.
> Interested parties include government agencies, corporations and industry groups, unions, the media, scientists, professional organisations, public interest groups, and individual citizens.

It is important to recognise that risk communication involves the multiple flows of information between scientists, decision-makers, the media and the public, and occurs at all stages from preliminary hazard assessment through to the development of zoning policies, the establishment of building codes and ordinances, and the issuing of forecasts and warnings. Improvements in risk communication will inevitably result in improved perceptions of risk and reductions in the levels of risk.

The key to good risk communication with the public involves the following.

- Limiting the use of technical jargon to a minimum and properly explaining any scientific terms that have to be used, for example, many lay-people do not understand probabilities and return periods or recurrence intervals.
- Fully explaining the basis of the conclusions and the uncertainties involved.
- Appreciating that individuals' perceptions of risk and resulting concerns are genuine and real and should not be dismissed as ignorant but should be carefully countered. For example, some slopes may well appear much more dangerous than others to the untrained eye or to those unaware of the details of landsliding and the intricacies of geology.

The need for good risk communication was recently brought into sharp focus by the decision of an Italian court (22 October 2012) to sentence six scientists and a former government official to six years in jail for manslaughter, owing to their failure to properly warn of the magnitude-6.3 earthquake that devastated the town of L'Aquila on 7 April 2009, killing 308 people. This decision initially caused consternation amongst geoscientists until clarification revealed that it was not the scientific assessments that were wrong, but that the group had provided 'inexact, incomplete and contradictory information' at a public meeting on 31 March 2009. At that meeting, the civil protection official had stated 'the scientific community tells us there is no danger, because there is an on-going discharge of energy (small shocks). The situation looks favourable'. None of the six scientists who were present at the meeting chose to correct this statement. The lessons to be learned from this unfortunate episode are generally similar to those that emerged from the adverse reactions to various claims contained in the Fourth Assessment Report of the IPCC (IPCC, 2007).

- The scientists who make the assessments need to be closely involved in writing the reports and making the public statements.
- If a publicist(s) or other spokesperson(s) is used, then one or more qualified scientists should be present to correct misleading or incorrect statements.

2.8. Landslide risk assessment procedures

It has been shown that risk assessment is a complex, multi-stage process that combines scientific investigation, expert judgement and human values. As hazard, exposure and vulnerability are dynamic in space and time, risk assessments may have to be periodically updated or even re-done in order to accommodate change. Therefore, the linear model portrayed in Figure 2.1 should, in many circumstances, be replaced with a circular model, where risk management is followed by risk monitoring, which reveals changes in risk levels or even the recognition of new risks, thereby resulting in the need for new risk assessments.

It is suggested that landslide risk assessment (LRA) should conform to the sequence presented in Figure 2.1, although the actual number of stages undertaken will depend upon the circumstances and contexts within which the risk assessment is undertaken. For example, the LRA for a coastal town built on pre-existing landslides prone to reactivation could involve greater risk evaluation and public participation than one undertaken for a dam site in an arid region or for an isolated construction site. Similarly, the LRA for the coastal town may involve much vaguer and more subjective risk estimation procedures than an investigation of cliff instability threatening a major hotel. However, the sequence of steps in the procedure should be the same, even though the level of detail, the investigatory effort and the techniques employed may vary from case to case. It is these varied approaches and techniques that are examined in detail in the following chapters.

Before embarking on the detail, however, it needs to be emphasised that the production of an LRA capable of withstand close scrutiny requires that the following points be properly addressed (based on DoE, 1995).

1. Scoping should be carried out carefully and reasons for limiting the areal extent of the risk assessment, restricting the time frame or excluding certain risks, should be clearly stated. The Australian Geomechanics Society (AGS, 2000) has suggested that before embarking on LRA it is important to ensure that the process is focused on relevant issues and that the limits or limitations of the analysis are recognised. This should involve setting clear objectives that define:
   - the site, being the primary area of interest
   - geographic limits involved in the study of processes that may affect the site
   - whether the analysis will be limited to addressing only property loss or damage, or will also include injury to persons, loss of life and other adverse consequences
   - the extent and nature of investigations that will be completed
   - the type of analysis that will be carried out
   - the basis for assessment of acceptable and tolerable risks.
2. The description of intention should be clear and precise.
3. Assumptions should be made explicit and recorded.
4. The nature of uncertainties should be made explicit and recorded.
5. The process of identifying hazards must be appropriate to the description of intention, but should also take account of the possibility of unintended or extremely rare events. Here the need is to 'think the unthinkable' so as to be aware of the potential for Black Swans and Perfect Storms, not because measures will be taken to minimise their likelihood or impact, but to show that the assessment has been complete.
6. Use should be made of historical studies and relevant analogues in establishing the nature of hazardous events.
7. If the magnitude of consequences cannot be estimated directly, then careful use should be made of comparable precedents.
8. Risk estimation will nearly always be judgemental, and the careful use of expert judgement is to be recommended.
9. The main purpose of risk evaluation should be to identify those risks that are considered to be unacceptable.

2.9. Risk assessment as a decision-making tool

Risk assessment techniques can be applied at all spatial scales of the decision-making process, from strategic planning to site evaluation. The nature of the risk assessment employed will generally vary according to the stage in the decision-making process, from a general indication of the threat across an area or region, to statements about the level of risk at a particular site. As mentioned earlier, Wong *et al.* (1997) have used the following terminology for the different applications:

- *Global* (i.e. *regional* or *broad-scale*) *risk assessment* determines the overall risk to a community posed by landslides, although the term *extensive risk* is increasingly coming into use. This level of approach can help define the significance of landslide hazards in relation to the overall risks faced by society and allow decision-makers to make a rational allocation of resources for landslide management.
- *Site-specific risk assessment* determines the nature and significance of the landslide hazards and risk levels at a particular site. This approach can help decision-makers decide whether the risk levels are acceptable and if risk reduction measures are required.

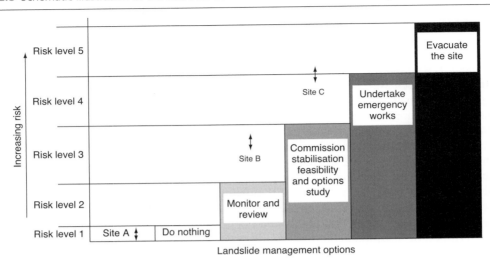

Figure 2.8 Schematic illustration of the link between risk levels and decision-making

It is important to appreciate that risk assessment is a decision-making tool, not a research tool. It can provide information of value at many levels, ranging from crude qualitative approximations to sophisticated quantitative analyses involving complex computations. The level of precision and sophistication required merely needs to be sufficient for a particular problem or context, so as to enable an adequately informed decision to be made. Thus, both qualitative and quantitative approaches have a role to play. Decisions will have to be made despite uncertainty, provided due acknowledgement is made of the limitations. In many instances, this may mean ensuring that the estimated level of risk is in the right *ballpark*. With reference to Figure 2.8, it may be more important to know within which of the range of risk levels a site or area falls, rather than a precise risk value. To illustrate this point, the assessed risk levels at a number of sites have been plotted on Figure 2.8. Site A clearly falls within risk level 1, suggesting that the most appropriate management strategy might be to do nothing. Site B has a higher estimated risk, indicating that an appropriate strategy would be to commission a feasibility and options study to determine the best ways of reducing the risk. The management decision at Site C may be more problematic, since the estimated risk appears to be close to the boundary between two risk levels. The decision maker(s) may choose to be cautious and evacuate the site or may seek more detailed evaluation of the risks before having to make a very difficult and politically sensitive decision.

2.10. Structure of the book

Having introduced the notion of risk and the main aspects of the risk assessment process, the remainder of this book will examine a range of approaches to landslide risk assessment using relevant examples from across the world. It is not intended to be a manual, but rather to appeal to all people interested in landslides and the problems they cause. The chapters have been arranged to cover the main stages in the risk assessment process as discussed in this chapter. Chapter 3 presents an introduction to the nature of landslide hazard, highlighting the importance of understanding the nature of the problem before trying to determine the risk. Qualitative and semi-quantitative risk assessment methods are then covered in Chapter 4, including the use of risk registers in the scoping or screening processes. However, as the main focus of the book is on quantitative risk assessment, the subsequent chapters are intended to provide an introduction to the key challenges that will need to be overcome by the practitioner.

- Chapter 5: the nature of uncertainty, predictability and probability.
- Chapter 6: estimating the probability of landsliding.
- Chapter 7: estimating the exposure of fixed and non-stationary elements in the danger zone.
- Chapter 8: estimating the vulnerability of the elements at risk to the landslide event.
- Chapter 9: estimating the consequences.
- Chapter 10: combining the hazard and consequence estimates to develop a measure of the risk.
- Chapter 11: the transition between risk assessment and risk management.
- Chapter 12: this final chapter is titled 'Future challenges' and gathers together discussions of some problem areas that could prove challenging in the years to come.

Each of Chapters 5–11 is accompanied by a series of examples to illustrate the different approaches that are available at each stage of the risk assessment process. However, as each site or problem will tend to be unique, these examples can only hope to give the reader an appreciation of how a particular issue *might* be addressed, rather than how it *should* be addressed.

There is a bias towards examples from the UK. This is perhaps inevitable, as both the authors are British. However, every attempt has been made to include examples from other landslide environments, drawing both on personal experience and on that of colleagues, as well as from the published literature.

The book is by no means exhaustive in its coverage, since this would be a monumental task. Thus, while all kinds of slope failure are considered, ranging from the small to the catastrophic and including volcanic landslides, submarine slides and earthquake-generated landslides, as well as more standard coastal and inland examples, the various individual types of landslide will not be examined in any great detail. There are many books on landslides that provide such information. This book is about landslide risk and how it can be assessed, and aims to bring together ideas and examples that illustrate the objectives, approaches, difficulties and shortcomings of landslide risk assessment.

By way of conclusion, it is important to re-emphasise comments made earlier regarding landslides, the risk-generating agents that are the focus of this book. As with all geohazards, it is easy to become engrossed in the complexities of landslide form and generation and in the details of investigated failures. Clearly, an understanding of landsliding is crucial to the formation of landslide hazard assessment, but it has to be remembered that this is only the first step in the development of a risk assessment. It is always important, therefore, to focus on the central issue, which is that landslide risk assessment is concerned with establishing the likelihood and extent to which slope failures could have an adverse impact upon humans and the things that humans value. Therefore, it is size, depth, suddenness, speed, run-out or travel distance, magnitude–frequency characteristics and secondary hazards that figure among the relevant parameters, together with the distribution of humans and the disposition, value and vulnerability of objects valued by humans. The fact that two potential landslides can be identical in all physical respects yet pose very different levels of risk owing to location shows that the agent is less crucial than the impact potential or consequences – in other words, emphasis should focus not on the *what* but on the *so what!*

REFERENCES

Adams J (1995) *Risk*. UCL Press, London.

AGS (Australian Geomechanics Society) (2000) Landslide risk management concepts and guidelines. *Australian Geomechanics* **35(1)**: 49–52.

BP (2008) *WREP's Comeback*. See http://www.bp.com/genericarticle.do?categoryId=9012063&contentId=7050346 (accessed 03/07/2013).

Bromhead EN (2005) Geotechnical structures for landslide risk reduction. In *Landslide Hazard and Risk* (Glade T, Anderson G and Crozier MJ (eds)). Wiley, Chichester, pp. 549–594.

Cvetkovich G and Lofstedt RE (eds) (1999) *Social Trust and the Management of Risk*. Earthscan, London.

DETR (Department of the Environment, Transport and the Regions) (2000) *Guidelines for Environmental Risk Assessment and Management – Revised Departmental Guidance*. Prepared with the Environment Agency and the Institute for Environment and Health. The Stationery Office, London.

DoE (Department of the Environment) (1995) *A Guide to Risk Assessment and Risk Management for Environmental Protection*. HMSO, London.

Douglas M and Wildavsky A (1983) *Risk and Culture: An Essay on the Selection of Technological and Environmental Dangers*. University of California Press, Berkeley, CA.

DTLR (Department for Transport, Local Government and the Regions) (2002) *Economic Valuation with Stated Preference Techniques: A Summary Guide. DTLR Appraisal Guidance*. DTLR, London.

Gould LC, Gardiner GT, De Luca DR, Tiemann AR, Dobb LW and Stolwijk JAJ (1988) *Perceptions of Technical Risks and Benefits*. Russell Sage Foundation, New York.

Hartford DND (2009) Legal framework considerations in the development of risk acceptance criteria. *Structural Safety* **31**: 118–123.

Holling CS (1979) Myths of ecological stability. In *Studies in Crisis Management* (Smart G and Stanbury W (eds)). Butterworth, Montreal, QC, pp. 97–109.

Holling CS (1986) The resilience of terrestrial ecosystems. In *Sustainable Development of the Biosphere* (Clark W and Munn R (eds)). Cambridge University Press, Cambridge, pp. 292–317.

Hood C and Jones DKC (eds) (1996) *Accident and Design: Contemporary Debates in Risk Management*. UCL Press, London.

HSE (Health and Safety Executive) (2006) *Guidance on Risk Assessment for Offshore Installations*. Offshore Information Sheet 3/2006. HSE, London.

IPCC (Inter-Governmental Panel on Climate Change) (2007) Summary for Policymakers. In *Climate Change 2007: The Physical Science Basis. Contribution of Working Group I to the Fourth Assessment Report of the Intergovernmental Panel on Climate Change* (Solomon S, Qin D, Manning M et al. (eds)). Cambridge University Press, Cambridge and New York, NY. See http://www.ipcc.ch/pdf/assessment-report/ar4/wg1/ar4-wg1-spm.pdf (accessed 03/07/2013).

IPCC (2012) Summary for Policymakers. In: *Managing the Risks of Extreme Events and Disasters to Advance Climate Change Adaptation. A Special Report of Working Groups I and II of the Intergovernmental Panel on Climate Change* (Field C, Barros V, Stocker TF et al. (eds)). Cambridge University Press, Cambridge and New York, NY, pp. 1–19. http://ipcc-wg2.gov/SREX/images/uploads/SREX-SPMbrochure_FINAL.pdf (accessed 03/07/2013).

Kasperson RE and Stallen PM (eds) (1991) *Communicating Risk to the Public*. Kluwer Academic Press, Dordrecht.

Kasperson RE and Kasperson JX (1996) The social amplification and attenuation of risk. *Annals of the American Academy of Political and Social Science* **545**: 95–105.

Lee EM and Charman JH (2005) Geohazards and risk assessment for pipeline route selection. In *Terrain and Geohazard Challenges Facing Onshore Oil and Gas Pipelines* (Sweeney M (ed.)). Thomas Telford, London, pp. 95–116.

Léone F, Aste JP and Leroi E (1996) Vulnerability assessment of elements exposed to mass movement: working towards a better risk perception. In *Landslides* (Senneset K (ed.)). Balkema, Rotterdam, vol. 1, pp. 263–268.

Lewis S and Hurst S (2005) Bow-tie: an elegant solution. *Strategic Risk*, November: pp. 8–10.

Lichtenstein S, Slovic P, Fischhoff B, Laymen M and Combs B (1978) Judged frequency of lethal events. *Journal of Environmental Psychology: Human Learning and Memory* **4**: 551–578.

Lofstedt RE and Frewer L (eds) (1998) *Risk and Modern Society*. Earthscan, London.

MAFF (Ministry of Agriculture, Fisheries and Food) (2000) *FCDPAG4 Flood and Coastal Defence Project Appraisal Guidance: Approaches to Risk*. MAFF Publications, London.

Pearce D, Atkinson G and Mourato S (2006) *Cost–Benefit Analysis and the Environment: Recent Developments*. Organisation for Economic Co-operation and Development (OECD), Paris, France.

Royal Society (1983) *Risk Assessment*. Royal Society, London.

Royal Society (1992) *Risk: Analysis, Perception and Management. Report of a Royal Society Study Group*. Royal Society, London.

Schwarz M and Thompson M (1990) *Divided We Stand: Redefining Politics, Technology and Social Choice*. Harvester Wheatsheaf, Hemel Hempstead.

Slovic P (2000) *The Perception of Risk*. Earthscan, London.

Slovic P, Fischhoff B and Lichtenstein S (1980) Facts and fears: understanding perceived risk. In *Societal Risk Assessment: How Safe is Safe Enough?* (Shwing R and Albers W (eds)). Plenum, New York, pp. 181–214.

Starr C (1969) Social benefit versus technological risk. *Science* **165**: 1232–1238.

Stern PC and Fineberg HV (eds) (1996) *Understanding Risk: Informing Decisions in a Democratic Society*. National Academy Press, Washington DC.

Thompson M, Ellis R and Wildavsky A (1990) *Cultural Theory*. Westview, Boulder, CO.

Vinnem JE (2007) *Offshore Risk Assessment: Principles, Modelling and Application of QRA Studies*, 2nd edn. Springer, London.

Wong HN, Ho KKS and Chan YC (1997) Assessment of consequence of landslides. In *Landslide Risk Assessment* (Cruden D and Fell R (eds)). Balkema, Rotterdam, pp. 111–149.

Landslide Risk Assessment
ISBN 978-0-7277-5801-9

ICE Publishing: All rights reserved
http://dx.doi.org/10.1680/lra.58019.047

Chapter 3
Landslide hazard

3.1. Introduction

To many people, the term 'landslide hazard' evokes images of sudden, dramatic and violently destructive events that cause loss of life and widespread devastation. In the UK, the 1966 Aberfan disaster fits this stereotype. A flowslide from a colliery spoil tip travelled down the valley side at 4.5 m per second, engulfing the Pantglas Infants' School; 144 people died, including 116 out of 250 pupils who had gathered for the morning assembly (Bishop *et al.*, 1969; Miller, 1974).

Huge, devastating landslides most certainly do occur. Amongst of the most recent was the August 2010 Gansu mudslide in Zhugqu County, China, when around 2×10^6 m^3 of mud and debris flattened three villages and killed over 1500 people. One of the greatest disasters of the last century occurred in Tadzhikistan in 1949, when a large earthquake (magnitude 7.5) triggered a series of debris avalanches and flows that buried 33 villages. Estimates of the death toll range from 12 000 to 20 000 (Wesson and Wesson, 1975). Between 23 000 and 25 000 people were killed in the town of Armero, Colombia by debris flows generated by the 1985 eruption of the Nevado del Ruiz volcano. The flows are believed to have been initiated by pyroclastic flows and travelled over 40 km from the volcano (Herd *et al.*, 1986; Voight, 1990). The 1970 Huascaran disaster destroyed the town of Yungay, Peru, killing between 15 000 and 20 000 people (Plafker and Ericksen, 1978). In this case, an offshore earthquake (magnitude 7.7) triggered a massive rock and ice avalanche from the overhanging face of a mountain peak in the Andes. The resulting turbulent flow of mud and boulders (estimated at $5-10 \times 10^7$ m^3) descended 2700 m, passing down the Rio Shacsha and Santa Valleys as a 30 m-high wave travelling at an average speed of 270–360 km/h in the upper 9 km of its path. Perhaps the fastest-moving known event was the Elm rockslide, which occurred in Switzerland in 1881. It is believed that quarrying operations undermined the mountain slope and caused a 10^7 m^3 landslide that moved across the valley towards the town of Elm at over 80 m/s; 115 people were killed (Buss and Heim, 1881; Heim, 1882, 1932).

These landslides, however, are examples of rare high-magnitude, low-frequency events (Figure 3.1) that occur with an estimated frequency of 40–45 per century. Not all have disastrous impacts on humans. For example, the 1965 Hope slide in British Columbia, Canada had a limited impact despite being one of the largest onshore failures in recorded history (4.7×10^7 m^3), while the conspicuous rock avalanche caused by the collapse of part of the summit of Mount Cook, New Zealand on 14 December 1991 (McSaveney *et al.*, 1992) also had minimal adverse consequences. However, growing global population and the spread of human activity together increase the potential for such events to produce adverse consequences and disastrous outcomes, because of the increasing exposure to hazard (see Chapter 1).

As a consequence, the published lists of major landslide disasters, which normally show fewer than 30 events since 1900 (see Jones, 1992b), and the apparent relative unimportance of landsliding in the

Figure 3.1 Hypothetical diagram to show the contemporary magnitude–frequency distribution of landslides divided into non-cost-inducing events, cost-inducing hazards and disasters, and the postulated relationships with average and total costs of impact. Note that both axes are logarithmic

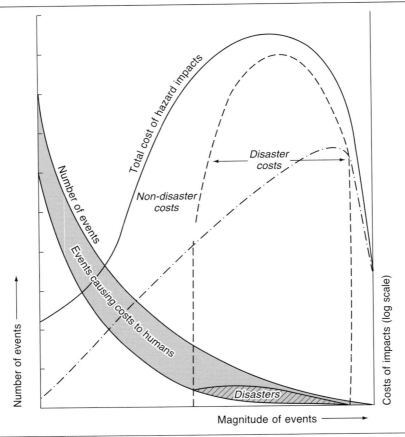

statistics produced on geohazard disasters – for example, there were 3006 deaths in landslide disasters over the period 1900–1976 (39 per year) out of a total of 4.8 million killed in all natural disasters (Red Cross, quoted in Crozier, 1986) – have done little to emphasise the significance of landsliding as a hazard. More recently, the Red Cross/Red Crescent (IFRCRCS, 1999), using a different threshold for disaster, quote an average of 14 landslide disasters per year for the decade 1988–1997 out of a total of 269 natural disasters per year, with average annual death tolls of 790 (out of 85 000), injured (267/65 000), affected (138 000/144 million) and made homeless (107 000/48 million), but even these figures must be seen as gross underestimates of both present and future significance, especially since landslide impacts are often subsumed under more conspicuous primary hazards such as earthquakes, volcanoes and tropical cyclones. The establishment of the Durham Fatal Landslide Database in 2002 as a focal part of the Durham International Landslide Centre should produce more realistic figures for landslide impacts in the future (for more details, see Chapter 9 and Petley (2012)).

It is crucial to recognise that the occurrence of conspicuous disasters is only a small part of the landslide story. Unfortunately, public perception of landslide hazard does not correspond to reality.

While the catastrophe potential of dramatic major failures periodically focuses attention on low-frequency (rare) events, the cumulative effects of ubiquitous minor failures tends to go little noticed. Indeed, the greatest total impact of landsliding is probably caused by medium-sized events, because of their greater frequency (Figure 3.1). Thus, McGuire *et al.* (2002) were correct in observing that 'landslides are the most widespread and undervalued natural hazard on Earth', although it has to be pointed out that an increasing proportion are actually quasi-natural or hybrid in origin owing to human activity (see Chapter 1). It is therefore important that proper attention be directed to evaluating the full range of potential landslide hazards for the purposes of risk assessment.

Over the last decade or so, the arena for landslide risk assessment has expanded dramatically to include the seabed. This is largely because of the increasing number of deep-water oil and gas fields that have been discovered and developed. The Ormen Lange field on the Norwegian margin, the Mad Dog and Atlantis fields lying downslope of the Sigsbee Escarpment in the Gulf of Mexico and the Raven field in the West Nile delta are examples of where the development has had to take account of landslide risk (e.g. Jeanjean *et al.*, 2003; Bryn *et al.*, 2004; Evans *et al.*, 2007). Hurricanes Camille in 1969, Ivan in 2004 and Katrina in 2005 triggered wave-induced mudslides in the Gulf of Mexico that damaged seabed pipelines (e.g. Gilbert *et al.*, 2007). Large submarine landslides may also generate tsunami. For example, an earthquake-triggered magnitude 7.1 submarine slide off the north coast of Papua New Guinea in 1998 generated a 15 m-high tsunami that destroyed three villages and killed more than 2000 people (Tappin *et al.*, 2008). It is still suspected that the 2004 Boxing Day tsunami that killed over 250 000 people may have partly been caused by a series of massive earthquake-triggered submarine landslides (e.g. Nadim and Locat, 2005).

3.2. Landslide mechanisms and type

Movements of ground or earth materials ranging from the very rapid to the extremely slow, can all be considered to be landslides because they all involve the 'movement of rock, debris or earth down a slope' (Cruden, 1991). The reference to 'down a slope' is important as it distinguishes landsliding from subsidence, although the two may often be intimately associated. Perversely, phenomena described as landslides are not restricted to the land (e.g. submarine slides) and do not necessarily involve sliding. Phenomena that could be described as landslides can involve six distinct mechanisms (Cruden and Varnes, 1996).

1 *Falling* involves the detachment of soil or rock from a steep face or cliff, along a surface on which little or no shear displacement occurs. The material then descends through the air by falling, before impact with the ground results in disintegration, with fragments subsequently bouncing and rolling some distance.
2 *Toppling* involves the forward rotation out of a slope of a mass of soil or rock about a point or axis below the centre of gravity of the displaced mass. The subsequent evolution of the displaced mass is similar to that of a fall.
3 *Spreading* or *lateral spreading* is the extension of a cohesive soil or rock mass over a lower deformable layer, combined with a general subsidence of the upper fractured mass into the softer underlying material. The surface of rupture is not a surface of intense shear and so is often not well defined. Spreading may also result from liquefaction or the flow and extrusion of a softer underlying layer of material.
4 *Flowing* is the turbulent movement of a fluidised mass over a rigid bed, with either water or air as the pore fluid (like, for example, wet concrete or running dry sand; see Hungr *et al.*, 2001). There is a gradation from flows to slides, depending on water content and mobility.

Figure 3.2 Landslide types

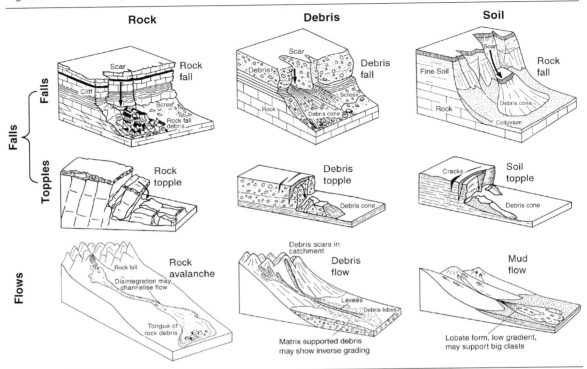

5 *Sliding* is the downslope movement of a soil or rock mass as a coherent body on surfaces of rupture or on zones of intense shear strain. Slides are characterised by the presence of a clearly defined shear surface at the contact between the moving mass and the underlying soil or rock.

6 *Composite and complex movements* take place when two or more types of displacement occur within the same landslide area. For example, sliding can be divided into rotational and translational groups depending on whether the basal shear surface is curved or planar, but in some cases rotational failures trend downslope into translational failures. Similarly, there are failures that represent a mixture of flowing and sliding, which are termed 'flow-slides'.

This classification has been found to apply to seabed landslides (e.g. Locat and Lee, 2002) as well as to major inland water bodies. However, the transformation of submarine slides and flows into *turbidity currents* appears to be unique to the marine environment. These are currents of rapidly moving, sediment-laden water moving down a slope through water; they are also known as *density currents* because the suspended sediment results in the current having a higher density than the clearer water into which it flows. The earthquake-triggered Grand Banks slide of 1929, for example, which occurred 280 km south of Newfoundland, generated a debris flow that transformed into a turbidity current that travelled over 1000 km at speeds ranging from 60 to100 km/h (Evans, 2001; Fine *et al.*, 2005). The turbidity current had an estimated flow thickness of several hundred metres and is believed to have flowed for up to 11 hours (Piper *et al.*, 1999). It severed 12 transatlantic telegraph cables.

The basic mechanisms outlined above combine with site factors such as topography, lithology, geological structure, hydrogeology, climate and vegetation to produce a remarkable diversity of

Figure 3.2 *Continued*

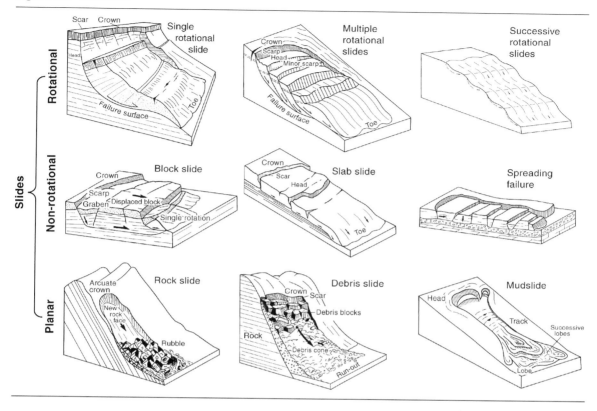

landslide types (Figure 3.2), a diversity that Terzaghi (1950) observed 'opens unlimited vistas for the classification enthusiast'. Indeed, numerous landslide classifications have been developed, most based on shape (morphology), the materials involved and some aspect of the principal mechanisms. The most widely accepted and comprehensive scheme in the English-speaking world is that of Hutchinson (1988). Simpler alternatives are those of Varnes (1978) (with modifications proposed in Cruden and Varnes, 1996) and the system developed by the EPOCH project for use in Europe (Dikau *et al.*, 1996) (Table 3.1). The overwhelming importance of local site conditions means that the landslide features observed on the ground are often difficult to classify, so great care is needed to ensure correct classification.

There are many books that describe the range of approaches to the investigation of landslides, including Brunsden and Prior (1984), Bromhead (1986, 1992) and Turner and Schuster (1996). Landslide investigation requires an understanding of the slope processes and the relationship of those processes to geomorphology, geology, hydrogeology, climate and vegetation (AGS, 2000). From this understanding, it should be possible to characterise the landslide threats within an area, including

- classifying the types of potential landsliding, the physical characteristics of the materials involved and the slide mechanisms
- assessing the physical dimensions of each potential landslide being considered, including the likely location on the slope and the areal extent and volume involved

Table 3.1 Classification of landslide types proposed by the EPOCH project

Type	Material		
	Rock	Debris	Soil
Fall	Rockfall	Debris fall	Soil fall
Topple	Rock topple	Debris topple	Soil topple
Slide (rotational)	Single (slump) Multiple Successive	Single Multiple Successive	Single Multiple Successive
Slide (translational, non-rotational)	Block slide	Block slide	Slab slide
Planar	Rock slide	Debris slide	Mudslide
Lateral spreading	Rock spreading	Debris spread	Soil (debris) spreading
Flow	Rock flow(sackung)	Debris flow	Soil flow
Complex (with run-out or change of behaviour downslope; note that nearly all forms develop complex behaviour)	For example, a rock avalanche	For example, a flow–slide	For example, a slump–earthflow

Note: A compound landslide is one that consists of more than one type, for example a rotational–translational slide. This should be distinguished from a complex slide, where more than one form of failure develops into a second form of movement – that is, a change in behaviour downslope by the same material.
Based on EPOCH (1993) and Dikau *et al.* (1996)

- assessing the likely triggering events
- estimating the resulting anticipated travel distances and velocities of movement
- addressing the possibility of fast-acting processes, such as flows and falls, from which people will find it more difficult to escape.

An area or a site may be affected by more than one type of landslide hazard; for example, deep-seated landslides on the site and rockfalls and debris flows from above the site. It is important to recognise that hazard assessment is not merely concerned with landsliding generated from within a site or area under consideration, but must also include landslides that might intrude into the site or area from elsewhere – hence the great emphasis placed on 'site and situation' by early applied geomorphologists (e.g. Brunsden *et al.*, 1975). For example, the ephemeral channel of the Rimac River in Peru passes through the densely populated suburbs of Lima on the almost flat coastal plain. Intense rainfall events in its upper catchment in the Andes cause devastating debris flows ('huaicos'), which can cause considerable damage in Lima, well away from the landslide source areas (Cascini *et al.*, 2005).

3.3. Landslide behaviour

The focus on what landslides 'look like' and how they are classified should not divert attention away from those aspects that influence landslide hazard and risk. The destructive intensity of a landslide is mainly related to kinetic parameters, such as velocity and acceleration, along with its dimensions and material characteristics (Léone *et al.*, 1996). In other words, 'it's not what it is but what it does that

Figure 3.3 Different stages of slope movement. (After Leroueil et al., 1996)

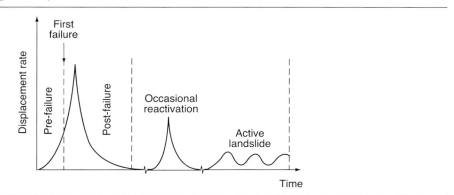

counts'. It follows that an understanding of the mechanical behaviour of a landslide at the different stages of its development is needed in order to properly support the development of a risk assessment.

Four stages of landslide movement can be defined (Figure 3.3) (Leroueil et al., 1996).

1. *Pre-failure movements* can involve small displacements that reflect the progressive development of shear surfaces, from isolated shear zones to continuous displacement surfaces. Soils generally behave as viscous materials, and can creep under constant stresses. This progressive creep is of considerable importance, since it provides forewarning of impending failure (i.e. a precursor), involving the development of tension cracks or minor settlement behind a cliff face, and bulging on the slope or at the slope foot. These movements equate with the late incubation stage of hazard outlined in Chapter 2, and it should be noted that the correct interpretation of precursors can reduce risk by stimulating monitoring and emergency action (e.g. public information, warnings or evacuation). Indeed, many major landslides are preceded by months to decades of accelerating creep (Voight, 1978), the major difficulty being determining exactly when the sudden major movement will occur. This transformation, sometimes called 'from creep to whoosh', is clearly portrayed by the failure of a hillslope above the Swiss village of Goldau in 1806, creating a $1-2 \times 10^7$ m^3 landslide that destroyed the village and killed 475 people. The last words of one of the victims were, according to Hsü (1978), 'For thirty years we have waited for the mountain to come, well, now it can wait until I finish stuffing my pipe'.

2. *Failure* occurs when the disturbing forces acting on the slope exceed the forces resisting failure (i.e. it corresponds to the 'trigger' in the model presented in Chapter 2). As disturbing forces increase owing, for example, to the over-steepening of a cliff by marine erosion or to the build-up of water pressure in the soil, deformations occur as shear strength is mobilised. For a non-brittle material, the shear strength will increase to an ultimate value and will then remain constant. However, for a brittle material, such as an over-consolidated clay, the shear strength will rise to a peak value and then decrease, as deformation continues, to a residual value. The residual strength of clay soils can be significantly less than the peak strength. When there is insufficient strength available to counter the disturbing forces, the slope or cliff will fail. Movement will occur until equilibrium is restored.
   Failure may involve the following:
   - *Peak strength failures*, in which the peak strength (the maximum stress that the material can withstand) is mobilised during failure. After failure, the shear resistance mobilised along the

shear surface decreases to a lower, residual value. Such slides are often characterised by large, rapid displacements; the velocity is proportional to the difference between the peak and residual strength values.

- *Progressive failure*, in which the fully softened strength or the residual strength is mobilised during first-time failure. The loss of strength is due to strain softening as a result of high lateral stresses developing within the slope. The in situ materials will expand to relieve this stress. If this expansion is large and the strains are concentrated at a particular horizon, for example on bedding planes, then peak strength may be exceeded and a shear surface formed along which strength is lost. This weakened surface can then lead to the development of a deep-seated slide as a shear surface propagates upwards towards the slope crest. Plastic clays (with plasticity index > 25%) are prone to strain softening and progressive failure. This style of landsliding generally involves slower and less dramatic movements than peak strength failures.

3 *Post-failure movements* include movement of the displaced mass from just after failure until it stops and are therefore equivalent to the 'initial' and 'primary' events of the hazard model outlined in Chapter 2. Some of the potential energy of the landslide (a function of slope height and geometry) is lost through friction as the material moves along the shear surface. The remainder is dissipated in the break-up and remoulding of the moving material and in accelerating it to a particular velocity (kinetic energy). In brittle materials, where there is a large difference between the peak and residual strengths, the kinetic energy can be very large, giving rise to long run-out landslides. Important mechanisms involved in the post-failure stage include the following.

- *Mass liquefaction* occurs when the soil structure suddenly fails without exerting its frictional shear resistance, for example as a result of rapid seismic loading. Channelised debris flows can develop as a result of the liquefaction of saturated stream channel or valley floor in-fills in response to rapid loading by hillside failures. Loess (wind-blown silt deposits, particularly extensive in China, where they are up to 280 m thick) is especially prone to liquefaction. For example, around 200 000 people were killed when a magnitude 8.5 earthquake in central China triggered numerous landslides in December 1920 (Close and McCormick, 1922), and at least twice that number are thought to have been buried by the Shansi earthquake of 1556, which killed an estimated 830 000 people (Eiby, 1980). The impact of one of the 1920 loess flows was graphically described by Close and McCormick, (1922):

> The most appalling sight of all was the Valley of the Dead, where seven great slides crashed into the gap in the hills three miles long, killing every living thing in the area except three men and two dogs. The survivors were carried across the valley on the crest of the avalanche, caught in the cross-current of two other slides, whirled in a gigantic vortex, and catapulted to the slope of another hill. With them went house, orchard, and threshing floor, and the farmer has since placidly begun to till the new location to which he was so unceremoniously transported.

Similar miraculous survivals are a feature of loess flows, but the prevailing consequences are more dire, as in 1988 when an earthquake (magnitude 5.5) southwest of Dushanbe, Tajikistan, triggered a series of landslides in a loess area. The slides turned into a massive mudflow (2×10^7 m^3), which travelled about 2 km across an almost flat plain (Ishihara, 1999). It buried more than 100 houses in Gissal Village under 5 m of debris; 270 villagers were killed or reported missing.

- *Sliding surface liquefaction*, which occurs when a shear surface develops in sandy soils and the grains are crushed or comminuted in the shear zone (Sassa, 1996). The resulting volume reduction causes excess pore pressure generation, which continues until the effective stress becomes small enough that no further grain crushing occurs. This mechanism is common in earthquake-triggered debris slides and flows, and can result in devastating large run-out landslides.

- *Remoulded 'quick-clay' behaviour* of sensitive Late Glacial and Post-Glacial marine clays of Scandinavia and eastern Canada (Bentley and Smalley, 1984; Torrance, 1987). Failure of a sensitive clay slope can result in the material remoulding to form a heavy liquid that flows out of the original slide area, often supporting rafts of intact clay. The affected area can quickly spread retrogressively, as the unsupported landslide backscar fails. The quick-clay slide that occurred in 1971 at St Jean Vianney, Quebec involved 6.9×10^6 m^3 of material, destroyed 40 homes and killed 31 people (Tavenas *et al.*, 1971).
- *Impact collapse flow–slides* can occur as a result of large cliff falls involving weak, high-porosity rock. The impact of the fall generates the excess pore pressures required for flow-sliding. Hutchinson (1988) has demonstrated that, on chalk coastal cliffs, the susceptibility to flow–sliding increases with cliff height. Falls from cliffs below 50 m high tend to remain at the cliff foot as talus. However, large falls from higher cliffs can generate flow-slides that travel up to 500 m.
- *Sturzstroms*, or high-speed flows of dry rockfall or rock-slide debris, can attain velocities of 30–50 m/s with enormous volumes of material ($1–16 \times 10^7$ m^3) (Hutchinson, 1988), as was displayed in the Huascaran (Peru) failure of 1970, described earlier. Sturzstroms are believed to involve turbulent grain flow, with the upward dispersive stresses arising from momentum transfer between the colliding debris (Hsü, 1975).
- *Turbidity currents* are currents of fast-flowing sediment-laden water moving down a sea or lake bed slope. The current moves because it has a higher density than the water through which it flows – the driving force of a turbidity current derives from its sediment, which renders the turbid water denser than the clear water above. Many submarine landslides (typically flows) transform into fast-moving turbidity currents. This transformation occurs through a variety of processes, including erosion of the dense landslide mass, breaking apart of the dense underflow, the breaking of internal waves and turbulent mixing (Felix and Peakall, 2006). The turbid water then rushes downward like an avalanche, picking up sediment and increasing in speed as it flows.

4 *Reactivation* is when part or all of a stationary, but previously failed, mass is involved in new movements, along pre-existing shear surfaces where the materials are at residual strength and are non-brittle. Reactivation can occur when the initial failed mass remains confined along part of the original shear surface. Such failures are generally slow-moving with relatively limited displacements associated with each new phase of movement, although there are circumstances when larger displacements can occur (Hutchinson, 1987). Reactivation is an episodic process, with phases of movement, often associated with periods of heavy rainfall or high ground-water tables, separated by inactivity (Figure 3.4). Reactivation also results in larger landslide masses becoming broken down into smaller units, a process sometimes called 'block disruption' (Brunsden and Jones, 1972).

An area can contain a *mosaic* of landslide features of different ages and origins, some of which may be subject to frequent reactivation. Indeed, the term *palimpsest* could be more appropriate, as older slides are often partially destroyed by overlying younger and younger slides. As a consequence, statements about landslide *age* can be both confusing and difficult to define: for example, does the age refer to the date of the original first-time failure(s) or a significant reactivation?

For areas where first-time failures are a regular occurrence, relatively young landslides can be dated with a reasonable degree of precision through their presence or absence on aerial photographs

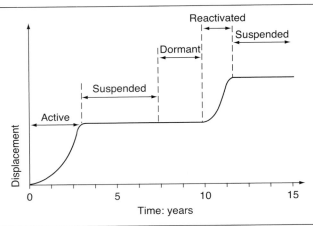

Figure 3.4 Landslide displacement in different states of activity (see Table 3.2 for definition of terms). (After Cruden and Varnes, 1996)

taken at different times. For example, the Natural Terrain Landslide Inventory in Hong Kong used vegetation cover and re-vegetation rates to provide an indirect measure of the date of shallow hillside failures. Evans and King (1998) report that in Hong Kong, landslide source areas start to re-vegetate within 2–3 years and reach 70% vegetation cover after 15–30 years. This allowed the following fourfold classification to be developed (Evans et al., 1997).

- Class A – totally bare of vegetation, assumed to be less than 2–3 years old.
- Class B – partially bare of vegetation, assumed to be between 2–3 and 30 years old.
- Class C – completely covered in grasses, assumed to be more than 15 years old.
- Class D – covered in shrubs and/or trees, assumed to be more than 25 years old.

The dating of older landslides can, however, prove extremely problematic. In some instances, it might be possible to use methods such as radiocarbon dating of organic deposits to define the inception of landsliding, as was achieved at Mam Tor, UK, where the first-time failure is believed to have been of Sub-Boreal age (around 3600 Radiocarbon Years ago), because the landslide materials have overridden tree roots dated at about 3200 Radiocarbon Years ago (Skempton et al., 1989). However, as Voight and Pariseau (1978) commented, 'the larger the mass movement (usually) the further back in time the event occurred, and in consequence, the more descriptive and less quantitative is our knowledge of the specifics of the event'. As a result, it can sometimes be more rewarding to focus on the timing of movements (i.e. the activity state of the landslide). Cruden and Varnes (1996) proposed six activity states (Table 3.2) that serve as a useful starting point. However, their definition of 'active' can be too restrictive and may become confused with 'suspended'. The failure to define 'inactive' is also problematic, and whether a landslide has to be capable of reactivation in whole or only in part to qualify for being classified as 'dormant' is unclear. As regards the term 'relict', it would appear better to add environmental to the list of conditions so as to include changes in vegetation and human influences. As a consequence, an alternative terminology is presented in Table 3.3, based partly on Jones and Lee (1994).

3.4. Potential for landsliding

The risk associated with landslide processes in an area is determined by the type of movements that can be expected to occur and their potential to produce adverse consequences. A wide variety of factors

Table 3.2 Landslide activity states

Activity state	Description
Active	Currently moving
Suspended	Moved within the last 12 months but not currently active
Dormant	An inactive landslide that can be reactivated
Abandoned	A landslide that is no longer affected by its original cause and is no longer likely to be reactivated. For example, the toe of the slide has been protected by a build-up of material, such as a floodplain or beach
Stabilised	A landslide that has been protected from its original causes by remedial measures
Relict	An inactive landslide developed under climatic or geomorphological conditions different from those at present

From Cruden and Varnes (1996)

(e.g. material characteristics, geological structure, pore water pressures, topography and slope angle) and causes (e.g. human activity, river and coastal erosion, weathering, seepage erosion, and high groundwater levels) are important in determining the occurrence of landsliding (e.g. Crozier, 1986). However, it is the presence or absence of pre-existing shear surfaces or zones within a slope that controls the character of landslide activity.

Table 3.3 Landslide activity states

Activity	Description
Active	Currently moving, or a currently unstable site such as an eroding sea-cliff or a site that displays a cyclical pattern of movement with a periodicity of up to 5 years
Suspended	Landslides and sites displaying the potential for movement, but not conforming to the criteria for 'Active' status
Dormant	A landslide or site that remains stable under most conditions, but may be reactivated in part or as a whole by extreme conditions
Inactive	A landslide or site of instability that is stable under prevailing conditions. Four sub-divisions can be identified *Abandoned*: a landslide that is no longer affected by its original cause and is no longer likely to be reactivated. For example, the toe of the slide has been protected by a build-up of material, such as a floodplain or beach *Stabilised*: a landslide that has been protected from its original causes by remedial measures *Anchored*: a landslide that has been stabilised by vegetation growth *Ancient*: an inactive landslide developed under climatic, environmental or geomorphological conditions different from those prevailing at present

Based on Jones and Lee, (1994)

Three main categories of landslide development have been recognised (Hutchinson, 1988, 1992):

- *First-time failures* of previously unsheared ground, often involving the mobilisation of the peak strength of the material. Such slides are often characterised by large, rapid displacements, particularly if there are significant differences between the peak and residual strength values.
- *Failures on pre-existing shear surfaces of non-landslide origin.* Probably the most important processes that have created such shears are flexural shearing during the folding of inter-bedded sequences of hard rocks and clay-rich strata, and periglacial solifluction.
- *Reactivation of pre-existing landslides,* where part or all of a previous landslide mass is involved in new movements, mainly along pre-existing shear surfaces.

This subdivision leads on to perhaps the simplest, yet most profound, concept in landslide hazard and risk studies, namely the subdivision of the land surface into (Hutchinson, 1992, 1995):

- areas where pre-existing landslides are present (the slid areas)
- areas which have not been affected by landsliding (the unslid areas).

In the former, there could be potential for reactivation; in the latter, new, first-time, landslides have to occur in previously unsheared ground.

The importance of this distinction between slid and unslid areas is that once a landslide has occurred, it can be made to move again under conditions that the slope, before failure, could have resisted. Thus reactivations can be triggered much more readily than first-time failures. For example, they may be associated with lower rainfall/groundwater level thresholds than first-time failures in the same materials. However, first-time failures may be more frequent than reactivations in those situations where the failed material is remobilised and transported beyond the landslide system or deposited at depths and angles that preclude further instability (e.g. Crozier and Preston, 1998).

3.5. Landslide velocity

Volume and speed combine to determine kinetic energy, which is sometimes equated with a landslide's destructive potential. Recorded landslide volumes range from around 1 m^3 up to the 1.2×10^{10} m^3 of the Flims landslide (Switzerland) and the 2×10^{10} m^3 of the Saidmarreh landslide (Iran), both of which have been exceeded by failures on young volcanoes (e.g. 2.6×10^{10} m^3 at Shasta, California). Speeds range from the imperceptible up to 400 km/h (Table 3.4).

Most slides tend to move at extremely slow to moderate speeds. Slides can prove exceedingly hazardous because of the profound disturbance (vertical and horizontal displacement) they cause to extensive tracts of ground. However, there are a number of slide types that, post failure, may develop into high-velocity events, including the following (Picarelli *et al.*, 2005; Hungr *et al.*, 2005).

- Flow-slides in saturated granular soils that liquefy as shearing continues, for example, loose constructed fills and mine waste and submarine slopes. Hunter and Fell (2003) identified a number of soil types that are prone to flow liquefaction and hence high-velocity landslides: very loose to loose clean sands; very loose to medium dense silty sands; silty sands with clay contents <10–20%; and sandy gravels and gravely sands with a trace of fines (<5–10%) at void ratios >0.3.
- Slides in sensitive clays or quick clays, for example the Attachie Slide on the Peace River, British Columbia, which developed in very sensitive glacio-lacustrine clays and silts (Fletcher *et al.*,

Table 3.4 Velocity classes for landslides

Velocity class	Description	Velocity: mm/s	Typical velocity	Nature of impact
7	Extremely rapid		5 m/s	Catastrophe of major violence; exposed buildings totally destroyed and population killed by impact of displaced material or by disaggregation of the displaced mass
		5×10^3		
6	Very rapid		3 m/min	Some lives lost because the landslide velocity is too great to permit all persons to escape; major destruction
		5×10^1		
5	Rapid		1.8 m/h	Escape and evacuation possible; structure, possessions and equipment destroyed by the displaced mass
		5×10^{-1}		
4	Moderate		13 m/month	Insensitive structures can be maintained if they are located a short distance in front of the toe of the displaced mass; structures located on the displaced mass are extensively damaged
		5×10^{-3}		
3	Slow		1.6 m/yr	Roads and insensitive structures can be maintained with frequent and heavy maintenance work, if the movement does not last too long and if differential movements at the margins of the landslide are distributed across a wide zone
		5×10^{-5}		
2	Very slow		16 mm/yr	Some permanent structures undamaged, or if they are cracked by the movement, they can be repaired
		5×10^{-7}		
1	Extremely slow			No damage to structures built with precautions

After WP/WLI (1995) and Cruden and Varnes (1996)

2002). After creeping for many decades, in 1973 the slide suddenly liquefied following a period of heavy rain and 7 Mm^3 of material descended a bedrock scarp at the foot of the slope and flowed across the Peace River floodplain.
- Slides in steep cut slopes in residual soils or colluvium. Rapid sliding is almost certain for slopes steeper than 35° (Picarelli *et al.*, 2005).
- Structurally controlled slides in strong rock can be extremely rapid owing to sudden loss of cohesion (e.g. Hungr and Evans, 2004). Many become fragmented and flow-like, forming extremely rapid rock avalanches (sturzstroms).
- Shallow debris slides on natural slopes with steep source areas (typically >25°, but these slides may occur on slopes of 18–20°). These failures typically occur during heavy rain. As fast movement occurs, soil situated downslope of the initial failure is over-ridden, liquefied by rapid undrained loading and incorporated in a growing debris avalanche (e.g. Sassa, 1985).

Flows tend to be rapid to extremely rapid events and can produce catastrophic events. This is not merely because flows have high kinetic energy but also because they move over the ground surface, thereby mobilising destructive force to overwhelm objects in their path, sometimes at very great distances from their source (travel distance), especially in the case of volcanic mudflows (lahars). For example, over 200 lahars occurred during the first rainy season after the June 1991 eruption of Mount Pinatubo in the Philippines (Pierson, 1999). The eruption had deposited 5–7 km^3 of pyroclastic flow deposits and 2×10^9 m^3 of tephra on the volcano flanks, filling many steep catchments with up to 220 m depth of material. Typical monsoon lahars in the region were 2–3 m deep and 20–50 m wide, with peak discharges of 750–1000 m^3/s. However, lahars triggered by intense typhoon rainfall moved at up to 11 m/s, with peak discharges reaching 5000 m^3/s. Lowland areas up to 50–60 km from the volcano were seriously affected (Pierson et al., 1992). River channels were infilled, leading to widespread flooding, and agricultural land was buried beneath several metres of sand and gravel.

Both slides and flows are generally considered more of a threat than falls, despite the fact that falls produce significant missiles, sometimes of enormous size. Although falls are confined to specific and somewhat restricted locations (namely, cliffs and crags), their occurrence can be the initiating event for the generation of destructive flows and slides downslope (e.g. Huascaran in 1970), illustrating how one type of landslide can evolve into other types of landsliding downslope during a single event sequence.

3.6. Landslide travel distance

Landslides can travel considerable distances beyond the original source area. For example, the 1903 Frank slide in southern Alberta, Canada involved a 3×10^7 m^3 rock slide from near the summit of Turtle Mountain that raced across the valley below, burying part of the town and killing at least 70 people. The mass of rock travelled about 4 km and covered 3 km^2 of the valley floor (Dowlen, 1903).

The travel distance of landslide debris is generally estimated in terms of a *travel angle* (defined as the slope of a line joining the tip of the debris to the crest of the main landslide source; also termed the *reach angle* α; Figure 3.5). The key factors that influence the travel distance include (Wong and Ho, 1996):

- slope characteristics, including slope height, slope gradient and the slope-forming materials
- failure mechanism, including collapse of metastable or loose soil structures leading to the generation of excess pore water pressures during failure, the degree of disintegration of the failed mass, the fluidity of the debris, the nature of the debris movement (e.g. sliding, rolling, bouncing or viscous flow) and the characteristics of the ground surface over which the debris travels (e.g. response to loading and surface roughness)
- the condition of the downhill slope, including, for example, the gradient of the deposition area, the existence of irregularities and obstructions, and the presence of well-defined channels.

Travel distances of historical landslides can be broadly defined from interpretation of aerial photograph or high-resolution satellite imagery or from geomorphological mapping (e.g. Griffiths et al., 2011; Hearn, 2011). Fell et al., (2008a,b) stressed that care should be taken when defining travel distance from the location of old landslide deposits because the source cannot always be accurately located and travel distance estimation may be subjected to significant error.

Empirical observations can be used to develop a relationship between travel distance and particular slope and material properties. Typical data on debris run-out for different mechanisms and scale of

Figure 3.5 Travel distance terminology. (From Hungr et al., 2005)

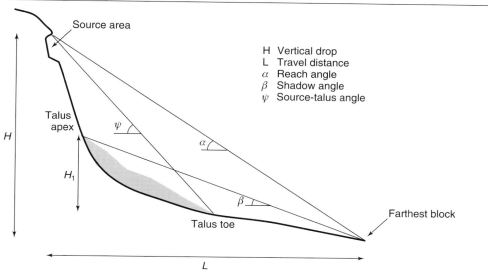

landslides in Hong Kong are given in Figure 3.6. For slides that break up, and in some cases become flows, the travel distance is usually estimated from the apparent friction angle and landslide volume (Figure 3.7) (Corominas, 1996; Finlay et al., 1999; Wong and Ho, 1996; Wong et al., 1997). However, caution needs to be exercised when extrapolating these relationships to different landslide settings and environments.

Figure 3.6 Landslide run-out: relationship between cut slope height and landslide run-out, Hong Kong. (After Wong et al. (1997)

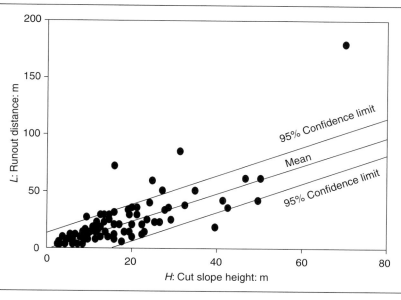

Landslide Risk Assessment

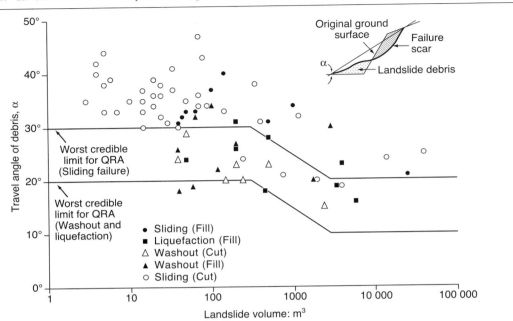

Figure 3.7 Landslide debris mobility data, Hong Kong. (After Ho et al., 2000)

A variety of regression equations have been developed that describe the relationship between the tangent of the travel/reach angle (tan $\alpha = H/L$) and the landslide volume V, generally in the form

$$\log \tan \alpha = A + B \log V$$

where A and B are constants. Values of A and B for different landslide types and flow-paths are presented in Table 3.5 (Corominas 1996; Hunter and Fell, 2003). However, Hungr et al. (2005) have stressed that care is needed when using these equations for predicting travel distance, since there can be significant scatter around the mean values and many landslides will travel beyond the calculated distance.

Table 3.5 Regression equation parameters for predicting landslide travel distance

Landslide type	Flow-path	A	B	R^2	Reference
Translational slides	Obstructed	−0.13	−0.06	0.76	Corominas (1996)
	Unobstructed	−0.14	−0.08	0.8	
Debris flows	Obstructed	−0.05	−0.11	0.85	
	Unobstructed	−0.03	−0.10	0.87	
Earthflows	Unobstructed	−0.22	−0.14	0.91	
All	Unconfined	0.77	0.09	0.71	Hunter and Fell (2003)
	Partly confined	0.69	0.11	0.52	
	Confined	0.54	0.27	0.85	

For rockfalls, most of the kinetic energy is dissipated on the first impact on a talus slope at the base of a cliff. The extent of bouncing and rolling of rockfalls can be defined by the shadow angle β, which is the angle of the line from the top of the talus apron or scree to the farthest block (Figure 3.5). The talus toe is used as the reference point beyond which the distance travelled by fallen blocks is determined. Evans and Hungr (1993) suggested that the minimum shadow angle should be 27.5°, although flatter angles have been observed where the talus or natural slope is relatively smooth, with β reaching 23–24° (Hungr et al., 2005). Zoning of rockfall hazard areas in Andorra la Vella, for example, used the regression equation (Copons 2004)

$$\log \tan \beta = 0.045 \log V - 0.223$$

Rockfall travel distance can be modelled using commercially available computer programs, such as the Colorado Rockfall Simulation Program (CRSP). This software uses slope and rock geometry and material properties, calculating falling rock bounce height, velocity and travel distance.

A range of analytical (numerical) methods are available to assess landslide travel distance, including sliding-block models, two-dimensional models that consider a typical profile of the slide and three-dimensional models that analyse the flow of a landslide over irregular terrain (Hungr et al., 2005). Among the most widely used single-body models is the 'sled' model proposed by Sassa (1988). In this model, it is assumed that all the energy losses during landslide movement will be dissipated through friction. The apparent angle (or coefficient) of friction, as determined in high-speed ring-shear tests, is used as a measure of the amount of friction loss that can occur during movement and flow of the debris. However, it should be noted that the deposition of the mass during motion can have a marked influence on run-out (Wong and Ho, 1996). As many landslides change mass during motion, the use of a sliding-block model in which there is a constant mass can lead to underestimation of the run-out. A number of models have been developed that take account of deposition during motion (e.g. Hungr and McClung, 1987; Hungr et al., 1984; Van Gassen and Cruden, 1989).

Under certain circumstances, landslides can travel unusually long distances (Hsü, 1975). These long-run-out *landslides* are mainly rockfalls and avalanches, although submarine slides display similar characteristics. The L/H ratio (i.e. run-out length to drop height) gives a measure of mobility, while the H/L ratio is an approximation of the coefficient of friction of sliding debris. Plots of H/L against run-out length L display a characteristic trend of decreasing H/L with increasing L, with small rockfalls characteristically having $H/L > 0.3$ (Collins and Melosh, 2003), while submarine slides have $H/L \ll 0.1$ (e.g. Hampton et al., 1996). Several explanations for this characteristic have been proposed (De Blasio, 2011). An interesting study of the 'landslides' on Saturn's icy moon, Iapetus (Singer et al. 2012), has concluded that its long-run-out features are the result of 'localized frictional heating of ice rubble such that sliding surfaces are slippery', thereby lending support to the hypothesis of landslide lubrication by basal melt layers advanced by De Blasio and Elverhoi (2008) and Weidinger and Korup (2009). This has implications for risk assessments in areas of snow-lie and frozen ground.

3.7. Nature of landslide hazards

The main types of threats include

- vertical displacement damage to structures and infrastructure, which occurs mainly at the expanding headwall parts of landslides or cliffs, except in the case of rotational failures, where it can occur almost anywhere over the entire length of the failure

- lateral displacement, which is a feature of rotational failures, although the piling-up of material against structures is a feature of all three mechanisms (i.e. falling, sliding and flowing)
- burial, which is a characteristic feature of flows, although it can occur below most types of large failure on steep slopes
- missile impacts and air-blast effects, which are mainly a feature of falls, although they can be associated with other types of major failures on steep slopes.

While landslide mechanism clearly play a dominant role in determining threat, it is clear that geological and topographic factors are also very important; that is, available potential energy increases the variety of threats.

Different parts of a landslide mass come under tension, compression, tilt, contra-tilt, heave and so on according to their location and the stresses involved. As a result, a particular asset impacted by a landslide may suffer damage from a number of different loads, depending on its location relative to the landslide. For example, in the case of a buried pipeline, full-bore rupture and a spill could occur as a result of the following processes.

1. *Lateral displacement*, with pipe rupture as a result of differential horizontal and/or vertical movement of the landslide mass. The potential for pipeline displacement is a function of landslide depth, the behaviour of the materials (e.g. plastic or block-type deformation), the speed of movement and the cumulative displacement that could occur over time.
2. *Spanning*, with pipe rupture as a result of removal of support along a significant length, due to landslide movement or scour associated with a debris flow evacuating material from beneath the pipeline. The potential for spanning is a function of the vertical displacement of the landslide mass, width and depth of the scour zone, or retreat of an eroding scar across the pipeline alignment.
3. *Loading*, with pipe rupture as a result of an imposed load (e.g. burial by landslide debris or the impact of falling or rolling rocks). This is a function of the depth of run-out of landslide debris, the size and height of fall of material onto the pipeline, or the velocity of missiles.
4. *Exposure and impact by landslide debris*. Exposure of a pipeline by flooding or scour can leave it vulnerable to the impact of large boulders mobilised by debris flows or large-run-out landslides that then pass along the channel. This sequence of events – flooding, scour and debris flow – can occur during the same storm event, making mountain stream crossings a particularly hazardous setting.

Five phases of landslide hazard can be recognised.

1. Hazards associated with pre-failure movements.
2. Hazards associated with the main phase(s) of movement.
3. Secondary hazards generated as a consequence of movement.
4. Post-event hazards as the system returns to stability.
5. Hazards associated with subsequent movements (i.e. reactivation).

3.7.1 Pre-failure hazards

During this phase, the actual hazard is usually restricted to minor falls, the formation of tension cracks and minor grabens, and minimal displacements that can affect infrastructure and buildings. However, people's perceptions of hazard are likely to be raised because of the uncertain threat of impending events of unknown magnitude, violence and timing, and this may result in alarm. This is the time

3.7.2 Main-phase hazards

In this phase, the main attributes that contribute to hazard are

- rate of onset (suddenness)
- volume involved in movement
- depth of movement
- areal extent of movement
- speed of movement
- displacement distance
- scale of surface distortion
- development of missiles
- destructive forces mobilised.

Any hazard assessment will need to establish the likely importance of these attributes in the context of any specific landslide scenario, especially as the phenomena that threaten human life are not necessarily the same as those that disrupt underground infrastructure. Four of the more important of these main-phase hazards are worth further elaboration at this point, with specific reference to the landslide portrayed in Figure 3.8 (see also Table 3.6):

1. *Loss of cliff-top land/cliff recession.* Active debris removal and undercutting at the landslide toe by marine or river erosion can generate repeated sequences of pre-failure movements, failure and reactivation events. Each first-time failure will involve the detachment of sections of cliff top or main scarp crown, leading to the progressive retreat of the cliffline/main scarp. Average coastal cliff-top recession rates of up to 2 m a year can occur on exposed soft-rock cliffs, although losses within an individual year can be much greater (Lee and Clark, 2002). Despite individual failures generally tending to cause only small amounts of cliff retreat, the cumulative effects can be dramatic. For example, the Holderness coastline of the UK has retreated by around 2 km over the last 1000 years, resulting in the loss of at least 26 villages listed in the Domesday survey of 1086; 7.5×10 Mm^3 of land has been eroded in the last 100 years (Pethick, 1996; Valentin, 1954). A review of the variety of approaches available for predicting cliff recession rates is presented in Lee and Clark (2002). The same processes operate on undercut or unprotected river cliffs and meander scars, although the process tends to be more episodic, reflecting the intervals between erosive floods, and the rates of retreat tend to be slower. On inland landslides, the process is even more episodic, often with significantly longer intervening periods of stability interrupted by head-scarp retrogression.

2. *Differential ground movement and distortion within the main body.* There is a close association between landslide type and the style of ground disturbance that can be experienced. For example, in the single rotational slide shown in Figure 3.8
    - vertical settlement and contra-tilt occur at the slide head
    - the downslope movement of the main body can generate significant lateral loads
    - differential horizontal and vertical settlement occurs between the individual blocks within the main body
    - compression and uplift (heave) occurs in the toe area, and some fixed objects may be displaced or buried.

Figure 3.8 Landslide morphology: (a) block diagram of an idealised rotational failure earthflow; (b) landslide features – see Table 3.6 for definitions of numbers. (After Cruden and Varnes, 1996; IAEG, 1990)

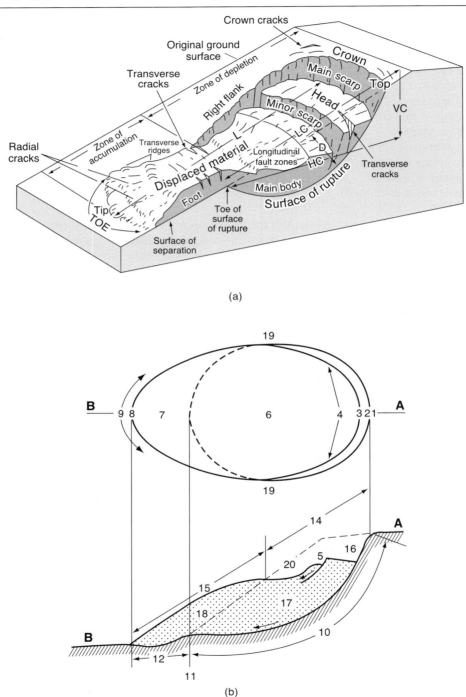

Table 3.6 Definitions of landslide features referred to in Figure 3.8

No.	Name	Definition
1	Crown	Practically undisplaced material adjacent to highest parts of main scarp
2	Main scarp	Steep surface on undisturbed ground at upper edge of landslide caused by movement of displaced material (13, stippled area) away from undisturbed ground – that is, it is visible part of surface of rupture (10)
3	Top	Highest point of contact between displaced material (13) and main scarp (2)
4	Head	Upper parts of landslide along contact between displaced material and main scarp (2)
5	Minor scarp	Steep surface on displaced material of landslide produced by differential movements within displaced material
6	Main body	Part of displaced material of landslide that overlies surface of rupture between main scarp (2) and toe of surface of rupture (11)
7	Foot	Portion of landslide that has moved beyond toe of surface of rupture (11) and overlies original ground surface
8	Tip	Point on toe (9) farthest from top (3) of landslide
9	Toe	Lower, usually curved, margin of displaced material of a landslide, most distant from main scarp (2)
10	Surface of rupture	Surface that forms (or that has formed) lower boundary of displaced material (13) below original ground surface (20)
11	Toe of surface of rupture	Intersection (usually buried) between lower part of surface of rupture (10) of a landslide and original ground surface (20)
12	Surface of separation	Part of original ground surface (20) now overlain by foot (7) of landslide
13	Displaced material	Material displaced from its original position on slope by movements in landslide; forms both depleted mass (17) and accumulation (18): stippled on Figure 3.8(b)
14	Zone of depletion	Area of landslide within which displaced material (13) lies below original ground surface (20)
15	Zone of accumulation	Area of landslide within which displaced material (13) lies above original ground surface (20)
16	Depletion	Volume bounded by main scarp (2), depleted mass (17) and original ground surface (20)
17	Depleted mass	Volume of displaced material (13) that overlies surface of rupture (10) but underlies original ground surface (20)
18	Accumulation	Volume of displaced material (13) that lies above original ground surface (20)
19	Flank	Undisplaced material adjacent to sides of surface of rupture (right and left is as viewed from crown)
20	Original	Surface of slope that existed before landslide took place ground surface

After Cruden and Varnes (1996)

In general, the amount of ground disturbance and the severity of the hazard will be greatest during first-time failure.
Other landslide features (generally hard-rock slope failures or flows) can simply remain as stable debris fans or boulder ramparts at the base of the original failure site. For many large-run-out landslides, the debris fan can have a very high long-term margin of stability against renewed

movement. For example, Hutchinson (1995) records that after the Mayunmarca sturzstrom, the emplaced debris was dry and had a factor of safety of at least 2.9 (see Chapter 5).

3. *Run-out*. The run-out of landslide debris can cause considerable damage. The distance the slide mass will travel and its velocity determine the extent to which the landslide will affect property and people downslope, and their ability to escape. The main hazards are lateral impact and burial.
4. *Impact of falling boulders or debris*. Falling rocks and boulders can present a major public safety issue in many areas, especially along road or railway cuttings or where rock cliffs form backdrops to bathing beaches. In addition, falling or bouncing rocks act as missiles and can cause considerable property damage and disruption to services, as well as harm to people.

3.7.3 Secondary hazards

Secondary hazards are geophysical phenomena generated as a consequence of a primary event. The situation is complicated in the case of landslides, as they can be both *primary hazards* and *secondary hazards* as generated by earthquakes, volcanic eruptions (e.g. lahars), tropical cyclones, floods and even primary landslides. Either way, the resultant landslides may result in *secondary impacts* that in certain circumstances can be even greater than those imposed by the primary event. The prime example involves large landslides, such as rock avalanches and rock slides, blocking narrow, steep-sided valleys and forming *landslide dams* (Schuster, 1986), the second concerns the potential for landslides to form *tsunamis*.

3.7.3.1 Landslide dams

Table 3.7 provides an indication of the scale of such landslide dams, which tend to be a feature of seismically active steep-relief mountain areas undergoing uplift and erosion (e.g. the Indus River catchment) or deeply dissected thick sequences of weakly consolidated sediments such as loess.

Landslide dams give rise to two important flood hazards:

1. *Upstream* or *back-water flooding* occurs as a result of impounding of water behind the dam leading to the relatively slow inundation of an area to form a temporary lake. In 1983, the small town of Thistle, Utah, was inundated by a 200 m-long, 50–60 m-deep lake that formed behind a landslide dam on the Spanish Fork River (Kaliser and Fleming, 1986). A similar lake that formed behind a rock avalanche dam in 1513 led to the drowning of the Swiss hamlets of Malvaglia and Semione (Eisbacher and Clague, 1984). The headstream valleys of many river systems, such as the River Indus, contain the remains of numerous sediment in-fills deposited behind former landslide dams.
2. *Downstream flooding* can occur in response to failure of a landslide dam. The most frequent failure modes are overtopping because of the lack of a natural spillway or breaching due to erosion. Failure of the poorly consolidated landslide debris generally occurs within a year of dam formation. The effects of the resultant floods can be devastating, partly because of their magnitude and partly because of their unexpected occurrence. For example, in the winter of 1840–1841, a spur of the Nanga Parbat Massif, in what is now Pakistan, failed during an earthquake, completely blocking the River Indus and causing the formation of a 60–65 km long lake. The dam, originally up to 200 m high, breached in early June 1841. The lake emptied in 24 hours, causing what has been called 'the Great Indus flood', during which hundreds of villages and towns were swept away. A Sikh army encamped close to the river about 420 km downstream was overwhelmed by a flood of mud and water estimated to be 25 m high; about 500 soldiers were killed (Mason, 1929). Failure of the Deixi landslide dam on the Min River, China in 1933 resulted in a wall of water that was 60 m high some 3 km downstream. The

Landslide hazard

Table 3.7 A selection of historic landslide dams

Landslide	Year	Dammed river	Landslide volume: 10^6 m^3	Dam height: m^3	Dam width: m^3	Lake length: km	Lake volume: 10^6 m^3	Dam failure
Slumgullion earth flow, USA	1200–1300	Lake Fork, Gunnison River	50–100	40	1700	3		No
Usay landslide, Tadzhikistan	1911	Murgab River	2–2.5	300–550	1000	53		Partial
Lower Gros Ventre landslide, USA	1925	Gros Ventre River	38	70	2400	6.5	80	Yes
Deixi landslide, China	1933	Min River	150	255	1300	17	400	Yes
Tsao-Ling rockslide, Taiwan	1941/42	Chin-Shui-Chi River	250	217	2000		157	Yes
Cerro Codor Sencca rockslide, Peru	1945	Mantaro River	5.5	100	580	21	300	Yes
Madison Canyon rockslide, USA	1959	Madison River	21	60–70	1600	10		No
Tanggudong debris slide, China	1967	Yalong River	68	175	3000	53	680	Yes
Mayunmarca rockslide, Peru	1974	Mantaro River	1600	170	3800	31	670	Yes
Gupis debris flow, Pakistan	1980	Ghizar River		30	300	5		No
Polallie Creek debris flow, USA	1980	East Fork Hood River	0.07–0.1	11	230		105	Yes
Thistle earth slide, USA	1983	Spanish River	22	60	600	5	78	No
Pisque River landslide, Ecuador	1990	Pisque River	3.6	56	60	2.6	3.6	Partial
Tunawaea landslide, New Zealand	1991	Tunawaea Stream	4	70	80	0.9	0.9	Yes
Rio Torro landslide, Costa Rica	1992	Rio Tooro	3	100	75	1.2	0.5	Yes
La Josefina rockslide, Ecuador	1993	Paute River	20–44	100+	500	10	177	Partial

From Schuster (1986) and Sassa (1999)

floodwaters had an average velocity of 30 km/h and reached the town of Maowen (58 km downstream) in two hours. The total length of valley flooded was 253 km and at least 2423 people were killed (Li Tianchi et al., 1986).

3.7.3.2 Tsunami generation

Tsunamis can be generated in a range of circumstances

- where fast moving or massive landslides descend into water
- where the slip plane of a major failure extends beneath a water body
- where a wholly submerged (submarine) landslide causes significant displacement.

Although traditionally associated with giant waves generated in marine environments, similar phenomena can develop in extensive bodies of freshwater.

When fast-moving landslides enter bodies of water, the impact can be devastating as a consequence of the generation of tsunamis. In 1792, a landslide off the flanks of Mount Unzen, Japan generated a tsunami in Ariake Bay that caused 14 500 deaths around the shoreline (McGuire, 1995). In July 1958, a major rockslide (of about 3×10^7 m^3) occurred on the margins of Lituya Bay, Alaska, triggered by a magnitude 7.5 earthquake on the Fairweather fault. A 30–200 m high tsunami wave was generated that inundated the shoreline of the Bay, with a run-up that probably reached up to 530 m high (Mader, 2002; Miller, 1960). Recent modelling of the event has indicated that the wave was caused by the impact of the slide moving at 110 m/s into 120 m of water.

Although the role of landslides in generating tsunamis has long been known, it is only recently that the true catastrophic potential has begun to be recognised (Keating and McGuire, 2000). Moore et al. (1989) claim that huge submarine landslides on the Hawaiian Ridge have produced tsunamis large enough to affect most of the Pacific margin. For example, Lipman et al. (1988) consider that the huge Alika slide from Hawaii Island (1.5–2 $\times$ 10^9 m^3, c. 105 000 BP) produced tsunamis with run-up heights up to 325 m on Lanai Island, more than 100 km away. Evidence of catastrophic wave erosion up to 15 m high along the New South Wales coast has also been linked to the collapse of portions of Mauna Loa volcano, over 14 000 km away (Bryant, 1991). Blong (1992) considers that the Pacific Basin may experience comparable submarine-landslide-generated tsunamis with a frequency of between 1 and 4 per 100 000 years, indicating a global frequency close to the higher figure. It should be noted, however, that some authorities favour an alternative explanation involving meteorite impacts (e.g. Jones, 1992a; Jones and Mader, 1996).

More recently, there is a growing appreciation that at least some of the greatest tsunamis appear to be related to the collapse of volcanoes. The eruption of Mount St Helens, USA in 1980 graphically illustrated two important points. First, landsliding can trigger a volcanic eruption (Lipman and Mullineaux, 1981) – so the blast effects, ashfalls and so on produced in 1980 could, quite properly, be interpreted as secondary hazards generated by the landslide. Second, volcanic edifices appear prone to major instability. Structural failure of volcanic edifices is now recognised to be ubiquitous, as can be illustrated by the observation that 75% of large volcanic cones in the Andes have experienced collapse during their lifetime (Francis, 1994).

The rate of collapse is unknown, with estimates ranging from four per century (Siebert, 1992) to more than double that number. How many collapses involve cones on coastlines or on volcanic islands is also unclear. However, about 70 gigantic slides have been identified around the Hawaiian Islands

(Moore *et al.*, 1994) and there is a growing consensus that such failures occur mainly during warm inter-glacial periods, such as the present, when global sea levels are high (McGuire *et al.*, 2002). For example, collapse of the western flank of the Cumbre Vieja volcano on La Palma in the Canary Islands is expected to generate a massive tsunami toward the coasts of Africa, Europe, South America, Newfoundland and possibly even the USA. It is predicted that within six hours of the failure, waves reaching 9 m would arrive in Newfoundland and 14–18 m high waves would hit the shores of South America. Nine hours after the collapse, crests reaching 9–21 m could surge onto the East Coast of the USA, where the floodwaters might extend inland several kilometres from the coast. Accurate estimates of the scale of economic loss are yet to be made, but are thought to be in the multi-trillion US dollar range (McGuire *et al.*, 2002).

The accumulations of sediment in the shallow marine environments associated with continental margins have long been known to suffer instability sufficient to generate tsunamis (Murty and Wigen, 1976). The classic example is the huge Storegga slide off the west coast of Norway (Bugge *et al.* 1987, 1988), which covers an area of 34 000 km², It involved the displacement of some 5.6×10^{12} m³ of sediment in three separate failures between 50 000 and 7200 BP. The first of two movements around 7200 BP is thought to have generated a tsunami on the coast of Scotland with a run-up height of about 4 m (Dawson *et al.*, 1988).

Landslides entering the more confined waters of big rivers, natural lakes or impounded reservoirs similarly displace large volumes of water in the form of waves and surges, which can have devastating local effects, as occurred during the Vaiont disaster in the Alpine region of north-east Italy. Filling of the reservoir in 1960 resulted in the opening of a 2 km long crack on the flanks of Mount Toc. During the next three years, the unstable slope moved downslope by almost 4 m until the evening of 9 October 1963, when a 2.5×10^8 m³ rockslide occurred. The slide entered the reservoir at around 25 m/s, sending a wave around 100 m high over the crest of the concrete dam. The flood wave destroyed five downstream villages and killed around 1900 people (Hendron and Patton, 1985; Kiersch, 1964).

The question as to whether or not tsunamis could also be formed in major inland spreads of water has recently been answered by Kremer *et al.* (2012), who reported that Lake Geneva suffered the equivalent of a tsunami in AD 563. The cause is claimed to be a historically established major rockfall at the eastern end of the lake, known as the Tauredunum Event, The evidence indicates that the fall generated a tsunami that reached a maximum height of 16 m on the northern shoreline. Interestingly, it was 13 m high at the present site of Lausanne and 8 m high at Geneva, thereby highlighting that it is not only urban areas located on coastlines and within fjords that are at risk from destructive tsunamis, but also those located on some inland lake shorelines.

3.7.4 Post-event hazards as the system returns to stability

Following a major landslide event, there is a period, sometimes extending for years, during which the landslide system attempts to achieve overall stability (akin to the aftershocks of an earthquake). During this time, there will be falls and small failures from the backscar, together with movements within the failed mass, including the fragmentation of some of the larger masses. There may be some retreat of the main backscar that could further threaten property and infrastructure, but otherwise the main threat is to curious people moving over the slide and unaware of the dangers.

3.7.5 Landslide reactivation hazards

Landslides that have involved shearing and remain confined within the original shear surface (e.g. clay slope or mudrock failures that have not run out beyond the shear surface) will remain prone to reactivation. The likely displacement during reactivation can be estimated as that required to change the geometry of the slide so that the average shear stress on the shear surface is not increased (Vaughan, 1995). The rate of displacement v can be determined from (Leroueil et al., 1996; Vulliet, 1986)

$$v = F(\sigma'_n, \tau)\tau$$

where F is a function of the normal effective stress σ'_n and the applied shear stress τ.

The hazards associated with the reactivation stage are similar to those experienced during the main phase of movement (i.e. the failure stage). Of particular significance is the differential ground movement and distortion as the failed mass slides along the pre-existing shear surface or surfaces. Extensive areas of landslide mantled terrain exist in many countries, especially in areas prone to neotectonic activity and/or that experienced repeated fluctuations from cold to temperate conditions during the Pleistocene age (i.e. the last 2.6 million years). In many cases, surficial erosion and deposition has largely concealed the existence of the slid material. As a consequence, the greatest inland landslide hazard can be associated with the unanticipated reactivation of pre-existing landslides following prolonged heavy rainfall or as a direct result of human activity, especially through loading and unloading of portions of already-slid ground and the disruption of natural drainage. For example, research has shown that many of the recent instability problems in South Wales, UK, appear to be associated with ancient landslides that have been reactivated by housing development or industrial activity (Sir William Halcrow and Partners, 1986, 1988). However, reactivation can also be generated by new first-time failures from adjacent un-slid ground and from the effects of basal erosion by rivers.

The classic case of reactivation concerns the construction of the A21 Sevenoaks Bypass in the UK, which had to be halted in 1966 because bulldozers cut through innocent-looking grass-covered lobes of material that proved to be the remains of periglacial landslides and solifluction lobes (Skempton and Weeks, 1976). The affected portion of the route had to be realigned at a (then) cost of £2 million. Since then, awareness of the potential problems and costs of reactivation has not only led to the widespread use of geomorphological and engineering geology mapping but has also contributed to the realignment or even shelving of several proposed roads in the UK.

There is evidence that there can be repeated periods of landslide reactivation, especially related to wet year sequences (e.g. Maquaire, 1994, 1997; Schrott and Pasuto, 1999). Lee and Brunsden (2000), for example, demonstrated that for the west Dorset coast in the UK, there was a direct relationship between wet year sequences and landslide activity. Wet year sequences were identified by calculating the cumulative number of years with effective rainfall (moisture balance) greater than the mean value of 319 mm (Figure 3.9). This value is set to zero every time it falls below the mean, on the assumption that the groundwater levels only become critical when they rise above the average level. On this basis, a number of distinct wet year sequences (of three years in length or more) were recognised: 1874–1877, 1914–1917, 1927–1930, 1935–1937, 1958–1960, 1965–1970, 1979–1982 and 1993–1998. There have thus been eight wet year sequences in 130 years, each with a duration of 3–6 years. As indicated on Figure 3.9, since the late 1950s, these wet year sequences have broadly coincided with the timing of major reactivations within the Black Ven–Spittles mudslide complex (before the 1950s, less attention was probably given to the recording of landslide events, other than the most dramatic failures).

In general, reactivation of pre-existing landslides present only a minor threat to life, since movements, when they occur, usually involve only slow to extremely slow displacements. Even when large

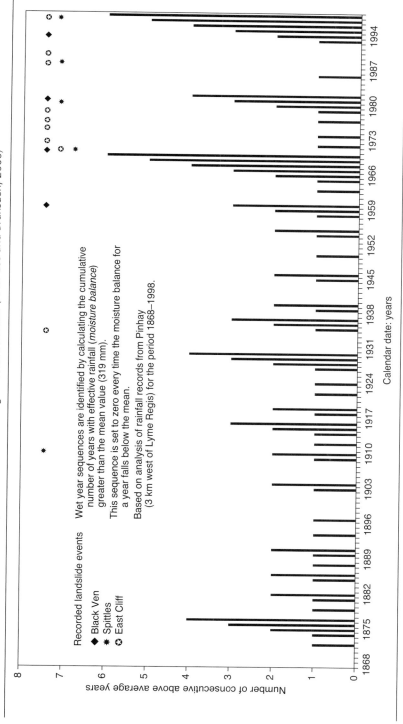

Figure 3.9 Lyme Regis, UK: number of years with above-average effective rainfall. (After Lee and Brunsden, 2000)

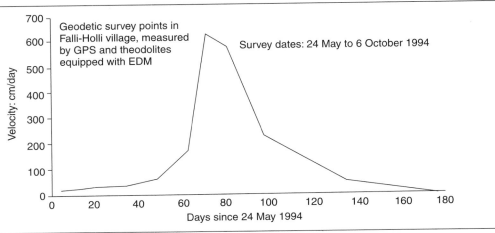

Figure 3.10 Landslide reactivation displacements: Falli-Holli village, Switzerland. (After Lateltin and Bonnard, 1999)

displacements occur, the rate of movement tends to be gentle and not dramatic, as was reported graphically for a slide near Lympne in Kent, UK, in 1725, where a farmhouse sank 10–15 m overnight, 'so gently that the farmer's family were ignorant of it in the morning when they rose, and only discovered it by the door-eaves, which were so jammed as not to admit the door to open' (Gostling, 1756).

However, the cumulative effects of episodes of slow movement can cause considerable damage to buildings, services and infrastructure. Damage due to slope instability can lead to expensive remedial measures or, where repair is considered uneconomic, the abandonment and loss of property. In 1994, for example, reactivation of a large landslide near Freiberg, Switzerland caused the destruction of 41 homes in the tourist village of Falli-Holli, with estimated losses of US$15 million (Lateltin and Bonnard, 1999). The maximum recorded displacements had been around 7 m per day (Figure 3.10). Reactivation of the Portuguese Bend landslide in the Palos Verdes Hills, California, USA resulted in $45 million of damage (1996 prices) to roads, homes and other structures between 1956 and 1959 (Merriam, 1960). Subsequent litigation awarded $41 million (1996 prices) to property owners from the County of Los Angeles, on the grounds that the problems had been initiated by road construction works (Vonder Linden, 1989).

3.8. Landslide intensity

The destructive intensity of a landslide is related to kinetic parameters, such as velocity and acceleration, the total and differential displacement, along with its dimensions (e.g. depth and volume (and thus mass m) of the moving material and depth of deposits after the movement ceases), and the material characteristics (Hungr, 1997; Léone et al., 1996). The maximum movement velocity v is a key factor in landslide destructiveness (Table 3.4 and 3.8). An important limit appears to lie around 5 m/s, approximately the speed of a person running away from a slide.

In first-time slides, the available potential energy (a function of slope height and geometry) is progressively dissipated into several components (Leroueil et al., 1996). *Frictional energy* moves the detached material over the shear surface. The amount of frictional energy required depends on the

Table 3.8 Examples of landslide velocity and damage

Landslide velocity class	Landslide name, location	Source	Estimated landslide velocity	Damage
7	Elm, Switzerland	Heim (1932)	70 m/s	115 deaths
	Goldau, Switzerland	Heim (1932)	70 m/s	457 deaths
	Jupile	Bishop (1973)	31 m/s	11 deaths, houses destroyed
	Frank, Canada	McConnell and Brock (1904)	28 m/s	70 deaths
	Vaiont, Italy	Mueller (1964)	25 m/s	1900 deaths by indirect damage
	Ikuta, Japan	Engineering News Record (1971)	18 m/s	15 deaths, equipment destroyed
	St Jean Vianney, Canada	Tavenas et al. (1971)	7 m/s	14 deaths, structures destroyed
6	Aberfan, Wales	Bishop (1973)	4.5 m/s	144 deaths, some buildings destroyed
5	Panama Canal	Cross (1924)	1 m/s	Equipment trapped, people rescued
4	Handlová, Slovakia	Zaruba and Mencl (1969)	6 m/day	150 houses destroyed, complete evacuation
3	Schuders, Switzerland	Huder (1976)	10 m/year	Road maintained with difficulty
	Wind Mountain, Washington, USA	Palmer (1977)	10 m/year	Road and railway require frequent maintenance, buildings adjusted periodically
2	Lugnez, Switzerland	Huder (1976)	0.37 m/yr	Six villages on slope undisturbed
	Little Smokey	Thomson and Hayley (1975)	0.25 m/yr	Bridge protected by slip joint
	Klosters, Switzerland	Haefeli (1965)	0.02 m/yr	Tunnel maintained, bridge protected by slip joint
	Fort Peck Spillway, Montana, USA	Wilson (1970)	0.02 m/yr	Movements unacceptable, slope flattened

From Cruden and Varnes (1996)

stress–displacement behaviour of the material (Figure 3.11). In elasto-plastic or ductile materials, almost all the potential energy is dissipated as frictional energy, resulting in low movement rates and small overall displacements. Any remaining potential energy is dissipated in the break-up and remoulding of the moving material (*energy of disaggregation*) and in accelerating it to a particular velocity (*kinetic energy*). However, in brittle materials, where there is a large difference between the peak and residual strengths, the kinetic energy can be very large, giving rise to long-run-out landslides. Figure 3.12 compares landslide volume and the normalised run-out distance (run-out length/slope height) for rock avalanches, quick clay failures and submarine slides. Run-out length generally increases with the volume of the failed mass, as the energy per unit volume increases with the slide height and hence its size.

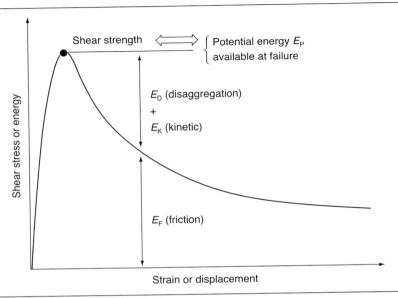

Figure 3.11 Schematic diagram showing the redistribution of potential energy after failure. (After Leroueil et al., 1996)

Landslide reactivation generally involves the sliding of a rigid block over a rigid base. The rate of displacement depends on a number of factors, including the local slope geometry, the total stresses and pore pressures induced by the thrust exerted by the soil mass, the rate of pore pressure dissipation, and fluctuations in the groundwater table (Leroueil et al., 1996). Highly variable rates of displacement are common, ranging from 10 m/month to less than 1 mm/month. Often, movements are progressive, starting in some sections and spreading downslope because of the thrust exerted by the moving mass. As a result, reactivated slides can behave like glaciers, with zones of tension and compression occurring at the same time in different parts of the slide.

The way in which the displaced materials behave as they move can have a significant impact on the degree of damage inflicted on assets. Taking a buried pipeline as an example, the potential for a rupture will be influenced by whether the movements involve the plastic deformation of clays around the pipe, as in some mudslides, or brittle-style block movements that cause significant lateral pressures to build up. Structures that happen to be located on a moving mass are damaged in proportion to their internal distortion (Cruden and Varnes, 1996). The extensive Lugnez landslide in Switzerland, for example, has moved down a 15° slope at up to 0.37 m/year since 1887. However, as the whole 25 km² landslide is moving as a single block without internal distortion, properties within the six villages on the landslide have been unaffected.

Rockfall intensity can be defined in terms of the kinetic energy KE, given by

$$KE = \tfrac{1}{2}mv^2$$

The velocity v of the falling rock (mass m) at the base of the cliff is

$$v = \sqrt{2gh}$$

where g is the acceleration due to gravity (9.81 m/s²) and h is the fall height.

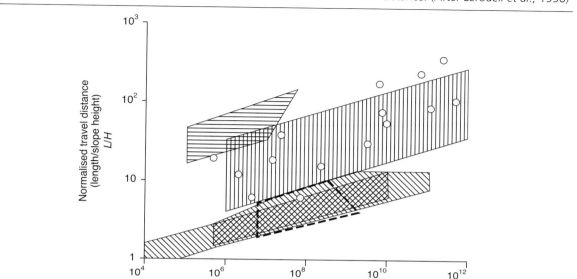

Figure 3.12 Relationship between landslide volume and normalised travel distance. (After Leroueil et al., 1996)

Landslide and debris flow intensity can also be considered to be a function of landslide mass m and maximum landslide velocity v. Table 3.9 presents the estimated intensity for a range of landslides based on these criteria (Cardinali et al., 2002). For a given landslide volume, fast-moving rockfalls have the highest landslide intensity, while rapidly moving debris flows exhibit intermediate intensity and slow-moving landslides have the lowest intensity.

Uzielli et al. (2008) proposed a general model for the intensity I (a dimensionless parameter, with values between 0 and 1) of a landslide or debris flow impacting on an element at risk:

$$I = K_s K_1 I_k + K_2 I_m$$

where K_s is the area affected/total area of interest is the spatial impact ratio, K_1 is the kinetic energy factor, K_2 is the kinematic intensity factor, I_k is the kinetic intensity parameter and I_m is the kinematic intensity parameter.

The kinetic (velocity-related) and kinematic (displacement-related) factors are defined by expert judgement and relate to the kinetic and magnitude characteristics of the landslide event in causing damage to

Table 3.9 Landslide intensity scale for different landslide types

Estimated volume: m³	Fast-moving landslide (rockfall)	Rapidly moving landslide (debris flow)	Slow-moving landslide (reactivated slide)
<0.001	Slight		
<0.5	Medium		
>0.5	High		
<500	High	Slight	
500–10 000	High	Medium	Slight
10 000–50 000	Very high	High	Medium
>500 000		Very high	High
≫500 000			Very high

After Cardinali et al. (2002)

particular elements at risk. For example, the damage caused by a slow-moving landslide is generally associated with ground displacement (i.e. the kinematics), whereas a rapidly moving landslide will cause damage because of the velocity of the impact (i.e. the kinetics). Possible values for these factors for different landslide types are presented in Table 3.10.

The kinetic energy parameter I_k is related to the landslide velocity, ranging from 0 for events moving at less than 16 mm/year (5×10^{-7} mm/s; extremely slow) to 1 for extremely rapid landslides (5×10^3 mm/s; 5 m/s). Between these two endpoints, the parameter is defined as

$$I_k = 0.1[\log_{10}(\text{velocity}) + 6.3]$$

The kinematic intensity parameter I_m ranges from 0 for structures not situated on the landslide mass to 1 for events where the actual displacement D_a is greater than the threshold displacement D_t that would cause very serious damage or loss of functionality of the structure. Between these points

$$I_m(\text{for } D_a/D_t < 0.5) = \frac{2D_a^2}{D_t^2}$$

$$I_m(\text{for } 0.5 \leq D_a/D_t \leq 1.0) = 1 - \frac{2(D_t-D_a)^2}{D_t^2}$$

Table 3.10 Values for kinetic energy and kinematic intensity parameters

Category	Landslide type	Kinetic energy factor K_1	Kinematic intensity factor K_2
Structures	Rapid	0.9	0.1
Structures	Slow	0.15	0.85
Persons	Rapid	0.75	0.25
Persons	Slow	1.0	0

From Uzielli et al. (2008)
Note: $K_1 + K_2 = 1$

For example, consider a rapid debris flow running out at 200 m/s towards structures located outside the source area

- from Table 3.10, the kinetic (velocity-related) and kinematic (displacement-related) factors are 0.9 and 0.1, respectively
- the kinetic energy parameter I_k is calculated from the velocity to be 0.86
- the kinematic intensity parameter I_m is assumed to be 0 because the structures are located outside the displaced mass
- the spatial impact ratio K_s is calculated from the area covered by the run-out (2880 m^2) and the total area occupied by the structures (4000 m^2), producing a value of 0.72.

The intensity for this debris flow is

$$I = K_s K_1 I_k + K_2 I_m$$
$$= 0.72 \times 0.9 \times 0.86 + (0.1 \times 0)$$
$$= 0.557$$

Uzielli *et al.* (2008) use this measure of intensity in combination with a measure of building 'susceptibility' (i.e. resistance to the landslide forces) to define physical vulnerability (see Chapter 8).

3.9. Landslide susceptibility and hazard zoning

Landsliding is a physical process/phenomenon. As was pointed out in Chapter 1, to become a *hazard*, the landslide process must pose a threat to humans and the things that humans value. The potential for adverse impact must be in the foreseeable future and relate to existing patterns of human occupancy and activity, as well as to planned developments or activities. In the case of the latter, the notion of *potential hazards* is acceptable, but the notion that all landslides are *potential hazards* because they may, at some stage in the future, interact adversely with human activity should be avoided as it is misleading.

These points are crucial to the appreciation of the distinction between *landslide susceptibility mapping* and *landslide hazard mapping*, especially as the terms 'susceptibility' and 'hazard' have tended to be used synonymously in the literature. *Landslide susceptibility* is the potential for landsliding to occur, while *landslide hazard* is the potential for landsliding to cause adverse consequences from a human perspective. The two are not the same, because of the 'human factor'. Thus, high susceptibility does not necessarily mean high hazard; indeed, landslide-prone remote areas may pose no threat to humans whatsoever. To convert susceptibility into hazard requires an understanding not only of the magnitude, character and frequency of landsliding, but also of the *probability* and *potential to cause perceived harm*. Unfortunately, this is not always attempted, and far too many studies that are presented as landslide hazard assessments are, in reality, landslide susceptibility assessments, focusing exclusively on the types of physical phenomena and merely *assuming* that there will be adverse consequences, or that some end user will re-interpret the results in terms of hazard.

The focus of landslide hazard assessment is mainly directed towards the following.

- *Prediction*, here defined as involving establishing the general likelihood and character of landsliding in an area. This can lead to the identification of areas with different levels of threat (i.e. landslide zonation or hazard zonation), which can be used to provide the spatial framework for establishing land use zonation plans, development controls and patterns of building regulations.

- *Forecasting*, defined as involving the identification of the location, timing, magnitude and character of individual landslides. This is a detailed, site-specific exercise, heavily reliant on monitoring and modelling. It is the necessary basis for the development of early-warning systems and the triggering of emergency action procedures.

Landslide zoning involves the preparation of landslide susceptibility and hazard maps, and is often carried out at the local government level for planning urban development and by state or federal governments for regional land use planning or disaster management planning (e.g. Cascini *et al.*, 2005). It may also be required by land developers, those managing recreational areas, or those developing major infrastructure developments, such as pipeline systems/networks, highways and railways.

The joint ISSMGE, IAEG and ISRM Technical Committee on Landslides and Engineered Slopes (JTC-1) has produced guidance on landslide zoning (Fell *et al.*, 2008a,b). It suggests that typical situations where landsliding can be an issue in land use planning include

- areas with a history of landsliding (e.g. the town of Ventnor on the Isle of Wight, UK, which is built on a large pre-existing landslide complex, and Hong Kong, where there are frequent shallow slides on steep, natural terrain)
- areas where the potential for landsliding could be expected because of topographic, geological and geomorphological conditions (e.g. steep slopes in weak rocks)
- sites where man-made slopes or structures could fail and initiate rapid landslides (e.g. loose silty, sandy fills, mine waste dumps on hillsides, and tailings dams)
- forestry works and clearance for agricultural land, where the resulting landsliding may cause environmental damage (e.g. on Vancouver Island in British Columbia).

Landslide susceptibility zoning involves defining the nature and spatial distribution of existing and potential landslides in an area (Fell *et al.*, 2008a). The approach usually involves developing an inventory of past landslides, together with an assessment of areas with a potential for future landsliding, so as to create a spatial grading into zones of differing frequency and intensity. Table 3.11 provides examples of landslide susceptibility terms that might be used in characterising potentially affected areas.

Table 3.11 Examples of landslide susceptibility mapping terms

Susceptibility term	Rockfalls: probability that rockfalls will reach the area, given that rockfalls occur from a cliff	Small slides on natural slopes: proportion of area in which small slides may occur	Large slides: proportion of area in which small slides may occur
High	>0.5	>0.5	>0.5
Moderate	>0.25 to 0.5	>0.25 to 0.5	>0.25 to 0.5
Low	>0.01 to 0.25	>0.01 to 0.25	>0.01 to 0.25
Very low	0 to 0.01	0 to 0.01	0 to 0.01

From Fell *et al.* (2008b)

Table 3.12 Landslide hazard terms recommended by Fell et al. (2008a,b)

Hazard term	Rockfalls: number per year per km of cliff or rock slope	Cut and fill slides: number per year per km of cut or fill	Small slides on natural slopes: number per year per km^2	Large slides; natural slopes: annual probability of sliding
Very high	>10	>10	>10	10^{-1}
High	1 to 10	1 to 10	1 to 10	10^{-2}
Moderate	0.1 to 1	0.1 to 1	0.1 to 1	10^{-3} to 10^{-4}
Low	0.01 to 0.1	0.01 to 0.1	0.01 to 0.1	10^{-5}
Very low	<0.01	<0.01	<0.01	$<10^{-6}$

Landslide susceptibility zoning clearly forms the essential basis for landslide hazard zoning. Only once the nature, magnitude, frequency and distribution of future landsliding have been determined is it possible to evaluate the threat to existing and planned human activity, in terms of the potential for damage and destruction of buildings, infrastructure, communication lines and so on, as well as the possibility of death and injury to humans, livestock and other animals. The results of such analyses are frequently presented in map form and the revealed spatial patterns are of value to both development planners and engineers.

The resulting hazard zoning maps should provide the spatial framework for the consideration of the threat posed by landsliding and the potential consequences; that is, the risk assessment process. However, to date, landslide hazard assessments have generally been restricted to determining the nature of the hazard (size and type of failure) and where it might occur (i.e. susceptibility), rather than the probability of occurrence and likely adverse consequences. This is because it is extremely difficult to determine the association between factors, such as potential triggering events, and landslide activity across a broad area. The landslide hazard terms recommended by Fell et al. (2008a,b) are presented in Table 3.12.

A wide range of approaches have been used to generate landslide susceptibility and hazard zoning. All are based on one or more of the following assumptions (Hearn and Griffiths, 2001):

1. The location of future slope failures or ground movements will largely be determined by the distribution of existing or past landslides; that is, known landslide locations will continue to be a source of hazard.
2. Future landslides or ground movements will occur under similar ground conditions to those pertaining at the sites of existing or past landslides; that is, the conditioning or controlling factors that have given rise to existing landslides can be ascertained and their distribution reliably mapped over wider areas. Together, these factors provide a reasonable indication of the relative tendency for slopes to fail.
3. The distribution of existing and future landslides can be approximated by reference to conditioning factors alone, such as rock type or slope angle.

The extent to which these conclusions will hold validity under conditions of global change driven by the twin forces of global warming and anthropogenic land cover change is open to question. Two basic methodologies can be recognised (Aleotti and Chowdhury, 1999; Carrara et al., 1998) (Figure 3.13):

Figure 3.13 Classification of landslide hazard assessment techniques. (Based on Aleotti and Chowdhury, 1999)

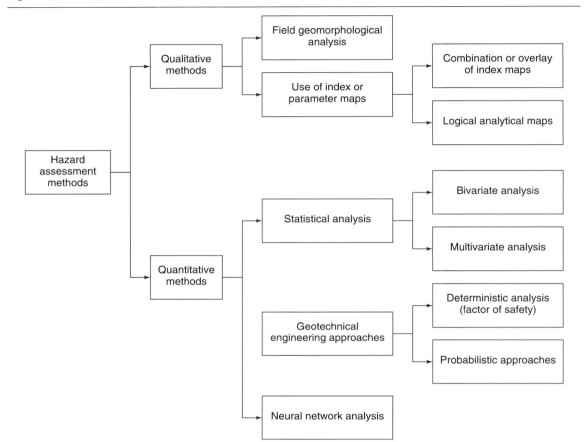

1 *Direct mapping approaches*, in which the hazard is defined in terms of the distribution of past landslide events. This is described by Soeters and van Westen (1996) as an experience-based approach, whereby the specialist evaluates direct relationships between landslides and their geomorphological and geological settings through the use of direct observations from a survey of existing landslide sites. The main types of direct mapping are as follows.
   - *Landslide inventories*. This is the most straightforward approach to landslide hazard mapping and involves compiling a database of pre-existing landslides within an area from aerial photographs, ground survey or historical records.
   - *Geomorphological analysis*, where the hazard is determined directly in the field by the geomorphologist, drawing on individual experience and the use of reasoning by analogy with similar sites elsewhere. As this approach relies on expert judgement, reproducibility can be a major issue.
   - *Qualitative map combination*. This involves the identification of key factors that appear to control the pattern of landsliding. Each of these factors is then assigned a weighted value, based on experience of the particular landslide environment. The weighted factor scores are then summed to produce hazard values that can be grouped into hazard classes. The exact

weighting of factors can be difficult to determine, and as Soeters and van Westen (1996) noted, 'insufficient field knowledge of the important factors prevents the proper establishment of the factor weights, leading to unacceptable generalisations'.

2   *Indirect mapping approaches*, where the results of the statistical analysis of a large number of parameters considered to influence landsliding are mapped so as to derive a predictive relationship between the terrain conditions and the occurrence of landsliding. Two different statistical approaches are used.

- *Bivariate statistical analysis*, in which each factor map (e.g. slope angle, geology or land use) is combined with the landslide distribution map and weighted values based on landslide densities are calculated for each parameter class (e.g. slope class, lithological unit or land use type). A variety of statistical methods have been applied to calculate the weighting values, including the *information value method* (Yin and Yan, 1988) and the *weight-of-evidence modelling method* (Spiegelhalter, 1986).
- *Multivariate statistical analysis*, in which all relevant factors are sampled either on a large-grid basis or in morphometric units. The presence or absence of landsliding within each of the units is also recorded. The resulting matrix is analysed using multiple regression or discriminant analysis (e.g. Carrara, 1983, 1988; Carrara *et al.*, 1990, 1991, 1992). Large datasets are needed to produce reliable results.

There has been a major expansion in the development of landslide hazard methodologies in the last decade or so, generally reflecting the greater access to geographic information system (GIS) technologies. Further details, examples and reviews of the various limitations of landslide hazard assessment methodologies can be found in a variety of sources, including Aleotti and Chowdhury (1999), Ardizzone *et al.* (2002), Carrara *et al.* (1999), Glade *et al.* (2005), Guzzetti (2005), Guzzetti *et al.* (1999, 2006), Hansen (1984), Hearn and Griffiths (2001), Karam (2005), Martha *et al.* (2012), Parise *et al.* (2012), Reichenbach *et al.* (2002), Soeters and Van Westen (1996), Van Westen (1993, 2004), Van Westen *et al.* (1999, 2000) and Varnes (1984). However, it is worth echoing the views of Carrara *et al.* (1999) regarding the use of GIS technologies for landslide hazard assessment.

- Computer-generated results are mistakenly considered to be more objective and accurate than products derived by experts in the conventional way through extensive field mapping.
- The use of GIS can result in the production of less accurate hazard maps if generated by users who are not experts in earth sciences.
- There has been an increased focus on the use of new computational techniques for landslide hazard assessment, and a declining interest in the collection of reliable data.

As Van Westen (2004) wrote, 'GIS has become an almost compulsory tool in landslide hazard and risk assessment, and it is the challenge to keep on using it as a tool, and not as an objective in itself'.

3.10. Hazard models

As has already been shown in Chapter 2, a key preliminary stage in landslide risk assessment is the identification of the different types of hazard present or credibly envisaged for a site or within an area, and to employ this information in the development of *hazard models*. The hazard models should classify the different types of threat and quantify their future frequency and magnitude. This can involve defining the different mechanisms and scales of failure, each with a corresponding frequency and impact potential. The hazard models should cover the full range of hazards at a level of detail that is consistent with the resolution of the available data and as determined by scoping (see Chapter 2). Consideration must be given to hazards originating off site or beyond the boundaries of the area under consideration, as well as within the immediate site or area, since it is possible for landslides both upslope and downslope to affect a site or location. The effects of proposed developments should also be considered, since these human-generated changes have the potential to alter the nature, scale, frequency and significance of future landslide hazards. The adequacy and appropriateness of the hazard models will greatly affect the accuracy of subsequent produced frequency and consequence assessments.

Each situation will be different because of variations in client requirements and purpose of study and the uniqueness of individual sites or areas. It follows that hazard models need to be individually designed to reflect site conditions and cannot be provided 'off-the-shelf'. Hazard models should focus on the following.

- *What could happen?* The nature and scale of the landslide events that might occur in the foreseeable future. Often, an important issue will be the way in which hazards develop, from incubation, via the occurrence of a triggering or initiating event, to the slope response and all possible outcomes.
- *Where could it happen?* The hazard model will need to provide a spatial framework for describing variations in hazard across a site or area. In many instances, this framework will be provided by a geological map, although a geomorphological map is preferable.
- *Why might such events happen?* That is, what are the circumstances associated with particular landslide events?
- *When events might happen?* That is, what are the timescales within which particular events can be expected to occur?

In Hong Kong, for example, the development of hazard models is a central element of natural terrain landslide hazard studies (e.g. Ng *et al.*, 2002). The general approach involves consideration of the site in the context of its regional geological and geomorphological settings, any man-made influences that may have modified this setting, and the history of landsliding in the area. Parry *et al.* (2006) describes how the hazard model identifies the controls on the location, type, magnitude, frequency and run-out characteristics of the site-specific landslide hazards. In Hong Kong, these models are typically based on:

- evaluating the natural terrain hazards that have occurred in the past, through aerial photograph interpretation (API) and field verification
- engineering geological and geomorphological mapping of terrain characteristics and interpreting how the landscape at the site has evolved.

Field evidence of past initiation and debris run-out behaviour can be used to establish specific models for the location, size, failure type and mobility mechanisms of potential failures. These hazard models are the input to the subsequent stage, for example establishing appropriate design criteria for mitigation works.

A hazard model, no matter how good it is, is not a risk assessment, merely an essential component of a risk assessment.

3.11. Uncertainty, assurance and defensibility

Understanding the nature and scale of potential landslide problems in an area provides the foundations for landslide risk assessment. Poor problem definition will lead to poor risk assessment.

However, it is clear that some degree of uncertainty will exist in the knowledge of any site or area, because of insufficient data, natural spatial variability and the difficulties in predicting the implications of future environmental change on slope stability (see Chapter 2). Fookes (1997) noted that the art of geological or geotechnical assessment is 'the ability to make rational decisions in the face of imperfect knowledge'. Decisions about landslide hazard almost always incorporate uncertainty to one degree or another, and there is usually a need to rely on judgement.

Recognition of uncertainty clearly has implications for how much effort should be spent defining the landslide hazard. There is no simple answer to this question, but a number of issues will be of relevance in reaching a decision.

1. *The need for adequate site investigation.* Experience has shown that shortfalls in data acquisition almost inevitably lead to the possibility that significant risks have been overlooked. For example, in a review of site investigation practice, the Ground Board of the Institution of Civil Engineers suggested that in any civil engineering or building projects, the largest technical and financial risk lies in the ground conditions (ICE, 1991). Figure 3.14 illustrates the estimated rate of acquiring information on a typical site. To achieve a robust and reliable understanding of ground conditions that support the development of a hazard model, it is advisable not to cut corners in the investigation process.

2. *The need to provide assurance to the client, decision-maker or regulator* that the hazard model delivers a reliable basis for risk assessment. In a perfect world, every landslide or potentially unstable slope would be thoroughly investigated. However, in reality, factors such as the urgency in finding a solution to a pressing problem, cost and the availability of suitable investigation equipment mean that compromises have to be reached between the effort involved and the degree to which the uncertainties are reduced. In other circumstances, the client, decision-maker or regulator may only be convinced that there is sufficient knowledge of the

Figure 3.14 Estimated upper and lower bounds of landslide information anticipated during the stages of a site investigation. (After Fookes, 1997)

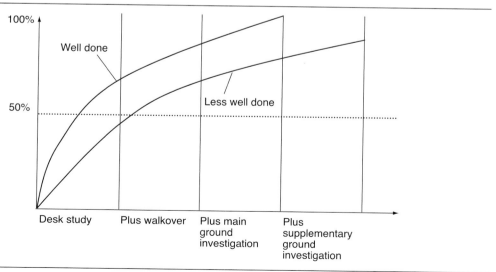

Figure 3.15 An illustration of how the degree of assurance in landslide hazard assessment might change with the intensity of investigation

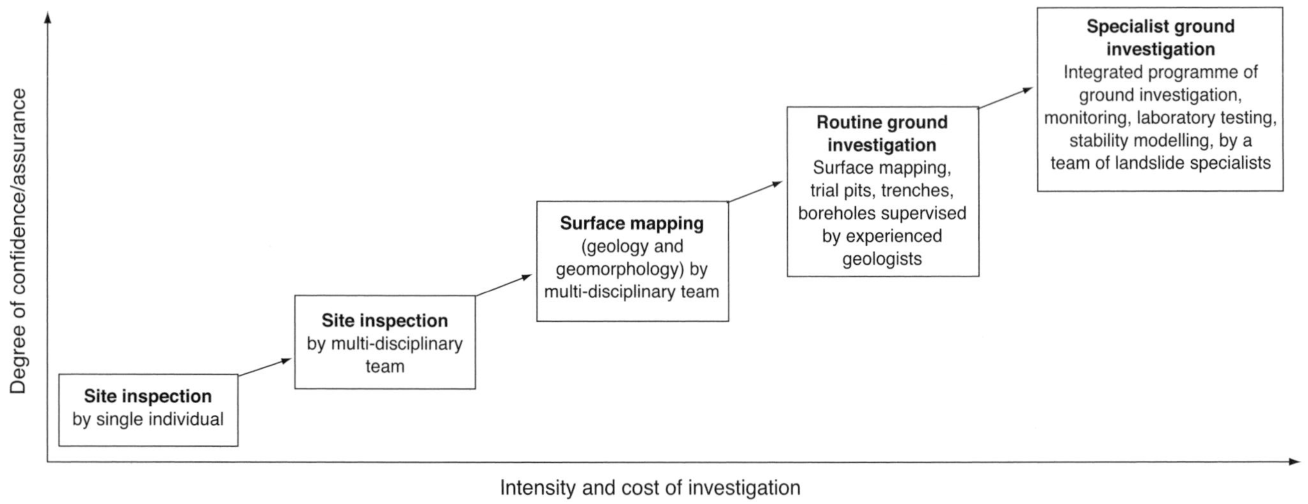

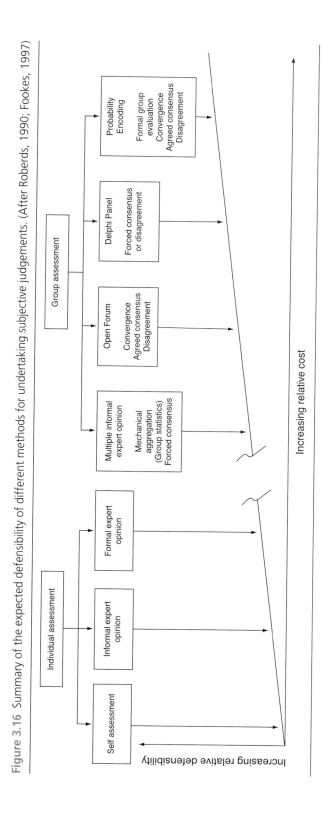

Figure 3.16 Summary of the expected defensibility of different methods for undertaking subjective judgements. (After Roberds, 1990; Fookes, 1997)

ground conditions and risks after an extensive borehole investigation. In reaching a decision as to the level of assurance that is appropriate to the particular assessment, a key factor should be the scale of consequences if the hazard model should prove to be inadequate or even wrong. Figure 3.15 illustrates how the degree of assurance might increase with increased investigation effort. For example, a simple site inspection of an apparently stable slope may be sufficient if the assets at risk are of low value and where there would be no threat to the public should the slope fail unexpectedly. However, if an unexpected failure would cause major economic loss or loss of life, it might be appropriate to undertake a more rigorous investigation to provide a higher level of assurance that the slope is safe. More detailed investigation may not significantly reduce the uncertainties, but often will result in a higher level of confidence about the judgements.

3   *The need to ensure that the judgements are defensible.* It is important for practitioners to appreciate that their judgements about risk may provoke considerable disagreement and controversy, especially if these judgements have implications for property values or the development potential of a site. In an increasingly litigious world, there will be a need for practitioners to demonstrate that they have acted in a professional manner appropriate to the circumstances. Figure 3.16 introduces the notion of increasing defensibility with increasing sophistication of the approach used in order to overcome the problems associated with subjective judgement (Chapter 5).

REFERENCES

Aleotti P and Chowdhury R (1999) Landslide hazard assessment: summary review and new perspectives. *Bulletin of Engineering Geology and Environment* **58(1)**: 21–44.

Ardizzone F, Cardinali M, Carrara A, Guzzetti F and Reichenbach P (2002) Impact of mapping errors on the reliability of landslide hazard maps. *Natural Hazards and Earth System Sciences* **2**: 3–14.

AGS (Australian Geomechanics Society) (2000) Landslide risk management concepts and guidelines. *Australian Geomechanics* **35**: 49–52.

Bentley S and Smalley IJ (1984) Landslips in sensitive clays. In *Slope Instability* (Brunsden D and Prior DB (eds)). Wiley, Chichester, pp. 457–490.

Bishop AW (1973) Stability of tips and spoil heaps. *Quarterly Journal of Engineering Geology* **6**: 335–376.

Bishop AW, Hutchinson JN, Penman ADM and Evans HE (1969) *Geotechnical investigation into the causes and circumstances of the disaster of 21 October 1966.* A selection of technical reports submitted to the Aberfan Tribunal 1–80, Welsh Office, HMSO, London.

Blong RJ (1992) Some perspectives on geological hazards. In *Geohazards: Natural and Man-made* (McCall GJH, Laming DJC and Scott SC (eds)). Chapman and Hall, London, pp. 209–216.

Bromhead EN (1986) *The Stability of Slopes.* Surrey University Press, London.

Bromhead EN (1992) *The Stability of Slopes.* 2nd edn. CRC Press, London.

Brunsden D and Jones DKC (1972) The morphology of degraded landslide slopes in south-west Dorset. *Quarterly Journal of Engineering Geology* **3**: 205–223.

Brunsden D and Prior DB (1984) *Slope Instability.* Wiley, Chichester.

Brunsden D, Doornkamp JC, Fookes PG, Jones DKC and Kelly JMN (1975) Large scale geomorphological mapping and highway engineering design. *Quarterly Journal of Engineering Geology* **8**: 227–253.

Bryant EA (1991) *Natural Hazards.* Cambridge University Press, Cambridge.

Bryn P, Kvalstad T, Guttormsen TR, Kjaerness PA, Lund JK, Nadim F, and Olsen J (2004) Storegga slide risk assessment. *OTC 16560, Offshore Technology Conference '04.* Houston, Texas.

Bugge T, Befring S and Belderson RH (1987) A giant three-stage submarine slide off Norway. *Geo-Marine Letters* **7**: 191–198.

Bugge T, Belderson RH and Kenyon NH (1988) The Storegga Slide. *Philosophical Transactions of the Royal Society A* **325**: 357–388.

Buss E and Heim A (1881) *Der bergsturz von Elm*. Wurster, Zurich.

Cardinali M, Reichenbach P, Guzzetti F, Ardizzone F, Antonini G, Galli M, Cacciano M, Castellani M and Salvati P (2002) A geomorphological approach to the estimation of landslide hazards and risks in Umbria, Central Italy. *Natural Hazards and Earth System Sciences* **2**: 57–72.

Carrara A (1983) Multivariate models for landslide hazard evaluation. *Mathematical Geology* **15**: 403–427.

Carrara A (1988) Landslide hazard mapping by statistical methods: a 'black box' approach. In *Workshop on Natural Disasters in European Mediterranean Countries*, Perugia, Consiglio Nazionale delle Ricerche, Perugia, 205–224.

Carrara A, Cardinali M and Guzzetti F (1992) Uncertainty in assessing landslide hazard and risk. *ITC Journal* **2**: 172–183.

Carrara A, Cardinali M, Detti R, Guzzetti F, Pasqui V and Reichenbach P (1990) Geographic information systems and multivariate models in landslide hazard evaluation. In ALPS 90 Alpine Landslide Practical Seminar, *Sixth International Conference and Field Workshop on Landslides*, Aug. 31–Sept. 12, Milan, Italy, Universita degli Studi de Mialno, 17–28.

Carrara A, Cardinali M, Detti R, Guzzetti F, Pasqui V and Reichenbach P (1991) GIS techniques and statistical models in evaluating landslide hazard. *Earth Surface Processes and Landforms* **16**: 427–445.

Carrara A, Guzzetti F, Cardinali M and Reichenbach P (1998) Current limitations in modeling landslide hazard. *Proceedings of IAMG '98* (Buccianti A, Nardi G and Potenza R (eds)), pp. 195–203.

Carrara A, Guzzetti F, Cardinali M and Reichenbach P (1999) Use of GIS technology in the prediction and monitoring of landslide hazard. *Natural Hazards* **20(2–3)**: 117–135.

Cascini L, Bonnard C, Corominas J, Jibson R and Montero-Olarte J (2005) Landslide hazard and risk zoning for urban planning and development. In *Landslide Risk Management* (Hungr O, Fell R, Couture R and Eberhardt E (eds)). Balkema, Rotterdam, pp. 199–236.

Close U and McCormick E (1922) Where the mountains walked. *National Geographic Magazine* **41(5)**: 445–464.

Collins GS and Melosh HJ (2003) Acoustic fluidization and the extraordinary mobility of sturzstroms. *Journal of Geophysical Research* **108**: 2473–2486.

Copons R (2004) Avaluació de la perillositat de caiguda de blocs a Andorra la Vella (Principat d'Andorra). PhD thesis, Universitat de Barcelona.

Corominas J (1996) The angle of reach as a mobility index for small and large landslides. *Canadian Geotechnical Journal* **33**: 260–271.

Cross W (1924) *Historical Sketch of the Landslides of the Gaillard Cut. Memoir 18*. National Academy of Sciences, Washington DC, pp. 22–43.

Crozier MJ (1986) *Landslides: Causes, Consequences and Environment*. Croom Helm, London.

Crozier MJ and Preston NJ (1998) Modelling changes in terrain resistance as a component of landform evolution in unstable hill country. *Lecture Notes in Earth Science* **78**: 267–284.

Cruden D (1991) A simple definition of a landslide. *Bulletin of the International Association of Engineering Geology* **43**: 27–29.

Cruden DM and Varnes DJ (1996) Landslide types and processes. In *Landslides: Investigation and Mitigation* (Turner AK and Schuster RL (eds)). Transportation Research Board, National Research Council, National Academy Press, Washington, DC, pp. 36–75. Special Report 247.

Dawson AG, Long D and Smith DE (1988) The Storegga Slides: evidence from eastern Scotland for a possible tsunami. *Marine Geology* **82**: 271–276.

De Blasio FV (2011) Landslides in Valles Marineris (Mars): a possible role of basal lubrication by sub-surface ice. *Planetary and Space Science* **59**: 1384–1392.

De Blasio FV and Elverhoi A (2008) A model for frictional melt production beneath large rock avalanches. *Journal of Geophysical Research* **113**: F02014.

Dikau R, Brunsden D, Schrott L and Ibsen M-L (1996) *Landslide Recognition.* Wiley, Chichester.

Dowlen WE (1903) The Turtle Mountain rock slide. *Engineering and Mining Journal* **77**: 10–12.

Eiby GA (1980) *Earthquakes.* Heinemann, London.

Eisbacher GH and Clague JJ (1984) *Destructive Mass Movements in High Mountains: Hazard and Management.* Geological Survey of Canada, Ottawa. Paper 84–16.

Engineering News Record (1971) Contrived landslide kills 15 in Japan. 187, 18.

EPOCH (1993) *Temporal Occurrence and Forecasting of landslides in the European Community.* European Community, Brussels. Contract 90 0025.

Evans TG, Usher N and Moore R (2007) Management of geotechnical and geohazard risks in the West Nile Delta. *Proc. 6th Intl. Offshore Site Investigation and Geotechnics Conference: Confronting New Challenges and Sharing Knowledge.* Society for Underwater Technology, London.

Evans NC and King JP (1998) *The Natural Terrain Landslide Study: Debris Avalanche Susceptibility.* Geotechnical Engineering Office, Hong Kong. GEO Technical Note TN 1/98.

Evans NC, Huang SW and King JP (1997) *The Natural Terrain Landslide Study: Phase III.* Geotechnical Engineering Office, Hong Kong. GEO Special Project Report SPR 5/97.

Evans SG (2001) Landslides. In *A Synthesis of Geological Hazards in Canada* (Brooks GR (ed.)). Geological Survey of Canada, pp. 43–79. Bulletin 548.

Evans SG and Hungr O (1993) The assessment of rockfall hazard at the base of talus slopes. *Canadian Geotechnical Journal* **30**: 620–636.

Felix M and Peakall J (2006) Transformation of debris flows into turbidity currents: mechanisms inferred from laboratory experiments. *Sedimentology* **53**: 107–123.

Fell R, Corominas J, Bonnard C, Cascini L, Leroi E and Savage W on behalf of the JTC-1 Joint Technical Committee on Landslides and Engineered Slopes (2008a) Guidelines for Landslide Susceptibility, Hazard and Risk Zoning. *Engineering Geology* **102**: 85–98.

Fell R, Corominas J, Bonnard C, Cascini L, Leroi E and Savage W on behalf of the JTC-1 Joint Technical Committee on Landslides and Engineered Slopes (2008b) Commentary: Guidelines for Landslide Susceptibility, Hazard and Risk Zoning. *Engineering Geology* **102**: 99–111.

Fine IV, Rabinovich AB, Bornhold BD, Thomson RE and Kulikov EA (2005) The Grand Banks landslide-generated tsunami of November 18, 1929: preliminary analysis and numerical modelling. *Marine Geology* **215**: 45–57.

Finlay PJ, Mostyn GR and Fell R (1999) Landslides: prediction of travel distance and guidelines for vulnerability of persons. *Proceedings of the 8th Australia/New Zealand Conference on Geomechanics*, Hobart. Australian Geomechanics Society, vol. 1, pp. 105–113.

Fletcher L, Hungr O and Evans SG (2002) Contrasting failure behaviour of two large landslides in clay and silt. *Canadian Geotechnical Journal* **39**:46–62.

Fookes PG (1997) Geology for engineers: the geological model, prediction and performance. *Quarterly Journal of Engineering Geology* **30**: 290–424.

Francis PW (1994) Large volcanic debris avalanches in the central Andes. In *Abstracts of the International Conference on Volcanic Instability on the Earth and Other Planets.* Geological Society, London.

Gilbert RB, Nodine MC, Wright SG, Cheon JY, Wrzyszcznski M, Coyne M and Ward EG (2007) Impact of Hurricane-induced mudslides on pipelines. *OTC 18983. Offshore Technology Conference '07*, Houston, TX.

Glade T, Anderson M and Crozier MJ (eds) (2005) *Landslide Hazard and Risk*. Wiley, Chichester.

Gostling Rev W (1756) Letter to *Gentleman's Magazine* 26, 160.

Griffiths JS, Lee EM, Brunsden D and Jones DKC (2011) The Cherry Garden Landslide, Etchinghill Escarpment, Southest England. In *Geomorphological Mapping: Methods and Applications* (Smith MJ, Paron P and Griffiths JS (eds)). Elsevier, Amsterdam, pp. 397–411.

Guzzetti F (2005) Landslide hazard and risk assessment. Dissertation, University of Bonn.

Guzzetti F, Carrara A, Cardinali M and Reichenbach P (1999) Landslide hazard evaluation: a review of current techniques and their application in a multi-scale study, Central Italy. *Geomorphology* **31(1–4)**: 181–216.

Guzzetti F, Galli M, Reichenbach P, Ardizzone F and Cardinali M (2006) Landslide hazard assessment in the Collazzone area, Umbria, Central Italy. *Natural Hazards and Earth System Sciences* **6**: 115–131.

Haefeli R (1965) Creep and progressive failure in snow, rock and ice. *Proceedings of the Sixth International Conference on Soil Mechanics and Foundation Engineering*. University of Toronto Press, Toronto 3, pp. 134–148.

Hampton MA, Lee HJ and Locat J (1996) Submarine landslides. *Reviews of Geophysics* **34**: 33–59.

Hansen A (1984) Landslide hazard analysis. In *Slope Instability* (Brunsden D and Prior DB (eds)). Wiley, Chichester, pp. 523–602.

Hearn GJ and Griffiths JS (2001) Landslide hazard mapping and risk assessment. In *Land Surface Evaluation for Engineering Practice* (Griffiths JS (ed.)). Geological Society, London, pp. 43–52. Engineering Group Special Publication 18.

Hearn GJ (ed.) (2011) *Slope Engineering for Mountain Roads*. Geological Society of London, pp. 103–116. Engineering Geology Special Publication 24.

Heim A (1882) Der bergsturz von Elm. *Zeitschrift der Deutschen Geologischen Gesellschaft* **34**: 74–115.

Heim A (1932) Bergsturz und Menschenleben. *Beiblatt zur Viierteljahrsschrift der Naturforschenden Gesellschaft in Zurich* **77**: 1–217. Translated by N Skermer (*Landslides and Human Lives*, BiTech Publishers, Vancouver, BC, 1989).

Hendron AJ Jr and Patton FD (1985) *The Vaiont Slide: a geotechnical analysis based on new geologic observations of the failure surface*. Waterways Experiment Station Technical Report, US Army Corps of Engineers, Vicksburg, MI.

Herd DG and the Comite de Estudios Vulcanologies (1986) The 1985 Ruiz volcano disaster. *Eos* **67**: 457–460.

Ho K, Leroi E and Roberts B (2000) Quantitative risk assessment: application, myths and future direction. In *Proceedings of the Geo-Engineering Conference, Melbourne Australia*. Publication 1, pp. 269–312.

Hsü KJ (1975) On sturzstroms – catastrophic debris stream generated by rockfalls. *Bulletin of the Geological Society of America* **86**: 129–140.

Hsü KJ (1978) Albert Heim: observations on landslides and relevance to modern interpretations. In *Rockslides and Avalanches – 1, Natural Phenomena* (Voight B (ed.)). Elsevier, New York, pp. 71–93.

Huder J (1976) *Creep in Bundner Schist*. Norwegian Geotechnical Institute, Oslo, pp. 125–153.

Hungr O (1997) Some methods of landslide hazard intensity mapping. In *Landslide Risk Assessment* (Cruden D and Fell R (eds)). Balkema, Rotterdam, pp. 215–226.

Hungr O and Evans SG (2004) The occurrence and classification of massive rock slope failure. *Felsbau* **22**: 16–23.

Hungr O and McClung DM (1987) An equation for calculating snow avalanche runup against barriers. In *Avalanche Formation, Movements and Effects*. International Association of Hydrological Sciences, London, pp. 605–611.

Hungr O, Corominas J and Eberhardt E (2005) Estimating landslide motion mechanisms, travel distance and velocity. In *Landslide Risk Management* (Hungr O, Fell R, Couture R and Eberhardt E (eds)). Balkema, Rotterdam, pp. 99–128.

Hungr O, Evans SG, Bovis MJ and Hutchinson JN (2001) A review of the classification of landslides of flow type. *Environmental and Engineering Geoscience* **3**: 221–238.

Hungr O, Morgan GC and Kellerhals R (1984) Quantitative analysis of debris torrents for design of remedial measures. *Canadian Geotechnical Journal* **21**: 663–677.

Hunter G and Fell R (2003) Travel distance angle for rapid landslides in constructed and natural slopes. *Canadian Geotechnical Journal* **40**: 1123–1141.

Hutchinson JN (1988) General Report: morphological and geotechnical parameters of landslides in relation to geology and hydrology. In *Landslides* (Bonnard C (eds)). Balkema, Rotterdam, pp. 3–95.

Hutchinson JN (1987) Mechanisms producing large displacements in landslides on pre-existing shears. *Memoir of the Geological Society of China* **9**: 175–200.

Hutchinson JN (1992) Landslide hazard assessment. In *Landslides* (Bell DH (ed.)). Balkema, Rotterdam, vol. 3, pp. 1805–1841.

Hutchinson JN (1995) The assessment of sub-aerial landslide hazard. In *Landslides Hazard Mitigation*. Royal Academy of Engineering, London, pp. 57–66.

IAEG (International Association of Engineering Geology) (1990) IAEG Commission on Landslides: Suggested nomenclature for landslides. *Bulletin of the International Association of Engineering Geology* **41**: 13–16.

ICE (Institution of Civil Engineers) (1991) *Inadequate Site Investigation*. Thomas Telford, London. Report by the Ground Board of ICE.

IFRCRCS (International Federation of the Red Cross and Red Crescent Societies) (1999) *World Disasters Report*. Edigroup, Chene-Bourg.

Ishihara K (1999) Liquefaction-induced landslide and debris flow in Tajikistan. In *Landslides of the World* (Sassa K (ed.)). Kyoto University Press, Kyoto, pp. 224–226.

Jeanjean P, Hill A and Taylor S (2003) The challenges of confidently siting facilities along the Sigsbee Escarpment in the Southern Green Canyon area of the Gulf of Mexico. Framework of integrated studies. *Proceedings of the Offshore Technology Conference*, Houston, TX, Paper 15156.

Jones AT (1992a) Comment on 'Catastrophic wave erosion on the southeastern coast of Australia: impact of the Lanai tsunami ca 105 ka?' *Geology* **20**: 1150–1151.

Jones AT and Mader C (1996) Wave erosion on the southeastern coast of Australia: tsunami propagation and modelling. *Australian Journal of Earth Science* **43**: 479–483.

Jones DKC (1992b) Landslide hazard assessment in the context of development. In *Geohazards: Natural and Man-made* (McCall GJH, Laming DJC and Scott SC (eds)). Chapman and Hall, London, pp. 117–141.

Jones DKC and Lee EM (1994) *Landsliding in Great Britain*. HMSO, London.

Kaliser B and Fleming RW (1986) The 1983 landslide dam at Thistle, Utah. In *Landslide Dams: Processes, Risk and Mitigation* (Schuster RL (ed.)). Geotechnical Special Publication No. 3, American Society of Civil Engineers, 59–83.

Karam KS (2005) Landslide hazards assessment and uncertainties. PhD thesis, Department of Civil and Environmental Engineering, Massachusetts Institute of Technology, Cambridge, MI.

Keating BH and McGuire WJ (2000) Island edifice failures and associated tsunami hazards. *Pure and Applied Geophysics* **157**: 899–955.

Kiersch GA (1964) Vaiont Reservoir disaster. *Civil Engineering* **34(3)**: 32–39.

Kremer K, Simpson G and Girardclos S (2012) Giant Lake Geneva tsunami in AD 563. *Nature Geoscience* **5**: 756–757.

Lateltin OJ and Bonnard C (1999) Reactivation of the Falli-Holli landslide in the Prealps of Freiburg, Switzerland. In *Landslides of the World* (Sassa K (ed.)). Kyoto University Press, Kyoto, pp. 331–335.

Lee EM and Brunsden D (2000) *Coastal Landslides of Southern England: Mechanisms and Management*. Post Conference Tour: Viii ISL Cardiff. Keynote Papers CD-ROM. Available from: e.bromhead@kingston.ac.uk.

Lee EM and Clark AR (2002) *Investigation and Management of Soft Rock Cliffs*. Thomas Telford, London.

Léone F, Aste JP and Leroi E (1996) Vulnerability assessment of elements exposed to mass movement: working towards a better risk perception. In *Landslides* (Senneset K (ed.)). Balkema, Rotterdam, vol. 1, pp. 263–268.

Leroueil S, Vaunat J, Picarelli L, Locat J, Lee H and Faure R (1996) Geotechnical characterisation of slope movements. In *Landslides* (Senneset K (ed.)). Balkema, Rotterdam, vol. 1, pp. 53–74.

Li Tianchi, Schuster RL and Wu Jishan (1986) Landslide dams in South-central China. In *Landslide Dams: Processes, Risk and Mitigation* (Schuster RL (ed.)). American Society of Civil Engineers, Reston, VA, pp. 146–162. Geotechnical Special Publication No. 3.

Lipman PW and Mullineaux D (eds) (1981) *The 1980 Eruptions of Mount St Helens*. US Geological Survey, Washington, DC. Professional Paper 1250.

Lipman PW, Normark WR, Moore JG, Wilson JB and Gutmacher SE (1988) The giant submarine Alika debris slide, Mauna Loa, Hawaii. *Journal of Geophysical Research* **93(B5)**: 4279–4299.

Locat J and Lee HJ (2002) Submarine landslides: Advances and challenges. *Canadian Geotechnical Journal* **39**: 193–212.

McConnell RG and Brock RW (1904) Report on the Great Landslide at Frank, Alberta. In *Annual Report for 1903*. Department of the Interior, Ottawa, Canada.

Mader CL (2002) Modelling the 1958 Lituya Bay mega-tsunami, II. *Science of Tsunami Hazards* **20(5)**: 241–250.

Maquaire O (1994) Temporal aspects of the landslides located along the coast of Calvados (France). In *Temporal Occurrence and Forecasting of Landslides in the European Community. Final Report* (Casale R, Fantechi R and Flageollet JC (eds)). European Commission, Brussels, vol. 1, pp. 211–234.

Maquaire O (1997) The frequency of landslides on the Normandy coast and their behaviour during the present climatic regime. In *Rapid Mass Movement as a Source of Climatic Evidence for the Holocene* (Matthews JA, Brunsden D, Frenzel B, Glaser B and Weiss MM (eds)). Gustav Fischer, Stuttgart, pp. 183–195.

Martha TR, van Westen CJ, Kerle N, Jetten VG and Kumar V (2012) Landslide hazard and risk assessment using semi-automatically created landslide inventories. *Geomorphology* **184**: 139–150.

Mason K (1929) Indus floods and Shyock glaciers: the Himalayan Journal. *Records of the Himalayan Club, Calcutta* **1**: 10–29.

McGuire WJ (1995) Volcanic landslides and related phenomena. In *Landslides Hazard Mitigation*. Royal Academy of Engineering, London, pp. 83–95.

McGuire WJ, Mason I and Kilburn C (2002) *Natural Hazards and Environmental Change*. Arnold, London.

McSaveney MJ, Chinn TJ and Hancox GT (1992) Mount Cook Rock Avalanche of 14 December 1991. *New Zealand. Landslide News* **6**: 32–34.

Merriam R (1960) The Portuguese Bend landslide, Palos Verdes, California. *Journal of Geology* **68**: 140–153.

Miller DJ (1960) *Giant Waves in Lituya Bay, Alaska*. US Government Printing Office, Washington, DC. Geological Survey Professional Paper 354-C.

Miller J (1974) *Aberfan – A Disaster and its Aftermath*. Constable, London.

Moore JG, Clague DA, Holcomb RT, Lipman PW, Normark WR and Torresan ME (1989) Prodigious submarine landslides on the Hawaiian Ridge. *Journal of Geophysical Research* **94(B13)**: 17465–17484.

Moore JG, Normark WR and Holcomb RT (1994) Giant Hawaiian landslides. *Annual Review of Earth and Planetary Sciences* **22**: 199–144.

Mueller L (1964) The rock slide in the Vaiont Valley. *Rock Mechanics and Engineering Geology* **2**: 148–212.

Murty TS and Wigen SO (1976) Tsunami behaviour on the Atlantic Coast of Canada and some similarities to the Peru Coast. *Royal Society of New Zealand Bulletin* **15**: 51–60.

Nadim F and Locat J (2005) Risk assessment for submarine slides. In *Landslide Risk Management* (Hungr O, Fell R, Couture R and Eberhardt E (eds)). Balkema, Rotterdam, pp. 321–333.

Ng KC, Parry S, King JP, Franks CAM and Shaw R (2002) *Guidelines for Natural Terrain Hazard Studies*. Geotechnical Engineering Office, Hong Kong. GEO Report 138.

Palmer L (1977) Large landslides of the Columbia River Gorge, Oregon and Washington. In *Landslides, Reviews in Engineering Geology* (Cruden DR (eds)). Vol. 3, Geological Society of America, Boulder, CO, pp. 69–83.

Parise M, Iovine GR, Reichenbach P and Guzzetti F (2012) Introduction to the special issue 'Landslides: forecasting, hazard evaluation, and risk mitigation' *Natural Hazards* **61**: 1–4.

Parry S, Ruse ME and Ng KC (2006) Assessment of natural terrain landslide risk in Hong Kong: an engineering geological perspective. *IAEG2006: Proceedings of the 10th IAEG International Congress*, Nottingham, Paper 299.

Pethick JS (1996) Coastal slope development: temporal and spatial periodicity in the Holderness cliff recession. In *Advances in Hillslope Processes* (Anderson MG and Brooks SM (eds)). Wiley, Chichester, pp. 897–917.

Petley DN (2012) Global patterns of loss of life from landslides. *Geology* **40**: 927–930.

Picarelli L, Oboni F, Evans SG, Mostyn G and Fell R (2005) Hazard characterization and quantification. In *Landslide Risk Management* (Hungr O, Fell R, Couture R and Eberhardt E (eds)). Balkema, Rotterdam, pp. 27–62.

Pierson TC (1999) Rainfall-triggered lahars at Mt Pinatubo, Philippines, following the June 1991 eruption. In *Landslides of the World* (Sassa K (ed.)). Kyoto University Press, Kyoto, pp. 284–289.

Pierson TC, Janda RJ, Umbal JV and Daag AS (1992) *Immediate and Long-term Hazards from Lahars and Excess Sedimentation in Rivers Draining Mt Pinatubo, Philippines*. US Geological Survey, Washington, DC. Water-Resources Investigations Report 92-4039.

Piper DJW, Cochonat P and Morrison ML (1999) The sequence of events around the epicenter of the 1929 Grand Banks earthquake: initiation of debris flows and turbidity current inferred from sidescan sonar. *Sedimentology* **46**: 79–97.

Plafker G and Ericksen GE (1978) Nevados Huascaran avalanches, Peru. In *Rockslides and Avalanches – 1 Natural Phenomena* (Voight B (ed.)). Elsevier, Amsterdam, pp. 277–314.

Reichenbach P, Carrera A and Guzzetti F (2002) Preface: assessing and mapping landslide hazards and risk. *Natural Hazards and Earth System Sciences* **2**: 1–2.

Roberds WL (1990) Methods for developing defensible subjective probability assessments. *Transportation Research Record* **1288**: 183–190.

Sassa K (1985) The mechanism of debris flows. *Proceedings, XI International Conference on Soil Mechanics and Foundation Engineering*, San Francisco, vol. 1, pp. 1173–1176.

Sassa K (1988) Geotechnical model for the motion of landslides. In *Landslides* (Bonnard C (ed.)). Balkema, Rotterdam, vol. 1, pp. 37–55.

Sassa K (1996) Prediction of earthquake induced landslides. In *Landslides* (Senneset K (ed.)). Balkema, Rotterdam, pp. 115–132.

Sassa K (ed.) (1999) *Landslides of the World*. Kyoto University Press, Kyoto.

Scheidegger AF (1973) On the prediction of the reach and velocity of catastrophic landslides. *Rock Mechanics* **5**: 231–236.

Schrott L and Pasuto A (eds) (1999) Temporal stability and activity of landslides in Europe with respect to climate change (TESLEC). *Geomorphology, Special Issue* **30(1–2)**.

Schuster RL (ed.) (1986) *Landslide Dams: Processes, Risk and Mitigation*. American Society of Civil Engineers, Reston, VA. Geotechnical Special Publication No. 3.

Siebert L (1992) Threats from debris avalanches. *Nature* **356**: 658–659.

Singer KN, McKinnon WB, Schenk PM and Moore JM (2012) Massive ice avalanches on Iapetus mobilized by friction reduction during flash heating. *Nature Geoscience* **5(8)**: 574–578.

Sir William Halcrow and Partners (1986) *Rhondda Landslip Potential Assessment*. Department of the Environment and Welsh Office, Cardiff.

Sir William Halcrow and Partners (1988) *Rhondda Landslip Potential Assessment: Inventory*. Department of the Environment and Welsh Office, Cardiff.

Skempton AW and Weeks AG (1976) The Quaternary history of the Lower Greensand escarpment and Weald Clay Vale near Sevenoaks, Kent. *Philosophical Transactions of the Royal Society, London* **A283**: 493–526.

Skempton AW, Leadbeater AD and Chandler RJ (1989) The Mam Tor landslide, North Derbyshire. *Philosophical Transactions of the Royal Society, London* **A329**: 503–547.

Soeters R and Van Westen CJ (1996) Slope instability recognition, analysis and zonation. In *Landslides: Investigation and Mitigation* (Turner AK and Schuster RL (eds)). National Research Council, National Academy Press, Washington, DC, pp. 129–177. Transportation Research Board, Special Report 247.

Spiegelhalter DJ (1986) Uncertainty in expert systems. In *Artificial Intelligence and Statistics* (Gale WA (ed.)). Addison-Wesley, Reading, MA, pp. 17–55.

Tappin DR, Watts P and Grilli ST (2008) The Papua New Guinea tsunami of 17 July 1998: anatomy of a catastrophic event. *Natural Hazards and Earth System Sciences* **8**: 243–266.

Tavenas F, Chagnon JY and LaRochelle P (1971) The Saint-Jean Vianney landslide: observations and eyewitness accounts. *Canadian Geotechnical Journal* **8**: 463–478.

Terzaghi K (1950) Mechanisms of landslides. In *Application of Geology in Engineering Practice* (Berkey Volume) (Paige S (ed.)). Geological Society of America, Boulder, CO, pp. 83–123.

Thomson S and Hayley DW (1975) The Little Smokey Landslide. *Canadian Geotechnical Journal* **12**: 379–392.

Torrance JK (1987) Quick clays. In *Slope Stability* (Anderson MG and Richards KS (eds)). Wiley, Chichester, pp. 447–474.

Turner AK and Schuster RL (eds) (1996) *Landslides: Investigation and Mitigation*. National Research Council, National Academy Press, Washington, DC. Transportation Research Board, Special Report 247.

Uzielli M, Nadim F, Lacasse S and Kaynia AM (2008) A conceptual framework for quantitative estimation of physical vulnerability to landslides. *Engineering Geology* **102**: 251–256.

Valentin H (1954) Der landverlust in Holderness, Ostengland von 1852 bis 1952. *Die Erde* **6**: 296–315.

Van Gassen W and Cruden DM (1989) Momentum transfer and friction in the debris of rock avalanches. *Canadian Geotechnical Journal* **26**: 623–628.

Van Westen CJ (1993) Application of geographic information systems to landslide hazard zonation. PhD dissertation, Technical University Delft.

Van Westen CJ (2004) Geo-information tools for landslide risk assessment – an overview of recent developments. In *Landslides, Evaluation and Stabilization. Proceedings of the 9th International*

Symposium on Landslides (Lacerda W, Ehrlich M, Fontoura S and Sayao A (eds)), Rio de Janeiro, pp. 39–56.

Van Westen CJ, Seijmonsbergen AC and Mantovani F (1999) Comparing landslide hazard maps. *Natural Hazards* **20**: 137–158.

Van Westen CJ, Soeters R and Sijmons K (2000) Digital Geomorphological landslide hazard mapping of the Alpago area, Italy. *International Journal of Applied Earth Observation and Geoinformation* **2(1)**: 51–59.

Varnes DJ (1978) Slope movement types and processes. In *Landslide: Analysis and Control* (Schuster RL and Krizek RJ (eds)). Transportation Research Board, National Research Council, Washington DC, pp. 11–33.

Varnes DJ (1984) *Landslide Hazard Zonation: a Review of Principles and Practice*. Engineering Geology Commission on Landslides and other Mass Movements on Slopes, UNESCO, Paris.

Vaughan PR (1995) Possible actions to help developing countries mitigate hazards due to landslides. In *Landslides Hazard Mitigation*. The Royal Academy of Engineering, London, pp. 114–122.

Voight B (ed.) (1978) *Rockslides and Avalanches – 1, Natural Phenomena*. Elsevier, Amsterdam.

Voight B (1990) The 1985 Nevado del Ruiz volcano catastrophe – anatomy and retrospection. *Journal of Volcanology and Geothermal Research* **42**: 151–188.

Voight B and Pariseau WG (1978) Rockslides and avalanches: an introduction. In *Rockslides and Avalanches – 1, Natural Phenomena* (Voight B (ed.)). Elsevier, Amsterdam, pp. 1–67.

Vonder Linden K (1989) The Portuguese Bend landslide. *Engineering Geology* **27**: 301–373.

Vulliet L (1986) Modelisation des pentes naturelles en movement. DSc thesis, Ecole Polytechnique Federale de Lausanne.

Weidinger JT and Korup O (2009) Frictionite as evidence of for a large Late Quaternary rockslide near Kanchenjunga, Sikkim Himalayas, India – Implications for extreme events in mountain relief destruction. *Geomorphology* **103**: 57–65.

Wesson CVK and Wesson RL (1975) Odyssey to Tadzhik – an American family joins a Soviet Seismological Expedition. *US Geological Survey Earthquake Information Bulletin* **7(1)**: 8–16.

Wilson SD (1970) Observational data on ground movements related to slope instability. *Journal of the Soil Mechanics and Foundations Division, ASCE* **96(SM4)**: 1521–1544.

Wong HN and Ho KKS (1996) Travel distance of landslide debris. In *Landslides* (Senneset K (ed.)). Balkema, Rotterdam, vol. 1, pp. 417–423.

Wong HN, Ho KKS and Chan YC (1997) Assessment of consequences of landslides. In *Landslide Risk Assessment* (Cruden D and Fell R(eds)). Balkema, Rotterdam, pp. 111–149.

WP/WLI (International Geotechnical Societies' UNESCO Working Party for World Landslide Inventory) (1995) A suggested method of a landslide summary. *Bulletin of the International Association of Engineering Geology* **43**: 101–110.

Yin KL and Yan TZ (1988) Statistical prediction models for slope instability of metamorphosed rocks. In *Landslides* (Bonnard C (ed.)). Balkema, Rotterdam, pp. 1269–1272.

Zaruba Q and Mencl V (1969) *Landslides and their Control*. Elsevier, Amsterdam.

Landslide Risk Assessment
ISBN 978-0-7277-5801-9

ICE Publishing: All rights reserved
http://dx.doi.org/10.1680/lra.58019.097

Chapter 4
Qualitative and semi-quantitative risk assessment

4.1. Introduction

Slope inspections and landslide studies have always involved some rudimentary form of risk assessment, although it may seldom have been recognised as such (e.g. Fell and Hartford, 1997). Informal assessments of risk have generally relied on the judgement and skill of experienced engineers, geologists and geomorphologists. Recognition of hazards, mapping of areas of current or past instability, creation of a ground model, and development of an understanding of the causes and mechanisms of failure have all been shown to be essential for making judgements about the significance of landslide problems within an area or at a particular site. As a consequence, decision-makers have often been able to act on specialists' advice without having recourse to the explicit quantification of risk. However, consistency between different specialists is difficult to achieve, both for successive inspections of the same slope and when comparing the relative significance of different landslide problems. But the world is changing, and both clients and consultants are becoming increasingly aware of the need to introduce ever more rigorous and systematic procedures to formalise the evaluation process and thereby enhance the 'openness', 'objectivity' and 'consistency' of such judgements.

A large proportion of this book is devoted to quantitative risk assessment methods. However, there are many instances when an estimate of risk in terms of economic impact or loss of life either cannot be realistically achieved because of constraints of time, resources and data availability or is simply not required (see Chapter 1). For example, decision-makers or agencies responsible for managing a large number of slopes may be aware that landsliding is an issue in a particular area, but only need advice on:

- which particular landslides present a significant or pressing threat to assets or the population. This can involve screening out those sites where landsliding is perceived to present only minor or negligible problems
- how to prioritise these significant sites so as to ensure that the most urgent problems are dealt with first.

One of the earliest examples of qualitative risk assessment is the Cut Slope Ranking System developed in the late 1970s by the Hong Kong Geotechnical Control Office (GCO) (Koirala and Watkins, 1988). A risk score was calculated for each of the 8500 cut slopes and retaining walls registered in the Hong Kong slope inventory. The system was used to assess whether individual slopes met the required safety standards and to prioritise slope improvements. Scores were assigned to a wide variety of slope characteristics and were used to calculate both an 'instability score' and a 'consequence score' for each slope (Table 4.1). The sum of the two scores yields a relative risk score (the 'total score').

Table 4.1 The Hong Kong Cut Slope Ranking System

Instability score			Consequence score		
Component		Score	Component		Score
e	Height H (m)	Soil slopes = $H \times 1$ Rock slopes = $H \times 0.5$ Mixed slopes = $H \times 1$	t	Distance to building, road or playground from toe of slope (m)	Buildings = Actual distance Roadways = Distance + 2 m Playground = Greater of actual distance or $\frac{1}{2}H$
f	Slope angle (rock)	90° = 10 >80° = 8 >70° = 5 >60° = 2 <60° = 0	u	Distance to building, road or playground from top of slope (m)	Buildings = Actual distance Roadways = Distance + 2 m Playground = Greater of actual distance or $\frac{1}{2}H$
f	Slope angle (others)	>60° = 20 >55° = 15 >50° = 10 >35° = 3 <35° = 0	v	Extensive slope at toe of slope	Extensive slope at top = 0.5 Extensive slope below = 20
g	Angle of slope above, or presence of road above	Slope >45° = 15 Slope >35° or major road = 10 Slope >20° or minor road = 5 Slope <20° = 0	w	Multiplier for type of property at risk at toe	Hospitals, schools, residential = 2 Factories, playgrounds = 1.5 Major roads = 1 Minor roads = 0.5 Open space = 0
i	Associated wall	Height of wall × 2	x	Multiplier for type of property at risk at toe	Hospitals, schools, residential = 2 Factories, playgrounds = 1.5 Major roads = 1 Minor roads = 0.5 Open space = 0
j	Slope condition	Loose blocks = 10 Signs of distress = 10 Poor = 5 Good = 0	y	Multiplier for risk factor	For densely populated area or where buildings may collapse = 1.25 Otherwise = 1

k	Condition of associated wall	Poor = 10 Fair = 5 Good = 0
l	Adverse jointing	Adverse joints noted = 5
m	Geology	Colluvium/shattered rock, thin soil mantle = 15 Thick volcanic soil = 10 Thick granitic soil = 5 Sound rock (massive) = 0
n	Water access: impermeable surface on and above slope	None = 15 50% (partial) = 8 Complete–poor = 5 Complete–good = 0
o	Ponding potential at crest	Ponding area at crest = 5
p	Channels	None, incomplete = 10 Complete, major cracks = 10 Complete = 0
q	Water carrying services within H of crest	Yes = 5 No = 0
r	Seepage	Heavy seepage, mid-height and above = 15 Heavy seepage near toe = 10 Slight seepage, mid-height and above = 5 Slight seepage near toe = 2

From Koirala and Watkins (1988)

Instability score = $\sum (e \text{ to } r)$

Consequence score = $y\{20w[1.5(e+i) - t1.5(e+i)] + (40x)[(e+i) - u(e+i) + (vx) + 2(e+i)]\}$

Since the 1970s, a variety of methods have been developed that combine measures of the relative likelihood of landsliding and the severity of the resulting consequences to produce a statement of landslide risk. These methods include

- risk registers
- relative risk scoring or rating
- risk ranking matrices
- failure modes, effects and criticality analysis (FMECA).

Two contrasting styles have developed: those based on expert judgement and those that are designed as 'expert systems' that can be used by a wide range of practitioners. Those schemes based on expert judgement are often developed specifically for a particular area or project and reflect limited available information on the landslide history, causal factors or impacts. As a result, there is a strong reliance on the experience of experts in developing subjective judgements about the likelihood of landsliding and the consequences. This type of approach is particularly useful in remote regions where there is a pressing need for prioritising the investment of limited landslide management resources. However, the subjectivity inherent in the assessment process and different perceptions about the meaning of what appear to be everyday terms such as 'high', 'medium' and 'low' can lead to confusion and a diversity of interpretations (e.g. Crozier and Glade, 2005).

Expert systems, by contrast, are designed to ensure that the reliance on the judgement of the practitioner in applying the scheme is minimised. This is only possible where considerable research effort has already been directed towards determining the relative significance of different slope parameters and assigning weighting factors. The work typically involves collation of historical landslide data, statistical analysis and numerical modelling. As Wong (2005) notes, this research effectively replaces the subjective judgement that would otherwise have had to be made by the individual practitioners in applying the scheme. While an expert system is generally more repeatable than the use of expert judgement, there are relatively few situations where there suitable landslide and slope performance data are available to justify the use of this type of approach. Indeed, examples are restricted to transportation routes in developed countries, such as the Rock Fall Hazard Rating System in Oregon, USA (Pierson *et al.*, 1990) and the Slope Risk Analysis System in New South Wales, Australia (Stewart *et al.*, 2002), or major urban centres with a history of landslide problems, such as Hong Kong (e.g. the Cut Slope Ranking System; see Wong (2005) and Wong and Ko (2006) for other examples).

4.2. Risk registers

A risk register is an *active* (i.e. continuously updatable) document that sets out *all the known risks (threats)* in an area or at a particular site, together with the decisions taken about how to monitor and manage them. The structure of the register should reflect the overall risk assessment and management process, concentrating on both the nature of the potential problems arising from different types and scales of slope failure and the recommended mitigation actions. It should be reviewed and updated as and when new risks (threats) are identified, or following the implementation of further studies or mitigation measures.

Risk registers, despite being a very basic or rudimentary form of risk assessment, can prove a useful tool in the screening and prioritisation of landslide problems at the early stages of a project. They can also provide a means of tracking landslide management decisions through a project and thereby help to ensure that low-priority issues are not ignored.

Example 4.1

The coastal cliffs of Scarborough's South Bay, UK, are 2 km long and developed in a sequence of Jurassic rocks and glacial tills. The cliffs were purchased by Scarborough Borough Council in the 1890s. At that time, the council undertook a programme of coast protection, landscaping and drainage as part of a range of slope treatment measures. The landscaped slopes have since become a major recreational resource of the town and are covered with an elaborate network of paths. Despite the stabilisation works, the cliffs (often sloping at 30–40° and over 50 m high) are regularly affected by slow, minor landslides that result in disrupted footpaths or small losses of cliff-top land. Of greater concern, however, were the results of geomorphological mapping and historical research undertaken following the unexpected and dramatic 1993 Holbeck Hall landslide, which revealed that other large and damaging landslides have occurred in historic times (Lee, 1999; Lee and Clark, 2000; Lee et al., 1998).

As a consequence of the Holbeck Hall landslide, a preliminary assessment of the risks posed by coastal instability was undertaken, based on geomorphological assessment and a thorough review of historical sources (Rendel Geotechnics, 1994). The South Bay cliffline was subdivided into eight separate cliff sections, following the recognition of *four previously failed major landslides*, separated by *four intact coastal slopes*; that is, cliffs unaffected by major failures (Figure 4.1). A range of failure scenarios were identified for each section and recorded in a risk register, two examples of which are shown in Table 4.2. Each failure type was evaluated in terms of likelihood, based on the historical frequency, and the expected scale of potential consequences to the current assets along and adjacent to the cliffline.

- *Small-scale shallow failures* of the coastal slopes might lead to slight to moderate damage to footpaths and other structures, but would not lead to cliff-top loss. This type of failure can be expected to occur somewhere within South Bay, on average, once every year.
- *Large failure* involving rapid cliff-top recession and run-out of debris across the beach could lead to total loss of the seawalls, coastal slope structures and cliff-top property within the affected area. The historical frequency of such failures of the intact steep slopes was estimated to be four events in 256 years (see Example 6.3 in Chapter 6).
- *Reactivation of pre-existing landslides* involving episodes of significant ground movement during wet winters.

Although this approach represented only a basic form of risk assessment (i.e. risk identification), it did provide a mechanism for ensuring that all instability risks on the Scarborough urban coastline were identified at an early stage of the landslide management process. Once the risks had been identified, they could be screened and prioritised. The development of the risk register facilitated the formulation of an action plan that included the following.

1. *A strategic coastal defence study*. The preliminary risk assessment recommended the preparation of an integrated cliff and foreshore management plan; this was subsequently developed as part of a strategic coastal defence study of the Scarborough urban frontage.
2. *Land use planning*. A planning guidance map that related the landslip potential and risk for the purposes of forward planning and development control was produced for the local planning authority.
3. *Monitoring and early warning*. A systematic observation and monitoring strategy was developed to provide the basis for early warning of potential problems in priority areas. The equipment employed included simple crack monitoring studs on buildings, inclinometers, piezometers and tilt meters.

Landslide Risk Assessment

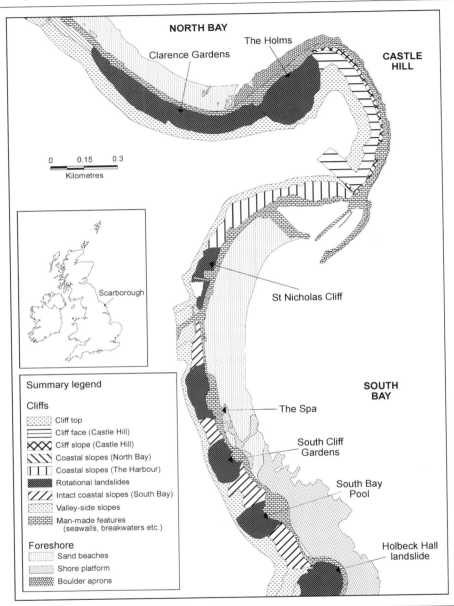

Figure 4.1 Example 4.1: summary geomorphological map of the Scarborough urban coast. (After Lee, 1999)

4.3. Relative risk scoring

In many instances, it may be inappropriate to evaluate risk in absolute terms because of the difficulties in assigning meaningful values for the hazard, the assets or elements at risk, and the possible adverse consequences. In such circumstances, it can be useful to assess the relative levels of the threat, or *relative risk*, to different sites posed by particular hazards, based on both factual data and subjective appraisal. A wide variety of schemes have been developed to do this, ranging from the expert systems

Qualitative and semi-quantitative risk assessment

Table 4.2 Example 4.1: part of the landslide risk register developed for Scarborough's South Bay (note that the action plan has been periodically reviewed and amended in the succeeding years)

Cliff section	Setting	Elements at risk	Landslide hazard	Likelihood of failure	Consequences	Recommended action plan (1994)	Actions (1995)
Holbeck Gardens	Intact, protected coastal cliff	Cliff footpaths Putting green Cliff-top property Seawall and promenade	Small-scale shallow translational failures on the cliff face	Failures could occur most years	Slight to moderate footpath damage Slide could expand and trigger a major landslide (unloading)	Council to monitor slopes and record failures; all failures to be treated as soon as possible	Consultant has been commissioned to implement action plan
			Large landslide involving rapid runout and loss of up to 100 m of cliff-top land	This type of failure could be expected in South Bay once every 100 years	Loss of cliff-top property and putting green, footpaths and seawall/promenade Potential for loss of life and injury because of rapid runout	Council to install inclinometers and piezometers; monitor slopes; establish contingency plan for emergency works; establish planning controls	Consultant has been commissioned to implement action plan
Spa Cliff	Previously failed (1737) major landslide	Footpaths The Spa Pavilions and Ocean Ballroom Cliff-top property Seawall and promenade	Rockfalls and small slides off the rear cliff face	Failures could be expected to occur during wet periods	Moderate to serious damage to cliff-top road and infrastructure	Council to install crack monitoring devices; monitor slopes and record failures; treat all localised failures as soon as possible; establish planning controls	Consultant has been commissioned to implement action plan
			Small-scale shallow translational failures on the coastal slope	Failures could occur most years	Slight to moderate footpath damage		
			Reactivation would be associated with slow, intermittent ground movement	Reactivation could be expected to occur during wet periods (say, 1 year in 10)	Slight to moderate and severe damage to the Pavilions, Ocean Ballroom and seawalls		

Adapted from Rendel Geotechnics (1994)

developed in Hong Kong for retaining walls, cut slopes and embankment fills (e.g. Wong, 2005) to methods that rely heavily on expert judgement. The value of relative risk assessment is that it can enable sites to be compared quickly and cheaply, thereby allowing early decisions to be made about where limited financial resources should be directed (e.g. Clark *et al.*, 1993; Hearn 1995; Wong, 2005).

The *relative risk scoring* approach utilises the basic definition of risk outlined in Chapter 1:

risk = probability (hazard) × adverse consequences

However, the hazard (i.e. landsliding) probability and adverse consequence elements in this equation are all represented by relative scores or rank values, with the risk being the product of these scores. For example, the probability of landsliding of a particular magnitude can be represented by a *hazard number*:

hazard number = hazard score × probability score

Similarly, the adverse consequences can be represented by a consequence value multiplied by the vulnerability of the assets or elements at risk:

adverse consequences = consequence value × vulnerability

The risk, expressed as a *risk number*, is calculated as

risk number = (hazard score × probability score) × (risk value × vulnerability)

The risk numbers produced can then be used to place each site within an arbitrarily defined scale of *risk classes*, thereby allowing some comparison between sites and so providing a basis for management decisions.

Example 4.2
A series of recent rockfalls and rockslides off the 50 m-high South Shore Cliffs, Whitehaven, UK, raised concerns over the safety of foreshore users. Boggett *et al.* (2000) described the use of a qualitative risk-scoring scheme to evaluate the problems and identify where emergency remedial works were required.

A detailed geomorphological and geotechnical mapping exercise was undertaken to collect information on the cliff conditions and failure mechanisms, which facilitated the division of the cliffline into five distinct sections, subdivided into morphological zones. Scoring scales were then developed for the four main components of the risk assessment: hazard magnitude H, *probability* of occurrence within the specified 10-year time frame P, the significance of the *elements at risk* or *risk value* R and the likely scale of adverse impact or *vulnerability* V, as shown in Table 4.3.

Relevant scores for each component were then allocated to each section of the cliff and multiplied together to yield a *risk number* R_N, as shown in Table 4.4, with a theoretical maximum of 300:

risk number = hazard × probability × risk value × vulnerability

The risk numbers for each cliff section are presented in Table 4.4. These scores were used to classify each section into one of five arbitrarily defined risk classes. A risk zonation plan was then produced

Qualitative and semi-quantitative risk assessment

Table 4.3 Example 4.2: South Shore Cliff, Whitehaven: qualitative risk assessment – hazard and risk scoring scheme (each column is a separate ranking scheme)

Number	Hazard H	Probability P (chance of occurrence in 10 years)	Risk value R	Vulnerability V
1	Small failure/erosion	Unlikely	Hardstanding areas not in use	Little or no effect
2	Moderate failure and occasional small falling blocks	Possible	Unoccupied building/public right of way (beach)	Nuisance or minor damage
3	Substantial failure and occasional large falling blocks	Likely	Roads/footpath	Major damage
4	Deep failure (>30 m) and large rockfall		Major structure/mine buildings	Loss of life
5	Major failure		Residential area	

From Boggett et al. (2000)

Table 4.4 Example 4.2: South Shore Cliff, Whitehaven – examples of cliff section risk assessment (see Figure 4.2 for locations)

Cliff subsection	Hazard score H (1–5)	Probability score P (1–3)	Risk value score R (1–5)	Vulnerability score V (1–4)	Risk R_N ($H \times P \times R \times V$) (4–300)	Risk class
1	4	2	3	3	72	IV
2	3	3	2	3	54	III
3	3	2	2	2	24	II
4	4	3	3	3	108	V
5	2	2	2	2	16	II

Risk classes:
$R_N > 100$ V   Highest
R_N 60–100   IV
R_N 30–60   III   Moderate
R_N 10–30   II
R_N 0–10   I   Lowest

Note: Each cliff section contains a number of different morphological zones
1 Shore platform
2 Lower cliff
3 Mid cliff
4 Upper cliff
5 Cliff top

From Boggett et al. (2000)

Figure 4.2 Example 4.2: Whitehaven cliffs: risk zonation plan. (After Boggett *et al.*, 2000)

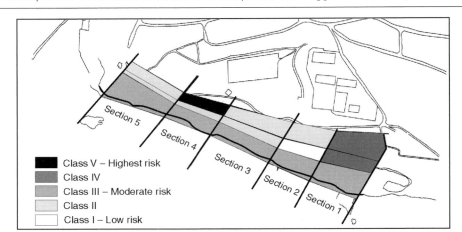

(Figure 4.2) that defines the key problem areas. The high-risk zone (section 4) coincides with the site of a large tension crack that was proceeding as a slow slide. As the slide develops, the likelihood of rockfall or topple increases, resulting in a high risk.

This is a classic example of the risk scoring technique. Nevertheless, it may be criticised for only having three divisions of probability and four of vulnerability. There is also no indication of 'degree of certainty', which often makes for a useful addition.

Example 4.3

Since 2003, Saskatchewan Highways and Transportation (Canada) has been using a risk scoring method to prioritise landslide sites along the road network for monitoring and remediation. The approach used is based on the Alberta Transportation Landslide Management System, which defines risk as

$$\text{risk} = \text{probability factor} \times \text{consequence factor}$$

The probability factor reflects the likelihood of a landslide occurring during the lifetime of a structure. The probability scores range from 1 to 20, depending on slope conditions and stability analysis results (Table 4.5). The consequence factor reflects the impact of a landslide on the transportation infrastructure or driver safety. Scores range from 1 to 10, and reflect public safety, the need for road closure and loss of infrastructure (Table 4.6). The risk score ranges from 1 to 200, and provides a means of identifying priority sites for response and management (Table 4.7).

Kelly *et al.* (2005) describe how the scoring system was used at 69 sites along the Saskatchewan provincial road network. After an initial inspection in 2004, probability and consequence scores were assigned to each site by members of an expert panel familiar with the sites. The risk scores varied from 1 to 160 and were used to assign a response level and management approach for each site. A total of 25 sites were classified as being either 'priority' or 'urgent'.

Table 4.5 Example 4.3: the Saskatchewan Highways and Transportation qualitative risk rating approach – the probability factor

Probability factor (PF)	Natural slopes	Engineered slopes
1	Geologically stable. Very low probability of landslide occurrence	$F > 1.5$ on basis of effective stress analysis with calibrated data and model.[a] Historically stable. Very low probability of landslide
3	Inactive, apparently stable slope. Very low probability of landslide occurrence or remobilisation	$1.5 > F > 1.3$ on basis of effective stress analysis with calibrated data and model.[a] Historically stable. Low probability of landslide
5	Inactive landslide with moderate probability of remobilisation. Moderate uncertainty level; or active slope with very slow constant rate of movement; or indeterminate movement pattern	$1.3 > F > 1.2$ on basis of effective stress analysis with calibrated data and model.[a] Minor signs of visible movement. Moderate probability of landslide
7	Inactive landslide with high probability of remobilisation, or additional hazards present. Uncertainty level high. Perceptible movement rate with define zones of movement	$1.2 > F > 1.1$ on basis of effective stress analysis with calibrated data and model.[a] Perceptible signs of movement, or additional hazards present. High probability of landslide
9	Active landslide with moderate, steady or decreasing rate of movement in defined shear zone	$F < 1.1$ on basis of effective stress analysis with calibrated data and model.[a] Obvious signs of ongoing slow to moderate movement
11	Active landslide with moderate, increasing rate of movement	Active landslide with moderate, increasing rate of movement
13	Active landslide with high rate of movement at steady or increasing rate	Active landslide with high rate of movement at steady or increasing rate
15	Active landslide with high rate of movement with additional hazards[b]	Active landslide with high rate of movement with additional hazards[b*]
20	Catastrophic landslide is occurring	Catastrophic landslide is occurring

From Kelly et al. (2005)
F = factor of safety
[a] If the described conditions for slope analysis are unknown or not met, increase probability factor by one category; e.g. if quality of data used in analysis is not known, increase PF from 1 to 3
[b] Additional hazards are factors that can greatly increase the rate of movement (eroding toe, groundwater etc.)

Example 4.4

The population of Baguio City on the Island of Luzon, Philippines, has grown from 128 000 in 1982 to over 250 000 in 2004, a rate of about 4% per annum. This growth has led to a rapid spread of the urban area onto marginal, unstable mountainous terrain, resulting in a significant increase in loss of life and

Table 4.6 Example 4.3: the Saskatchewan Highways and Transportation qualitative risk rating approach – the consequence factor

Consequence factor (CF)	Typical consequences
1	Shallow cut slopes where slide may spill into ditches or fills where slide does not have an impact on pavement or driver safety, maintenance issue
2	Moderate fills and cuts, not including bridge approach fill or headslopes, loss of portion of the roadway or slide onto road possible, small volume. Shallow fills where private land, water bodies or structures may be affected. Slides affecting use of roadways and safety of motorists, but not requiring closure of roadway. Potential rockfall hazard sites
4	Fills and cuts associated with bridges, intersectional treatments, culverts and other structures, high fills, deep cuts, historic rockfall hazard areas. Sites where partial closure of the road or significant detours would be a direct and avoidable result of a slide occurrence
6	Sites where closure of the road would be a direct and unavoidable result of a slide occurrence
10	Sites where the safety of public and significant loss of infrastructure facilities (e.g. a bridge abutment) or privately owned structures will occur if a slide occurs. Sites where rapid mobilisation of a large-scale slide is possible

From Kelly *et al.* (2005)

property damage caused by landslides. Saldivar-Sali and Einstein (2007) suggest that this situation had arisen, in part, because of inadequate and ineffective control over land development.

A qualitative risk rating tool was developed by Saldivar-Sali and Einstein (2007) to support land use planning in the area. The assessment of risk in the Baguio area involved defining map units based on bedrock geology and slope class. These units were subdivided to reflect variations in vegetation type.

Table 4.7 Example 4.3: the Saskatchewan Highways and Transportation qualitative risk rating approach – risk scores and management responses

Risk score	Response level	Management approach
>125	Urgent	Inspect at least once per year. Monitor instrumentation at least twice per year in the spring and fall. Investigate and evaluate mitigation measures
75–125	Priority	Inspect once per year. Monitor instrumentation at least once per year
27.5–75	Routine	Inspect every 3 years. Monitor instrumentation at least every 3 years with an increased frequency for selected sites as required
<27.5	Inactive	No set instrumentation monitoring or inspection schedule. Monitored and inspected as required in response to maintenance requests

From Kelly *et al.* (2005)

The hazard rating for each subunit was then multiplied by the land use and population factors to yield a risk rating (RR), defined as

RR = hazard rating × land use multiplier × population multiplier

The hazard rating was based on combinations of 'hazard contributing factors'; bedrock geology, slope gradient and vegetation type (Table 4.8). A range of scores were used for each hazard rating to allow site characteristics to be taken into account, reflecting relatively higher or lower hazard conditions. The rating is essentially a measure of landslide susceptibility rather than hazard.

The land use multiplier was used to reflect variations in the assets at risk within an area. Built-up areas are the highest-value land use (1.0) and include roads and highways, along with homes, industrial buildings and critical economic activities (Table 4.9). Multipliers less than 1 reduce the risk rating for non-built-up areas, reflecting differences in the value of the assets at risk. Grasslands are under-utilised land where future development is expected. Agricultural land is relatively highly scored (0.9) because it is a significant component of the local economy. Miscellaneous areas include mine pit sites, tailings ponds and reservoirs. Forest areas are largely undeveloped and hence have the lowest multiplier (0.8).

The population multiplier reflects the strong correlation between the population density in each local area (barangay) and the risk to people (Table 4.10). The multipliers range from 1 for the highest barangay populations (over 5000) to 0.75 for barangays with populations below 1000.

Table 4.11 presents the risk ratings for a selection of barangays in Baguio City. The approach is intended to assist engineers and land use planners in delineating land use and building constraints. Saldivar-Sali and Einstein (2007) also suggest that the method can serve as a basis for improving building codes, by allowing the regulations to reflect site-specific geological conditions rather than being generalised.

4.4. Risk ranking matrices

An alternative approach to risk rating involves the development of a risk matrix, in which a measure of the likelihood of a hazard occurring is matched against the increasing severity of consequences to provide a ranking of risk levels (see Figure 2.6). Although rankings are value judgements, experienced landslide specialists should be able to make realistic assessments of the likelihood of events and consequences, based on an appreciation of the landslide environment, together with knowledge of the particular site.

Tables 4.12 and 4.13 present typical scales of hazard likelihood and consequences that could be adapted to particular circumstances (AGS, 2000, 2007). *Relative risk levels* can then be assigned to different combinations of hazard and consequences (e.g. Table 4.14). Each relative risk level should mark a step up in the degree of threat and a change in the acceptability or tolerability of the risk (e.g. Table 4.15). Although the designation of risk levels can appear somewhat arbitrary, they do provide a framework for making comparisons between different sites within an area.

Example 4.5

Uncontrolled logging of forest land is known to result in an increase in landslide activity (e.g. Sidle *et al.*, 1985). VanDine *et al.* (2002) describe the use of risk matrices to assist the British Columbia Ministry of Forests plan timber harvesting operations at Perry Ridge, where local residents were

Landslide Risk Assessment

Table 4.8 Example 4.4: the Baguio City risk assessment – hazard ratings

Geological class	Vegetation type	Hazard rating	
VII	Pugo formation slopes >50%	IV	97–99
		III	95–96
		II	93–94
		I	91–92
VI	(Class not used)	IV	89–90
		III	87–88
		II	85–86
		I	83–84
V	Pugo formation slopes 0–50%	IV	81–82
		III	79–80
		II	77–78
		I	75–76
IV	(Class not used)	IV	73–74
		III	71–72
		II	69–70
		I	67–68
III	Zigzag formation (all slopes)	IV	65–66
	Baguio formation, slopes >31%, 0–8%	III	63–64
	Mirador limestone, slopes <18%	II	61–62
		I	59–60
II	Mirador limestone, slopes >31%	IV	57–58
		III	55–56
		II	53–54
		I	51–52
I	Baguio formation, slopes 9–30%	IV	49–50
	Mirador limestone, slopes 19–30%	III	47–48
	Klondyke formation (all slopes)	II	45–46
		I	43–44
0	Kennon limestone (all slopes)	IV	41–42
		III	39–40
		II	37–38
		I	35–36

From Saldivar-Sali and Einstein (2007)
Vegetation types: IV, none; III, broadleaf; II, grass or crop land/agricultural land; I, broadleaf mix or bushes/scrub
Geological types: Pugo formation = stratified sequence of basaltic and andesitic rocks; zigzag formation = conglomerates, sandstone and shale; Kennon limestone; Klondyke formation = conglomerates, tuffaceous sandstone, volcanic and tuff breccia, siltstone and mudstone; Mirador limestone; Baguio formation = pyroclastics

concerned about the resulting threats to people, property and water supply. The risk assessment undertaken was described as a 'consensual qualitative risk assessment', based on available information and empirical evidence, together with the experience and judgement of the three geohazard specialists involved in the project.

Table 4.9 Example 4.4: the Baguio City risk assessment – land use multipliers

Land use	Multiplier
Built up	1
Grasslands	0.95
Agriculture	0.9
Miscellaneous	0.85
Forest	0.8

From Saldivar-Sali and Einstein (2007)

Table 4.10 Example 4.4: the Baguio City risk assessment – population multipliers

Population/barangay	Multiplier
0–1000	0.75
1000–2000	0.8
2000–3000	0.85
3000–4000	0.9
4000–5000	0.95
>5000	1

From Saldivar-Sali and Einstein (2007)

Table 4.11 Example 4.4: hazard and risk ratings for selected barangays in Baguio City

Barangay	Hazard rating	Land use multiplier	Population multiplier	Risk rating
Engineer's Hill	65–66	1	0.85	55–56
Military cut-off	47–48	1	0.8	38
Camp 7	43–44	0.95	0.95	41–42
SLU–SVP	65–66	0.9	0.85	50
Bakakeng Central	65–66	1	0.95	62–63
Camp Sioco	49–50	1	0.8	39–40

From Saldivar-Sali and Einstein (2007)

Landslide Risk Assessment

Table 4.12 Indicative measures of landslide likelihood

Level	Descriptor	Description	Indicative annual probability[a]
A	Almost certain	The event is expected to occur	$> \approx 10^{-1}$
B	Likely	The event will probably occur under adverse conditions	$\approx 10^{-2}$
C	Possible	The event could occur under adverse conditions	$\approx 10^{-3}$
D	Unlikely	The event might occur under very adverse circumstances	$\approx 10^{-4}$
E	Rare	The event is conceivable, but only under exceptional circumstances	$\approx 10^{-5}$
F	Not credible	The event is inconceivable or fanciful	$< 10^{-6}$

From AGS (2000)

[a] '$\approx$' means that the indicative value may vary by say $\pm$ half of an order of magnitude, or more

Table 4.13 Indicative measures of consequence

Level	Descriptor	Description
1	Catastrophic	Structure completely destroyed or large-scale damage requiring major engineering works for stabilisation
2	Major	Extensive damage to most of structure, or extending beyond site boundaries, requiring significant stabilisation works
3	Medium	Moderate damage to some of structure, or significant part of site, requiring large stabilisation works
4	Minor	Limited damage to part of structure, or part of site, requiring some reinstatement/ stabilisation works
5	Insignificant	Little damage

From AGS (2000)

Table 4.14 Qualitative risk assessment matrix: levels of risk to property

Likelihood		Consequences to property				
		1 Catastrophic	2 Major	3 Medium	4 Minor	5 Insignificant
A	(almost certain)	VH	VH	H	H	M
B	(likely)	VH	H	H	M	L–M
C	(possible)	H	H	M	L–M	VL–L
D	(unlikely)	M–H	M	L	VL–L	VL
E	(rare)	L–M	L–M	VL–L	VL	VL
F	(not credible)	VL	VL	VL	VL	VL

From AGS (2000)

Table 4.15 Indicative risk level implications

Risk level		Examples of implications
VH	Very high risk	Extensive detailed investigation and research, planning and implementation of treatment options essential to reduce risk to acceptable levels; may be too expensive and not practical
H	High risk	Detailed investigation, planning and implementation of treatment options required to reduce risk to acceptable levels
M	Moderate risk	Tolerable provided that treatment plan is implemented to maintain or reduce risks. May be accepted. May require investigation and planning of treatment options
L	Low risk	Usually accepted. Treatment requirements and responsibility to be defined to maintain or reduce risk
VL	Very low risk	Acceptable. Manage by normal slope maintenance procedures

From AGS (2000)

The 76 km^2 area was subdivided into 32 catchments or hydrological units. A range of potential hazards were identified from previous studies, including flash floods, debris flows and landslides. The existing probability of each type of event was rated 'high', 'moderate', 'low', 'very low' or 'none', based on the past occurrence of the event, independent of its magnitude.

Consequences were rated 'high', 'moderate' and 'low', based on the elements at risk (people, property and water supply) and the inferred severity of the events that could affect those assets. Examples of potential consequences are presented in Table 4.16. The ratings for the three groupings of the elements at risk were not intended to be directly comparable.

Three risk matrices were developed, one for each of the elements at risk: people, property and water supply (Table 4.17).

Using a risk matrix for each of the 32 hydrological units, the position of the established past hazard–consequence relationship for each of the three recognised hazards was plotted; that is, between 3 and

Table 4.16 Example 4.5: ratings of potential consequences

	People	Property	Water supply
High	Death	Destruction of multiple residences	Destruction of multiple water intakes or very high increase in turbidity
Moderate	Serious injury	Destruction of single residence or damage to multiple residences	Destruction of single water intake or high increase in turbidity
Low	Minor injury	Damage to single residence	Damage to single water intake or moderate to low increase in turbidity

From VanDine et al. (2002)

Table 4.17 Example 4.5: risk matrices developed for different elements at risk

Hazard	Consequence to people			Consequence to property			Consequence to water supply		
	High	Moderate	Low	High	Moderate	Low	High	Moderate	Low
High	VH	H	M	VH	H	M	VH	H	M
Moderate	H	M	L	H	M	L	H	M	L
Low	M	L	VL	M	L	VL	M	L	VL
Very low	L	VL		L	VL		L	VL	

From VanDine et al. (2002)

9 of the 33 available cells contained information. The geohazard experts then assessed the likely relationships between hazard and adverse consequences that would exist following the commencement of logging activity, and the position of these *with logging* relationships were also plotted on the matrix. In those instances where the *with logging* case revealed a significant increase in estimated risk involving the identification of very high (VH), high (H) or moderate (M) risks, alternatives to logging or road building were suggested and landslide management recommendations were made.

This represents a good example of a fairly standard application of the risk matrix technique in the context of landsliding.

Example 4.6

Natural terrain landslides have caused fatalities, injuries and economic losses in Hong Kong (Wong et al., 2004), primarily because of the close proximity of dense urban development and steep hillslopes. The strategy for managing natural terrain landslide risk has been to avoid new development in vulnerable areas (e.g. Wong, 2003). Where this is not practicable, slope stabilisation or mitigation measures (e.g. debris flow barriers) are required by the Geotechnical Engineering Office (GEO). The management of landslide risks associated with natural terrain is directed towards identifying hazards and undertaking mitigation measures where necessary (Ng et al., 2002). The management principles vary slightly between new and existing development, with the key differences being as follows.

- *New development*: avoid sites that are subject to *severe* natural terrain hazards. Where hazards exist, undertake any necessary mitigation actions.
- *Existing development*: undertake urgent mitigation works where there exists an immediate and obvious danger. Where there is reason to believe that a dangerous situation could develop, mitigation should be undertaken when considered necessary.

A qualitative risk assessment approach was adopted to support the decision-making process for natural terrain developments. This approach, known as the design event approach, is used to identify the design basis (i.e. the design event magnitude) for the mitigation measures required to protect a development from natural terrain landslides on the adjacent hillsides (e.g. Ng et al., 2002; Wong, 2001). The required design event magnitude may be either a 'conservative' (notional annual probability = 0.01) or a 'worst credible' (notional annual probability = 0.001) event, depending on site conditions.

The appropriate design event for a particular site is determined using a risk matrix (Table 4.18). The approach follows the general principles of risk assessment, taking account of simple measures of

Table 4.18 Example 4.6: Hong Kong Design Event Approach – risk matrix

Susceptibility class		Consequence class				
		I	II	III	IV	V
A	Extremely susceptible. Notional annual probability ≥0.1	WCE	WCE	WCE	CE	N
B	Highly susceptible. Notional annual probability 0.1–0.01	WCE	WCE	CE	CE	N
C	Moderately susceptible. Notional annual probability 0.01–0.001	WCE	CE	CE	N	N
D	Low susceptibility. Notional annual probability <0.001	N	N	N	N	N

From Ng et al. (2002)
WCE = 'Worst credible event', with a notional annual probability of 0.001 (1 in 1000 years)
CE = 'Conservative event', with a notional annual probability of 0.01 (1 in 100 years)
N = Further study not required

hazard probability, exposure (i.e. proximity) and the elements at risk. However, the matrix is not a conventional one.

- The consequence class is not really a measure of consequences, but rather a measure of the elements at risk and their exposure and proximity.
- The susceptibility class is not really a measure of hazard to the *site*, but rather an indication of the current and recent activity state of the adjacent *hillslopes*.

Four landslide susceptibility classes are recognised, which broadly relate to the probability of failure of the surrounding hillslopes. The relevant class for a particular site is determined from aerial photograph interpretation, field mapping of evidence of past landsliding and other relevant information. Most natural terrain studies involve the development of a hazard model that identifies likely geomorphological and geological controls on the location, type, magnitude, frequency and run-out characteristics of the hazards concerned (e.g. Parry *et al.*, 2006). The notional ranges of probability of occurrence indicated for each category serve as yardsticks to aid judgement, in addition to the general guidance given in the assessment of relative susceptibilities. The consequences of failure are grouped into five classes based on the types of asset at risk and the proximity to the hillside (Table 4.19).

Risk levels are not expressed explicitly, but in terms of the design event requirement to provide an acceptable level of risk reduction. The design requirements have been benchmarked against 17 cases. Wong (2005) suggests that the approach is relatively easy to use and is favoured by many geotechnical practitioners in Hong Kong.

It should be noted that the design event approach is directed primarily towards hazard management rather than risk management. As a result, mitigation measures are based on event magnitude and not *at-site* risk. Sites with similar scales of hazard but contrasting exposed populations could have the same standard of protection (i.e. the design event), although the risk will be different.

4.5. The FMECA approach

Failure modes, effects and criticality analysis (FMECA) is a systematic approach to analysing how parts of a system, such as an engineered slope, might fail (the failure mode). For example, possible

Table 4.19 Example 4.6: Hong Kong Design Event Approach – consequence classes

Proximity	Facility group number				
	1, 2 Buildings and major infrastructure	3 Densely used open space, quarries, moderately used roads	4 Lightly used open space, low-volume roads	5 Remote areas and very low-volume roads	
Very close (e.g. if angular elevation from the site is ≥30°)	I	II	III	IV	
Moderately close (e.g. if angular elevation from the site is ≥25°)	II	III	IV	V	
Far (e.g. if angular elevation from the site is <25°)	III	IV	V	V	

From Ng et al. (2002)
For full details of assets included in facility group numbers, see Wong and Ho (1995)
The angular elevation is the angle between the horizontal and the line from the slope crest to the upslope boundary of the site

failure modes on a protected coastal cliff might include collapse of the seawall, leading to a renewal of marine erosion at the cliff foot, or blocked drains leading to locally high porewater pressures. Each failure mode can be assessed to determine the effects of that failure and to identify how critical it might be to the stability of the overall system. For example, seawall failure will result in an almost immediate renewal of cliff foot erosion and the re-establishment of active instability on the coastal slope. By contrast, failure of the drainage system is likely to have a less immediate or direct impact on the stability of the slopes. Nevertheless, drainage failure could lead to small-scale landslides that, under certain circumstances, may lead to the progressive decline in overall stability, resulting in an increase in the likelihood of a larger event.

The FMECA approach provides a structured framework for the qualitative analysis of the various components of a system, using engineering judgement to generate scores or rankings, rather than probabilities. The FMECA approach has been used as a risk assessment tool in the dam industry (e.g. Hughes et al., 2000; Sandilands et al., 1998). Lee (2003) has suggested that the approach may also be useful in the strategic monitoring of coastal slopes that have been stabilised by a combination of structural elements, including, for example, a concrete seawall at the cliff foot and slope drainage measures.

The approach involves the development and analysis of an LCI (location, cause, indicator) diagram for each slope (Figure 4.3). An LCI diagram sets out the individual constructed components of each man-made slope (seawall, drainage network, retaining structures, etc.) and how their lack of integrity

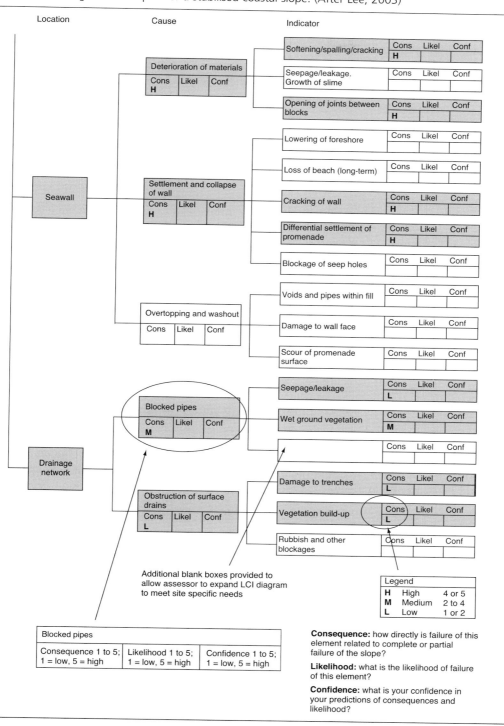

Figure 4.3 An LCI diagram developed for a stabilised coastal slope. (After Lee, 2003)

might contribute to the overall failure of the slope. Failure resulting from a range of possible causes (e.g. undercutting or high porewater pressures) and with different indicators (e.g. blocked drains or seepage) is considered by means *of indicator–cause pathways*. The level of detail presented in an LCI diagram should reflect the available knowledge about any potential indicator–cause pathway.

The assessment procedure involves scoring three key factors on a range of 1 to 5, for each *indicator–cause pathway*.

1. The *consequence* expressed in terms of how directly is failure of an element related to complete failure of the slope:

    1 = failure of element is unlikely to lead to failure of the slope
    5 = failure of element is highly likely to lead to failure of the slope.

    For example, failure of a seawall will almost inevitably lead to failure of the engineered slope, because marine erosion will begin to undercut the cliff foot. However, small-scale failures of engineered slopes resulting from the collapse of a gabion-basket retaining structure behind a footpath are a common feature on many coastal slopes, but rarely lead to failure of the whole slope.

2. The *likelihood* of failure of an element, ranging from 1 (low) to 5 (high).
3. The practitioner's *confidence* in the reliability of his or her predictions of the consequence and likelihood factors. The confidence score ranges from 1 (very confident) to 5 (little or no confidence). This score allows a measure of uncertainty to be included within the assessment. Table 4.20 presents a range of factors that should be considered when determining the confidence score. In some situations, many of these issues may prove insignificant and the term effectively will represent detectability; that is, the degree to which the potential failure mode can be detected prior to its occurrence. Increased slope monitoring can therefore result in a decrease in the confidence score by reducing the uncertainty about the ground conditions.

Considerable experience is required to develop and use an LCI diagram. It is important that the scores be the product of careful scrutiny, ideally by a group or panel of experts. Hughes *et al.* (2000) recommend that the process should be 'transparent' and the reasoning behind the allocation of each value should be clearly documented.

The results of the LCI diagram analysis are used to identify those structural elements that contribute most to the overall risk. A number of measures can be defined, including

Element score = consequence of failure × likelihood of failure

which provides a measure of the degree of risk associated with the failure of a particular element of the slope, such as a seawall. High scores indicate those elements where remedial measures may be needed to reduce the risk.

Criticality score = element score × confidence

This gives a measure of the hazard that a particular indicator-cause pathway creates for the slope. High criticality scores can reflect uncertainty in consequence and likelihood scores, highlighting the need for further investigation.

Table 4.20 The FMECA approach: key considerations for defining a confidence score in an LCI diagram

Issue	Comment
Detectability	The ease with which potential failure mechanisms can be detected before failure occurs, through the use of instrumentation, which is a function of the cost and resources required to monitor signs of pre-failure movement within different components
Construction quality	The quality of construction materials and the workmanship will vary between engineered slopes and between individual components of a slope. This can sometimes be readily identified and incorporated into the likelihood score. Sometimes, evidence of poor quality or bad workmanship may not be readily apparent. The confidence score should take account of any uncertainty regarding construction quality
Operational maintenance	Maintenance is essential for ensuring the continued integrity of the structures. The confidence score should take account of any uncertainty regarding the future maintenance programme actually being undertaken. For example, poorly funded or ad hoc programmes may be subject to significant change and are likely to be unreliable
Quality of records	A full record of the 'as-built' construction and operational maintenance is essential for a reliable assessment of structural performance. Good records do not reduce the likelihood of failure, but they increase the confidence in the allocated likelihood score
Incompleteness of knowledge	The confidence score should take account of any significant gaps in knowledge about the condition, behaviour and performance of the structures

After Hughes *et al.* (2000)

A measure of the relative risk associated with failure of particular elements of the slope can be established from the product of the criticality score and an impact score:

relative risk = criticality score × impact score

An *impact score* can be determined through the use of the types of scoring or ranking systems described earlier in this chapter. Hughes *et al.* (2000) present an expanded scoring framework for assessing the impact of dam failure that has been modified here to suit high-velocity landslide events (Table 4.21; note that further adjustments would be needed to make it appropriate for slow-moving landslides). The scores for each type of *economic impact* are combined to provide a single measure of impact for the site or area. This is achieved by adjusting each impact score by a weighting factor, and adding the adjusted scores; these factors are based on expert judgement and should be reviewed and modified accordingly, depending on the local circumstances.

Loss of life is estimated from the total number of people at risk as

loss of life = population at risk × exposure

For high-velocity landslides, the *exposure factor* will vary with the length of forewarning time and the ability of people to escape or be evacuated. Hughes *et al.* (2000) suggest that the vulnerability factor may range from 0.5 if there is little or no forewarning to only 0.0002 for a warning time of 90 minutes.

Table 4.21 FMECA approach: impact scoring system

	Score	Population at risk
Residential properties affected		
0	0	0
0–15	1	30
15–50	2	100
50–250	3	500
Estimate (>250)	4	2 × estimate
Non-residential: number of people affected		
0	0	0
0–150	1	150
150–500	2	500
500–1000	3	1000
Estimate (>1000)	4	Estimate
Infrastructure affected		
None	0	0
Minor roads	1	25
Major regional infrastructure	2	50
Major national infrastructure	3	100
Major international infrastructure	4	Estimate
Recreational sites: number of people affected		
0	0	0
0–10	1	10
10–50	2	50
50–100	3	100
Estimate (>100)	4	Estimate
Industrial sites		
None	0	N/A
Light industrial	1	N/A
Public health industries	2	N/A
Heavy industrial	3	N/A
Nuclear, petrochemical	4	N/A
Utilities		
None	0	N/A
Local loss of distribution	1	N/A
Local loss of distribution/supply	2	N/A
Regional loss of distribution/supply	3	N/A
Significant impact on national services	4	N/A
Agriculture/habitat site		
Uncultivated/grassland	0	N/A
Pasture	1	N/A
Widespread farming	2	N/A
Intensive farming/vulnerable habitat/monument	3	N/A
Loss of international habitat/monument	4	N/A

From Hughes *et al.* (2000)
N/A, not applicable

Table 4.22 FMECA approach: standard tables for calculating impact scores

Impact	Population at risk (PAR)	Exposure factor[a]	Total (PAR × exposure)
Residential property		0.5	
Non-residential property		0.5	
Infrastructure		0.5	
Recreation		0.5	
		Total loss of life	

Impact	Score	Weight	Total (score × weight)
Residential property		0.15	
Non-residential property		0.15	
Infrastructure		0.1	
Recreation		0.05	
Industrial		0.25	
Utilities		0.25	
Agriculture/habitats		0.05	
		Total score	

Impact	Score	Factor	Total (score × weight)
Economic impact		100	
Potential loss of life		1	
		Total impact score	

From Hughes *et al.* (2000)
[a] Exposure varies with forewarning

The economic impact scores are combined with the estimated loss of life to give an overall impact score (Table 4.22).

The relative risk also allows comparison between the same elements on different slopes.

Example 4.7

As described in Example 4.1, the coastal cliffs of Scarborough's South Bay consist of a sequence of previously failed major landslides, separated by *intact coastal slopes* (Figure 4.1). It has been recognised that the intact slopes have the potential for large-scale failure involving rapid cliff-top recession and run-out of debris. The FMECA approach has been used at this site to provide a repeatable methodology for the strategic monitoring of the condition of the slopes. The example below considers one of the intact slopes (Site A).

The cliff under consideration is around 2 km long, 55–60 m high and developed in a sequence of glacial tills (around 25–30 m thick) overlying Jurassic sedimentary rocks (predominantly sandstones and mudstones). The cliff face slopes at 30–32°, becoming steeper towards the base, where a near vertical mass concrete seawall, built around 1889–1893, together with a concrete promenade, provide protection from wave attack. The cliffs were landscaped and partially drained around the same time to improve the stability of the slopes and allow them to be opened for public use. A cliff-top road runs parallel with the cliffline, set back 80–140 m from the cliff edge. A row of large private houses and hotels lines the landward side of the road.

It has been recognised that there is a possibility of a major landslide at the site which could lead to the loss of cliff-top property. Following the development of a hazard model for the site, three scenarios were proposed that might lead to major landsliding.

- Scenario 1: the development of a major landslide caused by the expansion of the shallow landslides that occur on the cliff face, probably in response to a combination of prolonged heavy rainfall and blocked drainage.
- Scenario 2: the development of a major landslide caused by a combination of exceptionally high groundwater levels, progressive failure of the mudstones at the base of the cliff and the gradual deterioration of the slope drainage system.
- Scenario 3: the development of a major landslide due to failure of the seawall and renewed marine erosion.

The site was visited during 2002, and a number of features recorded that relate to the condition and performance of the slopes and coastal defences. The following were specific features of note.

- Tension cracks were present, crossing a number of footways, and probably indicative of the early stages of development of shallow landslides.
- The seawalls were in poor condition, and were considered to have a residual life of around 10 years (defined as the period over which the structure could be expected to perform as an acceptable coastal defence with routine repairs and maintenance).
- There was a high probability of seawall failure during storm events. The probability of major structural failure of the seawall was assessed for a range of failure mechanisms (Table 4.23), as part of a condition survey (visual inspection and review of damage records, etc.). The combined annual probability of failure is the sum of the probabilities of each of these mechanisms, and was estimated in this case to be 0.056 (1 in 18).

The documentation of the current condition and risk was based on the FMECA approach, and involved the following stages.

1. *Completion of an LCI diagram.* The scores entered into each of the boxes were assigned on the basis of observations and knowledge of the site conditions (Figure 4.4).

Table 4.23 Example 4.7: Site A, South Bay, Scarborough – estimated annual probability of structural failure of the seawall

Principal seawall failure mechanisms					
Loss of apron	Undermining of toe	Block plucking	Break-up of wall face	Overtopping and washout	Combined annual probability
0	0.0025	0.015	0.03	0.008	0.056

From High Point Rendel (2000)

Qualitative and semi-quantitative risk assessment

Figure 4.4 Example 4.8: the LCI diagram for an intact coastal slope, South Bay, Scarborough. (After Lee, 2003)

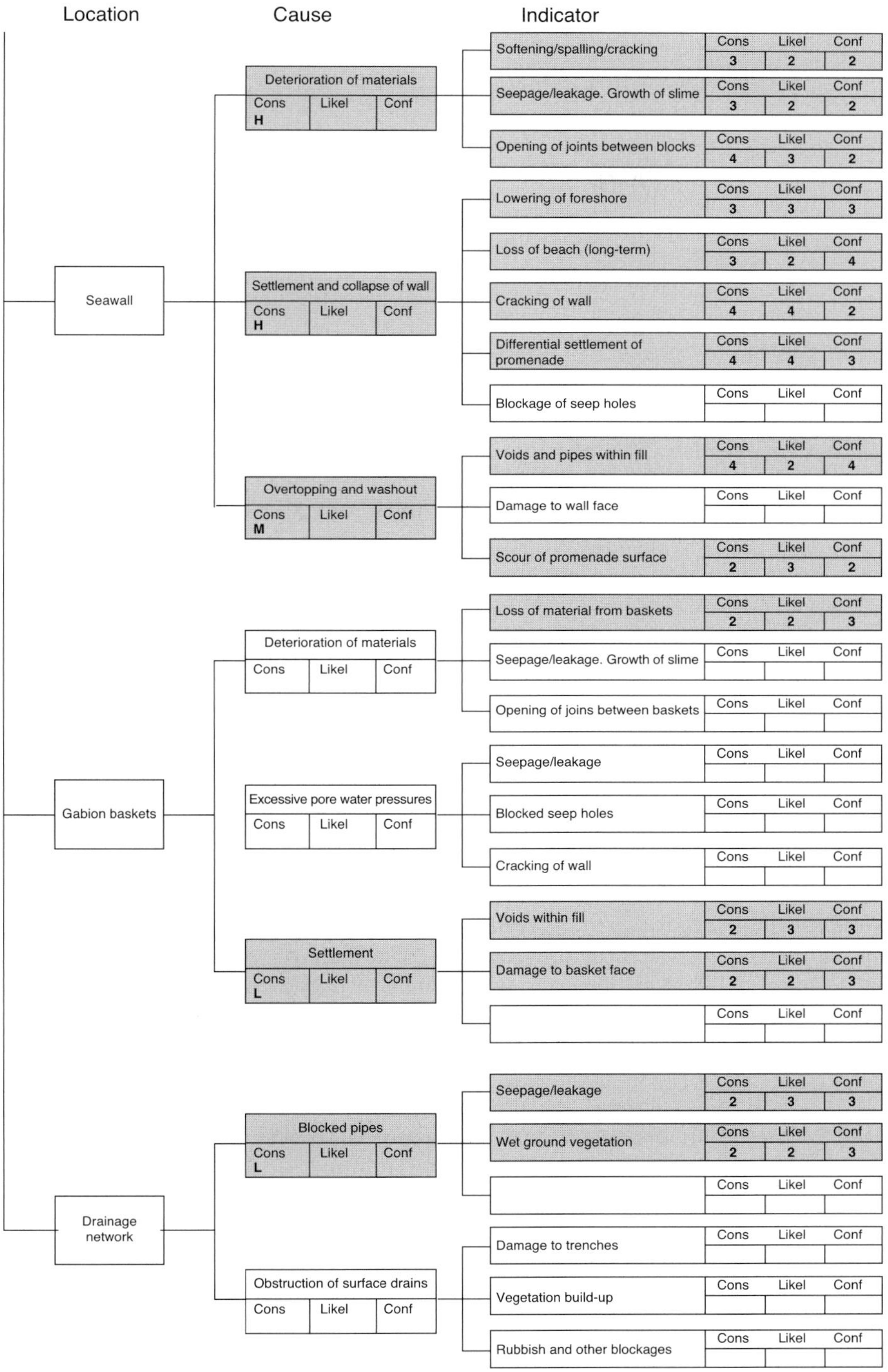

Location	Cause	Indicator	Cons	Likel	Conf
Seawall	Deterioration of materials (Cons H)	Softening/spalling/cracking	3	2	2
		Seepage/leakage. Growth of slime	3	2	2
		Opening of joints between blocks	4	3	2
	Settlement and collapse of wall (Cons H)	Lowering of foreshore	3	3	3
		Loss of beach (long-term)	3	2	4
		Cracking of wall	4	4	2
		Differential settlement of promenade	4	4	3
		Blockage of seep holes			
	Overtopping and washout (Cons M)	Voids and pipes within fill	4	2	4
		Damage to wall face			
		Scour of promenade surface	2	3	2
Gabion baskets	Deterioration of materials	Loss of material from baskets	2	2	3
		Seepage/leakage. Growth of slime			
		Opening of joins between baskets			
	Excessive pore water pressures	Seepage/leakage			
		Blocked seep holes			
		Cracking of wall			
	Settlement (Cons L)	Voids within fill	2	3	3
		Damage to basket face	2	2	3
Drainage network	Blocked pipes (Cons L)	Seepage/leakage	2	3	3
		Wet ground vegetation	2	2	3
	Obstruction of surface drains	Damage to trenches			
		Vegetation build-up			
		Rubbish and other blockages			

123

Table 4.24 Example 4.7: direct losses and impact scores

Direct losses	Impact score
<£10 000	1
£10 000–£100 000	2
£100 000–£1 million	3
£1 million–£10 million	4
>£10 million	5

From Lee (2003)

2 *Calculation of the impact score*. The estimated risk-free market values of the properties within the cliff-top area that might be affected by a major landslide were compiled with the assistance of a local estate agent. This yielded a value for potential direct losses of £1.8 million and an impact score of 4 (Table 4.24).
3 *Calculation of criticality and risk scores*. Criticality scores were calculated from the scoring attributed to elements in the LCI diagram. Rankings were attributed to the *criticality score*, the *consequence* × *likelihood* product and the *confidence score*.

The results of the FMECA analysis are presented in Table 4.25, and indicated the following:

- The highest risk scores were associated with differential settlement and cracking of the seawall and the possible presence of voids within the fill materials.
- The highest scores for *consequence* × *likelihood* were associated with the potential for differential settlement, cracking and opening up of joints within the seawall. These elements posed the greatest risk to the safety of the slope and needed to be the focus of prompt remedial action.
- High *confidence scores* (i.e. high uncertainty) were associated with the potential for beach loss and the occurrence of voids within the seawall fill materials. These were priority areas for further investigation. Resolution of the uncertainty could result in the consequence and likelihood scores going up or down.
- Calculation of risk scores for individual elements of this slope and others along the cliffline allowed a comparison between the different sites. For example, similar elements from adjacent slopes might have identical *consequence* × *likelihood* or *criticality* scores. However, a higher impact score would indicate that the risk generated by failure of the element at one site would be greater than at another. Priority needed to be given to addressing the higher-risk elements first.

4.6. Qualitative risk assessment: an easy option?

Qualitative methods are of value where the available resources or data dictate that more formalised quantitative assessment would be inappropriate or even impractical. Nevertheless, it remains essential that any qualitative risk assessment be based on a sound hazard model (see Chapter 3) and a good appreciation of the full range of possible outcomes (see Chapter 9). The use of qualitative methods should not be seen as an *easier option* than the quantitative methods approaches described in the following chapters. Indeed, there are a number of significant issues that can limit the reliability and/or usefulness of the qualitative approach unless carefully handled, including the following.

Table 4.25 Example 4.7: Site A, South Bay, Scarborough – risk summary table

Location	Cause/indicator	Criticality score	Criticality rank	Consequence × likelihood	C × L rank	Confidence score	Confidence rank	Risk score (impact[a] × criticality)
Seawall	Differential settlement	48	1	4 × 4 = 16	1 =	3	3 =	192
	Cracking of wall	32	2 =	4 × 4 = 16	1 =	2	8 =	123
	Voids and pipes within fill	32	2 =	4 × 2 = 8	5	4	1 =	128
	Lowering of foreshore	27	4	3 × 3 = 9	4	3	3 =	108
	Opening of joints	24	5 =	4 × 3 = 12	3	2	8 =	96
	Loss of beach	24	5 =	3 × 2 = 6	6 =	4	1 =	96
	Softening/spalling/cracking of materials; seepage/leakage: materials	12	9 =	3 × 2 = 6	6 =	2	8 =	48
	Scour of promenade surface	12	9 =	2 × 3 = 6	6 =	2	8 =	48
Gabion baskets	Voids within fill	18	7 =	2 × 3 = 6	6 =	3	3 =	72
	Damage to basket face	12	9 =	2 × 2 = 4	11 =	3	3 =	48
	Loss of material from basket	8	12	2 × 2 = 4	11 =	2	8 =	32
Drainage	Blocked pipes: wet ground; seepage/leakage	18	7 =	2 × 3 = 6	6 =	3	3 =	72

From Lee (2003)
[a]Impact score = 4; see Table 4.24

- The use of subjective scales to rank hazards and adverse consequences can be problematic, as perceptions of what actually constitutes a high or a low risk will vary considerably. This can lead to misunderstandings between professionals and unnecessary alarm among those the assessment is intended to inform.
- There is a need to ensure that the uncertainties associated with the identification of hazards and adverse consequences are fully documented and clearly conveyed to the users.
- Establishing whether the risk levels identified at a site are acceptable can be difficult. This can be particularly important where there is potential for loss of life or where there is a dispute between parties about the best way forward in managing landslide issues.
- Problems can arise in establishing an overall risk factor for a site where there might be multiple landslide hazards (e.g. rockfalls, debris flows and rotational slides), each with a different magnitude, frequency and range of potential adverse consequences.

As a final word, however, it is worth emphasising that 'the quality of a landslide risk assessment is related to the extent that the hazards are recognised, understood and explained which is not necessarily related to the extent to which they are quantified' (Powell, 2002). Attempts to quantify what are in effect qualitative judgements or intangibles can have the effect of assigning a spurious degree of accuracy to the assessment (de Ambrosis, 2002).

REFERENCES

AGS (Australian Geomechanics Society) (2000) Landslide risk management concepts and guidelines. *Australian Geomechanics* **35**: 49–52.

AGS (2007) Practice note guidelines for landslide risk management 2007. *Australian Geomechanics* **42(1)**: 63–114.

Boggett AD, Mapplebeck NJ and Cullen RJ (2000) South Shore Cliffs, Whitehaven – geomorphological survey and emergency cliff stabilisation works. *Quarterly Journal of Engineering Geology and Hydrogeology* **33**: 213–226.

Clark AR, Palmer JS, Firth TP and McIntyre G (1993) The management and stabilisation of weak sandstone cliffs at Shanklin, Isle of Wight. In *The Engineering Geology of Weak Rock* (Cripps JC and Moon CF (eds)). Geological Society, London, pp. 392–410. Engineering Group of the Geological Society Special Publication.

Crozier MJ and Glade T (1999) Frequency and magnitude of landsliding: fundamental research issues. *Zeitschrift fur Geomorphologie NF Suppl. Bd* **115**: 141–155.

de Ambrosis L (2002) Letters to the editor: re: landslide risk management concepts and guidelines. *Australian Geomechanics* **37**: 54–55.

Fell R and Hartford D (1997) Landslide risk management. In *Landslide Risk Assessment* (Cruden D and Fell R (eds)). Balkema, Rotterdam, pp. 51–108.

Hearn GJ (1995) Landslide and erosion hazard mapping at Ok Tedi copper mine, Papua New Guinea. *Quarterly Journal of Engineering Geology* **28**: 47–60.

High Point Rendel (2000) *The Holbeck–Scalby Ness Coastal Defence Strategy*. Scarborough Borough Council, Scarborough.

Hughes A, Hewlett H, Samuels PG, Morris M, Sayers P, Moffat I, Harding A and Tedd P (2000) *Risk Management for UK Reservoirs*. Construction Industry Research and Information Association (CIRIA), London.

Kelly AJ, Clifton AW, Antunes PJ and Widger RA (2005) Application of a landslide risk management system to the Saskatchewan highway network. In *Landslide Risk Management* (Hungr O, Fell R, Couture R and Eberhardt E (eds.)) Balkema, Rotterdam, pp. 571–580.

Koirala NP and Watkins AT (1988) Bulk appraisal of slopes in Hong Kong. In *Landslides* (Bonnard C (ed.)). Balkema, Rotterdam, vol. 2, pp. 1181–1186.

Lee EM (1999) Coastal Planning and Management: The impact of the 1993 Holbeck Hall landslide, Scarborough. *East Midlands Geographer* **21**: 78–91.

Lee EM (2003) Coastal change and cliff instability: development of a framework for risk assessment and management. PhD thesis, University of Newcastle upon Tyne.

Lee EM and Clark AR (2000) The use of archive records in landslide risk assessment: historical landslide events on the Scarborough coast, UK. In *Landslides: In Research, Theory and Practice* (Bromhead EN, Dixon N and Ibsen M-L (eds)). Thomas Telford, London, pp. 904–910.

Lee EM, Clark AR and Guest S (1998) An assessment of coastal landslide risk, Scarborough, UK. In *Engineering Geology: The View from the Pacific Rim: Proceedings of the 8th International Congress of the IAEG* (Moore D and Hungr O (eds)), Vancouver, pp. 1787–1794.

Ng KC, Parry S, King JP, Franks CAM and Shaw R (2002) *Guidelines for Natural Terrain Hazard Studies*. Geotechnical Engineering Office, Hong Kong. GEO Report 138.

Parry S, Ruse ME and Ng KC (2006) Assessment of natural terrain landslide risk in Hong Kong: an engineering geological perspective. *IAEG2006: Proceedings of the 10th IAEG International Congress: Proceedings of the 10th International Congress of the IAEG*. Nottingham, Paper 299.

Pierson LA, Davis SA and Van Vickle R (1990) *Rockfall hazard rating system implementation manual*. US Department of Transport, Federal Highway Administration Report No. FHWA-OR-EG-90-01.

Powell G (2002) Letters to the editor: discussion 'Landslide risk management concepts and guidelines'. *Australian Geomechanics* **37**: 45–53.

Rendel Geotechnics (1994) *Preliminary study of the coastline of the urban areas within Scarborough Borough: Scarborough Urban Area*. Scarborough Borough Council, Scarborough.

Saldivar-Sali A and Einstein HH (2007) A landslide risk rating system for Baguio, Philippines. *Engineering Geology* **91**: 85–99.

Sandilands NM, Noble M and Findlay JW (1998) Risk assessment strategies for dam based hydro schemes. In *The Prospect for Reservoirs in the 21st Century* (Tedd P (ed.)), Thomas Telford, London.

Sidle RC, Pearce AJ and O'Loughlin CL (1985) *Hillslope Stability and Land Use*. American Geophysical Union, Washington, DC.

Stewart IE, Baynes FJ and Lee IK (2002) The RTA guide to slope risk analysis version 3.1. *Australian Geomechanics* **37(2)**: 115–147.

VanDine DF, Jordan P and Boyer DC (2002) An example of risk assessment from British Columbia, Canada. In *Instability: Planning and Management* (McInnes RG and Jakeways J (eds)). Thomas Telford, London, pp. 399–406.

Wong HN and Ho KKS (1995) New priority classification system for soil cut slopes. Special Project Report No. SPR 6/95. Geotechnical Engineering Office, Civil Engineering and Development Department, Hong Kong.

Wong HN (2001) Recent advances in slope engineering in Hong Kong. *Proceedings of the 14th Southeast Asian Geotechnical Conference, Hong Kong* Vol. 1, pp. 641–659.

Wong HN (2003) Natural terrain management criteria – Hong Kong practice and experience. *Proceedings of the International Conference on Fast Slope Movements: Prediction and Prevention for Risk Mitigation*, Naples, Italy.

Wong HN (2005) Landslide risk assessment for individual facilities. In *Landslide Risk Management* (Hungr O, Fell R, Couture R and Eberhardt E (eds)) Balkema, Rotterdam, pp. 237–296.

Wong HN and Ko FWY (2006) *Landslide risk assessment: application and practice.* GEO Report No. 195. Geotechnical Engineering Office, Civil Engineering and Development Department, Hong Kong.

Wong HN, Ko FWY and Hui THH (2004) *Assessment of Landslide Risk of Natural Hillsides in Hong Kong.* Special Project Report No. SPR 5/2004, Geotechnical Engineering Office, Civil Engineering and Development Department, Hong Kong.

Landslide Risk Assessment
ISBN 978-0-7277-5801-9

ICE Publishing: All rights reserved
http://dx.doi.org/10.1680/lra.58019.129

Chapter 5
Introduction to probability and quantitative assessment

5.1. Uncertainty and probability
5.1.1 Uncertainty
Uncertainty is a characteristic feature of scientific studies because of the difficulty of achieving perfect knowledge. However, there are many views as to what constitutes uncertainty, some of which have already been discussed (see Chapter 1). Bradley (2011) suggested it was 'a catch-all term for all the ways one might fail to be certain about things'. The uncertainties associated with landsliding and its consequences are generally associated with

- *Randomness* (*aleatory uncertainty*, i.e. about games of chance), where the outcome is inherently unpredictable, like the roll of a set of dice. This type of uncertainty can describe the occurrence of landslide triggering events (e.g. rainfall or earthquake loading over a critical threshold value) that can be considered random in time and space. Aleatory uncertainty cannot be reduced by further study, since it expresses the inherent variability of a phenomenon.
- *Lack of knowledge* (*epistemic uncertainty*), where the uncertainty is associated with the level of understanding of a problem. These uncertainties are due to things we could in principle know but in practice do not. In many situations, the difficulty in estimating future landslide activity is due to a lack of knowledge about ground conditions, failure mechanisms and the circumstances that could lead to failure, rather than randomness. This is usually the case for rare or extreme events and the behaviour of complex pre-existing landslides. Epistemic uncertainty can be reduced with time as more data are collected and more research is completed, but it is unlikely ever to be eliminated. But the adverse consequence element of risk assessment suffers from the additional uncertainties of imperfect knowledge of future conditions.

It is arguable whether natural processes are random. For example, tropical revolving storm systems (e.g. cyclones, typhoons and hurricanes) are often treated as random (i.e. stochastic) processes. However, these storms occur in response to patterns of atmospheric change within the tropics, but, as current understanding of these regional weather patterns is limited (i.e. there is epistemic uncertainty), these storms are modelled as random processes because climatologists know too little about weather patterns to predict storm activity deterministically. The assumption of randomness is a convenience for modelling, not inherent in reality. Similarly, variability of soil properties may be modelled as if they arose from random processes rather than simply as a result of spatial variation that could be observed given enough ground investigation and testing.

Landslide specialists are familiar with uncertainty associated with the variability and complexity of the ground and groundwater conditions, potential failure mechanisms and the difficulty in characterising the engineering properties within and across a slope. This uncertainty influences the extent to which slope stability is predictable. For example, Ho and Lau (2010) describe how landslides have occurred on slopes in Hong Kong that were previously engineered to the required safety standards. These incidents highlight the difficulties in accurately predicting the performance of slopes developed in tropical residual soils due to uncertainties.

In landslide risk assessment, the uncertainties can be associated with the likelihood of both the threat (e.g. a rockfall from a road cutting) and its possible adverse consequences (e.g. the chance of a fatality given that the rockfall has occurred). Consider, for example, the impact of the rockfall on a passing vehicle. If the outcome (e.g. death of the driver) was the same every time a similar incident occurred, then the consequences would be entirely predictable. On the other hand, if the outcome of similar rockfalls varied with factors such as the nature of the impact on the vehicle, the make of the vehicle, the time of day and the number of occupants, then the consequences would clearly be more uncertain and less predictable.

5.1.2 Probability and probability rules

Probability is the branch of mathematical study focused on providing measures of uncertainty or the 'degree of predictability'.

Probability theory is a logical construct based on a limited number of rules or axioms.

- *Axiom 1*: the probability P of an event E is a non-negative number between 0 and 1.
- *Axiom 2*: the probability P of an event E that will *certainly* occur is 1.0:

$$P(E) = 1$$

It follows that $P(E) = 0$ indicates that the event will *certainly not* happen.

The relationship between probability and uncertainty is illustrated in Figure 5.1. There is no uncertainty at 0 or 1, because the outcome is certain, but uncertainty exists throughout the rest of the range and is at its greatest at 0.5 when it is not possible to differentiate between the relative likelihoods of the two outcomes (i.e. 'event' or 'non-event') occurring. Away from this point, the uncertainty decreases as the outcomes become more certain. Probabilities of 0.75 and 0.25 display the same level of uncertainty.

- *Axiom 3:* the probability of two mutually exclusive events (i.e. they share no points in common) is the sum of their probabilities:

$$P(A \cup B) = P(A) + P(B)$$

in which $\cup$ indicates the union of the two events.

For example, if the losses caused by landslide events were to be classified on an interval scale <£1 million, £1–5 million, £5–10 million and >£10 million, then these events would be mutually exclusive, since the loss can only fall within one class. This axiom can be illustrated using a simple Venn diagram (Figure 5.2(a)). Each set (i.e. circle) represents an outcome (e.g. a type of landslide), with the overall sample space (i.e. the entire diagram) defining all possible events.

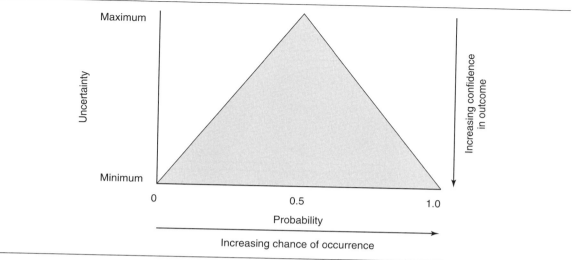

Figure 5.1 The relationship between probability and uncertainty. (After Lee, 2009)

For potentially unstable slopes there are two possible outcomes – failure or no failure (for pre-existing landslides, the equivalent outcomes are movement or no movement). These two outcomes are mutually exclusive and collectively exhaustive (i.e. they encompass all possibilities), so the probability of 'an event' occurring, together with the probability of 'no event' occurring, must equal 1.0 (Figure 5.2(b)):

$P(\text{event}) + P(\text{no event}) = 1$

This can also be expressed as

$P(\text{event}) = 1 - P(\text{no event})$

In Figure 5.2(a), the events were mutually exclusive and there was no overlap between the two sets. However, some events are not mutually exclusive and the sets may overlap (Figure 5.2(c)). For example, a landslide-related fatality may occur in a road cutting as a result of either a rockfall (R) or debris flow (DF), both of which could occur during the same rainstorm or earthquake; in this case, the fatality will occur as a result of only one of the events. If the occurrence of one event makes it neither more nor less probable that the other occurs, then the rockfall and debris flow are statistically independent events.

In this case, the overall probability of a fatality will need to take account of the overlap on the Venn diagram, because when $P(R)$ and $P(DF)$ are added, the probability of the intersection ($\cap$) is added twice. The intersection is defined as

$P(R \cap DF) = P(R) \times P(DF)$

To compensate for that double addition, the intersection needs to be subtracted:

$P(\text{fatality, R or DF}) = P(R) + P(DF) - P(R \text{ and } DF)$

Figure 5.2 Venn diagrams to explain probability concepts. (a) Mutually exclusive events: $P(A \cup B) = P(A) + P(B)$. (b) Mutually exclusive and collectively exhaustive events: $P(\text{event}) + P(\text{no event}) = 1$. (c) Non-mutually exclusive events: $P(\text{fatality, R or DF}) = P(R) + P(DF) - P(R \cap DF)$. (d) Conditional probability: $P(B|A) = P(A \cap B)/P(A)$. (e) Conditional probability: $P(B|A) = P(B)/P(A)$. (f) Conditional probability: $P(D|C) = P(D)/P(C)$

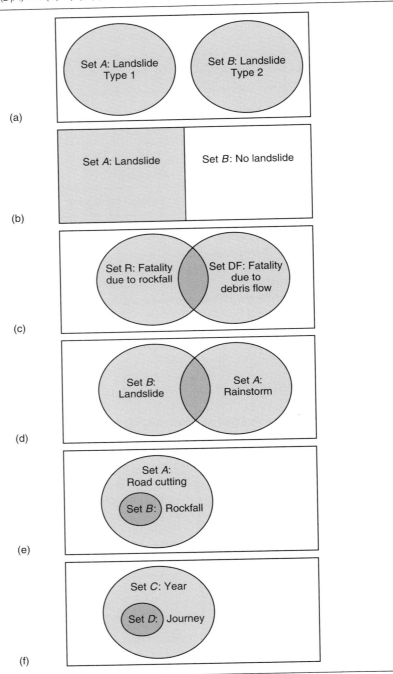

which is

$$P(\text{fatality, R or DF}) = P(R) + P(DF) - [P(R) \times P(DF)]$$

or

$$P(\text{fatality, R or DF}) = 1 - \{[1 - P(R)] \times [1 - P(DF)]\}$$

For example, if the probabilities of a fatality in the hypothetical road cutting due to a rockfall and a debris flow were 0.2 and 0.3, respectively, then

$$P(\text{fatality, R or DF}) = P(R) + P(DF) - [P(R) \times P(DF)]$$

$$= 0.2 + 0.3 - (0.2 \times 0.3)$$

$$= 0.44$$

$$P(\text{fatality, R or DF}) = 1 - \{[1 - P(R)] \times [1 - P(DF)]\}$$

$$= 1 - [(1 - 0.2) \times (1 - 0.3)]$$

$$= 0.44$$

A similar issue would arise if a landslide event could be initiated by two or more potential triggering events, such as an earthquake, rainstorm or toe erosion. The basic formula for calculating the probability of an outcome arising from three or more different events is

$$P(\text{outcome}) = 1 - \{[1 - P(\text{event 1})] \times [1 - P(\text{event 2})] \times [1 - P(\text{event 3})]\}$$

Many events are dependent on the occurrence of preconditioning factors and initiating events. This gives rise to conditional probability (i.e. the events are conditional on the occurrence of these factors). For example, the conditional probability that a landslide (event B) occurs in response to an intense rainstorm (event A) is defined as (Figure 5.2(d))

$$P(B|A) = \frac{P(A \cap B)}{P(A)}$$

The vertical line '|' denotes 'given', as in *given that the triggering event (A) has occurred*. $P(A)$ represents the probability of the rainstorm and $P(A \cap B)$ the probability that both the rainstorm and the landslide occur (i.e. the overlap between sets A and B). Note that the occurrence of the rainstorm does not necessarily lead to generation of the landslide. This equation can be rewritten to define the joint probability of A and B as

$$P(A \cap B) = P(B|A) \times P(A)$$

This is known as the *multiplication rule for joint probability* and provides the arithmetic framework for calculating the probability along a conditional sequence of events using event trees (see Section 5.4.2). The probability of each branch is the joint probability of the events that it contains.

For example, the probability of damage to a building sited on a debris flow pathway can be represented by a simple conditional probability model:

$$P(\text{damage}) = P(\text{landslide}) \times P(\text{hit}|\text{landslide}) \times P(\text{damage}|\text{hit})$$

where:

- $P(\text{landslide})$ is a measure of the expected likelihood of a debris flow along the channel; for example, $P(\text{landslide}) = 0.25$.
- $P(\text{hit}|\text{landslide})$ is the probability of a 'hit' *given that the debris flow event occurs*; for example, $P(\text{hit}|\text{landslide}) = 0.01$.
- $P(\text{damage}|\text{hit})$ is the probability of damage *given that a 'hit' has occurred*. It is a measure of the chance (0 to 1) that the building would be damaged by the event; for example, $P(\text{damage}|\text{hit}) = 0.5$.

Using this simple model,

$$P(\text{damage}) = 0.25 \times 0.01 \times 0.5$$

$$= 0.00125$$

Where more than one branch of an event tree produces the same outcome (e.g. the likelihood of a fatality), then the total probability of this outcome is the union of the relevant branch probabilities:

$$P(\text{fatality}) = P(\text{branch 1}) + P(\text{branch 2})$$

For example, a series of mutually exclusive and collectively exhaustive volumetric classes could be used to model rockfall activity along a road cutting: $<100 \text{ m}^3$, $100\text{–}500 \text{ m}^3$ and $>500 \text{ m}^3$. Each event could cause a fatality, and is represented by a different branch along an event tree. The analysis reveals that the outcome probability increases with the event size, from 0.01 to 0.1 and 0.5, respectively:

$$P(\text{fatality}) = P(\text{branch 1}) + P(\text{branch 2}) + P(\text{branch 3})$$

$$= 0.01 + 0.1 + 0.5$$

$$= 0.61$$

In Figure 5.2(e), the set B is a subset of set A, with every element of B also an element of A. In this instance, the conditional probability of B given the occurrence of A is

$$P(B|A) = \frac{P(A \cap B)}{P(A)}$$

As the set B lies entirely within set A, then $P(A \cap B)$ is the same as $P(B)$ and

$$P(B|A) = \frac{P(B)}{P(A)}$$

If, for example, a 1 m wide rockfall (B) occurs somewhere within a 1000 m-long road cutting (A), then the conditional probability of its occurrence at a particular point (e.g. a bus shelter or a stationary vehicle) is

$P(B|A) = 1/1000$ (expressed in m)

$\quad\quad\quad = 0.001$

If a vehicle is making a single journey through the cutting, then the time spent in the cutting can be described as the overlap between the journey time (D) and a larger set of time such as a year (C; Figure 5.2(f)):

$$P(D|C) = \frac{P(C \cap D)}{P(C)}$$

As the journey time set D lies entirely within the larger set C, then $P(C \cap D)$ is the same as $P(D)$. Thus, if the vehicle travels through the 1000 m long cutting at 30 km/h, then

$P(D|C)$ = length of cutting (m)/speed (m/h)

$\quad\quad\quad = 1000/30\,000$

$\quad\quad\quad = 0.033$ hours

If C is assumed to be a whole year, then

$P(D|C) = 0.033/(365 \times 24)$

$\quad\quad\quad = 0.0000038$

$\quad\quad\quad = 3.80\text{E-}06$ years

These last two examples for the road cutting together define the exposure of a vehicle to a rockfall event. This involves a combination of being in the 'wrong place' (i.e. in the danger zone) at the 'wrong time' (i.e. the occupancy period within the danger zone):

$P(\text{wrong place and time}) = P(\text{wrong place}) \times P(\text{wrong time})$

$\quad\quad\quad\quad\quad\quad\quad\quad\quad\quad\quad = P(B|A) \times P(D|C)$

$\quad\quad\quad\quad\quad\quad\quad\quad\quad\quad\quad = 0.001 \times 3.80\text{E-}06$

$\quad\quad\quad\quad\quad\quad\quad\quad\quad\quad\quad = 3.80\text{E-}09$

5.1.3 The meaning of probability

The probability axioms and theorems define the mathematics of probability, but do not address the meaning of the numbers that are generated. Probability can be used to describe uncertainties associated with the random occurrence of events over time; for example, it is known that a flood event will occur, but not when. However, it can also be used to describe uncertainties arising from an incomplete knowledge or understanding of the circumstances that could lead to an event. For example, there is uncertainty as to whether an event will occur, although it is believed that it could.

These two examples of contrasting interpretations of probability reflect the existence of a crucially important dichotomy.

- *Frequentist*: the relative frequency of occurrence of an event in a number of repetitions of the experiment is a measure of the probability of that event. The probability value is seen as being objective because it exists in the real world and is, in principle, measurable by doing the experiment.
- *Degree of belief (Bayesian)*: the probability of an uncertain event is the quantified measure of the practitioner's belief or confidence in the outcome, according to their state of knowledge at the time. Probability under this interpretation is often described as subjective because it is a person's degree of belief rather than a feature of the real world.

These fundamentally different interpretations lead to what has been called the 'Janus-faced' nature of probability, after the Roman god with two faces (Hacking, 2006). Christian (2004) suggested that in the 'frequentist' interpretation, probability is inherent in the state of nature and that it is the role of the analyst to estimate it, whereas in the 'degree of belief' interpretation, probability is in the mind and it is the analyst's role is to elicit this belief.

5.1.4 The frequentist interpretation

The 'frequentist' interpretation defines an event's probability as the limit of its relative frequency in a large number of trials. The probability of an event is a function of the number of favourable outcomes, such as throwing a '6' with a single die, compared with the total number of *possible* outcomes (there are six sides to the die):

$$P(\text{event}) = \frac{\text{number of favourable outcomes}}{\text{total number of possible outcomes}}$$

$$= \tfrac{1}{6} = 0.1667$$

In most situations, it is not possible to define probability in such precise terms, because the chance of each outcome is not equal and may be dependent on a complex range of other factors. Therefore, an alternative way of expressing probability is in terms of the frequency that a particular outcome occurs during a particular number of trials or experiments:

$$P(\text{event}) = \frac{\text{number of favourable outcomes}}{\text{total number of trials}}$$

There are a limited number of outcomes when throwing dice. The same is true of landslides. The probability distribution for landslide activity on a potentially unstable slope is shown in Figure 5.3. As described above, the two possible outcomes – event or no event – are mutually exclusive and collectively exhaustive, so

$$P(\text{event}) + P(\text{no event}) = 1$$

$$P(\text{event}) = 1 - P(\text{no event})$$

$$P(\text{event}) = 1 - [1 - P(\text{event})]$$

It is common practice to use units of time, such as years, as individual trials. The chance of a landslide occurring in a particular year is therefore expressed in terms of an *annual probability*. Thus, the

Figure 5.3 Discrete events

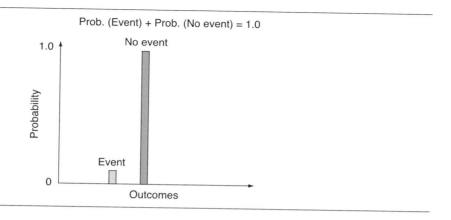

frequentist probability of a landslide is

$$\text{annual probability (landslide)} = \frac{\text{number of recorded landslides}}{\text{length of record period in years}}$$

For example, if records for a particular stretch of cliffline reveal three landslides in the past 300 years, then

$$\text{annual probability} = \frac{3}{300}$$

$$= 0.01, \text{ or } 1\%$$

It is important to appreciate that a 1% annual probability of occurrence also means a 99% probability of non-occurrence in any year. However, an annual probability of 0.01 (1 in 100 or 1%) does not mean that an event will occur once in a 100 years or every 100th year.

In the 'frequentist' interpretation, the historical record (i.e. the past) is a guide to the future. The future probability can be revealed by statistical analysis of time series data (i.e. the data do the work). This assumes that the number of times any particular event has occurred in a large number of trials or years (i.e. its *relative frequency*) converges to a limit as the number of repetitions or years increases – that is, the longer the time period, the closer the revealed frequency is to the 'true' value. The relative frequency that it is based upon is the result of actual observations of how things behave, whether or not we understand the causes of the behaviour. It is widely used to produce spatial representations of frequency or probability of atmospheric hazards, such as tropical revolving storms (hurricanes, typhoons, etc.) and tornado paths on the continental USA.

An essential prerequisite is a data series of landslide events for the area of interest. For example, an inventory of landslide features on the Brooks Peninsula, Vancouver Island, was compiled from the interpretation of 1 : 15 000 to 1 : 31 680 scale aerial photographs flown between 1950 and 1996 (Guthrie and Evans, 2004a). A total of 136 debris flows and debris slides were identified as having occurred over the 46-year period, indicating a landslide frequency of

relative frequency = (number of recorded landslides)/(time period)

$$= 136 \text{ landslides}/46 \text{ years}$$

$$= 2.95 \text{ landslides/year}$$

For the surveyed area (286 km^2), the landslide frequency is 0.01 landslides per km^2 per year.

However, the 'frequentist' approach can only be strictly applied if uniformitarian principles apply both over the relevant period of the past *and* through the present into the future. In other words, 'the past can be a guide to the future' only if the controls on conditions have remained relatively uniform, both over the past and into the future.

The relative frequency approach underpins the application of quantitative risk assessment (QRA) in sections of the oil and gas industry. For example, the 6th Report of the European Gas Pipeline Incident Data Group (EGIG, 2005) provides statistics on the incident frequency (from pin-hole crack to rupture) for cross-country pipelines in Europe between 1970 and 2004: a total exposure of 2.77×10^6 km-years; this is a measure used to express the size of the pipeline network and the number of years it has been in operation (e.g. if a 10 km pipeline had been in operation for 10 years, this would be the equivalent of an exposure to threats for 100 km-years). The overall incident frequency is 0.4 incidents per 1000 km per year. The major causes of incidents are external interference (50%), construction defect or material failure (17%), and corrosion (15%). The frequency of pipeline ruptures due to ground movement is 0.028 per 1000 km per year, of which 52% have been caused by landslides (0.015 per 1000 km per year). This value of 0.015 per 1000 km per year might be used as the expected rupture frequency due to landslides in a pipeline risk assessment, regardless of the terrain. A similar database for oil pipelines in western Europe is maintained by CONCAWE ('Conservation of Clean Air and Water in Europe' – the oil companies' European association for environment, health and safety in refining and distribution), and indicates a long-term spill frequency of 0.52 per 1000 km per year, of which 2.3% were caused by landslides (0.012 per 1000 km per year) (CONCAWE, 2007).

However, the EGIG and CONCAWE statistics relate to incidents across all types of terrain and hence potentially underestimate the rupture rate in landslide-prone terrain. BP has assembled a database for landslide ruptures in tropical Andean mountains (mainly Colombia; see Table 5.1) (Sweeney, 2005; Sweeney *et al.*, 2005). For older pipelines in these mountains, the rupture rate of 2.8 per 1000 km per year is nearly 200 times more frequent than the EGIG and CONCAWE 'all terrains' figure. Even for pipelines built with the benefit of modern geo-engineering best practice, the frequency is 0.33 per 1000 km per year – that is, a rate that is over 20 times greater than the EGIG and CONCAWE 'all terrains' rupture frequency. The comparable figure for the recently commissioned Camisea natural gas liquids pipeline in Peru is 2.7 per 1000 km per year (four events over 500 km of mountainous or hilly terrain in three years since 2004) (Lee *et al.*, 2009).

These reported rupture frequencies highlight an inherent problem with the relative frequency approach, where either the datasets are so generalised as to be almost meaningless or they are very site-specific. The relative frequency interpretation precludes any extrapolation to situations beyond the boundaries of the data series. Under a strict relative frequency interpretation, it is simply not possible to make a judgement as to where a pipeline in southern China is more likely to fall in the

Table 5.1 Landslide rupture frequencies on Andean pipelines

Pipeline	Year built	Landslide ruptures	Difficult terrain length: km	Frequency: ruptures per 1000 km per year
Cano Limon–Covenas	1985	9	220	2.56
Central Llanos	1987	5	190	1.88
Transecuatoriano	1972	25	264	3.27
Trans Panama	1982	1	60	0.88
Apiay–Bogata	1988	1	104	0.74
Oleoducto de Colombia	1991	0	116	0
Ocensa I	1994	0	60	0

From Sweeney et al. (2005)

range of Andean rupture statistics, or what sub-population of the European data set to use as an analogy. If there is no database of landslide related ruptures for an area, then it is not possible to generate rigorous, objective estimates of rupture frequency.

The landslide history within the Nile Delta illustrates a common problem with the 'frequentist' approach (Lee, 2009). It is one of the world's largest deltas, with a submarine fan of about 100 000 km^2 extending northwards into the Mediterranean Sea, and is a major area of hydrocarbon exploration (e.g. Moore et al., 2007). The delta has been subject to cycles of rapid sediment deposition, erosion and episodic submarine landslide activity for at least the last 250 000 years (e.g. Loncke 2002). Within the last 30 000 years, there has been a sequence of major landslide events that have generated large turbidity currents. This event history has been captured as a series of large turbidite deposits on the abyssal plain, the Herodotus Basin (Reeder et al., 1998).

Around 500 km^3 of turbidite sediment has been generated from the West Nile delta over the period 30 000 to 6500 BP (calendar years; see Table 5.2) (Reeder et al., 1998, 2002). This sequence requires

Table 5.2 West Nile delta turbidites recorded in the Herodotus Basin

Years BP: thousand years BP, adjusted ^{14}C dates	Event name	Herodotus Basin turbidite volume: km^3
7.0	a	0.1
7.5	b	6
8.0	c	80
8.5	Debrite	2.4
10.5	d	126
11.9	e	25.2
12.6	f	0.1
14.8	g	72
>28.8	o	190

From Reeder et al. (1998, 2002)

at least nine major landslide events since 30 000 BP, including at least two extremely large landslide events (>100 km³ volume; 190 km³ at 30 000 BP and 126 km³ at 9000 BP). Using this data series suggests that the historical frequency of events over the last 30 000 years has been around 1 event per 3000 years.

However, over this period, significant changes have occurred to the environmental controls on landslide activity in the West Nile delta, suggesting that the 1 per 3000 years historical frequency should not be used as a basis for estimating the future probability. A simple conceptual model that highlights how the West Nile delta landslide regime is influenced by broader environmental controls in presented in Figure 5.4. The main controls are as follows:

- The *sediment load of the Nile* influences the sedimentation rate on the delta slope through the influence on the frequency of hyperpycnal (high-density river discharges that flow along the seabed as turbidity currents) and hypopycnal (low-density river discharges that create suspended sediment plumes) flow activity (e.g. Mulder and Syvitski, 1995). Since the commissioning of the Aswan High Dam (AHD) in 1964, the downstream discharge has been reduced to 50% of 19th-century flows and 64% of early 20th-century flows. Irrigation and evaporation downstream of Aswan result in water discharges to the sea of about 5 km³/year (i.e. 10% of water released from the AHD: 55 km³/year). The contemporary sediment input to the coastal zone is probably less than 5×10^6 m³/year, compared with over 100×10^6 m³/year in the 19th century.
- The *sea level relative to the shelf edge* determines the shelf width and water depth across the shelf, which control the rate of offshore seabed sediment transport across the shelf and the loadings (e.g. wave-induced) on the delta slope sediments.
- The *pore pressure regime within the delta slope sediments* is a function of the sedimentation rate and permeability of the materials, and controls the rate of landslide activity.

Both the sediment load and sea level have changed significantly over time, resulting in variations in sedimentation rates, the pore pressure regime and the frequency of landslide events. A number of discrete time periods can be defined within the last 30 000 years within which differing combinations of environmental controls have profoundly influenced the seabed process regime and the susceptibility of the delta slopes to landslide activity:

- In the Weichselian (Last Glacial) Low Stand (30 000 – 15 000 BP) there were very high sediment loads and low relative sea-levels, rapidly falling to −120 m.
- From the Weichselian Termination to the Mid Holocene (15 000 – 6000 BP), there were very high sediment loads and low relative sea levels, rapidly rising from −120 m.
- From the Mid Holocene to the twentieth century (6000 BP to pre-AHD), there were low sediment loads and high relative sea levels, slowly rising to the present level.

The conditions over the next 100 years or so are expected to be controlled by negligible Nile sediment loads (post construction of the AHD) and relative sea levels rising by around 1 m. Very slow sedimentation rates are expected, with a very low, dissipating pore pressure regime. As a result, major landslide activity could be expected to be significantly lower than that experienced during the Weichselian Termination to Mid Holocene period. The probability of future major landsliding is therefore likely to be considerably lower than that suggested by the 1/3000 frequency produced by the historical record. However, this conclusion does not follow from a strict application of relative frequency statistics, since it draws on the use of judgement and knowledge of the controls on landslide activity.

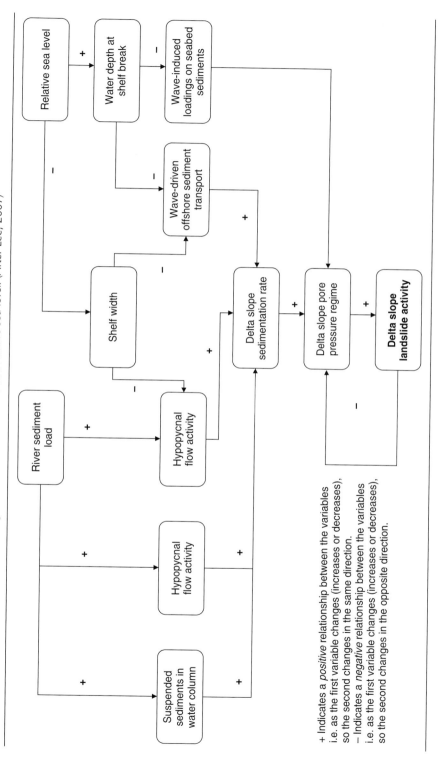

Figure 5.4 Submarine landslide influence diagram: river sediment loads and relative sea level. (After Lee, 2007)

5.1.5 The 'degree of belief' interpretation

In the 'degree of belief' interpretation, understanding the circumstances associated with landslide events – either past or hypothetical – is the key to the future. 'Degree of belief' probability is a measure of an individual's belief that a particular event will occur, given their state of knowledge at that time. This approach is widely used in QRA to evaluate unique or 'one-off' events for which there is no obvious historical precedent and also in areas where there has been a history of landsliding but no time series data exist. In contrast to the 'frequentist' approach, there is no minimum requirement for landslide data.

For example, on a section of the protected glacial till cliffs on the Whitby coast in the UK, the occurrence of a landslide event is conditional on seawall failure (e.g. Lee, 2009). The probability of seawall failure was estimated to be 0.01, based on an assessment of seawall condition made in accordance with the Environment Agency's National Sea and River Defence Survey's manual (EA, 1999). This manual provides a grading system for each type of defence on a scale of 1 to 5. The descriptions for each defence type are typically based on the scale of defects, the requirement for maintenance or remedial works, the urgency of work required, and a quantification of the defects in terms of length, area and height. The manual provides descriptions and includes photographic examples to help give each asset a condition score.

A 'degree of belief' probability value is simply the quantified expression of engineering or geotechnical judgement. There is no 'true' 'degree of belief' probability value; it is a function of the particular individual's or group's state of knowledge and beliefs. Different individuals or groups may make different estimates, based on the same evidence or knowledge.

The 'degree of belief' interpretation, somewhat confusingly, includes both logical (i.e. objective) and subjective views of probability (e.g. Gillies 2000). According to the logical interpretation, it is possible to derive a clear and indisputable probability value through rational evaluation of the available information. In this view, there is only one 'degree of belief' that would be rational to hold. Given the same information and knowledge, all experts would be expected to come to the same conclusion about the probability of an event (i.e. probability under this interpretation is subjective because it is a person's 'degree of belief' rather than a feature of the world, it is, however, objective in the sense that there is only one 'degree of belief' that it is rational to hold). To most clients with experience of employing experts, this might appear as a utopian ideal. However, as the applied geosciences rely heavily on induction, interpretation and judgement, the objectivist approach is not widely applicable in estimating landslide probability.

In the subjective interpretation, equally rational experts may hold different beliefs, and hence there is no single 'correct' probability value, even though a single figure may be produced by consensus. This is the real world for most applied geoscientists, with probability being a statement of belief about uncertainty.

5.1.6 Probability interpretations for landslide risk assessment

Probability calculus does not recognise the distinction between 'frequentist' and 'degree of belief' interpretations, and both are used interchangeably in everyday life. However, within the context of QRA, the existence of contrasting routes to estimating probability values leads to considerable disagreement over their relative importance. To many, the mathematical rigour of statistical analysis of data sets far outweighs the judgemental, non-repeatable basis of subjective probability. Others consider that mindless number-crunching comes a poor second to the application of knowledge and personal experience. The reality is that the application of QRA for landslides needs to accommodate both interpretations.

Both the relative frequency and 'degree of belief' interpretations have been used to provide a measure of the uncertainty of future landslide activity (e.g. Lee, 2009). This is because landslide problems generally include both aleatory and epistemic uncertainties. However, the balance between these sources uncertainty tends to vary with the nature of the landslide environment. In Hong Kong, for example, landslide activity on natural terrain tends to be dominated by frequent first-time failures, triggered by heavy rainfall (over 8000 landslides in the last 50 years within an area of 640 km^2) (Evans and King, 1998; Evans et al., 1997; King, 1997, 1999). The main uncertainty is associated with the random nature of the landslide process over time (i.e. the occurrence of triggering events) and space (i.e. the varying susceptibility of the underlying geology and slope classes). This type of landslide activity lends itself to statistical analysis, provided that there is an adequate time series of past events and no major influences of environmental change.

The landslide environment of Great Britain is very different, dominated as it is by ancient pre-existing landslides (Jones and Lee, 1994). These features are believed to be largely inherited from the later Pleistocene, when repetitive fluctuations from glacial to interglacial conditions resulted in periods of partial permafrost melt and enhanced freeze–thaw activity, leading to widespread landsliding on valley sides and flanks of hills and along escarpments (pre-existing slides probably account for over 90% of the 7533 inland landslides recorded by the National Landslide Review; Jones and Lee, 1994). Little is known about the ground conditions or reactivation behaviour of the vast majority of these landslides, and hence it is an environment dominated by epistemic uncertainty. Estimates about the annual probability of reactivation generally rely heavily on judgement and tend to be subjective.

Uncertainties associated with the consequences of landslides are usually epistemic, since it is rare to have access to databases that relate landslide properties such as volume or speed of movement to damage. Although standard curves that link floodwater depth to property damage have been developed to assist the estimation of flood risks (e.g. Penning-Rowsell et al., 2003), this is not an approach that appears to have been practical for landslide risk. Assessments of the probability of a fatality occurring if a vehicle is hit by a falling rock are not based on repeated experiments using crash test dummies or datasets of past incidents (including non-fatal accidents). The vulnerability of an individual to a rockfall is usually a 'degree of belief' probability, a judgement that is typically bounded by the event size that nobody could survive, $P(fatality) = 1$, and one that is too small to have any chance of causing death, $P(fatality) = 0$.

5.2. Probability distributions
5.2.1 The binomial distribution (yes/no events with a given probability)
Although the annual probability of a discrete event may be relatively low (e.g. the 0.01 probability of seawall failure at Whitby), over time the likelihood of the event occurring increases. This relationship between event probability and time can be modelled using a binomial experiment (Bernoulli trial). This is a statistical experiment that, in its simplest form, can be used to predict the outcome of tossing a coin and has the following properties.

- The experiment consists of repeated trials.
- Each trial can result in just two possible outcomes: heads or tails (i.e. yes/no).
- The probability of success is constant: 0.5 on every coin-tossing trial.

- The trials are independent – that is, getting heads on one trial does not affect whether heads occurs on the other trials.

The model can be used to determine the probability of getting heads (the event) during a sequence of coin-tossing trials, as follows:

$$P(\text{event}) \text{ in } n \text{ trials} = 1 - [1 - P(\text{event})]^n$$

$$= 1 - (1 - 0.5)^n$$

In two trials there will be a 75% chance of getting heads, in five trials a 97% chance and in ten trials a 99.9% chance.

If the occurrence or non-occurrence of landslides is assumed to be essentially the same as tossing a coin, then the binomial model can be used to calculate the probability of a single event, whose annual probability is known, occurring in a defined period of time (years). The Whitby seawall can be taken as an example.

- The experiment consists of n repeated trials. Each year in a time period is considered to be a separate trial – that is, 100 years would represent 100 trials.
- Each trial can result in just two possible outcomes: a failure or a non-failure.
- The probability of failure is the same on every trial and is defined by the annual probability (0.01).
- The trials are independent in that the outcome on one trial does not affect the outcome on other trials – this is the reverse of the 'gambler's fallacy', where the probability of one event somehow reflects the previous outcomes.

Thus, for a 20-year period,

$$P(\text{event}) \text{ in } 20 \text{ years} = 1 - (1 - 0.01)^{20}$$

$$= 1 - (0.99)^{20}$$

$$= 1 - 0.8179$$

$$= 0.1821 \text{ or } 18\%$$

This indicates an 82% probability that an event will not occur in the 20-year period. But note that the probability of occurrence rises from 0.01 (1%) for a single year to 0.095 (9.5%) for 10 years and 0.18 (18%) for a 20-year period, even though the annual probability has remained the same. The longer the time period, the greater the likelihood of an event occurring, although there is clearly not a linear relationship between the two. Recalculating the above equation for periods of 50 years and 100 years reveals that the probability of a 0.01 annual probability event occurring has risen to 39.5% and 63.4%, respectively, the latter clearly showing that the '100-year event' does not necessarily occur in any 100-year period.

The accumulation of probability over time is known as the *cumulative probability* (Figure 5.5). This can help define an annual probability by working back from a judgement of the expected time period over

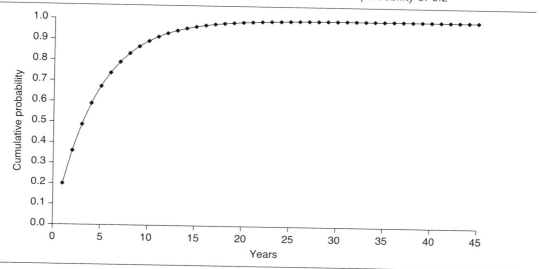

Figure 5.5 Cumulative probability over 45 years for an event with an annual probability of 0.2

which a single failure would be expected to have occurred. For example, if a slope is judged to have an estimated 95% chance of failure in the next 14 years, it is possible to 'track back' to yield an estimated annual probability of 0.2, as demonstrated in Table 5.3, using the equation

annual $P(\text{event}) = 1 - (1 - \text{probability event in } x \text{ years})^{1/n}$

$$= 1 - (1 - 0.95)^{1/14}$$

$$= 0.2$$

Table 5.3 shows that there is an 100% chance of failure within 45 years and that the probability of occurrence in any year gets progressively smaller as the years pass, despite the annual probability remaining constant at 0.2. This is because the probability of an event occurring in any year needs to take into account the possibility that the event has actually occurred in the previous years – that is, it is a 'one-off' event that may have already happened. Thus, the annual probability for a one-off event occurring in year 2 (and subsequent years) needs to be modified as follows:

probability of failure in year n = annual probability of failure

× (cumulative P failure not occurred by year n)

The binomial experiment can also be very useful in modelling the impact of repeated events on assets at risk. For example, a 5 m-long bench has been placed on a seaside promenade at the base of an unstable cliffline. Local authority records suggest a minimum rockfall frequency of five incidents per year along the 1000 m long cliffline. Each rockfall can be represented by a separate trial with two possible outcomes: impact with the bench or no impact. The probability of one or more impacts is related to the probability of the rockfall in a specific trial hitting the bench (i.e. the bench being in the 'wrong place') and the number of falls per year. The probability of the bench

Table 5.3 The probability of an event with an annual probability of 0.2 (1 in 5) occurring over a 45-year time period

Year n	Annual probability	Cumulative probability[a]	Probability of occurrence[b] in year n	Probability not occurred[c] in n years
1	0.2	0.2000	0.2000	0.8000
2	0.2	0.3600	0.1600	0.6400
3	0.2	0.4880	0.1280	0.5120
4	0.2	0.5904	0.1024	0.4096
5	0.2	0.6723	0.0819	0.3277
6	0.2	0.7379	0.0655	0.2621
7	0.2	0.7903	0.0524	0.2097
8	0.2	0.8322	0.0419	0.1678
9	0.2	0.8658	0.0336	0.1342
10	0.2	0.8926	0.0268	0.1074
11	0.2	0.9141	0.0215	0.0859
12	0.2	0.9313	0.0172	0.0687
13	0.2	0.9450	0.0137	0.0550
14	0.2	0.9560	0.0110	0.0440
15	0.2	0.9648	0.0088	0.0352
16	0.2	0.9719	0.0070	0.0281
17	0.2	0.9775	0.0056	0.0225
18	0.2	0.9820	0.0045	0.0180
19	0.2	0.9856	0.0036	0.0144
20	0.2	0.9885	0.0029	0.0115
21	0.2	0.9908	0.0023	0.0092
22	0.2	0.9926	0.0018	0.0074
23	0.2	0.9941	0.0015	0.0059
24	0.2	0.9953	0.0012	0.0047
25	0.2	0.9962	0.0009	0.0038
26	0.2	0.9970	0.0008	0.0030
27	0.2	0.9976	0.0006	0.0024
28	0.2	0.9981	0.0005	0.0019
29	0.2	0.9985	0.0004	0.0015
30	0.2	0.9988	0.0003	0.0012
31	0.2	0.9990	0.0002	0.0010
32	0.2	0.9992	0.0002	0.0008
33	0.2	0.9994	0.0002	0.0006
34	0.2	0.9995	0.0001	0.0005
35	0.2	0.9996	0.0001	0.0004
36	0.2	0.9997	0.0001	0.0003
37	0.2	0.9997	0.0001	0.0003
38	0.2	0.9998	0.0001	0.0002
39	0.2	0.9998	0.00004	0.0002
40	0.2	0.9999	0.00003	0.00013
41	0.2	0.9999	0.00003	0.00011
42	0.2	0.9999	0.00002	0.00009
43	0.2	0.9999	0.00002	0.00007
44	0.2	0.9999	0.00001	0.00005
45	0.2	1.0000	0.0000	0.0000

[a]Cumulative probability calculated as follows: probability(n years) $= 1 - (1 - 0.2)^n$
[b]Probability of occurrence in year n calculated as probability(year n) = cumulative probability(year n) − cumulative probability(year $n - 1$)
[c]Probability not occurred in n years calculated as probability(no event, n years) $= 1 -$ cumulative probability(n years)

being in the wrong place is then

$$P(\text{wrong place}) = \text{length of bench/length of cliffline}$$

$$= 5/1000$$

$$= 0.005$$

and the probability of a rockfall impacting the bench is

$$P(\text{impact}) = 1 - [1 - P(\text{wrong place})]^{\text{number of falls/year}}$$

$$= 1 - (1 - 0.005)^5$$

$$= 0.024$$

The binomial model can be expanded into a *binomial series* to predict the number of favourable outcomes x (i.e. landslides) in n trials in which the probability P remains constant in each trial, where the binomial probability $b(x; n, P)$ (the probability that an n-trial binomial experiment results in *exactly* x successes, when the probability of success on an individual trial is P) is given by

$$b(x; n, P) = {_nC_x} \times P^x \times (1 - P)^{n-x}$$

where ${_nC_x}$ is the binomial coefficient, stated as 'n choose x', or the number of possible ways to choose x 'successes' from n observations, which is given by

$$_nC_x = \frac{n!}{x! \times (n - x)!}$$

where ! indicates the factorial of the integer (e.g. $3! = 3 \times 2 \times 1 = 6$).

As an example, suppose a die is tossed five times. What is the probability of getting exactly two '4's? In this instance, the number of trials $n = 5$, the number of successes $P = 2$, and the probability of success on a single trial is 1/6 or 0.167. The binomial coefficient is then

$$_5C_2 = \frac{5!}{2! \times (5 - 2)!}$$

$$= \frac{120}{2 \times 6}$$

$$= 10$$

Therefore, the binomial probability is

$$b(2; 5, 0.167) = {_5C_2} \times (0.167)^2 \times (1 - 0.167)^{5-2}$$

$$= 10 \times (0.167)^2 \times (0.833)^3$$

$$= 0.161$$

Consider the situation where the promenade bench described earlier is protected by a small rock catch fence that traps the small to medium-sized rockfalls, but not the larger ones. The probability that a rockfall is trapped by the fence is 0.3. If five rockfalls occur at this site in a year, the probability that at most four of the five rockfalls are trapped (i.e. the number trapped ranges from 1 to 4) can be calculated by computing four individual probabilities (for $x = 1$ rockfall out of 5 trapped, $x = 2$ rockfalls, $x = 3$ rockfalls, $x = 4$ rockfalls). In this instance, the number of trials $n = 5$, the number of successes $P \leq 4$. The binomial coefficient in this example will vary with the number of successes. For the case of one success

$$_5C_1 = \frac{5!}{1! \times (5-1)!}$$

$$= \frac{120}{1 \times 24}$$

$$= 5$$

For the case of two successes, the coefficient becomes

$$_5C_2 = \frac{5!}{2! \times (5-2)!}$$

$$= \frac{120}{2 \times 6}$$

$$= 10$$

And so on to four successes, where the coefficient is 5.

The probability that up to four rockfalls are trapped by the rock fence is the sum of the probabilities for the four cases (one success, two successes, three successes and four successes):

$$b(x \leq 4; 5, 0.3) = b(x = 1; 5, 0.3) + b(x = 2; 5, 0.3) + b(x = 3; 5, 0.3) + b(x = 4; 5, 0.3)$$

$$= \left[5 \times 0.3^1 \times (1-P)^4\right] + \left[10 \times 0.3^2 \times (1-P)^3\right]$$

$$+ \left[10 \times 0.3^3 \times (1-P)^2\right] + \left[5 \times 0.3^4 \times (1-P)^1\right]$$

$$= 0.36 + 0.308 + 0.13 + 0.028$$

$$= 0.82$$

For given values of n and p, the set of probabilities of this form for $x = 0, 1, 2, \ldots, n-1$ is called a *binomial probability distribution*. Figure 5.6 presents a chart that can be used to predict the probability of an event occurring at least x times in n years (Gretener, 1967). For example, there is a 95% chance that an event with an estimated annual probability (P_0) of 0.01 (1/100) will occur at least once in 300 years ($nP_0 = 3$), at least five times in 900 years ($nP_0 = 9$) and at least ten times in 1600 years ($nP_0 = 16$).

5.2.2 The Poisson distribution (events that occur independently with a given rate)

The binomial distribution applies when it is known how often an event occurs and how often it does not occur. In other words, the total possibilities are known. But for natural phenomena such

Figure 5.6 Probability $P_{n,x}$ of a rare event to occur at least x times in n years. (After Gretener, 1967)

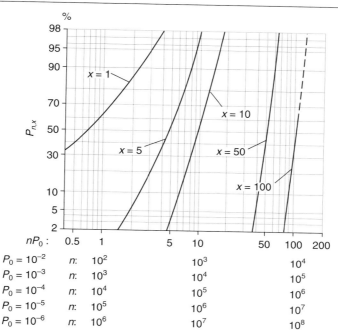

as landslides, while it may be possible to count the number of landslides that have occurred in a particular area during a particular period of time, it is not possible to count how many landslides did not occur. Similarly, while the binomial distribution deals with individual and discrete trials or lengths of time, it may be better to consider landsliding as a phenomena occurring in continuous time during which there could be any number of events. Both of these difficulties point to limitations of the applicability of the binomial distribution. An alternative is the *Poisson probability distribution*, a distribution widely used in predicting earthquakes and volcanic eruptions (and hence sometimes called the *model of catastrophic events*).

The general formula for the Poisson distribution is

$$P(n) = \frac{(\lambda t)^n \, e^{-\lambda t}}{n!}$$

where n is the number of predicted occurrences; t is the exposure period; $e = 2.71828\ldots$ is the base of natural logarithms; and λ is the *a posteriori* probability of occurrence, which is equivalent to 'P' in the binomial formula and has units that are the inverse of those of t (so if t is measured in years, then λ has units of year^{-1}, that is, 'per year').

Thus, so long as the *a posteriori* probability is known and conditions in the future are considered to be the same as those of the past, then the probability of an event occurring within differing time periods into the future can be calculated. For example, if studies of a stretch of cliff have revealed nine rockfalls in the past 80 years ($\lambda = 0.1125$), then the probability of one rockfall occurring in the next 25 years is

$$P(1) = \frac{(0.1125 \times 25)^1 \, e^{-0.1125 \times 25}}{1!}$$

$$= (2.8125 \times 0.060054)/1$$

$$= 0.168901 \text{ or } 17\%$$

The probability of two rockfalls is

$$P(2) = \frac{(0.1125 \times 25)^2 \, e^{-0.1125 \times 25}}{2!}$$

$$= \frac{7.91016 \times 0.060054}{2}$$

$$= 0.23752 \text{ or } 24\%$$

The probability of no (zero) rockfalls is

$$P(0) = \frac{(0.1125 \times 25)^0 \, e^{-0.1125 \times 25}}{0!}$$

$$= (1 \times 0.060054)/1$$

$$= 0.06 \text{ or } 6\%$$

It follows that the probability of no more than two events occurring in the 25-year time period is

$$P(\leq 2) = P(0) + P(1) + P(2)$$

$$= 0.168901 + 0.23752 + 0.06$$

$$= 0.466 \text{ or } 46\%$$

The Poisson distribution can be used in another way if the average number of events per unit time is known – that is, per month, per season, per year, per decade, per century, etc. In this case the Poisson expansion is used:

$$e^{-\lambda}, \quad \lambda e^{-\lambda}, \quad \frac{\lambda^2 e^{-\lambda}}{2!}, \quad \frac{\lambda^3 e^{-\lambda}}{3!}, \quad \frac{\lambda^4 e^{-\lambda}}{4!}, \quad \frac{\lambda^5 e^{-\lambda}}{5!}, \quad \ldots$$

where λ is the average value per unit time.

For example, if records for a stretch of coastal cliffs reveal the occurrence of 140 rockfalls in 100 years, then substituting the average of 1.4 per annum for λ in the above expansion reveals

$$e^{-1.4} = 0.2466 \text{ probability of no rockfalls in any one year}$$

$1.4 \, e^{-1.4} = 0.3452$ probability of one rockfall in any one year

$$\frac{(1.4)^2 \, e^{-1.4}}{2!} = 0.2417 \text{ probability of two rockfalls in any one year}$$

$$\frac{(1.4)^3 \, e^{-1.4}}{3!} = 0.1127 \text{ probability of three rockfalls in any one year}$$

$$\frac{(1.4)^4 \, e^{-1.4}}{4!} = 0.0395 \text{ probability of four rockfalls in any one year}$$

$$\frac{(1.4)^4 \, e^{-1.4}}{5!} = 0.0110 \text{ probability of five rockfalls in any one year}$$

and so on.

The general form of this expansion is

$$P(n) = \frac{\lambda^n \, e^{-\lambda}}{n!}$$

where λ is the probability or rate of occurrence and n is the number of events.

The Poisson distribution is a very versatile tool, which can be used in numerous ways. For example, suppose that a hypothetical region is subject to a range of natural hazards, including landslides, debris flows and rockfalls. All these events occur according to a Poisson process with mean rates λ of 1 in 5 years ($\lambda = 0.2$), 1 in 10 years ($\lambda = 0.1$) and 3 in 9 years ($\lambda = 0.33$), respectively. If it is assumed that the different types of event occur independently, then the chance of no events occurring in a single year can be calculated using the first equation in the above sequence:

$$P(n = 0) = e^{-0.2} \times e^{-0.1} \times e^{-0.33}$$

$$P(n = 0) = 0.818 \times 0.905 \times 0.716$$

$$= 0.53 \text{ or } 53\%$$

In the same hypothetical region, suppose that tropical storms can trigger numerous rockfalls from road cuttings, making driving particularly hazardous. Records indicate that, on average, one rockfall-related traffic accident occurs somewhere along the entire 500 km of the highway system during a storm. It is assumed that these incidents can be modelled by a Poisson process. The probability that at least one accident (i.e. ≥ 1) will occur during a storm along a particular 20 km-long section of highway can be calculated from the probability of no events occurring:

$$P(n \geq 1) = 1 - e^{-(20/500)}$$

$$= 1 - e^{-0.04}$$

$$= 1 - 0.960$$

$$= 0.039$$

By way of another hypothetical example, consider the case where historical records over a 450-year period indicate that major, deep-seated landslides have occurred on a cliffline with an average frequency of one event per 100 years ($\lambda = 0.01$). After years of inactivity, three events occur in the same year: what is the chance of this happening? Assuming a Poisson distribution, the probability of this sequence of events occurring is

$$P(n = 3) = (0.01)^3 \times e^{-0.01}/3!$$

$$= 1.65\text{E-}07 \text{ or about 1 in 6 million}$$

The Poisson distribution can also be used to calculate the probability of an event occurring in a particular time period and the average waiting time before the next event in a sequence. For example, it is estimated that the rate of major landslide activity on an area of seabed is three events over 170 000 years ($\lambda = 0.0000176$). The probability that one event (n) will occur in the next 100 years can be calculated using the cumulative distribution function:

$$P(n = 1) = 1 - e^{-\lambda t}$$

$$= 1 - e^{-(0.0000176 \times 100)}$$

$$= 1 - e^{-0.00176}$$

$$= 1 - 0.998237$$

$$= 0.001763$$

The average waiting time is the time at which there is a 50% probability of there being an event. In this example, it is around $t = 40\,000$ years (Figure 5.7):

$$P(n = 1) = 1 - e^{-(0.0000176 \times 40\,000)}$$

$$= 0.506$$

Figure 5.7 Example of the Poisson distribution ($\lambda = 0.0000176$). The average waiting time is the time at which there is a 50% probability of there being an event. Here, it is around $t = 40\,000$ years

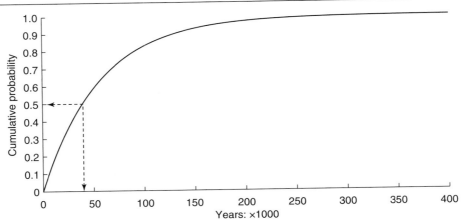

However, the Poisson distribution is 'memory-less' and the amount of time since the last event contains no information about the amount of time until the next event.

Tables 5.4(a) and 5.4(b) provide a numerical comparison between the Poisson model and the binomial model. The two models generate very similar results, except for relatively high probabilities and short

Table 5.4 Comparison between results generated by (a) the Poisson model and (b) the binomial model using the same annual probability and time periods

(a)

Annual probability	Time period					
	1	5	10	25	50	100
1	0.632	0.993	1.000	1.000	1.000	1.000
0.5	0.393	0.918	0.993	1.000	1.000	1.000
0.2	0.181	0.632	0.865	0.993	1.000	1.000
0.1	0.095	0.393	0.632	0.918	1.000	1.000
0.05	0.049	0.221	0.393	0.713	0.993	1.000
0.02	0.020	0.095	0.181	0.393	0.918	0.993
0.01	0.010	0.049	0.095	0.221	0.632	0.865
0.005	0.005	0.025	0.049	0.118	0.393	0.632
0.002	0.002	0.010	0.020	0.049	0.221	0.393
0.001	0.001	0.005	0.010	0.025	0.095	0.181
0.0005	0.000	0.002	0.005	0.012	0.049	0.095
0.0002	0.000	0.001	0.002	0.005	0.025	0.049
0.0001	0.000	0.000	0.001	0.002	0.010	0.020
					0.005	0.010

(b)

Annual probability	Time period					
	1	5	10	25	50	100
1	1.000	1.000	1.000	1.000	1.000	1.000
0.5	0.500	0.969	0.999	1.000	1.000	1.000
0.2	0.200	0.672	0.893	0.996	1.000	1.000
0.1	0.100	0.410	0.651	0.928	0.995	1.000
0.05	0.050	0.226	0.401	0.723	0.923	0.994
0.02	0.020	0.096	0.183	0.397	0.636	0.867
0.01	0.010	0.049	0.096	0.222	0.395	0.634
0.005	0.005	0.025	0.049	0.118	0.222	0.394
0.002	0.002	0.010	0.020	0.049	0.095	0.181
0.001	0.001	0.005	0.010	0.025	0.049	0.095
0.0005	0.000	0.002	0.005	0.012	0.025	0.049
0.0002	0.000	0.001	0.002	0.005	0.010	0.020
0.0001	0.000	0.000	0.001	0.002	0.005	0.010

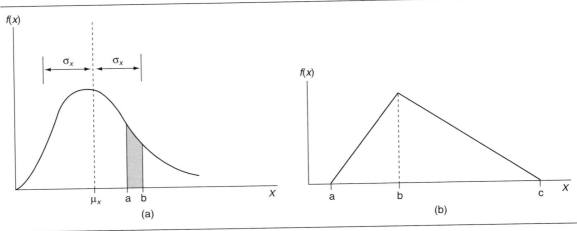

Figure 5.8 Probability density functions: (a) the probability that the variable x is between the values a and b is given by the shaded area; (b) a triangular distribution in which the value b represents the 'best estimate', and a and c represent the upper and lower limits of the range. (After Wu et al., 1996)

periods of time. For example, when the annual probability is 0.5 and the specified time is 1 year, the cumulative probability is equal to 50% using the binomial model, whereas it is equal to 39.3% using the Poisson model. The difference between the two models becomes negligible for lower annual probabilities and longer time periods. This can be significant in those situations where the landslide risk over short time periods (less than 25 years) is dominated by events with high annual probabilities.

5.2.3 Continuous probability distributions

A number of parameters, such as recession rates or soil shear strength, can be regarded as random variables with continuous probability distributions (Figure 5.8). These parameters can have an almost infinite number of possible values as compared with the binomial case, where there are only two possible outcomes: failure or no failure.

A continuous probability distribution differs from a discrete probability distribution in several ways. The probability that a continuous random variable will assume a particular value is almost zero. This may seem odd, but for a continuous variable there is an uncountable number of values that it can take, and this number is so large that the probability of any single value occurring is almost zero. The equation used to describe a continuous probability distribution is called a probability density function (pdf). All probability density functions satisfy the following conditions.

- The random variable y is a function of x: $y = f(x)$.
- The value of y is greater than or equal to zero for all values of x.
- The total area under the curve of the function is equal to 1.0.

Random variables are described by measures of central tendency (i.e. mean and median) and measures of variability (i.e. standard deviation and variance).

The most widely used distribution is the so-called *normal distribution* (or *Gaussian distribution*), which takes the form of a symmetrical bell-shaped curve.

The value of the random variable y is defined by the *normal* equation:

$$y = \frac{1}{(2\pi\sigma)^{1/2}} \exp\left(-\frac{(x-\mu)^2}{2\sigma^2}\right)$$

where x is a normal random variable, μ is the mean and σ^2 is the square of the standard deviation.

The shape of the distribution is defined by the mean and the standard deviation: the mean determines the centre of the probability distribution and the standard deviation determines the dispersion around the mean. When the standard deviation is large, the curve is short and wide; when the standard deviation is small, the curve is tall and narrow. Figure 5.9 highlights the relationship between the mean and the standard deviation: 68% of the values will lie within 1 standard deviation of the mean, 95% within 2 standard deviations, 99.7% within 3 standard deviations and 99.99% within 4 standard deviations. The standard normal distribution is a special case of the normal distribution, and is the distribution that occurs when a normal random variable has a mean of 0 and a standard deviation of 1.

In many instances, the probability distribution is not a bell-shaped curve but concentrated (i.e. skewed) to the right (negative) or left (positive). Skewed distributions are common when mean

Figure 5.9 The normal distribution, indicating the relationship between the mean μ and the standard deviation σ

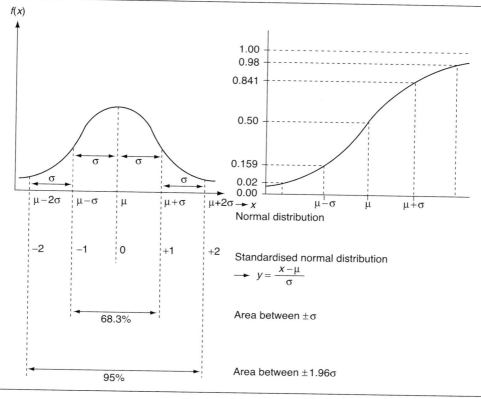

values are low, variances large and values cannot be negative: for example, the distribution of mineral resources in the Earth's crust. Such skewed distributions often closely fit the lognormal distribution (e.g. Johnson et al., 1994).

A lognormal distribution is a continuous probability distribution of a random variable whose logarithm is normally distributed. If X is a random variable with a normal distribution, then $Y = \log X$ has a log-normal distribution. If X has a log-normal distribution, then a log–log plot of the distribution will be nearly a straight line for a large portion of the body of the distribution. The log-normal distribution is the distribution of a random variable that takes only positive real values. Log-normal distributions have been observed in many systems, including ecology, human medicine and the environment.

Other common distributions include the exponential, gamma and Gumbel (e.g. Ang and Tang, 1984; Benjamin and Cornell, 1970; Hahn and Shapiro, 1967; Montgomery and Runger, 1994). For example, exponential functions have the form

$$f(x) = b^x$$

where b is called the *base* and x is called the *exponent* (or *power*), so the exponential function for the base of natural logarithms is

$$f(x) = \exp(x) = e^x$$

If $b > 1$, the function continuously increases in value as x increases. A special property of exponential functions is that the slope of the function also continuously increases as x increases.

A logarithmic function is the inverse of an exponential function: if $f(x)$ is the logarithm of x to base b,

$$f(x) = \log_b x$$

then, by definition,

$$x = b^{f(x)}$$

Again, we have the special case of the natural logarithm (ln) with $b = e$, and

$$f(x) = \log_e x = \ln x$$

The logarithmic distribution function is widely used in electronics, earthquake analysis and population prediction. It can also be used to model the deterioration of slope stability (or change in probability of failure) over time. For example, along a section of the protected glacial till cliffs on the Whitby coast, the probability of slope drainage system failure and a resulting landslide event was assumed to increase logarithmically with time as the system deteriorates, from an initial value (0.01) up to an expected 25-year residual life, at which point the probability of failure is 0.99. The change in probability P with the remaining drainage system 'life' T years is represented by

$$P = e^{a + b \ln T}$$

The constants a and b are defined as

$$a = \ln(P \text{ in final year})$$
$$= \ln(0.99)$$
$$= -0.01005$$

$$b = \frac{\ln(P \text{ in 1st year}) - \ln(P \text{ in final year})}{\ln(\text{residual life})}$$
$$= \frac{\ln(0.01) - (-0.01005)}{\ln(25)}$$
$$= \frac{-4.6 - (-0.01005)}{3.219}$$
$$= -1.427$$

For each year t up to year 25, the probability of failure is calculated as

$$P = e^{a + b \ln(25 - t)}$$

Thus, for year 10,

$$P = e^{-0.01005 - 1.427 \ln 15}$$
$$= e^{-0.01005 - 1.427 \times 2.71}$$
$$= e^{-3.877}$$
$$= 0.02$$

The results are displayed in Figure 5.10, which shows the probability of failure in each year and the cumulative probability of failure until year 24.

5.2.4 Fat-tailed distributions and power laws

In some instances, events may occur that are many standard deviations away from the mean value of a normal distribution and thus might be expected to have an exceptionally low probability. Nordhaus (2011) presents the example of prices on the US stock exchange, which fell 23% on 19 October 1987. Over the period 1950–1986, the daily standard deviation of price change was 1%. If stock market prices followed a normal distribution, then a 5% change would be expected, on average, every 14 000 years (5 standard deviations, i.e. 5 times 1%), while a 23% fall (23 standard deviations, i.e. 23 times 1%) might be expected once in the life of the universe! In another example, the oil price in the early 1970s was relatively stable, with a standard deviation of around 5% in a month. However, in 1973, there was a price rise that, using a normal distribution, would represent 37 standard deviations.

Large deviations from the mean are also known to occur for many other phenomena, including floods, earthquakes and landslides. Around 30 000 people were killed in Vargas State, Venezuela,

Figure 5.10 Whitby Cliffs – probability of a landslide event following drainage failure: (a) probability of failure in each year; (b) cumulative probability of failure until year 24 (logarithmic increase over time)

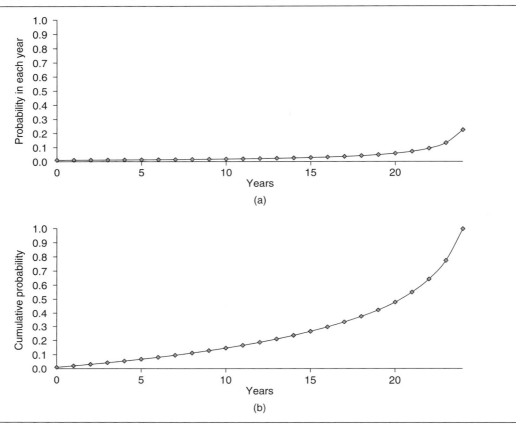

on 15 December 1999, when torrential rain triggered widespread flash floods and debris flows (Wieczorek et al., 2001). The neighbourhood of Los Corales was buried under 3 m of debris, and many homes in the region were swept away to the ocean. The daily rainfall over the period 14–16 December had been 120, 410 and 290 mm. Prior to the December storm, there had been no recorded daily rainfall in excess of 140 mm. Standard statistical analysis of the available rainfall time series (1961–1999) indicated that a daily rainfall total of 410 mm should be expected only once in several million years (Süveges and Davison, 2012). This has led the event to be used as an example of what are termed 'Dragon Kings' – events that lie well beyond the normal magnitude–frequency relationships developed for an area and impossible to predict from simple extrapolation (e.g. Sornette, 2009; Sornette and Ouillon, 2012).

The reality is, of course, that these 'surprise' events do not fit the normal distribution, and they have a higher probability than is suggested using that distribution. Such extreme events are often termed 'tail events'. These are rare events that, from past experience, should happen extremely infrequently. They are indicative of 'fat-tailed' or 'heavy tailed' distributions, such as the Pareto-type (power-law)

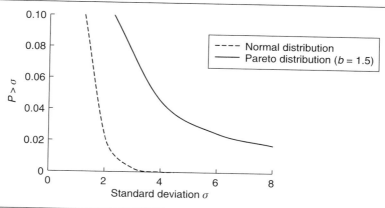

Figure 5.11 Illustration of the tails of a normal distribution and a Pareto distribution with a scale parameter $b = 1.5$. Each curve shows the probability that the variable will be greater than the number of standard deviations (sigma). (After Nordhaus, 2011)

distribution. The probability P of an event can be defined as

$$P = aX^{-b}$$

where X is the variable, a is a constant that ensures that the area under the curve adds up to 1.0, and b is a 'shape parameter' (the tail index) that defines the shape of the curve and reflects the importance of tail events. If b is small (<2), then the distribution is fat-tailed. If b is large, then the distribution resembles a normal distribution.

The difference between the normal and a Pareto distribution is shown in Figure 5.11. For the normal distribution, the probability curve is close to zero at ± 3 standard deviations, whereas in the Pareto distribution ($b = 1.5$) the probability remains significantly high at ± 8 standard deviations.

The earthquake frequency–magnitude distribution is usually described by the Gutenberg–Richter law (Gutenberg and Richter, 1949):

$$\log N(M) = a - bM$$

where $N(M)$ is the number of earthquakes with magnitude $\geq M$, and a and b are constants. This can be rewritten as

$$N(M) = 10^{a-bM} = 10^a 10^{-bM} = \text{constant} \times 10^{-bM}$$

from which it can be seen that as M increases, the frequency decreases exponentially. Although this relationship between frequency and magnitude is exponential, the Gutenberg–Richter law is in fact a power law when expressed in terms of earthquake amplitude A, since the magnitude is related to $\log A$, so

$$N = \text{constant} \times 10^{-bM} = \text{constant} \times 10^{-b \log(A/A_0)} = \text{constant} \times 10^{-b \log A}$$

$$\text{constant}' \times 10^{\log A^{-b}} = \text{constant}' \times A^{-b}$$

Landslide Risk Assessment

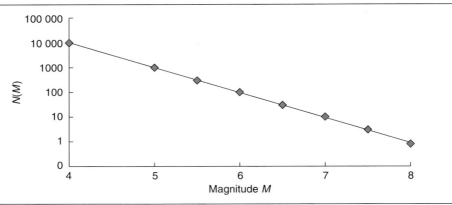

Figure 5.12 Gutenberg–Richter relationship for global earthquakes between 1904 and 2000. $N(M)$ is the number of earthquakes per year with magnitude $\geq M$

(A_0 is a constant that is part of the definition of the magnitude). The distribution is scale-invariant: when the number of events of magnitude M or greater is compared with the number of events of magnitude $M+1$ or greater, the difference is always the same, regardless of the value of M.

When plotted on a semi-logarithmic plot, with a logarithmic scale on the vertical (N) axis, the earthquake distribution is linear. The a value (the intercept of the line with magnitude zero) indicates the seismic activity. The b value is (minus) the slope of the frequency–magnitude distribution, and describes the relative size distribution of earthquakes. A higher b value indicates a relatively larger proportion of small events, and vice versa.

This distribution can be used to make a probabilistic hazard forecast by rewriting the Gutenberg–Richter law in terms of the probability of a target magnitude M_t or larger event:

$$P(M > M_t) = \frac{10^{a-bM}}{dT}$$

where dT is the observation period (e.g. 24 years). Taking $a = 5.1$ and $b = 0.96$, an event with $M \geq 6$ or larger has an annual probability of about 0.009 (1 in 111 years).

Figure 5.12 shows that the Gutenberg–Richter relationship for global earthquakes between 1904 and 2000 (here $N(M)$ is the number of earthquakes per year with magnitude $\geq M$; Kanamori and Brodsky, 2004). Every year approximately one earthquake with $M \geq 8$ occurs somewhere on the Earth, and there are ten events with $M \geq 7$ events and 100 with $M \geq 6$.

5.2.5 Landslide magnitude–frequency distributions

Landslide magnitude–frequency curves define the probabilistic nature of landslide size, occurrence and frequency, and hence can be an important component of risk assessment (Guthrie et al., 2008). It has been widely reported that landslide magnitude–frequency distributions can be described by an inverse power-law equation (e.g. Dai and Lee, 2001; Guzzetti et al., 2002; Martin et al., 2002; Stark and Hovius, 2001; Sugai et al., 1994). For example, Hungr et al. (1999) suggested that distributions of rockfalls from homogeneous areas could be fitted by a power law. The approach was used by

Dussauge-Peisser et al. (2002) to develop simple hazard models for cliff sites in France (Grenoble and Val d'Arly Gorge) and the USA (Yosemite Valley) based on the analysis of relatively long-duration rockfall inventories. They concluded that, above a given volume, the occurrence of rockfall sizes corresponded to a simple power law, regardless of the period of observation:

$$n(V) = aV^{-b}$$

where V is the rockfall volume, $n(V)$ is the number of events per year with a volume equal or greater than V, and a and b are constants. The constant b (which controls the shape of the distribution) is independent of geological setting and is fairly constant at around 0.45 (with a range of 0.41–0.46), whereas the constant a (defined as n_{100}, which represents the annual number of events greater than 100 m^3) varies with the site. This coefficient reflects the level of activity along the cliffline, and varied from 0.62 (Grenoble) to 2.16 (Yosemite Valley). When n_{100} was normalised by dividing by the surface area of each cliffline that was a potential rockfall source, values ranged from 0.26/10 km^2 (Grenoble) to 19.45/10 km^2 (Arly gorges).

Guthrie and Evans (2004a) developed a magnitude–frequency relationship for landslides on the Brooks Peninsula (a 286 km^2 study area) on the west coast of Vancouver Island, British Columbia, Canada. A total of 136 debris flows and debris slides were recorded from aerial photography for the period 1950–1996. The landslides ranged in size from 500 m^2 to 115 000 m^2, with a mean size of almost 9300 m^2. A magnitude–cumulative frequency curve was generated for these 136 identified landslides (Figure 5.13(a)), showing the number of landslides above a given size threshold occurring in a year. The relationship was defined by a power law:

$$y = 2 \times 10^6 x^{-1.5576}$$

where y is the cumulative frequency per year and x is the area of the slide in m^2.

The cumulative annual probability of an event of a particular magnitude M being equalled or exceeded is calculated as (Figure 5.13(b))

$$P(\geq M) = \frac{m}{n+1}$$

where n is the number of years in the time series and m is the rank order of the event magnitude.

The points on the cumulative frequency curve (Figure 5.13(a)) represent the frequency that a landslide over a particular magnitude is equalled or exceeded within the 46-year time series. For example, the smallest event (500 m^2) is equalled or exceeded 136 times:

$$\text{frequency}(\geq 500 \text{ m}^2) = 136/(46+1) = 2.89 \text{ events/year}$$

For the largest event:

$$\text{frequency}(\geq 115\,000 \text{ m}^2) = 1/(46+1) = 0.02 \text{ events/year}$$

A similar relationship was defined for landslides triggered by a single storm in Loughborough Inlet, on the British Colombia coast (Guthrie and Evans, 2004b). A total of 101 landslides (ranging from around 1000 m^2 to 409 000 m^2) were documented across 370 km^2 following a rainstorm that swept the British Columbia coastline on 18 November 2001. For landslides larger than 10 000 m^2, the

Landslide Risk Assessment

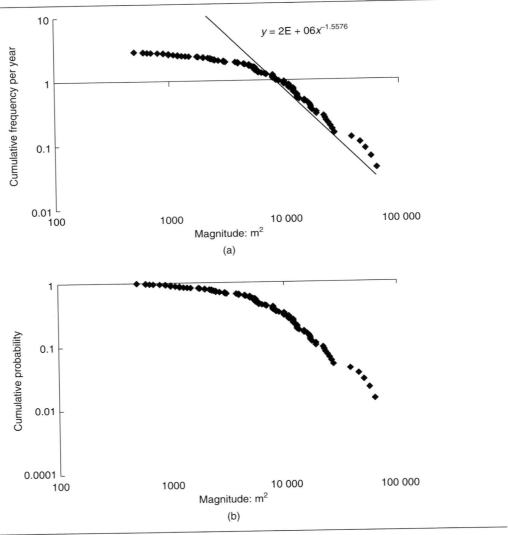

Figure 5.13 Magnitude–cumulative frequency curve for landslides in the Brooks Peninsula area (1950–1996). (After Guthrie and Evans, 2004a)

magnitude–frequency relationship was defined by

$$y = 5 \times 10^4 x^{-1.24}$$

A power-law distribution has also been used to describe the magnitude–frequency relationship of submarine landslides. For example, on the 12 000 km² carbonate platform off the northern coast of Puerto Rico, ten Brink et al. (2006) have shown that the cumulative number of 160 slope failures (0.07–20 km³ in volume) recorded from the study of detailed bathymetric data could be described by

$$n(L) = 26 V^{-0.64}$$

where $n(L)$ is the number of landslides exceeding volume V (km³). This relationship suggests that the cumulative number of events only becomes 1.0 with a landslide over 160 000 km³:

$$n(L) = 26 \times (160^{-0.64}) = 1.01$$

This event is over four times larger than the largest recorded feature (20 km³), suggesting that there may be physical limits to the maximum landslide size in this region (ten Brink *et al.*, 2006).

Very similar relationships were defined by Chaytor *et al.* (2009) for submarine landslides (up to 2410 km³ in volume) on the US Atlantic margin and the volume distribution of landslide sources from the Storegga landslide complex off the western coast of Norway (see also Haflidason *et al.*, 2005; Issler *et al.*, 2005): for the US Atlantic Margin,

$$n(L) = 35.809 V^{-0.37666}$$

and for Storegga,

$$n(L) = 39 V^{-0.44}$$

However, there is evidence that landslide frequency–magnitudes actually follow a power law for only a truncated portion of the entire distribution (Figure 5.14). The curves often show a flattening of their respective distributions at smaller sizes, and hence are not predictable by a single simple equation. For example, the power law for Brooks Peninsula predicts 125 events per year for volumes equal or greater than 500 m²:

$$y = 2 \times 10^6 x^{-1.5576}$$
$$= 2\,000\,000 \times 500^{-1.5576}$$
$$= 125/\text{year}$$

Figure 5.14 The conceptual relationship between landslide magnitude and frequency. (After Guthrie *et al.*, 2008)

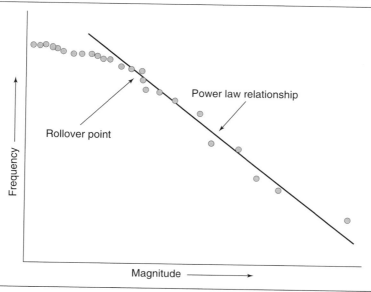

This is over 40 times the observed frequency. This departure of the data from a power law is known as the rollover, and has generated considerable discussion about probable cause (e.g. Guthrie and Evans, 2005; Guthrie et al., 2008). This trend has also been noted for submarine landslides, where the power law only appears to apply to a truncated portion of the data set (e.g. Stark and Hovius, 2001).

It has been suggested that rollover is due to under-representation of smaller events within the landslide data sets because of the difficulties in detecting landslides below a certain size (so-called censoring or biasing of the data; e.g. Hungr et al., 1999; Stark and Hovius, 2001). Brardinoni et al. (2003) attempted to quantify the missing data due to censoring in the temperate rain forest on the west coast of Canada. They concluded that small landslides may account for up to 85% of the number and 30% of the volume at aerial photograph scales of 1 : 12 000 to 1 : 15 000.

However, in many cases, it appears that the rollover point does not coincide with the minimum size of consistently mapped landslide (e.g. Guzzetti et al., 2002; Martin et al., 2002). For the Brooks Peninsula magnitude–frequency curve (Figure 5.13), the rollover occurs at approximately 10 000 m^2, considerably in excess of the minimum consistently resolvable size of 500 m^2 (Guthrie and Evans, 2004a).

It is now considered likely that there may be a physical explanation for the rollover, with the nature of the landscape providing limits to the size of landslides that can be generated (e.g. Guthrie and Evans, 2004a, 2004b; Hungr et al., 2008). For example, for large events, the number of available locations in a landscape that could support major failure events is clearly restricted by the topography, and may decay by a power law. Smaller events may also be restricted by the nature of the terrain and cause flattening of the distribution curve. In British Colombia, for example, debris slides tend to be initiated on the till and colluvium mantled steep middle and upper slopes and then flow to a lower slope position, such as a stream or valley bottom. The nature of the terrain ensures that potential landslide source areas and flow pathways tend to be in a particular size range, imposing a lower limit on the events that are likely to occur.

A number of alternative distributions have been used to describe landslide distributions over the full range of magnitudes, which overcome the problem of rollover. These include the following (Figure 5.15).

- *A double Pareto distribution* (Stark and Hovius 2001). This comprises two power law distributions, one for the part of the data set below the rollover threshold and the other for the larger events (e.g. Reed and Jorgensen, 2005). Guthrie and Evans (2004a,b) used this distribution to describe the data sets at Brooks Peninsula and Loughborough Inlet. The double Pareto model predicts the majority of the landslide data well, but less well at both tails. For landslides <630 m^2 at Brooks Peninsula and <1000 m^2 in Loughborough Inlet, there appear to be fewer small landslides than predicted by the curve.
- *The log-normal distribution.* Chaytor et al. (2009) demonstrated that the observed volumes of the identified failure scars on the US Atlantic margin, when plotted as a cumulative number on a logarithmic scale, show a very good fit to a lognormal distribution ($R^2 = 0.985$), with a standard deviation σ and sample mean μ of log volume of 2.27 and 6.60, respectively. This distribution is a good fit with the observed pattern of landsliding, with a relatively narrow dominant landslide size away from which the number of landslides falls off for both smaller and larger sizes. Log-normal distributions have also been found for the areas of landslides in Kashmir (Dunning et al., 2007) and for volumes of deposits of prehistoric turbidity currents in Italy (Talling et al., 2007).

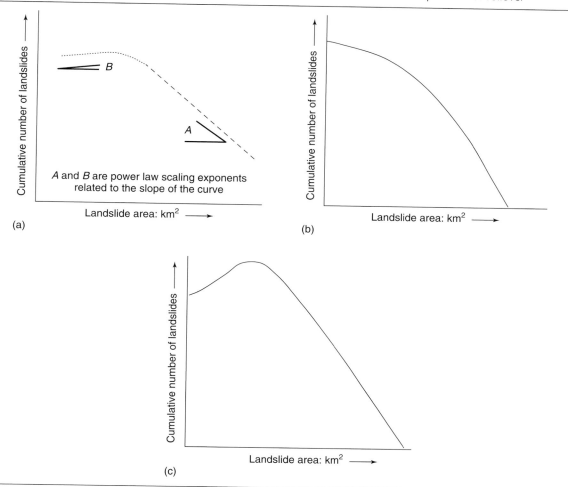

Figure 5.15 The (a) double Pareto, (b) exponential and (c) inverse gamma distributions that have been used to describe landslide distributions over the full range of magnitudes and overcome the problem of rollover

- *The gamma distribution.* Malamud *et al.* (2004) analysed well-documented landslide inventories of the 1994 Northridge earthquake, the 1998 Hurricane Mitch-related rainfall in Guatemala, and the rapid snowmelt in Umbria (Italy) in 1997, and found that the probability density functions of landslide area to be in good agreement with a truncated inverse gamma distribution. This distribution matches the observed pattern of frequent small sizes, a power law decrease in the number of larger events, and an exponential decrease in smaller events.

It appears likely that different distributions suit different landslide environments, reflecting variations in controlling factors such as the nature of the terrain (e.g. the availability of landslide source areas) and the geological setting (e.g. the occurrence and distribution of landslide-prone strata or materials). Those sites where landslide area is best matched by a gamma distribution are all from tectonically active mountain regions – environments characterised by very dense assemblages of landforms, including steep-sided gullies, incised mountain stream channels and narrow spurs. This results in a full range of slope aspects. The geology is typically complex, with highly variable structure limiting the spatial extent of landslide-prone strata. The landslide size distribution is a function of this setting, with many separate small slides of a particular size range concentrated along the stream and gully network. The maximum size tends to be limited by the availability of large potential source areas.

The log-normal distribution of landslides on the US Atlantic margin reflects a very different environment. The seabed is an almost uniform, gently sloping plain that has been divided into sections by submarine canyons. The geology is simple, with long-term sedimentation producing layers of gently dipping beds. As a result, potential source areas are very extensive and almost uniform. This gives rise to a relatively narrow dominant landslide size range, with few smaller or larger events.

5.2.6 Return periods and exceedence probability

The average time between two events that equal or exceed a particular size (magnitude) is known as the return period or recurrence interval, and is normally denoted by T_r. Thus, an event that is expected to be equalled or exceeded on average every n years has a return period of n years. Exceedence probability is the probability that a certain event value (e.g. landslide magnitude) is going to be exceeded.

The return period is usually computed from a continuous series of recorded events as

$$\text{return period } T_r = \frac{\text{number of events equal to or exceeding a certain defined magnitude}}{\text{number of years of continuous records}}$$

This allows the annual probability to be calculated as

$$\text{annual probability } P = \frac{1}{T_r}$$

For events that occur on a regular basis, such as spring thaw debris flows within a defined catchment, it may also be possible to use a simple form of *extreme event analysis* similar to that used for flood predictions. This requires continuous and complete records of events over a number of years, so that the size of the maximum magnitude event for each year is known. These extreme events are then ranked and the return period calculated using the so-called Weibull formula (Riggs, 1968):

$$\text{return period } T_r = \frac{\text{number of years in series } (n) + 1}{\text{rank number of event } (m)}$$

Thus, if there are 49 years of records, then the largest event (rank number 1) is the 50-year event (annual probability 0.02), and the second largest event is the 25-year event, and so on. Note that for the x year event, the size of the event will be *equalled or exceeded* on average once in every x years. The flood magnitude/return period relationship for the River Severn data (Table 5.5) is presented in Figure 5.16 as an example.

The Gumbel distribution is often used to model flood frequency–magnitude (e.g. Stedinger et al., 1993). The Gumbel EV1 distribution, for example, is (see Table 5.5)

$$F(Q) = \exp\left[-\exp\left(-\frac{x-u}{\alpha}\right)\right]$$

Introduction to probability and quantitative assessment

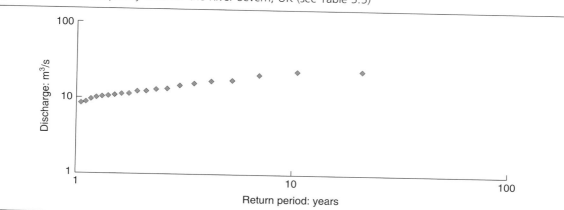

Figure 5.16 Flood frequency data for the River Severn, UK (see Table 5.5)

Table 5.5 Flood frequency data for the River Severn, UK (see Figure 5.16 for the flood return period plot)

Year	Discharge Q: m³/s	Probability P(x) discharge is equalled or exceeded (Gumbel EV1 distribution)[a]	Rank	Return period (Weibull formula): years
1951	10.4	0.766	16	1.31
1952	10.9	0.717	14	1.50
1953	10.6	0.747	15	1.40
1954	10.1	0.794	17	1.24
1955	8.5	0.917	20	1.05
1956	9.6	0.838	18	1.17
1957	23.7	0.033	1	21.00
1958	13.1	0.493	9	2.33
1959	23.1	0.040	2	10.50
1960	17	0.202	5	4.20
1961	11.4	0.666	12	1.75
1962	14.8	0.343	7	3.00
1963	12.3	0.573	11	1.91
1964	20.7	0.076	3	7.00
1965	13.4	0.464	8	2.63
1966	17.5	0.178	4	5.25
1967	12.4	0.562	10	2.10
1968	15.8	0.271	6	3.50
1969	8.8	0.898	19	1.11
1970	11.3	0.676	13	1.62

[a]Using Excel, column 3 is $P(x) = 1 - \text{EXP}(-\text{EXP}(-(Q-u)/\alpha))$; u and α are calculated as shown in the text

where $F(Q)$ is the probability that an annual flood will equal or be less than the discharge Q. The mean μ and variance v are estimated from the data series. These values are then used to derive the parameters α and u:

$$\alpha = \frac{\sqrt{6v}}{\pi}$$

$$u = \mu - \gamma\alpha$$

where $\gamma = 0.5772\ldots$ is the Euler–Mascheroni constant. The probability of exceedence of the EV1 mean annual flood is

$$P(m) = 1 - F(m)$$

The binomial and Poisson models (see above) can also be used to determine the exceedence probability. Using these discrete event models, the exceedence probability is the probability of experiencing one or more landslides (n) during time t:

$$P(n) = 1 - (1 - P)^t \quad \text{(the binomial model)}$$

$$P(n) = 1 - e^{-\lambda t} \quad \text{(the Poisson model)}$$

For example, a seawall protecting a coastal cliff is designed against the 200-year combination wave and storm surge level, but it is intended to operate for only 50 years. The probability that conditions exceeding the design value will occur during the first year of operation is

$$P = 1/\text{return period}$$

$$= 1/200$$

$$= 0.005$$

The probability that the seawall will not be subject to conditions exceeding the design value during its lifetime can be calculated from the annual probability of exceedence (i.e. 0.005) and non-exceedence (0.995). Using the binomial model, the probability of exceedence over 50 years is

$$P = 1 - (1 - P)^n$$

$$= 1 - (1 - 0.005)^{50}$$

$$= 0.22$$

Hence, the probability of non-exceedence is $1 - 0.22 = 0.78$.

Using the Poisson distribution, the probability of non-exceedence is

$$P = 1 - (1 - e^{-\lambda t})$$

$$= 1 - (1 - 0.778)$$

$$= 0.78$$

Table 5.6 Approximate event return periods and probabilities associated with different combinations of exceedence probability (%) and time periods (using the Poisson distribution)

Time period	Probability of exceedence of x% within time period		
	10%	5%	2%
50 years	1/500 event (0.002)	1/1000 event (0.001)	1/2500 event (0.0004)
100 years	1/1000 event (0.001)	1/2000 event (0.0005)	1/5000 event (0.0002)

In earthquake studies, the ground motion hazard at a particular location is often expressed along the lines of 'the probability of exceedence of 5% in 50 years' – that is, the ground motion (in m/s^2) that has a probability of 0.05 of being exceeded within 50 years (λ). This can be calculated using the Poisson distribution as follows.

The probability of an occurrence is 5% (0.05).

The probability of $n = 0$ (zero) occurrences is

$$P(0) = 1 - 0.05 = 0.95$$

$$P(0) = e^{-n}$$

where $n = 50\lambda$ is the number of exceedences in t years.

Therefore

$$e^{-n} = 0.95$$

$$n = -\ln(0.95) = 0.05129$$

As $n = 50\lambda$,

$$\lambda = 0.05129/50$$

$$= 0.0010258 \text{ (or } 1/975, \text{ approximately } 1/1000)$$

Approximate 5% and 10% exceedence probabilities for different time periods are presented in Table 5.6.

The relationship between probability (expressed as a percentage), return period and the length of the period under consideration is shown in Table 5.7. This indicates that an event with a return period of 100 years (the 100-year event) *will not* have a 100% probability of occurring in a period of 100 years – the true figure is 65%. It also follows that 100-year events do not occur 100 years apart or once every 100 years, although it is possible for there to be more than one occurrence in any period of a 100 years, and even occurrences in successive years. Note should also be taken of the fact that events with very long return periods have only a small likelihood of occurring during the design life of a building or structure, or within the planning horizons of most organisations. Thus, the 500-year event has only

Table 5.7 Percentage probability of the N-year event occurring in a particular period (Binomial model)

Number of years in period	N = average return period: years							
	5	10	20	50	100	200	500	1000
1	20	10	5	2	1	0.5	0.2	0.1
5	67	41	23	10	4	2	1	0.5
10	89	65	40	18	10	5	2	1
30	99	95	79	45	26	14	6	3
60	–[a]	98	95	70	31	26	11	6
100	–	99.9	99.4	87	65	39	18	9
300	–	–	–	99.8	95	78	45	26
600	–	–	–	–	99.8	95	70	45
1000	–	–	–	–	–	99.3	87	64

[a]Dashes indicate that the percentage probability is >99.9

an 11% chance of occurring during the lifetime of a building (taken here as 60 years), while the probability of a 1000-year event occurring is 6% (Table 5.7), thereby highlighting the problem of planning for extreme events.

It is important to stress that return-period statistics are extremely sensitive to the length and quality of the historical record. For example, Benson (1960) demonstrated that to achieve 95% reliability on the return period of a 50-year event required 110 years of continuous good quality records; such lengthy data sets are rare (Table 5.8). It must also be stressed that the computation of return periods assumes uniformitarianism – that is, past conditions are the same as those of the present and into the future. As landsliding is particularly sensitive to human activity and environmental change, it is necessary to treat computed return periods with great scepticism. Indeed, all *a posteriori* probabilities need to be used with great care.

Table 5.8 The length of the historical record required to estimate return period events with 95% and 80% reliability

Return period	Length of record in years required to deliver reliability of the return period estimate	
	95% reliable	80% reliable
2.33	40	25
10	90	38
25	105	75
50	110	90
100	115	100

After Benson (1960)

5.3. Judgement and the use of experts
5.3.1 Expert judgement

The absence of landslide time series or damage data sets in most regions of the world means that it is generally not practical to expect that estimates of landslide probability will be 'objective' truths, supported by 'hard facts'. In most cases, the estimates of probability will be based on a combination of frequency statistics (where available), calculations of conditional probability, and expert judgement, which involves combining use of the available data, knowledge and experience. This can be illustrated by returning to the simple model introduced earlier to estimate the probability of damage to a building sited along a debris flow pathway:

$$P(\text{damage}) = P(\text{landslide}) \times P(\text{hit}|\text{landslide}) \times P(\text{damage}|\text{hit})$$

$P(\text{landslide})$ is likely to be a 'degree of belief' probability. Although in some cases the estimate might be based on landslide frequency, judgement is required to determine whether the data set is a representative sample or appropriate to the specific site under consideration, and, if so, what methods should be used to analyse the data (see Chapter 6).

$P(\text{hit}|\text{landslide})$ will usually be a conditional probability, taking account of the area of the landslide zone, the size of the asset at risk and the area or volume of the landslide (see Chapter 7).

$P(\text{damage}|\text{hit})$ will typically be a 'degree of belief' probability unless statistics are available that relate landslide event size to damage (see Chapter 8).

In most cases, 'degree of belief' probabilities derived from expert judgement will play a key role in probability assessment.

Experts are people with specialist skills or knowledge in a particular field (e.g. landsliding or the consequences of landsliding). Judgement is the forming of an estimate or conclusion from the information available at that time, described as the *'intelligent use of experience'* (Einstein, 1991). It is the means by which evidence is recognised, supporting evidence compiled, conflicting evidence reconciled, and evidence of all kinds weighed according to its perceived significance (Vick, 2002). This evidence includes field and experimental data, along with personal experience and the lessons learnt by the broader scientific or engineering community from past events (analogues). Judgement is specific to the individual and hence is inherently subjective, but should recognise that both statistical data and detailed knowledge of site conditions are important in order to address all of the elements contributing to the uncertainty in the occurrence of landslides and their consequences.

The use of judgement allows extrapolation beyond the boundaries of the data series. As the Inter-Governmental Panel on Climate Change (IPCC) (2006) stated:

> using only classical likelihoods delimits the boundaries of knowledge to a small set of findings based either on the historical record or from models rigorously calibrated to historical data. This makes it difficult to provide meaningful scientific advice to policy makers on a large number of questions where history is unlikely to be a suitable guide for the future.

However, as Mark Twain wrote (Focht, 1994):

> good judgement comes from experience. And where does experience come from? Experience comes from bad judgement.

The use of expert judgement has a long tradition in geotechnical practice, where the available field and experimental data are often limited. In this context it is worth considering Casagrande's (1965) view of calculated risk as the use of imperfect knowledge, guided by judgement and experience, to estimate the probable ranges for all pertinent quantities that enter into the solution of a problem.

Judgement uses inductive reasoning that goes from the specific to the general by means of associative inference or analogy (Vick 2002). The process cannot prove that a conclusion is true, but it can establish whether it is likely to be true. This is not a comfortable position for many geotechnical engineers, for whom deductive reasoning from first principles is a highly valued approach. Factors of safety are often treated as real, objective truths because they are derived from geotechnical theory. To them, judgement is seen as ill-defined, subjective and non-repeatable. These positions reflect a long-standing duality in geotechnical engineering: the contrasting nature of theory and practice.

Theory and practice are simply two different approaches for assessing geotechnical problems: engineering science and engineering practice (Peck, 1979). The application of soil mechanics theory is considered by many to yield conclusions that are objective truths: if the slope stability model is sound, then the results derived from the model are statements of fact. However, many of the developers of soil mechanics theory also appreciated the importance of judgement. Towards the end of his life, Terzaghi wrote (cited in Goodman, 1999):

> I produced my theories and made my experiments for the purpose of establishing an aid in forming a correct opinion and I realized with dismay that they are still considered by the majority as a substitute for common sense and experience.

Engineering practice relies on lessons learnt from past precedent, case histories, analogies and experience, rather than just first principles. Its core principles cannot be codified, and it is essentially a skill learnt through experience. In this approach, calculated values for the factor of safety are empirical tools or useful indications of stability that can aid understanding of a slope problem (Duncan, 1996a,b; Morgenstern, 1995). However, the experience from one site cannot be simply transferred to another, as ground conditions can vary significantly between sites. Theory and analysis provide the framework for applying judgement (Vick, 2002).

For many, both theory and practice have a role in assessing geotechnical problems:

> The fact that equilibrium analyses of slope stability involve assumptions and limitations does not mean they are valueless. It means that they cannot be used without understanding and judgement. (Duncan, 1996a)

However, it is important to stress that the use of judgement presents a range of challenges that need to be addressed in the probability assessment process, including bias and a tendency for experts to be overconfident in their estimates.

Landslide experts willingly acknowledge the uncertainty that is inherent in predictions of slope stability, and most are familiar with probability theory as a notation for expressing their uncertainty. Nonetheless, there is an ample body of literature, building on Tversky and Kahneman (1974), demonstrating that, when asked to make judgements under conditions of uncertainty, humans tend to adopt heuristics (i.e. experience-based techniques for problem-solving) and biases. These heuristics and biases mean that their judgements may prove to be distorted reflections of their state of knowledge or uncertainty. The most common sources of bias include the following.

- *Availability bias*: basing judgements on outcomes that are more easily remembered. An expert who has been involved in the investigation of many fatal landslide incidents may overestimate the actual likelihood of this type of event.
- *Anchoring and adjustment bias*: focusing on a particular value within a range and making insufficient adjustments away from it in constructing an estimate. An example is the application of adjustment factors to long-term landslide frequencies that take account of significant recent environmental changes, as in the case of landsliding in the West Nile delta described earlier.
- *Cognitive distortion*: when the joint probability of combined events is not consistent with their separately estimated values. An example is the classic 'Linda problem' (Tversky and Kahneman, 1983). Linda is 31 years old, single, outspoken and very bright. She majored in philosophy. As a student she was deeply concerned with issues of discrimination and social justice, and also participated in antinuclear demonstrations. Which of these alternatives is more probable: (a) Linda is a bank teller or (b) Linda is a bank teller and active in the feminist movement? The selection of (b) illustrates cognitive distortion. It cannot be probabilistically correct, as 'feminist female bank tellers' is a subset of 'female bank tellers'. Unless all female bank tellers are feminists, then there must be more female bank tellers than feminist bank tellers.

  In the same vein, a tropical storm triggers an open hillslope landslide in southern China. Which is more likely, a debris slide or a debris slide that transforms into a channelised debris flow?
- *Representative bias*: this involves basing judgements on limited data and experience without fully considering other relevant evidence. For example, greater weight may be given to recent laboratory test results and stability analyses than to the known historical record of events and the performance history of the slopes. There may also be a tendency for an assessor to underestimate the uncertainty associated with values of parameters produced by laboratory testing, such as shear strength.

  The use of bent trees as an indicator of slope movement is another example. Although there may be bent trees on a landslide, bent trees are not confined to landslide settings. The probability of a bent tree (BT) given that there is a landslide (L) is not the same as the probability of a landslide (L) given the presence of a bent tree (BT) – that is, $P(BT|L)$ is not the same as $P(L|BT)$.

  Another example is the 'Jack problem' (Kahneman and Tversky, 1982). Jack is a 45-year-old man. He is married and has four children. He is generally conservative, careful and ambitious. He shows no interest in political and social issues and spends most of his free time on his many hobbies, which include home carpentry, sailing, and mathematical puzzles. Jack was selected at random from a sample of engineers and lawyers. What is the chance that he is an engineer? The first group of subjects was told that the sample consisted of 30 engineers and 70 lawyers, the second group 70 engineers and 30 lawyers. However, the mean estimate of both groups was a 50% probability of Jack being an engineer. Both groups ignored the base-rate frequencies provided and based their views on a perception of a stereotype engineer.

  In landslide studies, it is not uncommon for some geotechnical engineers to place considerably more reliance on the results of stability analysis than on simple field observations. For example, the computed factor of safety for modelled circular slip surface might be around 1.2, but field evidence clearly indicates that the site is actively unstable with a factor of safety around 1.0.
- *Motivational bias*: where the assessor's perception is influenced by non-technical factors. For example, if the objective is to demonstrate a benefit:cost ratio of greater than 1.0 for a scheme, then the landslide scenarios, timing of losses and consequences may all be overstated. In other circumstances, an assessor may be over-cautious and consciously overstate the likelihood of failure so as to avoid underestimating the chance of a major event.

Landslide Risk Assessment

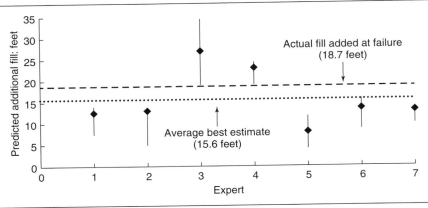

Figure 5.17 Seven experts' estimates of additional height of fill to cause failure of the I-95 embankment. (After Christian, 2004)

Experience has shown that experts tend to be too confident in their estimates and underestimate the uncertainties (e.g. Christian, 2004; Vick, 2002). Vick (2002) suggests that overconfidence increases as the task of probability estimation becomes more difficult (the 'hard–easy' effect). The greatest overconfidence occurs for questions about which there is no knowledge (Lichtenstein et al., 1982). The less people know of a subject, the less they recognise their lack of knowledge (Vick, 2002).

The problem of overconfidence can be illustrated with an example presented in Focht (1994). An embankment had been constructed on soft clay in 1969 near Boston in the USA. The road project was later abandoned, and in 1974 a symposium at the Massachusetts Institute of Technology was held to predict the height of additional fill that would cause the embankment to fail. Seven acknowledged experts agreed to make predictions based on the available geotechnical data and to provide a range within which the expert's confidence of the failure was 50% (50% confidence limits). Fill was placed incrementally over a 300-foot section, and failure occurred when 18.7 feet of fill had been added. The results of the experts' predictions are shown in Figure 5.17; the diamonds indicate each individual expert's best estimate and the lines their ranges. In no case did the amount of fill required to initiate failure fall within an expert's 50% confidence limits. However, the average of the seven experts' best estimates was 15.6 feet, and the confidence limits around this group estimate would have encompassed the actual result. This clearly reveals the benefits to be gained from having panels of experts when dealing with difficult or potentially dangerous situations (see below).

Christian (2004) identified a further problem with the use of experts: it is difficult to predict an expert's performance on the basis of credentials and experience. Some experts are good at synthesising the available evidence and making a rational judgement, whereas others appear to be less successful. For example, nine engineers of varying experience and qualification were asked to design six transmission tower footings resistant against uplift (Kondziolka and Kandaris, 1996). These footings were built and then tested to failure. A comparison of the engineers' predictions is presented in Figure 5.18, and shows the ratio of the predicted design capacity P and the actual failure load Q: $P/Q = 1$ indicates the exact prediction of the actual result (the diamonds indicate the average for each expert over all six footings and the lines the range of predicted values). In all cases, the predicted capacity is less than the observed. The two best performers (experts 1 and 2) had 14 and 30 years of experience, respectively, but no master's degree or PhD. The third best performer (expert 3) had limited

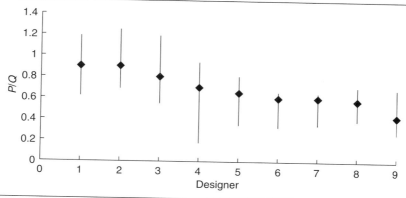

Figure 5.18 Ratio of the predicted uplift capacity P to measured uplift Q for six transmission tower foundations evaluated by nine designers. Each black diamond represents the average of six values of P/Q for each designer. The vertical bar represents the range of each designer's results. (After Christian, 2004)

experience (8 years) and no higher degrees. The worst performer (expert 9) had 22 years of experience. Experts 4 and 7 were the most highly qualified performers, with master's degrees and PhDs. An important lesson from this example is that neither experience nor credentials are necessarily good measures of an expert's potential performance.

5.3.2 The need for expert panels and their operation

It is now widely recognised that formal protocols and procedures are required to aid the elicitation of expert judgements (e.g. IPCC, 2000; SSHAC, 1997). Ideally, the assessment process should involve a group of experts, rather than single individuals, as this facilitates the pooling of knowledge and experience, as well as limiting bias (e.g. Roberds, 1990; Vick, 2002). Group consensus about a judgement is clearly desirable despite the increased cost. However, there may be significant differences of opinion between team members. It is necessary to attempt to resolve these differences of opinion, but they may persist, so that the following outcomes can result.

- *Convergence*: a single assessment is determined that expresses the common belief of all individuals in the group as expressly agreed to by every group member.
- *Consensus*: a single assessment is determined, although the assessment may not reflect the exact views of each individual. The consensus assessment may be a compromise derived from the individual assessments of group members but without the express agreement of the individuals concerned (forced), or the group may expressly agree to it for a particular purpose (agreed).
- *Disagreement*: multiple assessments are determined where convergence or consensus on a single assessment is not possible (e.g. owing to major differences of opinion).

In general, convergence is the most desirable outcome because it is defensible, but it may be difficult to achieve. Agreed consensus (i.e. with the concurrence of the group) is slightly less defensible, but is also less difficult to achieve. Forced consensus, without concurrence of the group, may be difficult to defend, but it is easy to achieve. Disagreement may be the most difficult outcome to use because it involves more than one view, but it is defensible.

The most widely used approaches to develop convergence or agreed consensus are the *open forum* and the *Delphi panel*. The open forum relies on the open discussion between group members to identify and resolve the key issues related to the landslide problems. The results can, however, be significantly distorted by the dynamics of the group, such as domination by an individual because of status or personality.

The Delphi panel is a systematic and iterative approach to achieving consensus, and has been shown to generally produce reasonably reproducible results across independent groups (e.g. Foster, 1980; Linstone and Truoff, 1975). The Delphi method was developed at the RAND Corporation in the early 1950s as a spin-off of an Air Force-sponsored research project, 'Project Delphi', designed to anticipate an optimal targeting of US industries by a hypothetical Soviet strategic planner.

Each individual in the group (panel) is provided with the same set of background information and is asked to conduct and document (in writing) a self-assessment. These assessments can then be used in one of three ways. They can be given to an independent individual who produces an average interpretation, which is then returned to each individual within the group for further comment and so on until consensus is achieved. Alternatively, the independent individual analyses and compares responses and goes back to the individuals concerned with comments, requesting revision or explanation as to why extreme or unusual opinions are held – a process of controlled feedback that is continued until there is agreement. The third approach is for assessments to be simply provided anonymously to the other assessors, who are encouraged to adjust their assessment in light of the peer assessment. Irrespective of the process adopted, it is typical for individual assessments to converge and for iterations to be continued until consensus is achieved. As the Delphi technique maintains anonymity and independence of thought, it precludes the possibility that any one member of the panel may unduly influence any other.

Useful lessons can be learnt from the way the US nuclear industry has tackled this issue for probabilistic seismic hazard assessment (PHSA). PHSA is used to estimate the likelihood that levels of ground motion will be exceeded in a given time period. It provides input to seismic safety assessment for many engineered structures. The main stages are as follows (Panel on Seismic Hazard Analysis, 1988).

1. Identification of seismic source zones around the area of interest. Each zone is assumed to have a uniform earthquake occurrence.
2. Specification of a time-rate for earthquake occurrence for each seismic source zone, with the distribution of earthquakes up to a limiting maximum magnitude.
3. Specification of an attenuation function that defines how ground motions decay with distance away from the source.
4. Probability analysis that integrates overall earthquake magnitudes and distances, and sums over all source zone contributions to obtain the expected exceedence frequency of every possible ground motion per unit time.

Typically, the method requires the use of both the 'frequentist' and 'degree of belief' approaches to probability assessment. Historical records are used to define the earthquake magnitude–frequency relationship. However, there often remain significant uncertainty associated with, for example, the maximum earthquake magnitude, the fault rupture characteristics (e.g. strike–slip, reverse or normal) and rupture frequency, which usually requires the use of 'weighting factors' derived from expert judgement. Multiple interpretations are often possible owing to the large uncertainties in all the available data. This can lead to a lack of consensus between experts, resulting in disagreement on the selection of ground motion for design at a given site (e.g. Savy et al., 2002).

The Senior Seismic Hazard Analysis Committee was established to review the PSHA process and consider ways of improving the use of experts (SSHAC, 1997). They found that most problems were caused by procedural rather than technical difficulties, especially the way in which experts were used (Budnitz et al., 1997). As a result, the SSHAC set out procedural guidance for the use of experts. Among the key points made by the committee are

- Expert judgement is used to represent the informed scientific community's state of knowledge.
- It is impractical and unnecessary to engage an entire scientific community in any meaningful interactive process.
- Decision-makers must always rely on a smaller, but representative, set of experts.
- An expert panel is a sample of the overall expert community.
- A technical integrator should act as the expert 'pollster' of that community, responsible for capturing efficiently and quantitatively the community's degree of consensus or diversity.

The committee believed that 'regardless of the scale of the PSHA study, the goal remains the same: to represent the centre, the body, and the range of technical interpretations that the larger informed technical community would have if they were to conduct the study' (Budnitz et al., 1997).

The recommended approach views the expert panel as a team, with a technical facilitator/integrator (TFI) as the team leader, working together through a series of structured interviews and workshops to arrive at a composite representation of the knowledge of the group and the broader technical community (Figure 5.19). To achieve this composite representation, the committee recommended a

Figure 5.19 The Senior Seismic Hazard Analysis Committee procedure for the use of an expert panel and TFI. (After SSHAC, 1997)

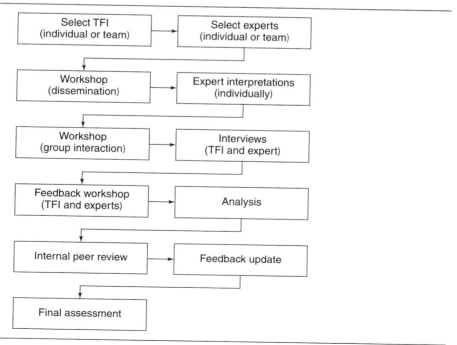

blending of judgemental and mathematical methods – that is, both degree of belief and relative frequency approaches. A trial application of the TFI procedures demonstrated that the approach can lead to an unusual degree of agreement among experts through thorough discussion of the available data, and through interaction between the experts (Savy et al., 2002). This, together with the focusing effect of the TFI, leads to narrower margins of variation without any coercion. The committee also recommended independent peer review, both throughout (participatory) and towards the end of the assessment (late stage).

A simplified version of this procedure was adopted by BP's Geohazard Assessment Team (GAT), which has undertaken seabed geohazard (especially landslide) risk assessments to support decision-making on deep-water field development (e.g. Evans et al., 2007; Moore et al., 2007). The GAT comprises specialists from a range of disciplines, including geophysics, geology, geomorphology, engineering geology, geotechnical engineering and risk analysis.

The assessment process involves the GAT team members convening in a series of structured workshops where the available information is presented and challenged. Within the workshop context, the individual specialists have a responsibility to both challenge the geohazard models developed from the available information and discuss the full range of possible interpretations that might be present in the broader scientific and engineering community. The workshops are led and supervised by a TFI, who is responsible for developing as much of a consensus as possible.

Over the years, many agencies have made use of expert elicitation, including the IPCC (e.g. Moss and Schneider, 2000), the European Environmental Agency (Meozzi and Iannucci, 2006) and the US Environmental Protection Agency (EPA, 2005). Various approaches have been developed, with the common aim of producing and documenting a credible and traceable account of the probability assessment process. The IPCC have advocated the use of the so-called Stanford/SRI protocol (Spetzler and Staël von Holstein, 1975) for use in the elicitation of expert judgements about the uncertainties in emissions. The procedure involves an analyst (elicitor) and a representative panel of experts, and can be divided into five separate phases.

1. *Motivating the experts*: this involves developing rapport with the experts and explaining and justifying the basic idea of probabilistic assessment. This phase also should include a systematic search for motivational bias – that is, any reason why an expert might say something that does not represent his or her true beliefs.
2. *Structuring the problem*: the objective of this phase is to arrive at a clear and unambiguous definition of the quantity to be assessed. Careful work may be needed to unearth unspoken assumptions that experts may be making but that are not actually stated in the problem.
3. *Conditioning*: the objective of the conditioning phase is to get all the experts ready to think about their judgement and avoid cognitive biases. Each expert has to talk about how he or she will go about making the judgement, what data will be used, and how. Alternatives should be discussed while the analyst looks carefully for possible cognitive biases so that a method is arrived at that minimises them.
4. *Encoding*: in this phase, each expert gives his or her assessment. When asking for the spread of some variable, the common bias is for experts to give a spread that is too narrow. The analyst can counter this to some extent by first asking each expert to choose extreme values (i.e. eliciting extreme values), and then asking each expert to describe conditions that might lead to results outside those extremes (i.e. extreme assessment). As a consequence, each expert then gets the chance to think again.

5 *Verifying*: the objective of this phase is to verify that the assessments obtained truly reflect the experts' views. Their assessments should be discussed with them individually to make sure that the estimates correctly represent their beliefs.

Many geoscientists tend to be sceptical about using expert elicitation methods. Often, these are not perceived as reliable or rigorous scientific methods, in contrast to those used in empirical studies. As a consequence, the results may be considered to be inherently less accurate. This can reflect a lack of knowledge about formal expert elicitation, or a disproportionate trust in the quality and representativeness of empirical data. By making judgements explicit and transparent, as is done in formal expert elicitation, criticism should in fact decrease instead of increase. Knol *et al.* (2010) have suggested that, by designing expert elicitations in a structured way, they begin to resemble the design of a scientific experiment and that this could create increasing trust in the results among sceptical geoscientists. The procedures can also provide a mechanism for greater transparency in the estimation process, justified through adequate documentation, thereby allowing any reviewer to be able to trace the reasoning behind particular estimates. However, there always remains the possibility that experts outside the group, third-party interests or members of the public will view the process as being biased and not reflecting their views.

A formal expert elicitation can, however, be a time-consuming and resource-intensive activity (van der Sluijs *et al.*, 2004). The entire procedure from defining a study and selecting experts to staging expert meetings, reconciling differing views, analysis and reporting can easily stretch over several months or even years (e.g. Hora and Iman, 1989). The decision to conduct a formal elicitation will undoubtedly depend on the price the client is willing to pay for more rigorous and defensible results.

More informal procedures can, of course, be developed. For example, a field visit was undertaken by an expert team (a combination of geomorphologists, engineering geologists and geotechnical engineers) to inspect the landslide risks along a crude oil pipeline (the pipeline crosses a number of landslide features that have been classified as high, medium and low risk). At the conclusion of the visit, the individual team members were asked separately to estimate the total landslide-related pipeline rupture risk of along the pipeline (they were not forewarned of this task). Each team member was given a pre-prepared sheet and asked to conduct and document (in writing) a self-assessment of the following question.

- What is the chance (expressed as a percentage) that there will be at least one pipeline rupture caused by landslides in the next 25 years?

Separate estimates were required for four management/resource cases

- Do nothing – that is, there should be no further investment in landslide monitoring and management activity.
- Maintain present management/resources – that is, continue with the current level of landslide monitoring and management.
- Improve present management – that is, provide additional staff and resources for landslide monitoring and management.
- High-risk site stabilisation/avoidance – that is, re-route to avoid the high-risk sites.

The results are presented in Figure 5.20 and Table 5.9, along with an agreed consensus value for each case. The objective of the assessment was to establish each individual's view of the total risk

Figure 5.20 Pipeline inspection team: expert panel views of the probability of rupture under different management scenarios. Black triangles indicate the consensus view

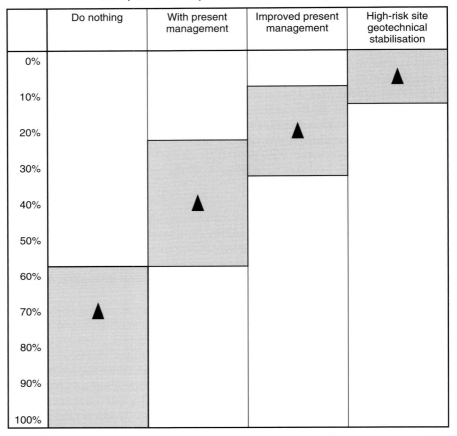

and the relative significance of high-, medium- and low-risk landslides over the 25-year 'lifetime' period of the pipeline. The results highlighted the degree of consensus between team members of their perception/understanding of the landslide risks and also provide an indication of the anticipated effectiveness of different management approaches.

The 'do nothing' case determines the total risk

$$\text{total risk} = P(\text{high}) + P(\text{moderate}) + P(\text{low})$$

The 're-routing at high-risk site' case determines the contribution from moderate- and low-risk sites:

$$\text{risk} = P(\text{moderate}) + P(\text{low})$$
$$= 5\%$$

Table 5.9 Pipeline risk example: estimated total 'lifetime' risk levels for different management/resource cases

	Do nothing	With present management	Improved present management	High-risk site re-routing
Expert 1	65–90% (77.5%)	35–55% (45%)	15–25% (20%)	0–10% (5%)
Expert 2	100%	50%	10%	5%
Expert 3	70%	40%	25%	10%
Expert 4	100%	50%	30%	10%
Expert 5	60%	25%	10%	5%
Consensus value	70%	40%	20%	5%

This also yields the expected contribution from high-risk sites

$$P(\text{high}) = \text{total risk} - [P(\text{moderate}) + P(\text{low})]$$
$$= 70\% - 5\%$$
$$= 65\%$$

The consensus 'lifetime' probability estimate was used to generate the annual probability of rupture. For example, if the high-risk sites are judged to have, collectively, an estimated 65% chance of causing pipeline rupture in the next 25 years, then it is possible to 'track back' to yield an estimated annual probability using a simple version of the binomial distribution

$$\text{annual } P(\text{event}) = 1 - [1 - P(\text{event in } n \text{ years})]^{1/n}$$
$$= 1 - [1 - ((1 - 0.65)^{1/25}$$
$$\approx 0.04$$

This example provides an indication of how the expert elicitation procedure can be adapted to form part of a field assessment of landslide issues. The key principles were the establishment of a team of experts, reviewing and discussing the available evidence as a group, the elicitation of the individual's views, and then group discussion to try and establish a consensus view. The results supported the development of a landslide management strategy that includes re-routing to avoid high- and medium-risk sites, and the automatic monitoring of low-risk sites.

5.4. Probability assessment and reliability methods

Probabilities and probabilistic methods are useful because of our ignorance of the true future frequency of events or the values of different parameters, such as shear strength. A common problem is that while it is possible to say with certainty that a landslide event will occur, it is not possible to say exactly when it will occur. It may be possible to estimate the annual probability of a particular event (known as *prediction*), but not what year it will occur in (known as *forecasting*). Probabilistic methods (predictions) are essentially sophisticated sensitivity tests in which single data values are replaced by probability distributions that cover all possible values or outcomes.

In general, probability assessment for landslide events and their consequences can involve one or more of the following types of methods (e.g. Christian, 2004).

- The use of available statistics or databases.
- Conceptual modelling using influence diagrams, event trees and fault trees.
- Reliability methods.

5.4.1 The use of landslide statistics

Statistical analysis of data series can yield important conclusions about the future occurrence of uncertain events. To a large degree, this is a 'black box' approach, with the focus solely on the outputs from a slope system (i.e. the landslide events) and with no, or limited, concern as to the nature and mechanics of the system. Typically, the statistics will be in the form of a landslide frequency over time within an area, with limited information about the causes and landslide types. An assumption will be made that historical frequency/year is the same as annual probability. The main issue to be considered is whether the statistics are representative of the area of interest, in terms both of space (e.g. the area of interest may actually lie outside the area covered by the statistics) and of time (e.g. the current or future environmental conditions may be different from those operating within the period covered by the statistics, so uniformitarian principles cannot be applied).

Suitable time series data sets are rare. Hong Kong is one of the few regions with an extensive database of natural terrain landslide events. It was developed because of concerns for the safety of new development sites, which led the Geotechnical Engineering Office to undertake a programme of region-wide landslide hazard mapping – the Natural Terrain Landslide Study (NTLI) (Evans and King, 1998; Evans et al., 1997; King, 1997, 1999). An inventory of landslide features and areas of intense gullying was compiled from the interpretation of high-level aerial photographs (1:20 000 to 1:40 000 scales) taken in 1945, 1964 and annually from 1972 to1994 (excluding 1977). Most parts of Hong Kong have appeared on between 20 and 23 sets of photographs.

A total of 26 780 natural terrain landslides were identified within an area of 640 km^2, mostly debris slides or debris avalanches within weathered rock or overlying colluvium. Although the majority of recorded landslides were relatively small features, a significant proportion (15%) exceeded 20 m in width. Of the total of 26 780 slides, 8804 were described as having occurred within the last 50 years (i.e. 'recently,'). Thus, the relative frequency of recent landsliding is

relative frequency = (number of recorded landslides)/(time period)

= 8804 landslides/50 years

= 176 landslides/year

For the surveyed area (630 km^2), the landslide frequency works out as 0.28 landslides per km^2 per year.

These average statistics smooth out the effects of geology and slope angle in controlling the distribution of landsliding across Hong Kong. As a result, subpopulations (i.e. samples) of the data set were generated for different combinations of geological unit and slope class, as derived from a 1:20 000-scale digital terrain model. Table 5.10 presents the landslide frequency statistics for a selection of the bedrocks and reveals landslide frequencies of 1–34 events/year, with slope class subpopulations varying from 0 to >10 events/year. Normalisation of the subpopulations allows the underlying characteristics of the data sets to be compared by dividing the frequency statistics by the area covered by each subpopulation. The resulting normalised landslide frequencies for the selected bedrocks range from around 0.08 to 1 event per km^2 per year, with slope class subpopulations varying from 0 to 1.4 events per km^2 per year.

Table 5.10 Hong Kong Natural Terrain Landslide Inventory: relative frequency statistics for selected bedrocks

Slope class: deg	Coarse-grained granite and pegmatite				Lai Chi Chong Formation[a]				Trachydacite, dacite and rhyolite lava			
	Area: km^2	Recent slides (<50 years old)	Landslide frequency: events/year	Density of recent slides: per km^2 per year	Area: km^2	Recent slides (<50 years old)	Landslide frequency: events/year	Density of recent slides: per km^2 per year	Area: km^2	Recent slides (<50 years old)	Landslide frequency: events/year	Density of recent slides: per km^2 per year
0–5	0.47	1	0.02	0.04	0.19	13	0.26	1.40	2.52	7	0.14	0.06
5–10	0.57	–	–	–	0.25	18	–	–	3.51	21	–	–
10–15	0.82	1	0.02	0.02	0.42	18	0.36	0.86	5.68	53	1.06	0.19
15–20	1.36	2	0.04	0.03	0.66	32	0.64	0.97	9.69	117	2.34	0.24
20–25	2.13	3	0.06	0.03	1.20	45	0.9	0.75	16.62	265	5.3	0.32
25–30	2.87	12	0.24	0.08	1.96	113	2.26	1.15	19.37	580	11.6	0.60
30–35	2.24	11	0.22	0.10	1.49	52	1.04	0.70	13.81	430	8.6	0.62
35–40	1.00	10	0.20	0.20	0.66	23	0.46	0.70	7.58	175	3.5	0.46
40–45	0.29	3	0.06	0.21	0.18	4	0.08	0.44	2.90	45	0.9	0.31
45–50	0.08	3	0.06	0.79	0.04	–	–	–	0.89	13	–	–
50–55	0.03	2	0.04	1.43	0.03	–	–	–	0.26	2	–	–
55–60	0.01	–	–	–	0.01	–	–	–	0.09	–	–	–
>60	0.01	–	–	–	0.02	–	–	–	0.08	–	–	–
Total	11.86	48	0.96	0.08	7.08	318	6.36	0.90	82.99	1708	34.16	0.41

From Evans and King (1998)

[a] The Lai Chi Chong Formation comprises coarse ash tuff, rhyolite lava, mudstone and siltstone

These relative frequency statistics provide an excellent framework for estimating the probability of landsliding in different geological and slope classes across Hong Kong. However, it is unclear how useful they would be for an assessment of landslide risk along a hypothetical gas pipeline route over the border in mainland China. Although the terrain and geological conditions are similar, there has been a very different land use history. Also, the statistics do not reveal which landslide events (e.g. width, depth or velocity characteristics) could cause pipeline damage, or in which topographic settings (e.g. ridgelines, gully heads or valley-side slopes) would the pipeline be most likely to fail.

5.4.2 The use of conceptual models, event trees and fault trees

Conceptual models, event trees and fault trees are techniques for describing the varying sequence of events that might be involved in the generation of different magnitudes of landslides and their consequences. They can be viewed as 'grey box' methods, involving a conceptual understanding of the nature and mechanics of the system, including responses to triggering events. They are particularly useful in situations where the probability assessment process is faced by

- insufficient databases of landslide activity or its consequences and hence limited quantitative understanding of the long-term behaviour of the slopes in the area of interest or the triggering processes
- limited understanding of the relevant processes operating on the slopes, with a consequent lack of reliable models to predict future slope behaviour.

One approach to overcoming these problems is to adopt a conceptual modelling approach. This involves identifying the nature of the interactions between different components within a physical system. For example, although the workings of a carburettor can be explained in isolation, its function and performance can only be understood within the spatial context of a larger system, namely the car engine. So, in the case of landsliding, the focus is on the interactions within and between components, such as a slope or a catchment (i.e. how the slope or catchment 'works'). The focus is directed towards gaining understanding in the context of the behaviour of broader-scale physical systems (i.e. a 'top-down' approach) rather than simply the cumulative product of site specific interactions between the imposed stresses and the strength of the materials (i.e. a 'bottom-up' approach), as used in stability analysis.

At a very simple level, a slope can be viewed as a physical system, with energy inputs (e.g. rainfall and earthquake events) and outputs (e.g. sediment transport by landslides and surface erosion). Slope failure can occur in response to the occurrence of an initiating event, such as heavy rainfall, undercutting or a large earthquake. However, some potential trigger events may not actually result in failure because the system state has not been sufficiently 'prepared' for failure (e.g. the antecedent groundwater levels may be very low). This sequence can be represented as a conceptual model that forms the basis for a simple conditional probability model:

$P(\text{landslide}) = P(\text{trigger}) \times P(\text{landslide}|\text{trigger})$

The uncertainty about whether the potential triggering event will actually initiate a landslide is accommodated in the expression $P(\text{landslide}|\text{trigger})$, and will reflect the system state at the time of the event (the pore pressure distribution, strength of the materials etc.).

Many landslide problems are more complex and cannot be represented by this type of simple model. For example, a small industrial estate has been located in a former quarry site. Part of the estate is protected by a wire rock fence. The estate comprises a combination of light structures, used by small firms making electronic components, car repair businesses and service companies, and adjacent parking lots. Following a severe winter, there was significant rockfall activity, raising concerns about the risks to the businesses and, in particular, their staff and customers.

There is a range of possible pathways that a rockfall event could take once it has become detached from the quarry face. Examples include

- breaking up at the cliff foot and coming to rest before the low barrier
- becoming trapped by the rock fence
- occurring on a section without a rock fence and coming to rest in open space
- occurring on a section without a rock fence and hitting a parked vehicle
- occurring on a section without a rock fence and hitting a building.

Of concern is the possibility of a rockfall hitting a load-bearing wall of a building, causing it to collapse, and leading to a fatality. This problem can be broken down into a series of logical steps and the chance of that outcome calculated by using a conditional probability model:

$$P(\text{fatality}) = P(\text{rockfall}) \times P(\text{reaches slope foot}|\text{rockfall}) \times P(\text{misses barrier}|\text{reaches slope foot})$$

$$\times P(\text{building hit}|\text{misses barrier}) \times P(\text{hits load-bearing wall}|\text{building hit})$$

$$\times P(\text{building collapse}|\text{hits load-bearing wall}) \times P(\text{fatality}|\text{building collapse})$$

Building this model requires information from a variety of different sources.

- $P(\text{rockfall})$ might be based on historical records or expert judgement.
- $P(\text{reaches slope foot}|\text{rockfall})$ is the probability that the rockfall reaches the slope foot, given that it has occurred. This judgement needs to take account of the presence of natural obstacles, such as trees or boulders at the base of the quarry face, which would prevent the rock fall reaching the rock fence or the estate boundary.
- $P(\text{misses barrier}|\text{reaches slope foot})$ is the probability that the rockfall misses the rock fence given that it has reached the slope foot. This will involve an assessment of the chance that the rockfall occurs along a section without a rock fence.
- $P(\text{building hit}|\text{misses barrier})$ is the probability that the rockfall hits a building, given that it has missed the rock fence. This will involve a combination of modelling rockfall travel distances and the chance of hitting a building rather than open space.
- $P(\text{hits load-bearing wall}|\text{building hit})$ is the probability that the rockfall hits a load-bearing wall, given that it has hit a building. This will involve a review of the building design and layout.
- $P(\text{building collapse}|\text{hits load-bearing wall})$ is the probability that the building collapses, given that the rockfall has hit a load-bearing wall. This will involve an understanding of the vulnerability of a particular type of structure to the impact of a rockfall of a particular size travelling at a particular velocity.
- $P(\text{fatality}|\text{building collapse})$ is the probability that a fatality occurs, given that the building has collapsed. This will involve an estimate of the chance that the building was occupied at the time of the event and the vulnerability of a person to the effects of a collapsing building.

However, this is only one of a number of event sequences that might lead to a fatality at this site. For example, visitors sitting in their vehicles in the parking lot might be hit by a falling rock, workers in their offices might be hit by rocks going through windows, and so on. Event trees are an effective method of tracing all the possible sequences (Figure 5.21). They are a specific form of branching logic diagram that allows all likely sequences of events, or combinations of scenarios, arising from an initial event to be mapped as a branching network. They use inductive analysis to 'decompose' complex problems into simpler sequences of initiating events, responses and outcomes, addressing the basic question 'what would happen if ...'?

Event tree analysis was developed in the 1960s as a form of decision analysis in the nuclear industry. The value of decision trees is that they provide a graphical insight into the way particular consequences might arise and can be used as a template for probability assessment. There is a long tradition of use in the assessment of dam safety (e.g. Hartford and Baecher, 2004), and they have proved valuable in analysing landslide risks, both on 'natural' and on engineered slopes (e.g. Ho et al., 2000; Hsi and Fell, 2005; Lacasse and Nadim, 2008, Lacasse et al., 2008; Lee and Moore, 2007; Lee et al., 2000).

Event trees are typically organised left–right, and comprise the following (Figure 5.22).

- The occurrence of an initiating event, such as a rockfall or a triggering event such as large earthquake or prolonged period of heavy rainfall.
- Branches mapping out all the alternative pathways that could develop following the initiating event. While each branch will be unique, they all link back to the initiating event.
- Nodes that act as transitions from one position along a pathway to one or more alternative pathways. Typically, the nodes define binary (Yes/No) alternative pathways (e.g. will a particular landslide respond to a triggering event?).
- Terminal points (leaves), the end of the braches, which define a unique end-state that is conditional on the preceding events.

The pathway options at each node should be collectively exhaustive (i.e. they should encompass all possibilities). As a result, the sum of the probabilities of moving along any of the pathway options at a node will equal 1.0:

$$P(A_1) + P(A_2) = 1$$

$$P(A_2) = 1 - P(A_1)$$

The probability of any one of the terminal points (i.e. the final outcomes: T_1, etc.) will be the conditional probability of all the nodes along the pathway, starting with the initiating event (IE):

$$P(T_1) = P(IE) \times P(A_1) \times P(B_1) \times P(C_1)$$

The total probability of a particular outcome (e.g. fatality) that could arise from more than one pathway through the event tree will be the sum of the terminal point probabilities, since these events are mutually exclusive:

$$P(\text{fatality}) = P(T_1) + P(T_5)$$

The combined probabilities of all the terminal points (e.g. T1 to T8) should equal the probability of the initiating event at the start of the event tree:

$$\sum P(T1 \text{ to } T8) = P(IE)$$

Introduction to probability and quantitative assessment

Figure 5.21 Example event tree: rockfalls in an industrial estate

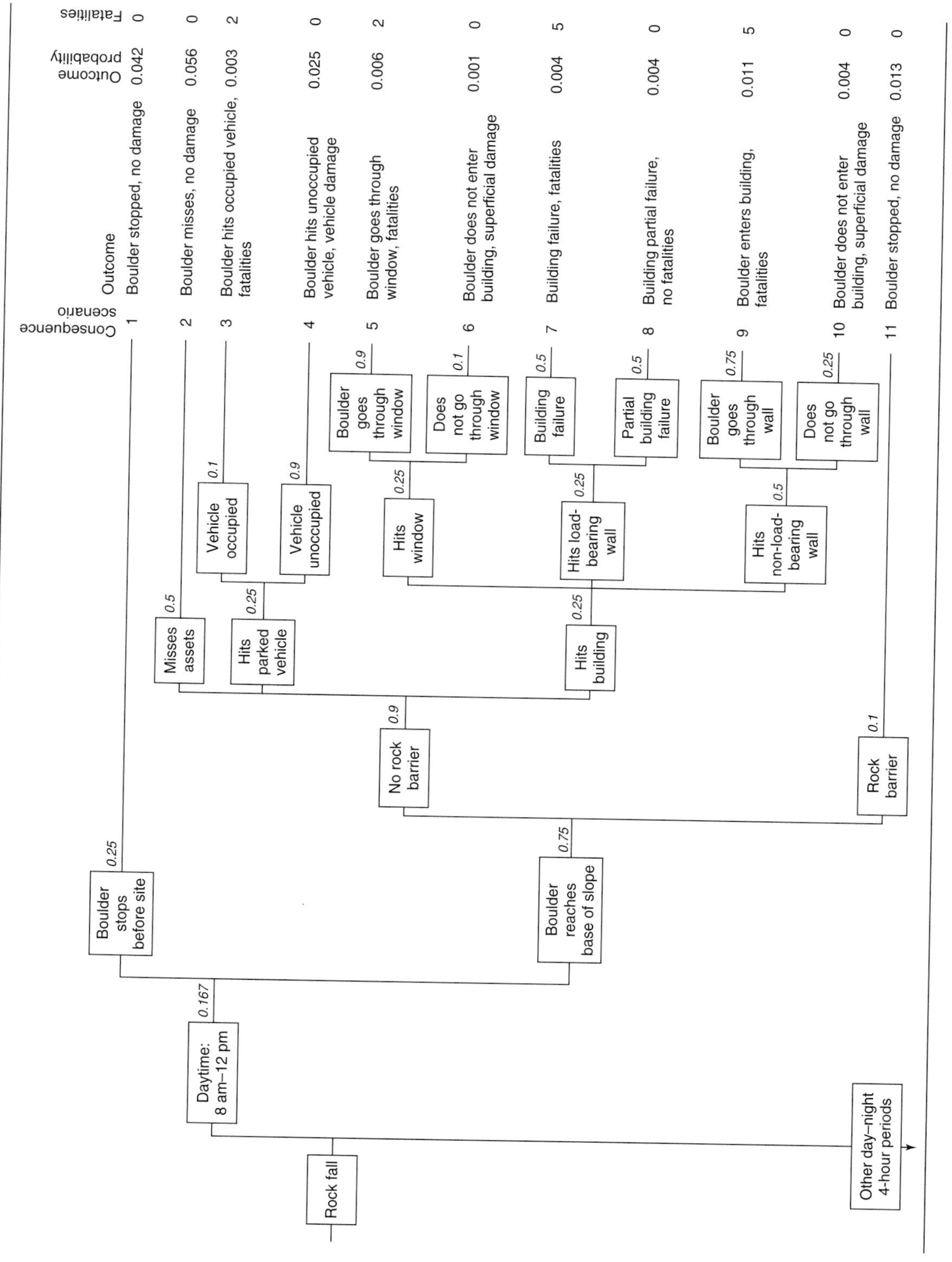

Consequence scenario	Outcome	Outcome probability	Fatalities
1	Boulder stopped, no damage	0.042	0
2	Boulder misses, no damage	0.056	0
3	Boulder hits occupied vehicle, fatalities	0.003	2
4	Boulder hits unoccupied vehicle, vehicle damage	0.025	0
5	Boulder goes through window, fatalities	0.006	2
6	Boulder does not enter building, superficial damage	0.001	0
7	Building failure, fatalities	0.004	5
8	Building partial failure, no fatalities	0.004	0
9	Boulder enters building, fatalities	0.011	5
10	Boulder does not enter building, superficial damage	0.004	0
11	Boulder stopped, no damage	0.013	0

Landslide Risk Assessment

Figure 5.22 The structure of an event tree: dam overtopping

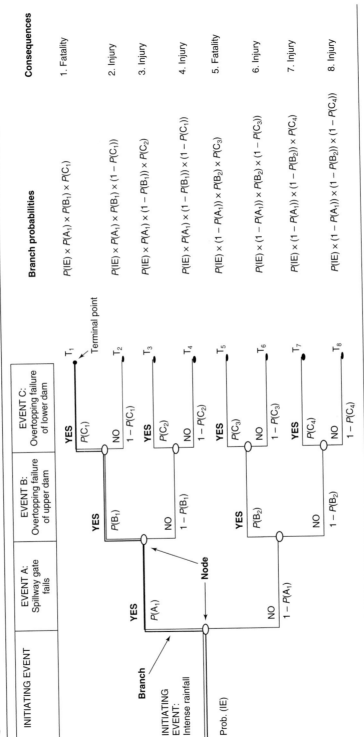

Typically, engineering judgement and experience play a key role in event tree analysis. Indicative probability values are often used to achieve consistency in the estimation of the branch probabilities (e.g. Lacasse and Nadim, 2011; Lacasse *et al.*, 2010).

- Virtually impossible: probability = 0.001. The event can be ruled out with almost complete confidence owing to known physical conditions or processes.
- Very unlikely: probability = 0.01. The possibility cannot be entirely ruled out on the basis of physical or other reasons.
- Unlikely: probability = 0.1. The event is unlikely, but it could happen.
- Completely uncertain: probability = 0.5. There is no reason to believe that one outcome is any more or less likely than the other to occur.
- Likely: probability = 0.9. The event is likely, but it may not happen.
- Very likely: probability = 0.99. The event is highly likely, but may not happen, although one would be surprised if it did not happen.
- Virtually certain: probability = 0.999. The event can be described and specified with almost complete confidence owing to known physical conditions or processes.

There is no unique way to represent a problem using an event tree. Developing an event tree is an inductive process, and hence the clarity and scope of a tree will reflect the skill, experience and insight of the experts involved. Influence diagrams are a useful first step in characterising landslide problems, providing a compact framework from which event trees can be constructed. They can be used to illustrate relationships between initiating events, system conditions, system responses and outcomes.

An influence diagram for a situation where renewed recession of a currently protected cliffline could occur as a result of seawall failure is presented in Figure 5.23. During severe storm conditions, extreme rainfall might trigger a landslide that could displace the wall or cause structural damage by loading it. The wall could also be under threat from the extreme wave loadings resulting from a combination of high storm waves and a storm surge (i.e. extremely high water levels). One approach to model this

Figure 5.23 Influence diagram: seawall failure

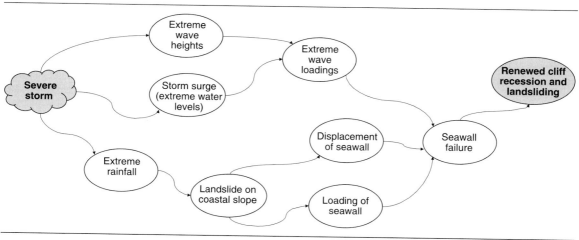

Landslide Risk Assessment

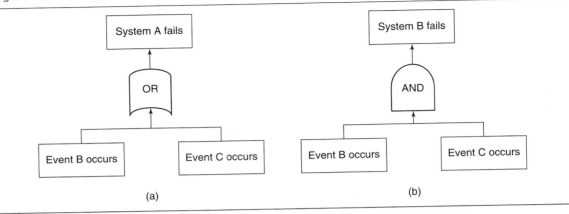

Figure 5.24 Fault tree analysis: general fault tree structure

situation would be to develop two separate event trees: one for the landslide-related scenario (scenario A) and the other for the extreme wave loadings (scenario B). As these are not independent events (they could occur during the same storm):

$$P(\text{seawall failure, A or B}) = 1 - \{[1 - P(A)] \times [1 - P(B)]\}$$

An alternative approach is to identify a particular consequence scenario (e.g. a fatality caused by a rockfall hitting a vehicle) and use so-called 'backward logic' to establish how the accident could happen – that is, the causal relationships that lead to the accident, undesired event or 'top event'. The combinations of factors needed to generate a particular consequence can be identified through the use of fault trees (e.g. Fussell, 1973, 1976; Kumamoto and Henley, 1996).

Fault tree analysis was developed in 1961 by the Bell Telephone Laboratory team to evaluate the reliability of missile launch control systems. The method has since become an integral part of safety assessment for a wide range of sectors, including the nuclear, chemical and oil and gas industries.

Fault trees are based on deductive analysis to identify all the possible ways of producing a single outcome (i.e. 'how could it happen'?). Working back from the top of the tree, the required preconditions are set down in progressively more detail. With reference to Figures 5.24 and 5.25, fault trees comprise the following.

- The top event: this is the failure of a system to perform its function; for example, a retaining wall to provide support to a slope.
- Primary or basic events (shown as open circles): these are the basic faults, conditions or failures that lead onto the top event (e.g. drainage failure in front of a retaining wall).
- Fault events (shown as rectangles): these are lower-level faults. Fault events receive inputs from and provide outputs to a logic gate.
- Logic gates: these define the relationships between events leading to the top event. There are two main types: the OR gate and the AND gate. In Figure 5.24(a), system A fails if one or both

Figure 5.25 Fault tree analysis: general principles

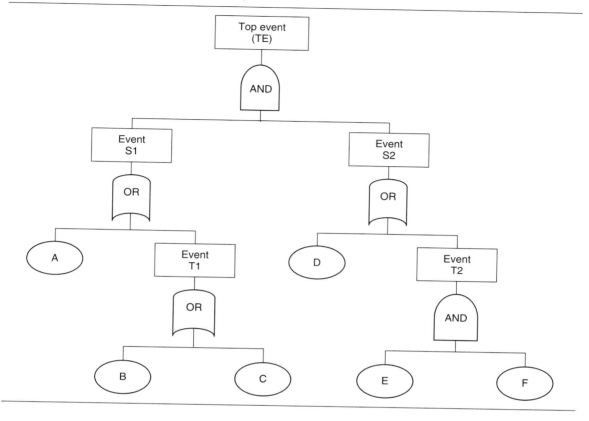

primary events B or C occur. Both events B and C have to happen in combination in Figure 5.24(b) for the top event to occur.

In Figure 5.25, as both events S1 AND S2 are required to cause the top event (TE), the probability of the top event is calculated as the intersection (∩) or overlap between events S1 and S2:

$P(TE) = P(S1) \times P(S2)$

As event S1 can be caused by either primary event A OR fault event T1, it is calculated as

$P(S1) = P(A) + P(T1)$

Event T1 is caused by B OR C, and is calculated as

$P(T1) = P(B) + P(C)$

Hence, the probability of event S1 becomes

$P(S1) = P(A) + P(B) + P(C)$

The logic is similar for event S2. However, event T2 is caused by a combination of both primary events E AND F, and is calculated as

$$P(T2) = P(E) \times P(F)$$

Hence, the probability of event S2 becomes

$$P(S2) = P(D) + [P(E) \times P(F)]$$

The probability of a precondition being met can be derived from historical statistics (e.g. the failure frequency of early warning systems) or through the use of expert judgement. Fault tree analysis has the capability of providing useful information concerning the likelihood of a top event or accident and the means by which such an event could occur. Efforts to improve slope safety can be focused and refined using the results of the analysis.

Fault tree analysis has been used in dam safety assessments (Figure 5.26), but has proved difficult to use in practice (Hartford and Baecher, 2004). This is because, unlike a piece of mechanical equipment, it is difficult to subdivide a dam into a set of individual components and then unequivocally link failures among a subset of the dam components to subsequent failures of others. For similar reasons, fault trees are not a widely used method of estimating landslide risk.

5.4.3 The use of reliability methods

Reliability can be defined as the probability that a structure (e.g. a retaining wall) or system (e.g. a slope) performs its required function (i.e. it does not fail) for a specified time period under stated conditions:

reliability = 1 − probability of failure

Reliability theory is well established in civil engineering, and the methods provide a means of assessing the performance of existing structures or slopes and quantifying the uncertainties associated with proposed structures (e.g. Baecher and Christian, 2003; Reeve, 2010). However, reliability methods require estimates of the variance of significant parameters and hence more data than conventional deterministic models. This can be expensive and will also require expert judgement. Nadim (2006) suggests that the additional cost tends to make reliability methods more useful for major projects than routine work.

Reliability methods take account of geotechnical uncertainties through analytical models to obtain probabilistic descriptions of the performance of a structure or slope (e.g. Christian, 2004). For example, the performance (or survival) function describes the probability that a variable X takes on a value greater than a number x. Thus, the survival function $G(x)$ of a structure could be defined as the probability that the material strength X is greater than an imposed load x:

performance function $G(x) = P(X > x) = 1 - F(x)$

where $F(x)$ is the cumulative probability distribution for the random variable X (in this case material strength).

The failure rate (or hazard function) is the frequency with which an engineered system fails, expressed for example in failures per hour. It can be defined in terms of the performance function (i.e. the

Introduction to probability and quantitative assessment

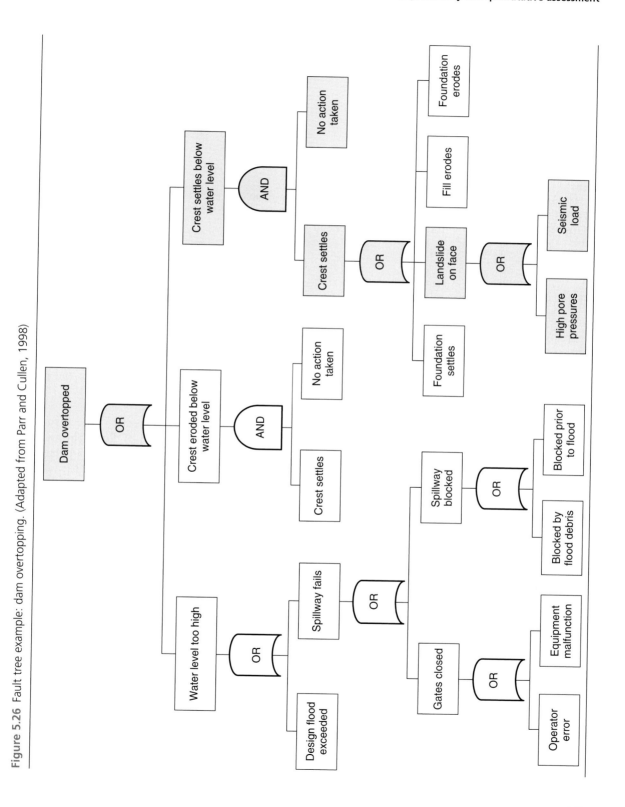

Figure 5.26 Fault tree example: dam overtopping. (Adapted from Parr and Cullen, 1998)

probability of no failure before time t):

$$\text{failure rate} = \frac{f(t)}{1 - F(t)}$$

where $f(t)$ is the 'time to failure distribution' (i.e. the failure density function).

The failure probability over time t is

$$G(t) = 1 - e^{-(\frac{t}{b})^a}$$

For example, on a series of railway cuttings, the slope drainage pipe sections are known to have a Weibull failure density (a two-parameter distribution, with $a = 2$ and $b = 300\,000$). The probability that a pipe section fails within 5 years (i.e. 43 830 hours) is

$$G(t) = 1 - e^{-(43\,830/300\,000)^2}$$
$$= 1 - 0.98$$
$$= 0.02$$

Traditional approaches to analysing slope stability are essentially *deterministic*, in that parameters are represented by single values (e.g. a shear strength value or year when the event will occur). Confidence in the results can be improved by varying the values by sensitivity testing. But sensitivity tests do not incorporate the likelihood of a particular value, beyond the engineer's judgement about the range within which the true value is expected to lie.

In probabilistic stability analysis, the probability of failure is defined with respect to a performance function $G(x)$, which represents slope stability as a function of a combination of slope parameters, each with a continuous probability distribution (i.e. random variables; note that x denotes a particular combination of values). For slopes, the key random variables are usually the shear strength parameters and the pore water pressure ratio (Chowdhury and Flentje, 2011). Conventional (i.e. deterministic) stability analysis models generally serve as the basis for probabilistic analyses (e.g. Nadim et al., 2005).

The probability that the factor of safety falls below 1.0 is determined from the frequency with which the input values would fall below the value that yields a factor of safety of 1.0 in the stability model. Slopes with the same factor of safety could have significantly different failure probabilities because of the uncertainties associated with the ground conditions (Figure 5.27) (Nadim, 2006). A low factor of safety does not necessarily correspond with a high probability of failure, and vice versa (Nadim et al., 2005). Indeed, the relationship between the factor of safety and probability of failure depends on the uncertainties in load and resistance.

The performance (or survival) function can also be expressed in terms of the factor of safety of a slope:

performance function $G(x) = $ factor of safety $- 1$

As a result, it is possible to use the performance function $G(x)$ to define the stability state:

$G(x) > 0$ 'safe' conditions

$G(x) \leq 0$ 'unsafe' or 'failure' conditions

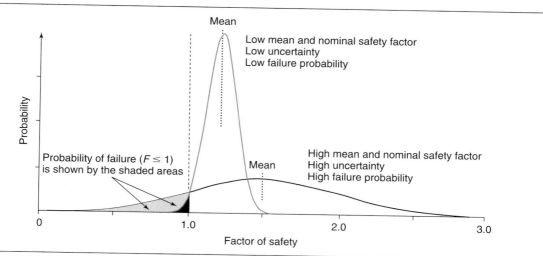

Figure 5.27 Stability analysis: comparison of two cases with different factors of safety and uncertainty. (After Nadim, 2006)

The FORM approach (first-order reliability method) (Ang and Tang, 1984) provides approximations for the mean and standard deviation of the factor of safety as a function of the mean and standard deviations of the various input parameters and their correlations.

The probability of failure can be estimated using the reliability index β:

$$\beta = \frac{\text{mean of } G(x)}{\text{standard deviation of } G(x)}$$

The reliability index is a measure of the reliability of an engineering system (e.g. a man-made slope) that reflects both the mechanics of the problem and the uncertainty in the input variables. The index is defined in terms of the expected mean value and the standard deviation of the performance function. It enables comparisons to be made about the reliability of different slopes without having to calculate absolute probability values. It can be regarded as providing a measure of the confidence in the ability of a slope to perform in a satisfactory manner – that is, not to fail.

The reliability index can also be expressed as

$$\beta = \frac{\text{mean factor of safety} - 1}{\text{standard deviation of factor of safety}}$$

The mean factor of safety minus 1 measures the difference between the mean factor of safety and a factor of safety F of 1.0 – that is, it indicates the margin of stability (Wu et al., 1996). By dividing this value by the standard deviation of F, the margin of stability becomes relative to the uncertainty in F.

The distribution function for the factor of safety must be assumed before the probability of failure can be estimated. If the reliability index is assumed to be the number of standard deviations by which the expected value of a normally distributed performance function $G(x)$ exceeds zero, then the probability

Table 5.11 Reliability index and probability of failure for various slopes

Mean factor of safety F	Standard deviation of F	Reliability index β	Probability of failure[a]
1	0.01	0.0	0.5
1.1	0.01	10.0	0
1.2	0.01	20.0	0
1.3	0.01	30.0	0
1.4	0.01	40.0	0
1.5	0.01	50.0	0
1	0.05	0.0	0.5
1.1	0.05	2.0	0.02275
1.2	0.05	4.0	3.17E-05
1.3	0.05	6.0	9.9E-10
1.4	0.05	8.0	6.66E-16
1.5	0.05	10.0	0
1	0.1	0.0	0.5
1.1	0.1	1.0	0.158655
1.2	0.1	2.0	0.02275
1.3	0.1	3.0	0.00135
1.4	0.1	4.0	3.17E-05
1.5	0.1	5.0	2.87E-07
1	0.2	0.00	0.5
1.1	0.2	0.50	0.308538
1.2	0.2	1.00	0.158655
1.3	0.2	1.50	0.066807
1.4	0.2	2.00	0.02275
1.5	0.2	2.50	0.00621
1	0.3	0.00	0.5
1.1	0.3	0.33	0.369441
1.2	0.3	0.67	0.252492
1.3	0.3	1.00	0.158655
1.4	0.3	1.33	0.091211
1.5	0.3	1.67	0.04779
1	0.4	0.00	0.5
1.1	0.4	0.25	0.401294
1.2	0.4	0.50	0.308538
1.3	0.4	0.75	0.226627
1.4	0.4	1.00	0.158655
1.5	0.4	1.25	0.10565

Expanded, after Chowdhury and Flentje (2003)
[a]Using Excel, for example, simply insert the function NORMSDIST($-\beta$) to obtain the probability value

of failure can be approximated by the cumulative standard normal distribution Φ evaluated at $-\beta$ (see Table 5.11):

probability of failure $\approx \Phi(-\beta)$

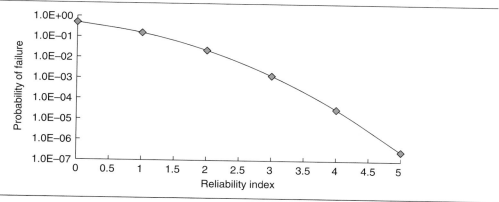

Figure 5.28 Relationship between reliability index and the probability of failure. (After Nadim, 2006)

Note that the standard normal distribution is widely tabulated and is available as a built-in function on many spreadsheet programs.

The relationship between probability of failure and the reliability index is illustrated in Figure 5.28, for a standard deviation of F of 0.1. 'Failure' in this context does not necessarily mean a catastrophic landslide (e.g. with large ground displacement), but may include what might be described in other contexts as pre-failure movement and development of tension cracks (e.g. Duncan, 2000). Thus, in general, the probability of failure may actually be regarded as the 'probability of unsatisfactory performance'.

The choice of distribution can, however, have a significant impact on the estimates of the probability of failure. Baecher (1987) demonstrated that for a slope with a mean factor of safety of 2 and a standard deviation of 0.4, the following probabilities of failure could be derived

- 2×10^{-3}, using a normal distribution
- 2×10^{-4}, using a lognormal distribution
- 2×10^{-4}, using a gamma distribution.

Probabilistic stability analysis is not for the faint-hearted, and involves a level of statistical competence that is beyond the scope of this book. Interested readers are therefore advised to refer to some of the detailed papers that address this subject in theoretical and practical terms, before moving further down this route (e.g. Alén, 1996; Ang and Tang, 1984; Chowdhury *et al.*, 2010; Li, 1991, 1992; Mostyn and Fell, 1997; Mostyn and Li, 1993; Nadim *et al.*, 2005; Nguyen and Chowdhury, 1984; Rosenbleuth, 1975, 1981; Vanmarcke, 1977, 1980; Yu and Mostyn, 1996).

REFERENCES

Alén C (1996) Application of a probabilistic approach in slope stability analysis. In *Landslides* (Senneset K (ed.)). Balkema, Rotterdam, pp. 1137–1148.

Ang A HS and Tang WH (1984) *Probability Concepts in Engineering Planning and Design*, vol. II. *Decision, Risk and Reliability*. Wiley, Chichester.

Baecher G (1987) *Statistical Quality Control of Engineered Embankments*. US Army Engineer Waterways Experiment Station, Vicksburg, MI.

Baecher GB and Christian JT (2003) *Reliability and Statistics in Geotechnical Engineering*. Wiley, Chichester, UK.

Benjamin JR and Cornell CA (1970) *Probability, Statistics and Decision for Civil Engineering.* McGraw-Hill, New York.

Benson MA (1960) *Characteristics of Frequency Curves Based on a Theoretical 1000 Year Record.* US Geological Survey Water Supply, Reston, VA. Paper 1543-A.

Bradley S (2011) *Scientific Uncertainty: A User's Guide.* Grantham Research Institute on Climate Change and the Environment, London. Technical Paper No. 56.

Brardinoni F, Slaymaker HO and Hassan MA (2003) Landslide inventory in a rugged forested watershed: a comparison between air-photo and field survey data. *Geomorphology* **54**: 179–196.

Budnitz RH, Apostolakis G, Boore DM, Cluff LS, Coppersmith KJ, Cornell CA and Morris PA. (1997) *Recommendations for Probabilistic Seismic Hazard Analysis: Guidance on Uncertainty and Use of Experts.* US Nuclear Regulatory Commission, Washington, DC. NUREG/CR-6372.

Casagrande A (1965) The role of the 'calculated risk' in earthwork and foundation engineering. *Journal of Soil Mechanics and Foundation Engineering* **91**(4): 1–40.

Chaytor JD, ten Brink US, Solow AR and Andrews BD (2009) Size distribution of submarine landslides along the U.S. Atlantic margin and its implication to tsunami hazards. *Marine Geology* **264**: 16–27.

Chowdhury R and Flentje P (2003) Role of slope reliability analysis in landslide risk management. *Bulletin of Engineering Geology and Environment* **62**: 41–46.

Chowdhury R, Flentje P and Bhattacharya G (2010) *Geotechnical Slope Analysis.* Balkema, Rotterdam.

Chowdhury R and Flentje P (2011) Practical reliability approach to urban slope stability. *11th International Conference on Applications of Statistics and Probability in Civil Engineering.* ICASP, 1–5, University of Woolongong.

Christian JT (2004) Geotechnical Engineering Reliability: How well do we know what we are doing? *Journal of Geotechnical and Geoenvironmental Engineering* **130**: 985–1003.

Conservation of Clean Air and Water in Europe (CONCAWE) (2007) *Performance of European Cross-country Oil Pipelines: Statistical Summary of Reported Spillages in 2005 and Since 1971.* Special Task Force on Oil Pipeline Spillages (OP/STF-1), CONCAWE Oil Pipelines Management Group, Brussels.

Dai FC and Lee CF (2001) Frequency-volume relation and prediction of rainfall-induced landslides. *Engineering Geology* **59**: 253–266.

Duncan JM (1996a) Soil slope stability analysis. In *Landslides: Investigation and Mitigation* (Turner AK and Schuster RL (eds)). National Academy Press, Washington DC, pp. 337–371. Transportation Research Board Special Report 247.

Duncan JM (1996b) State-of-the-art: limit equilibrium and finite-element analysis of slopes. *Journal of Geotechnical Engineering* **122**(7): 557–596.

Duncan JM (2000) Factor of safety and reliability in geotechnical engineering. *Journal of Geotechnical and Geoenvironmental Engineering* **126**: 307–316.

Dunning SA, Mitchell WA, Petley DN, Rosser NJ and Cox NJ (2007) Landslides predating and triggered by the 2005 Kashmir earthquake: rockfall to rock avalanches. *Geophysics Research Abstracts* 9, 06376.

Dussauge-Peisser C, Helmstetter A, Grasso J-R, Hantz D, Desvarreux P, Jeannin M, and Giraud A (2002) Probabilistic approach to rock fall hazard assessment: potential of historical data analysis. *Natural Hazards and Earth System Sciences* **2**: 15–26.

EA (Environment Agency) (1999) *National Sea and River Defence Surveys – Condition Assessment Manual. A Guide to the Visual Condition Assessment of Sea and River Defences.* Environment Agency, Peterborough.

European Gas Pipeline Incident Data Group (EGIG) (2005) *Gas Pipeline Incidents. 6th Report of the European Gas Pipeline Incident Data Group*. European Gas pipeline Incident Data Group, London. Document No. EGIG 05-R-0002. Available from: http://www.egig.nl.

Einstein HH (1991) Observation, quantification and judgement: Terzaghi and engineering geology. *Journal of Geotechnical Engineering* **117(11)**: 1772–1778.

EPA (Environmental Protection Agency) (2005) *Guidelines for Carcinogen Risk Assessment*. Risk Assessment Forum, Washington, DC. EPA/630/P-03–001B.

Evans NC and King JP (1998) *The Natural Terrain Landslide Study: Debris Avalanche Susceptibility*. Geotechnical Engineering Office, Hong Kong. GEO Technical Note TN 1/98.

Evans NC, Huang SW and King JP (1997) *The Natural Terrain Landslide Study: Phase III*. Geotechnical Engineering Office, Hong Kong. GEO Special Project Report SPR 5/97.

Evans TG, Usher N and Moore R (2007) Management of geotechnical and geohazard risks in the West Nile Delta. *Proceedings of the 6th International Offshore Site Investigation and Geotechnics Conference: Confronting New Challenges and Sharing Knowledge*. Society for Underwater Technology (SUT), London.

Focht J (1994) Lessons learned from missed predictions. *Journal of Geotechnical Engineering*, **120(10)**: 1653–1683.

Foster HD (1980) *Disaster Planning*. Springer-Verlag, New York.

Fussell JB (1973) A formal methodology for fault tree construction. *Nuclear Science Engineering* **52**: 421–432.

Fussell JB (1976) Fault tree analysis: concepts and techniques. *Proceedings of the NATO Advanced Study Institute on Generic Techniques in Systems Reliability Assessment* (Henley E and Lynn J (eds)). Noordhoff Publishing, Leyden, pp. 133–162.

Gillies D (2000) *Philosophical Theories of Probability*. Routledge, New York.

Goodman R (1999) *Karl Terzaghi: The Engineer as Artist*. American Society of Civil Engineers, Reston, VA.

Gretener PE (1967) The significance of the rare event in geology. *The American Association of Petroleum Geologists Bulletin* **51(11)**: 2197–2206.

Gutenberg B and Richter F (1949) *Seismicity of the Earth and Associated Phenomena*. Princeton University Press, Princeton, NJ.

Guthrie RH and Evans SG (2004a) Analysis of landslide frequencies and characteristics in a natural system, coastal British Columbia. *Earth Surface Processes and Landforms* **29**: 1321–1339.

Guthrie RH and Evans SG (2004b) Magnitude and frequency of landslides triggered by a storm event, Loughborough Inlet, British Columbia. *Natural Hazards and Earth System Sciences* **4**: 475–483.

Guthrie RH and Evans SG (2005) The role of magnitude–frequency relations in regional landslide risk analysis. In *Landslide Risk Management* (Hungr O, Fell R, Couture R and Eberhardt E (eds)). Balkema, Rotterdam, pp. 375–380.

Guthrie RH, Deadman PJ, Cabrera AR and Evans SG (2008) Exploring the magnitude-frequency distribution: a cellular automata model for landslides. *Landslides* **5**: 151–159.

Guzzetti F, Malamud BD, Turcotte DL and Reichenbach P (2002) Power-law correlations of landslide areas in central Italy. *Earth and Planetary Science Letters* **195**: 169–183.

Hacking I (2006) *The Emergence of Probability A Philosophical Study of Early Ideas about Probability, Induction and Statistical Inference*. Cambridge University Press, Cambridge.

Haflidason H, Lien R, Sejrup HP, Forsberg CF and Bryn P (2005) The dating and morphometry of the Storegga Slide. *Marine and Petroleum Geology* **22**: 123–136.

Hahn GJ and Shapiro SS (1967) *Statistical Methods in Engineering*. Wiley, New York.

Hartford DND and Baecher GB (2004) *Risk and Uncertainty in Dam Safety*. Thomas Telford, London.

Ho KKS and Lau JWC (2010) Learning from slope failures to enhance landslide risk management. *Quarterly Journal of Engineering Geology and Hydrogeology* **43**: 33–68.

Ho K, Leroi E and Roberds B (2000) Quantitative risk assessment: application, myths and future direction. *Proceedings of the Geo-Eng Conference*, Melbourne, Publication 1, pp. 269–312.

Hora S and Iman R (1989) Expert opinion in risk analysis: the NUREG-1150 methodology. *Nuclear Science and Engineering* **102**: 323–331.

Hsi JP and Fell R (2005) Landslide risk assessment of coal refuse emplacement. In *Landslide Risk Management* (Hungr O, Fell R, Couture R and Eberhardt E (eds)). Balkema, Rotterdam, pp. 525–532.

Hungr O, Evans SG and Hazzard J (1999) Magnitude and frequency of rock falls along the main transportation corridors of southwestern British Columbia. *Canadian Geotechnical Journal* **36**: 224–238.

Hungr O, McDougall S, Wise M and Cullen M (2008) Magnitude–frequency relationships of debris flows and debris avalanches in relation to slope relief. *Geomorphology* **96**: 355–365.

IPCC (2000) Quantifying uncertainties in practice. In *IPCC Good Practice Guidance and Uncertainty Management in National Greenhouse Gas Inventories*. Intergovernmental Panel on Climate Change, Geneva, ch. 6.

IPCC (2006) Guidance notes for lead authors of the IPCC fourth assessment report on addressing uncertainties. Appendix in Manning 2006. *Advances in Climate Change Research* **2**: 13–21.

Issler D, De Blasio FV, Elverhoi A, Bryn P and Lien R (2005) Scaling behaviour of clay-rich submarine debris flows. *Marine and Petroleum Geology* **22**: 187–194.

Johnson NL, Kotz S and Balkrishan N (1994) *Continuous Univariate Distributions*. Wiley, New York.

Jones DKC and Lee EM (1994) *Landsliding in Great Britain*. HMSO, London.

Kahneman D and Tversky A (1982) Subjective probability: a judgement of representativeness. In *Judgement under Uncertainty: Heuristics and Biases* (Kahneman D, Slovic P and Tversky A (eds)). Cambridge University Press, Cambridge, pp. 32–47.

Kanamori H and Brodsky EE (2004) The physics of earthquakes. *Reports on Progress in Physics* **67**: 1429–1496.

King JP (1997) *Natural Terrain Landslide Study: The Natural Terrain Landslide Inventory*. Geotechnical Engineering Office, Hong Kong. GEO Report No. 74.

King JP (1999) *Natural Terrain Landslide Study: The Natural Terrain Landslide Inventory*. Geotechnical Engineering Office, Hong Kong. GEO Technical Note TN 1/97.

Knol AB, Slottje P, van der Sluijs JP and Lebret E (2010) The use of expert elicitation in environmental health impact assessment: a seven step procedure. *Environmental Health* **9**: 19.

Kondziolka RE and Kandaris PM (1996) Capacity predictors for full scale transmission line test foundations. *Uncertainty in the geologic environment: From theory to practice*. American Society of Civil Engineers, Reston, VA, pp. 695–709.

Kumamoto H and Henley EJ (1996) *Probabilistic Risk Assessment and Management for Engineers and Scientists*. Institute of Electrical and Electronics Engineers Press, New York.

Lacasse S and Nadim F (2008) Landslide risk assessment and mitigation strategy. Invited lecture, state-of-the-art. *First World Landslide Forum, Global Landslide Risk Reduction, International Consortium of Landslides*, Tokyo, pp. 31–61.

Lacasse S and Nadim F (2011) Learning to live with geohazards: from research to practice. *ASCE GSP 22: Proceedings, Geotechnical Risk Assessment and Management*, pp. 64–116.

Lacasse S, Eidsvik U, Nadim F, Høeg K and Blikra LH (2008) Event Tree analysis of Åknes rock slide hazard. *IV Geohazards Québec, 4th Canadian Conference on Geohazards*, Quebec, pp. 551–557.

Lacasse S, Nadim F and Kalsnes B (2010) Living with Landslide Risk. *Geotechnical Engineering Journal of the SEAGS & AGSSEA* **41**(4).

Lee EM (2007) Landslide risk assessment: the challenge of estimating landslide probability. In *Engineering Geology in Geotechnical Risk Management*. Hong Kong Regional Group of the Geological Society of London, Hong Kong, pp. 13–31.

Lee EM (2009) Landslide risk assessment: the challenge of estimating the probability of landsliding. *Quarterly Journal of Engineering Geology and Hydrogeology* **42**: 445–458.

Lee EM and Moore R (2007) Ventnor Undercliff: development of landslide scenarios and quantitative risk assessment. In *Landslides and Climate Change: Challenges and Solutions* (McInnes R, Jakeways J, Fairbank H and Mathie E (eds)). Balkema, Rotterdam, pp. 323–334.

Lee EM, Audibert JME, Hengesh JV and Nyman DJ (2009) Landslide-related ruptures of the Camisea pipeline. system, Peru. *Quarterly Journal of Engineering Geology and Hydrogeology* **42**: 251–259.

Lee EM, Brunsden D and Sellwood M (2000) Quantitative risk assessment of coastal landslide problems, Lyme Regis, UK. In *Landslides: In Research, Theory and Practice* (Bromhead EN, Dixon N and Ibsen ML (eds)). Thomas Telford, London, pp. 899–904.

Lee YF, Lee DH and Chi YY (2009) A probabilistic method for evaluating slope failure by reliability index. *Journal of the Chinese Institute of Civil and Hydraulic Engineering* **21**(1): 91–103 (in Chinese).

Li KS (1991) *Reliability Index for Probabilistic Slope Analysis*. Department of Civil Engineering, University College, Australia Defence Force Academy, University of New South Wales, Sydney. Research Report No. R101.

Li KS (1992) A point estimate method in calculating statistical moments. *Journal of Engineering Mechanics* **118**: 1506–1511.

Lichtenstein S, Fischoff B and Phillips L (1982) Calibration of probabilities: the state of the art to 1980. In *Judgement under uncertainty: heuristics and biases* (Kahneman D, Slovic P and Tversky A (eds)). Cambridge University Press, Cambridge, pp. 306–334.

Linstone HA and Truoff M (eds) (1975) *The Delphi Method; Techniques and Applications*. Addison-Wesley, Reading, MA.

Loncke L (2002) Le delta profond du Nil: structure et evolution depuis le Messinien (Miocene terminal). PhD thesis, Universite P. et M. Curie, Paris.

Malamud BD, Turcotte, DL, Guzzetti F and Reichenbach P (2004) Landslide inventories and their statistical properties. *Earth Surface Processes and Landforms* **29**: 687–711.

Martin Y, Rood K, Schwab JW and Church M (2002) Sediment transfer by shallow landsliding in the Queen Charlotte Islands, British Columbia. *Canadian Journal of Earth Science* **39**: 189–205.

Meozzi PG and Iannucci C (2006) Facilitating the Development of Environmental Information into Knowledge: Government Agency Perspectives to Improve Policy Decision-making. *4th International Conference on Politics and Information Systems, Technologies and Applications*.

Montgomery DC and Runger GC (1994) *Applied Statistics and Probability for Engineers*. Wiley, New York.

Moore R, Usher N and Evans T (2007) Integrated Multidisciplinary Assessment and Mitigation of West Nile Delta Geohazards. *Proceedings of the 6th International Offshore Site Investigation and Geotechnics Conference: Confronting New Challenges and Sharing Knowledge*.

Morgenstern NR (1995) The role of analysis in the evaluation of slope stability. *Proceedings of the 6th International Symposium on Landslides*. Balkema, Rotterdam, pp. 1615–1629.

Moss R and Schneider SH (2000) Uncertainties in the IPCC TAR: recommendations to lead authors for more consistent assessment and reporting. In *Guidance papers on the cross cutting issues of the third assessment report of the IPCC* (Pachauri R, Taniguchi T, Tanaka K (eds)). World Meteorological Organization, Geneva, pp. 33–51.

Mostyn GR and Fell R (1997) Quantitative and semiquantitative estimation of the probability of landsliding. In *Landslide Risk Assessment* (Cruden D and R. Fell R (eds)). Balkema, Rotterdam, pp. 297–315.

Mostyn GR and Li KS (1993) Probabilistic slope analysis – state-of-play. In *Probabilistic Methods in Geotechnical Engineering* (Li KS and Lo S-CR (eds)). Balkema, Rotterdam, pp. 89–109.

Mulder T and Syvitski JPM (1995) Turbidity currents generated at river mouths during exceptional discharges to the World Oceans. *Journal of Geology* **103**: 285–299.

Nadim F (2006) Challenges to geo-scientists in risk assessment for sub-marine slides. *Norwegian Journal of Geology* **86**: 351–362.

Nadim F, Einstein H and Roberds W (2005) Probabilistic stability analysis for individual slopes in soil and rock. In *Landslide Risk Management* (Hungr O, Fell R, Couture R and Eberhardt E (eds)). Balkema, Rotterdam, pp. 63–98.

Nguyen VU and Chowdhury RN (1984) Probabilistic study of spoil pile stability in strip coal mines – two techniques compared. *International Journal for Rock Mechanics, Mining Science and Geomechanics* **21**: 303–312.

Nordhaus WD (2011) The Economics of Tail Events with an Application to Climate Change. *Review of Environmental Economics and Policy* **5(2)**: 240–257, http://dx.doi.org/10.1680/lra.58019.202 10.1093/reep/rer004.

Panel on Seismic Hazard Analysis (1988) *Probabilistic seismic hazard analysis*. National Academy Press, Washington, DC.

Parr NM and Cullen N (1998) Risk management and reservoir maintenance. *Journal of the Institution of Water and Environmental Management* **2**: 587–593.

Peck R (1979) Liquefaction potential: science versus practice. *Journal of the Geotechnical Engineering Division* **105(GT5)**: 393–398.

Penning-Rowsell E, Johnson C, Tunstall S, Tapsell S, Morris J, Chatterton J, Coker A and Green C (2003) *The Benefits of Flood and Coastal Defence: Techniques and Data for 2003*. Flood Hazard Research Centre, Enfield.

Reed W and Jorgensen M (2005) The double Pareto-lognormal distribution – a new parametric model for size distribution. *Communications in Statistics* **34**: 1733–1753.

Reeder MS, Rothwell RG, Stow DAV, Kahler G and Kenyon NH (1998) Turbidite flux, architecture and chemostratigraphy of the Herodotus Basin, Levantine Sea, south-eastern Mediterranean. In: *Geological Processes on Continental Margins: Sedimentation, Mass Wasting and Stability* (Stoker MS, Evans D and Cramp D (eds)). Geological Society, London, pp. 19–41. Special Publication 129.

Reeder MS, Stow DAV and Rothwell RG (2002) Late Quaternary turbidite input into the eastern Mediterranean basin: new radiocarbon constraints on climate and sea-level control. In *Sediment Flux to Basins: Causes, Controls and Consequences* (Jones SJ and Frostick LE (eds)). Geological Society, London, pp. 267–278. Special Publication 191.

Reeve D (2010) *Risk and Reliability: coastal and hydraulic engineering*. Spon Press, London.

Riggs HC (1968) Frequency curves. In *Techniques of Water Resources Investigations of the United States*. US Geological Survey, Reston, VA, ch. A-2.

Roberds WL (1990) Methods for developing defensible subjective probability assessments. *Transportation Research Record* **1288**: 183–190.

Rosenbleuth E (1975) Point estimates for probability moments. *Proceedings of the National Academy of Science of the USA* **72(10)**: 3812–3814.

Rosenbleuth E (1981) Two point estimates in probabilities. *Applied Mathematical Modelling* **5**: 329–335.

Savy J, Boissonnade A, Mensing R and Short S (1993) *Eastern U.S. Seismic Characterization Update*. Lawrence Livermore National Laboratory, Livermore. Report No. UCRL-ID-115111.

Sornette D (2009) Dragon-Kings, Black Swans and the prediction of crises. *International Journal of Terraspace Science and Engineering* **2(1)**: 1–18.

Sornette D and Ouillon G (2012) Dragon-kings: mechanisms, statistical methods and empirical evidence. *European Physical Journal Special Topics* **205**: 1–20.

Spetzler C and Staël von Holstein C (1975) Probability encoding in decision analysis. *Management Science* **22**: 340–358. http://dx.doi.org/10.1680/lra.58019.203 10.1287/mnsc.22.3.340.

SSHAC (1997) *Recommendations for Probabilistic Seismic Hazard Analysis: Guidance on Uncertainty and Use of Experts*. Lawrence Livermore National Laboratory, Livermore. Report No. NUREG/CR-6372.

Stark CP and Hovius N (2001) The characterization of landslide size distributions. *Geophysical Research Letters* **28**: 1091–1094.

Stedinger JR, Vogel RM and Foufoula-Georgiou E (1993) Frequency analysis in extreme events. In *Handbook of Hydrology* (Maidment D (ed.)). McGraw-Hill, New York.

Sugai T, Ohmori H and Hirano M (1994) Rock control on magnitude-frequency distribution of landslides. *Transactions of the Japanese Geomorphological Union* **15**: 233–251.

Süveges M and Davison AC (2012) A case study of a 'Dragon-King': the 1999 Venezuelan catastrophe. *European Physical Journal Special Topics* **205**: 131–146.

Sweeney M (2005) Terrain and geohazard challenges facing onshore oil and gas pipelines: historic risks and modern responses. In *Terrain and geohazard challenges facing onshore oil and gas pipelines* (Sweeney M (ed.)). Thomas Telford, London, pp. 37–51.

Sweeney M, Gasca AH, Garcia-Lopez M and Palmer AC (2005) Pipelines and Landslides in Rugged Terrain: a Database, Historic Risks and Pipeline Vulnerability. In *Terrain and Geohazard Challenges Facing Onshore Oil and Gas Pipelines* (Sweeney M (ed.)). Thomas Telford, London, pp. 641–660.

Talling PJ, Amy LA and Wynn RB (2007) New insight into the evolution of large-volume turbidity current: comparison of turbidite shape and previous modeling results. *Sedimentology* **54**: 737–769.

ten Brink US, Giest EL and Andrews BD (2006) Size distribution of submarine landslides and its implication to tsunami hazard in Puerto Rico. *Geophysics Research Letters* **33**: L11307.

Tversky A and Kahneman D (1974) Judgements under uncertainty: heuristics and biases. *Science* **185**: 1124–1131.

Tversky A and Kahneman D (1983) Extensional vs intuitive reasoning: the conjunction fallacy in probability judgement. *Psychological Review* **90(4)**: 293–315.

Van der Sluijs JP, Janssen PHM, Petersen AC et al. (2004) *RIVM/MNP Guidance for Uncertainty Assessment and Communication: Tool Catalogue for Uncertainty Assessment*. Copernicus Institute & RIVM, Utrecht/Bilthoven. Report No. NWS-E-2004-37.

Vanmarcke EH (1977) Probabilistic modelling of soil profiles. *Journal of the Geotechnical Engineering Division of the ASCE* **103**: 1227–1246.

Vanmarcke EH (1980) Probabilistic stability analysis of earth slopes. *Engineering Geology* **16**: 29–50.

Vick SG (2002) *Degrees of Belief: Subjective Probability and Engineering Judgement*. American Society of Civil Engineers, Reston, VA.

Wieczorek GF, Larsen MC, Eaton LS, Morgan BA and Blair JL (2001) *Debris-flow and Flooding Hazards Associated with the December 1999 Storm in Coastal Venezuela and Strategies for Mitigation*. US Geological Survey, Reston, VA. Open File Report 01-0144.

Wu TH, Tang WH and Einstein HH (1996) Landslide hazard and risk assessment. In *Landslides Investigation and Mitigation* (Turner AK and Schuster RL (eds)). National Academy Press, Washington, DC, pp. 106–120. Transportation Research Board Special Report 247.

Yu YF and Mostyn GR (1996) An extended point estimate method for the determination of the probability of failure of a slope. In *Landslides* (Senneset K (ed.)). Balkema, Rotterdam, pp. 429–433.

Chapter 6
Estimating the probability of landsliding

6.1. Introduction

What is the chance that a landslide will occur? To answer this question, it is necessary to appreciate the circumstances that might lead to a landslide and to know something of the history of landsliding in the area. The ultimate cause of all landsliding is the downward pull of gravity. The stress imposed by gravity is resisted by the strength of the materials forming the slope. A stable slope is one where the resisting forces are greater than the destabilising forces and therefore can be considered to have a margin of stability. By contrast, a slope at the point of failure has no *margin of stability*, since the resisting and destabilising forces are approximately equal. The quantitative comparison of these opposing forces gives rise to a ratio known as the 'factor of safety' F:

$$\text{factor of safety } F = \frac{\text{resisting forces}}{\text{destabilising forces}} = \frac{\text{shear strength}}{\text{shear stress}}$$

The factor of safety of a slope at the point of failure is assumed to be 1.0.

When a slope fails, the displaced material moves to a new position so that equilibrium can be re-established between the destabilising forces and the strength of the material. Landsliding therefore helps change a slope from a less stable to a more stable state with a margin of stability. No subsequent movement will occur unless the slope is subject to processes that, once again, affect the balance of opposing forces. In many inland settings, landslides can remain inactive or dormant for thousands of years. However, on the coast and along active river cliffs, erosion continues to remove material from the cliff foot, reducing the margin of stability, and promotes further recession.

As slope movements are the result of changes that upset the balance of forces, so that those offering resistance to movement are exceeded by those producing destabilisation, the stability of a slope can be viewed in terms of its ability to withstand potential changes. A slope is

- *stable* when the margin of stability is sufficiently high to withstand all transient forces in the short to medium term (i.e. hundreds of years), excluding excessive alteration by human activity
- *marginally stable* when the balance of forces is such that the slope will fail at some time in the future in response to transient forces attaining a certain magnitude
- *actively* when transient forces produce continuous or intermittent movement.

On any slope, the margin of stability will vary through time in response to weathering, basal erosion, loading and fluctuations in groundwater levels, due to natural and human influences. It will usually rise to a peak immediately after a landslide event and then decline progressively to lower levels as basal erosion or other slope processes (e.g. weathering and human activity) affect the slope stability

Figure 6.1 An example event sequence involved in the generation of landslide risk to a pipeline

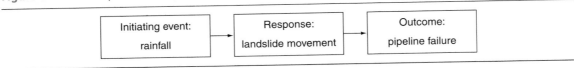

(Brunsden and Lee, 2004). This perspective makes it possible to recognise two categories of *causal factors* that are active in promoting failure:

- *Preparatory factors* work to make the slope increasingly susceptible to failure without actually initiating landsliding (e.g. the long-term effect of erosion at the base of a slope or weathering). This is the equivalent to the incubation phase described in Chapter 2.
- *Triggering factors* actually initiate landslide events (e.g. rainstorm events).

The combination of triggering and preparatory factors also provides a framework for modelling landslide scenarios. The occurrence of landslides involves an inter-related sequence of events driven by (Figure 6.1) the following.

1. An *initiating* or *triggering event* (e.g. an earthquake, high groundwater levels or construction activity).
2. The slope *response*, controlled by the nature and stability state of the slope as determined by the preparatory factors, together with propagating conditions such as high groundwater levels or progressive removal of toe support by stream or gully undercutting.
3. Subsequent *outcomes* determined by the style of landslide behaviour, the topography, and the locations of human-valued assets relative to the landslide movement and their vulnerability.

As discussed in Chapter 3, the development of a hazard model provides a sound framework for estimating the probability of landsliding. Investigation efforts should be directed towards addressing the following questions.

1. What is the condition of the contemporary slopes within the area of interest?
   - Is there evidence of active instability (e.g. tension cracks, settlement or heave) and associated damage to structures?
   - Are there pre-existing weaknesses within the slope (e.g. pre-existing shears of non-landslide origin or thin clay bands)?
   - Are existing slope stabilisation or erosion control measures showing signs of distress and could their failure initiate landsliding?
2. What type of landslide events can be expected to occur in the area of interest?
   - Are these events the product of minor pre-failure movements, first-time failure, post-failure movement or reactivation?
   - What are the likely mechanisms of failure (e.g. falling, toppling, spreading, flowing or sliding) and landslide types?
   - What is the range of event sizes that could occur (including the maximum credible event size)?
   - What velocities and travel distances could be expected (see Chapter 3)?
3. How frequently have landslides occurred in the past?
   - How many events have occurred over a particular time period?

- If active cliffs are involved, what is the rate of recession, over both the short term and the longer term?
- What is the magnitude–frequency distribution of past events?
- Is there any evidence that the frequency of landsliding has been increasing (e.g. as a result of human influence on slopes or climate change) or decreasing (e.g. owing to climate change, sea-level fall or reforestation) over a particular time period of the relevant past?

4 When did past landslide events take place?
- Is there evidence for the timing of past landslide activity?
- Can dates be determined for all recorded landslide events or only a sample?

5 Can past landslide events be associated with a particular triggering factor, such as rainfall or earthquakes?
- Are there reliable records of rainfall or earthquakes that cover the time period represented by the landslide inventory?
- Is there a trend in the frequency or timing of the triggering factors, or can the records be considered to be stationary?

6.2. Approaches to estimating landslide probability

As described in Chapter 5, there are two fundamentally different probability interpretations: the *frequentist* and the *degree of belief* interpretations. Despite their limitations, both interpretations can be used to estimate the probability of future landslide activity (e.g. Lee, 2009).

- The *frequentist* approach involves the statistical analysis of an inventory of landside events that have occurred in a particular region or area over a specific time period. This approach can be seen to be objective. However, it takes no account of site-specific geomorphological or geological conditions across the area, changes in environmental controls over the time period covered by the inventory or specialist experience gained from working in similar areas. Essentially, this is a 'black box' approach where the focus is solely on the statistical analysis rather than understanding the circumstances associated with landslide events. It can be undertaken by a statistician and requires no input from landslide specialists.
- The *degree of belief* approach is viewed as being subjective, but involves making a judgement based on the available information and the experience of the specialist(s) involved in the assessment. The focus tends to be on understanding the underlying causes of landsliding, including the controlling, preparatory and triggering factors. Clearly, it requires specialist knowledge of landsliding.

It would be misleading to suggest that the decision about which approach to use is a philosophical choice between objective statistics or subjective judgement. In most cases, the approach will reflect a range of more practical issues, including the availability of information and the skills and experience of the people undertaking the assessment. In many situations, an absence of information on landslide activity is the main problem facing the engineer or landslide specialist. This is particularly so for developments in remote regions, such as the Baku-Supsa oil pipeline in Georgia (Lee and Charman, 2005). Here, the choice was between the use of expert judgement or no probabilistic assessment. The decision to proceed with a probabilistic assessment was made by the client as it enabled them to directly compare the landslide hazard with other threats to the pipeline. In this instance, the specialists were able to build on the limited local record of landslide activity by applying their experience of landslide behaviour from similar temperate mountain environments, so as to generate an estimate of the probability of landslide-related pipeline rupture.

In other situations, the approach to estimating probability will reflect the experience and skills of the specialist or research team involved. For example, probabilistic stability analysis is an interesting and challenging analytical tool for experienced geotechnical engineers, but it is not an approach that others, including many geomorphologists, would naturally turn to. Thus, while the Norwegian Geotechnical Institute (NGI) have pioneered the use of probabilistic stability analysis on many seabed projects (e.g. Nadim *et al.*, 2003, 2005), the geomorphologists and engineering geologists of BP's Geohazard Assessment Team have favoured the use of a combination of historical frequency analysis and expert judgement to estimate landslide probability in the initial stages of the West Nile delta development (e.g. Moore *et al.*, 2007). However, as geotechnical engineers were subsequently added to the BP team, probabilistic stability analysis was introduced to estimate the threat to seabed infrastructure from shallow translational planar failures (Dimmock *et al.*, 2012).

Academic research interests are often different from those of the commercial sector when it comes to landslide risk assessment. Over the past decade, there has been considerable research focused on the application of statistically based methods for estimating landslide probability, such as the characterisation of magnitude–frequency curves in British Colombia (e.g. Guthrie and Evans, 2004a, 2004b; Guthrie *et al.*, 2008) or for submarine landslides (e.g. Chaytor *et al.*, 2009; ten Brink *et al.*, 2006) and the use of Poisson models to generate region-wide estimates of landslide exceedance probability (e.g. Coe *et al.*, 2000; Guzzetti, 2005; Guzzetti *et al.*, 2005). All these studies have been based on detailed and comprehensive landslide inventories that were either available prior to the research or collected as part of the research. Little academic landslide research appears to have been directed towards the application of degree of belief approaches or the use of techniques such as event tree or fault tree analysis, or the issues associated with using expert judgement. In the few instances where these topics have been studied, the impetus has come from client-funded research and development to address practical problems, such as modelling the economic losses associated with the reactivation of coastal landslides (e.g. Lee and Moore, 2007; Lee and Sellwood, 2002; Lee *et al.*, 2000).

By contrast, in the commercial world, landslide risk assessment is driven by client needs. The people undertaking the probability assessment do so because the client requires the information to support some aspect of their decision-making process. Incomplete knowledge of landslides, tight schedules and limited resources are the reality for most projects. Faced with these challenges, projects are often more about making the best use of the available information and experience than having to choose between one particular approach over another. In many cases, the lack of a comprehensive landslide inventory will rule out most statistical methods. The absence of detailed subsurface information and representative laboratory test results will rule out probabilistic stability analysis. What is often left is a combination of expert judgement, based on the available landslide statistics and knowledge, plus experience gained from other projects and sites. The resultant judgements are inevitably subjective, but recognise that both statistical data and knowledge of site conditions are important to address all of the elements contributing to the uncertainty in the occurrence of landslides.

The examples presented in this chapter can be grouped together into three broad categories.

- *Statistical methods*, including the use of databases of incidents caused by landslides, the analysis of landslide inventories (historical frequency analysis, the use of magnitude–frequency curves and the estimation of exceedance probability using the Poisson and binomial models) and the use of statistical models to simulate coastal cliff recession. These are, in essence, 'black box' approaches where the focus is on the occurrence of events through time, with no or limited concern for the nature and mechanics of the landslide process.

- *Conceptual models*, all of which involve incorporating experience and judgement into a framework provided by landslide statistics. They can be viewed as 'grey box' methods, involving a conceptual understanding of the nature and mechanics of the system, including responses to triggering events. The examples range from the use of empirical adjustment factors to modify area-wide landslide statistics to take account of local conditions, linking landslide probability to the frequency of triggering events and the use of event trees to model the landslide process.
- *Reliability methods*, in which a number of examples of the use of probabilistic stability analysis are presented. These methods take account of the uncertainties in geotechnical parameters (e.g. shear strength, pore water pressures and loads) and use stability models to generate probabilistic estimates of the reliability of a slope. It is tempting to view them as 'white box' models where all of the processes are known in detail and taken into account by the model. The reality is, however, that simplifications and assumptions have to be made. As Nadim *et al.* (2005) wrote, 'to characterise the uncertainties in a soil property, one needs to combine, in addition to the actual data, knowledge about the quality of the data, knowledge on the geology, and most importantly engineering judgement'.

While the statistical methods are clearly frequentist in approach, both the conceptual modelling and reliability methods involve a combination of frequency statistics and degree of belief. This combination is a pragmatic solution to fairly typical problem of imperfect knowledge. The output from this mixed approach is clearly not a 'true' probability in the relative frequency sense, but what the landslide specialists believe to be the 'best estimate' value, given their state of knowledge and experience. The value cannot be expected to be fixed in stone, as it will change as more information becomes available to reduce the uncertainties.

It is also important to be aware of the practical limitations of statistical methods. It not possible to use these methods to yield a probability for events with no historical precedent (e.g. a major coastal landslide triggered by the failure of a seawall; Lee and Sellwood, 2002) or for unique cases (e.g. the collapse of the flank of Mount St Helens in May 1980; Voight *et al.*, 1981). By their very nature, such 'Black Swans' and 'Dragon Kings' (see Chapter 5) are outliers from normal statistical distributions and not usually represented in historical data sets. In these cases, the statistical approach is of limited value without the use of expert judgement or conceptual modelling.

6.3. Statistical methods: the use of incident databases

Generic accident or incident databases are widely used to estimate the probability of a hazard in many industries, including the oil and gas, petrochemical and nuclear sectors. For example, the US Pipeline and Hazardous Materials Safety Administration (PHMSA) is responsible for protecting the American public and the environment by ensuring the safe and secure movement of hazardous materials to industry and consumers by all transportation modes, including the nation's 2.3 million mile pipeline system. Incidents that result in unintentional releases from hazardous liquid pipelines are reported to the PHMSA, which maintains a database of pipeline incident reports (available online at http://primis.phmsa.dot.gov/comm/reports/safety/psi.html).

PHSMSA (2010) provides baseline frequencies for a range of geological hazards (Table 6.1). Between 2008 and 2010, the major causes of incidents were material or equipment failure (38%) and corrosion (21%). The frequency of pipeline ruptures due to landslides is 0.0000123 per mile per year (0.00000764 per km per year). This value can be used as the expected rupture frequency due to landslides in a pipeline risk assessment, regardless of the terrain, as shown for the TransCanada Keystone Pipeline in Example 6.1.

Table 6.1 PHSMA baseline incident frequencies

Hazard	Incident frequency per mile per year
Corrosion	2.90E-04
Excavation damage	1.22E-04
Materials and construction	3.00E-04
Hydraulic event	1.47E-04
Ground movement	1.23E-05
Washout and flooding	1.14E-05

From Keystone (2009)

In other situations, incident databases can provide a useful preliminary indication of the levels of landslide threat that might be expected in an area. On other occasions, as shown in Example 6.2 from Papua New Guinea, it may be possible to use the incident data from one region to address a problem in a remote region where there has been no previous development or performance history.

Industry databases provide records of damage incidents caused by a variety of factors, including landslides. In doing so, they provide an indication of the probability of the 'top event' (e.g. pipeline rupture as a result of landsliding) rather than the probability of landsliding. However, landslide incidents are simply statistics in these datasets, and little needs to be known about the geological conditions that apply to them. Indeed, in some cases, the statistics have probably been used by risk analysts without any discussion with landslide experts.

Many geologists are sceptical about the value of performance databases because, in their mind, they are so generalised as to be almost meaningless at a site level. This is certainly true, but they do have value, especially at the feasibility stage of long linear projects in remote regions when a relatively quick, high-level statement on landslide risk is needed to support the appraisal of various project options.

Example 6.1: The TransCanada Pipeline

The TransCanada Keystone Pipeline is a proposed crude oil pipeline designed to transport 900 000 barrels per day of crude oil from facilities at Hardisty, Alberta, Canada, southward to the Port Arthur and east Houston areas of Texas, USA. It was recognised that a major landslide event could be possible along mountainous sections of the proposed route. As a result, landslide potential was a factor that was addressed during the routing stage. For example, crossings of mountain areas were selected where landslide potential was considered minimal. In addition, the potential for landslide activity would be monitored during operations through regular aerial and intermittent ground patrols and through landowner awareness programmes.

As part of the environmental impact statement prepared for the project, a pipeline incident frequency and spill volume analysis was undertaken for the entire system (Keystone 2009). The main threats included corrosion, excavation damage, materials and construction, hydraulic events, ground movement (landslides), and wash-out and flooding. The analysis was based on publicly available historical incident data collected from the US PHMSA incident reports (see Table 6.1).

The US-wide baseline landslide frequencies from the PHMSA (0.0000123 per mile per year) were adjusted to take account of variations in landslide hazard along the route. The assessment of landslide

Table 6.2 Example 6.1: Keystone pipeline project – landslide susceptibility classes and adjustment factors

Landslide susceptibility	Adjustment factor	Miles exposed
Low	0.1	1230.2 (1979 km)
Moderate	0.8	71.3 (115 km)
High	1.0	370.2 (595 km)

From Keystone (2009)

hazard was based on the US Geological Survey (USGS) national-scale landslide susceptibility mapping (Radbruch-Hall et al., 1982), and expert judgement was used to generate the adjustment factors used for each susceptibility class (Table 6.2). Unfortunately, no information was provided by Keystone (2009) to explain the rationale behind selecting these adjustment factors.

For each state crossed by the route, an overall weighted adjustment factor was developed based on the relative length of pipeline that passes through land characterised by the three different landslide susceptibility classes. The results of the assessment are presented in Table 6.3.

Example 6.2: Papua New Guinea

A desk-based review of potential oil pipeline corridors from the Highlands Region of Papua New Guinea (PNG) to the south coast was undertaken to support the initial project planning. One of the corridor options involved routing a 30 km section of the pipeline through a remote highland area developed in very weathered volcanic soils (e.g. pyroclastic and lahar deposits), deeply dissected by a dense network of incised stream valleys, canyons and gullies (50–200 m deep with side-slope angles in the range of 35– 60°). The region is seismically active and has a humid rainforest climate. Given the nature of the terrain, it was expected that this section of corridor would be exposed to a wide range of landslide threats, including reactivation of major, pre-existing landslide complexes,

Table 6.3 Example 6.1: Keystone pipeline – estimated landslide incident frequencies by US state

State	Length: miles	Weighted adjustment factor	Incident frequency per mile per year
Montana	282.3	0.44	6.6E-06
South Dakota	312.8	0.71	1.0E-05
Nebraska	255.2	0.18	3.5E-06
Total Segment 1	850.3	0.46	6.9E-06
Nebraska	2.5	1.0	1.3E-05
Kansas	211.1	0.16	3.3E-06
Oklahoma	82.4	0.1	2.5E-06
Total Segment 2	296.0	0.15	3.1E-06
Oklahoma	154.9	0.17	3.3E-06
Texas	323.3	0.25	4.3E-06
Total Segment 3	478.2	0.22	34.0E-06

From Keystone (2009)

shallow slides on steep slopes (triggered by heavy rainfall and/or strong earthquakes) and channelised debris flows.

An indication of the expected level of landslide threat to the pipeline within this area was required in order to make comparisons with other geohazard risks along alternative corridor options (e.g. karst collapse, active faulting and river bed scour). However, little was known about the nature and distribution of landsliding and, unlike the Keystone example above, there was no relevant database of pipeline rupture incidents. In addition, the dense rainforest cover obscures the underlying slope morphology, thereby preventing the reliable mapping of landslides from satellite imagery. As a consequence, the approach adopted was to identify an analogous situation for which statistics on the landslide-related pipeline rupture rate have been reported.

It was concluded that the conditions in the area of interest in PNG were broadly similar to those found along the Coca River, on the flanks of the Reventador Volcano in Ecuador. This is where the Trans-Ecuador oil and Poliducto propane pipelines suffered multiple landslide-related ruptures during two earthquakes on 5 March 1987 ($M_s = 6.1$ and 6.9) (Schuster, 1991). The pipelines were located in a shared right of way, partly in sidelong ground and partly in the floodplain of the Coca River, where they crossed numerous tributary valleys.

Most of the pipeline damage was caused by flash flooding of the Coca River, partly due to the failure of a series of landslide dams created when debris flows spread across the valley floor. Flooding encroached upon the pipeline alignment, resulting in severe scouring and removal of sections of the entire pipeline right of way. In at least four locations, the pipelines were severed by direct debris flow impact. The total loss of revenue before the reconstructed line began its service in August 1987 was estimated at nearly $800 million. Approximately 40 km of the oil pipeline had to be reconstructed, at a cost of around $50 million.

Given its similarities with the Coca River area, the volcanic highland section in PNG might be expected to present landslide risk levels comparable to those in the high-landslide-susceptibility terrain found in parts of the Andes. As described in Chapter 5, BP has assembled a database for landslide-related pipeline ruptures in tropical Andean mountains (Sweeney et al., 2005). This database includes statistics for the Trans-Ecuador oil pipeline (Transecuadorian), which has had a rupture frequency of 3.27 per 1000 km per year (see Table 5.1). This frequency was used to generate a high rupture rate of 3.0 ruptures per year over the 30 km section through similar conditions in PNG. Alternative corridors are longer, but reduce the pipeline exposure to steep, landslide prone terrain and avoid the areas with high concentrations of pre-existing landslides. By adopting one of these, it was considered possible that the landslide risk could be reduced to levels comparable to those of modern, geo-engineered pipelines in the Andes (that is, 0.3 rupture per 1000 km per year; see Table 5.1).

It is clear that incident history data from Andean pipelines cannot be expected to provide precise and reliable rupture frequency statistics for a pipeline in PNG. However, the statistics probably do highlight the relative geohazard risk levels between different corridor alternatives. The absolute levels of risk specific for the PNG pipeline could only be estimated from more in-depth geotechnical analysis of the corridors.

6.4. Statistical methods: historical frequency assessment

It is increasingly recognised that there exist serious problems in using historical data to predict future landslide activity, because of the non-linear relationship that exists between triggering factors such as rainfall and landslide activity, the variable effects of human development on vegetation cover, land use and water balance, and the uncertain consequences of climate change (Jones, 1992). Nevertheless, it is a commonly held view that the historical frequency of landsliding in an area can provide an indication of the future probability of such events. The approach relies on the assumption that the proportion of times any particular event has occurred in a large number of trials (i.e. its *relative frequency*) converges to a limit as the number of repetitions increases. This limit is called the *probability of the event*. There is no need for the number of different possible outcomes to be finite. This is essentially a statistical approach depending on a long period of reliable and continuous record.

Probability and frequency are fundamentally different measures and should not be used interchangeably. For example, in a game of chance (e.g. rolling dice), there are a finite number of different possible outcomes, which are assumed to be equally likely. The frequency of any event is then defined as the proportion of the total number of possible outcomes for which that event does occur. Thus, the probability of throwing a '6' with a single die is 1 in 6 (0.167). However, an individual may roll the die 100 times on three successive occasions and record 25 successful throws with a '6' on the first, a '10' on the second and a '5' on the third. In each case, the probability based on historical frequency would be different from the classical probability of 1 in 6. Indeed, it is possible for 100 throws of the die to yield no '6' at all at one extreme and a hundred '6's at the other, although the probability of the latter is *vanishingly small*, thereby drawing attention to the fact that *improbable* (i.e. extremely low probability) does not mean *impossible* (cannot happen).

In the context of landsliding, therefore, obtaining annual probability data from historical records requires the following.

- Ascertaining the extent to which environmental change during the period under consideration (the past as determined by historical records and the future as determined by risk assessment scoping) can be dismissed as insufficiently important to invalidate the basic uniformitarian assumption that 'the past is a valid guide to the future'.
- Ensuring that only that portion of the historical record which is complete and accurate is used, while employing every means available to make the usable record as long as possible.
- Undertaking steps to overcome the problems of sample size, as discussed above, by using every available means of increasing the sample population.

Example 6.3: Scarborough Cliffs, UK

As described in Example 4.1, the 2 km-long cliffline in South Bay, Scarborough, UK can be subdivided into eight separate sections dominated by two contrasting geomorphological units: large, pre-existing landslides and intervening intact (i.e. unfailed) steep slopes. The recent history of landsliding in both of these settings was established through a search of journals, prints, reports, records and local newspapers (held on microfiche) archived at the Scarborough local library and Admiralty charts held at the Hydrographic Office, Taunton (Lee and Clark, 2000).

The earliest reported major landslide in South Bay is the 1737/38 failure at the site of the present-day Spa (Figure 6.2). Descriptions of the slide can be found in Schofield (1787) and Whittaker (1984), among others, along with a number of artists' illustrations. During this event, an acre of cliff top land (205 m by 33 m) sank 15.5 m, complete with cattle grazing on it. This was accompanied by 5.5–6.4 m of toe heave at the cliff foot and on the beach, creating a bulge around 25 m broad and 90 m in length.

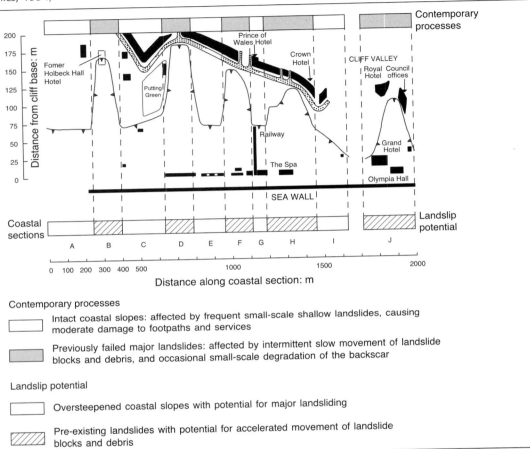

Figure 6.2 Example 6.3: summary of cliff instability hazards, South Bay, Scarborough. (After Lee, 1999; Rendel Geotechnics, 1994)

Little is known out about the timing of the two other major pre-existing slides on South Cliff (the South Bay Pool and South Cliff Gardens landslides; sections D and F in Figure 6.2), other than that they both appear to have fresher morphology (and hence are probably younger) than the Spa landslide. As they are shown on the earliest reliable map of this coastline (an Admiralty chart of 1843), they probably occurred in the period 1738–1842. The most recent event was the 1993 landslide, which led to the destruction of Holbeck Hall, a large cliff-top hotel (Clark and Guest, 1994; Clements, 1994; Lee, 1999).

The historical frequency of failure of the intact steep slopes was estimated to be four events in 256 years (i.e. one event in 64 years). Thus, the annual probability of failure, P, of any one of the eight original intact slopes was estimated to be

$$P(\text{section } s) = 4/(8 \times 256) = 0.00195 \ (1 \text{ in } 512)$$

The time series generated through the archive search allowed an estimate to be made of the frequency of particular types of major landslide events along the cliffline. A degree of caution is needed here, since the construction of seawalls along the foot of the entire cliffline (around the 1890s) has significantly altered the environmental controls on landslide activity. The cliffs are no longer subject to marine erosion at their base, which would promote failures by causing undercutting and removing debris from the landslide toes. The 1993 Holbeck Hall landslide demonstrated that, despite these defences, major landslides can and still do occur (Clark and Guest, 1994; Lee, 1999; Lee et al., 1998a). The estimation of landslide probability from historical frequency needs to consider whether the conditions that generated the pre-1842 landslides have remained valid following the construction of the seawalls and will continue to be valid in the future. If they are, then the time series and subsequent modelling provide reasonable predictions for decision-making. However, if changes in environmental conditions have been profound, then great uncertainty surrounds the validity and usefulness of the results.

Example 6.4: Argillite Cut, British Columbia, Canada

In 1982, a rockfall landed on a car that was stuck in traffic in a large cutting (the 'Argillite Cut') on British Columbia Highway 99, killing a woman and disabling her father. The history of rockfall activity was used by Bunce et al. (1997) to provide an estimate of the probability of potentially hazardous incidents within the 476 m-long cutting. Records of rockfall incidents had been collected by the British Columbia Ministry of Transportation and Highways since 1952. Before 1988, however, incidents were only recorded during inspections by Ministry geotechnical and engineering staff. Systematic recording began in 1988 with the appointment of Capilano Highway Services Ltd as maintenance contractors. Between November 1988 and December 1992, there were nine incidents when a rock was found on the road, indicating that the average number of rockfalls in the 4.12-year period was about 2.2 per year.

Only falls larger than 0.15 m diameter appear to have been recorded, and Bunce et al. (1997) suggest that there may have been other occasions when private citizens, police and the maintenance contractor's employees removed rockfall debris without any record being made. The assessment of rockfall frequency was improved by the mapping of rockfall marks on the highway surface. These marks tend to be distinctive features and are often isolated, sharply defined, circular to angular, concave depressions. A total of 60 rockfall impact marks were identified along a 393 m stretch of road that had been resurfaced 4.75 years prior to the study. Some of these marks were interpreted as multiple impacts from a single event, suggesting that only 35 events had occurred that might have posed a threat to motorists. The number of rockfalls within the whole of the cutting was estimated as

$$\text{number of rockfalls} = \frac{\text{length of Argillite Cut}}{\text{length of resurfaced section}} \times \text{number of rockfall incidents}$$

$$= \frac{476 \times 35}{393} = 42.39$$

$$= 8.925 \text{ incidents/year} \quad \text{(over the 4.75-year period)}$$

The historical records and impact marks suggest a frequency of rockfalls of between 2.2 and 8.9 incidents a year. These figures can be used to provide an estimate of the probability of rockfall activity. However, as the probability cannot exceed 1.0, it is more appropriate to estimate the probability for a shorter unit of time than a year, such as a month. The risk can then be calculated on a monthly basis and then multiplied by 12 to yield an annual figure (see Chapter 10):

$$\text{risk} = (\text{monthly probability} \times \text{consequences}) \times 12$$

If the annual mean number of events is 2.2, then the monthly mean x is 2.2/12 (i.e. 0.18). Therefore, assuming the Poisson distribution (see Chapter 5), the probability of at least one event in a month, P, is

$$P = 1 - P(\text{no event in the month})$$
$$= 1 - e^{-x} = 1 - e^{-2.2/12} = 0.167$$

This is a minimum figure. If the estimated frequency of rockfalls based on impact marks is used (8.925 per annum or 0.7438 per month), then the probability of one or more events per month rises to 0.525.

In the Argillite Cut example, Bunce *et al.* (1997) used the annual frequency of rockfalls as the *number of trials* in a binomial model to determine the probability that rockfalls would hit one or more vehicles in a year (see Example 10.13).

6.5. Statistical methods: the use of landslide magnitude–frequency curves

Landslide magnitude–frequency curves can be viewed as a form of hazard model that represents the probability distribution of landslides of different sizes occurring through time in a region. This type of analysis requires a complete and reliable inventory of historical landslides within a region or a pre-defined subset within the region (e.g. colliery spoil tips in the south Wales coalfield) that includes a measure of the event magnitude. This measure could be the volume of displaced material, the area of the landslide, the dimensions of the failure scar, the length of run-out and so on. Unfortunately, there are few complete landslide inventories worldwide (Guzzetti *et al.*, 2005).

As described in Chapter 5, there are many documented examples of landslide magnitude–frequency distributions approximating an inverse power law relationship:

$$n(\geq M) = aM^{-b}$$

where M is the event magnitude and $n(\geq M)$ is the number of landslides per year greater than or equal to magnitude M.

If it is assumed that the historical frequency per year is the same as future frequency per year or annual probability, then this distribution can form the basis for estimating the probability of landslide events of a particular magnitude. However, most landslide magnitude–frequency curves follow a power law for only a truncated portion of the entire distribution, showing a flattening of the distribution at smaller sizes (rollover; see Chapter 5). For the portion that does fall under a power law, the exponent b (i.e. the slope of the line) is similar for many data sets worldwide (often between -1.3 and -1.8) (Guthrie, 2009).

The recurrence interval RI for events equalling or exceeding a particular magnitude can be calculated using the Weibull equation (see Chapter 5):

$$\text{RI} = \frac{n+1}{m}$$

where n is the number of years in the time series and m is the rank order of the event magnitude (i.e. run-out distance).

The cumulative annual probability of an event of a particular magnitude M being equalled or exceeded is calculated as

$$P(\geq M) = \frac{m}{n+1}$$

This is the equivalent to 1/RI.

However, this is essentially a statistical approach to estimating probability, and there are a number of important issues that need to be considered when undertaking magnitude–frequency analysis.

- Whether or not the magnitude–frequency curve is representative of the entire region covered by the landslide inventory. The locations of future landslides are assumed to be randomly distributed across the region and not biased by geological and landscape controls or limited by the availability of sites where landslides could actually occur. For example, in some areas, the density of large landslide features in the landscape will increase until all the potential sites have been utilised. In this instance, the historical magnitude–frequency curve may overestimate the future probability of large events.
  In other areas, landslides of a particular magnitude may be concentrated along a specific landform, such as an escarpment, or a particular geological outcrop or structural setting. This can also introduce spatial bias into the data set – that is, the inventory does not really reflect a single region-wide power law relationship, but a series of separate inventories each with a limited range of event magnitudes.
- Whether the magnitude–frequency curve is representative of the entire time period covered by the landslide inventory. This implies that the occurrence of landslides has been random over time and not clustered around a limited number of periods when environmental conditions were especially conducive to slope failure. Such clustering can introduce temporal bias into the data set, with the magnitude–frequency curve being the product of a number of discrete landslide episodes, each with different magnitude–frequency characteristics, and therefore unlikely to be representative of the near future.
- Whether the full range of potential landslide magnitudes is captured by the inventory. Inventories derived from historical aerial photography or incident records may cover too short a time period to identify the maximum event size that might occur in a region. For example, in February 2006, the village of Guinsaugon, on Leyte Island in the Philippines, was destroyed by a 1.5×10^7 m^3 rockslide-debris avalanche that ran out a horizontal distance of 3800 m over a vertical distance of 810 m (Evans et al., 2007). Over 1100 people perished in the event. An event of this magnitude was unexpected in this region of the Philippines. The landslide was initiated along a splay of the active Philippine Fault, and subsequent field mapping revealed the presence of older rockslide-debris avalanche deposits on the valley floor parallel to the fault.

Examples 6.5 and 6.6 illustrate the potential problems with this approach. In the South Wales flow slide example, it is unclear what the time period covered by the analysis should be. It might be the period over which spoil tipping was undertaken at those collieries where flow slides occurred. It could have been the period of tipping at all South Wales collieries, including those where flow slides did not occur. There are also uncertainties over whether the analysis reveals a coalfield-wide relationship between flow slide occurrence and run-out distance, or whether this relationship is only applicable to those sites where flow slides actually occurred. Also, do the results mean that the run-out magnitude–frequency relationship describes the flow slide potential at each individual spoil tip, irrespective of topographic setting and tip conditions? This appears unlikely.

The West Nile delta example highlights a different problem. In this example, a sedimentary record of major turbidite deposits over the last 30 000 years is used to generate a mega-landslide magnitude–frequency relationship for the delta. However, environmental conditions have changed dramatically over this period and events have occurred in distinct clusters in time rather than randomly through time. As a result, the long-term magnitude–frequency relationship could provide a very poor indication of landslide probability under current conditions.

Example 6.5: Flow slides in the South Wales Coalfield, UK

The industrial development of the South Wales Coalfield in the UK took place mainly from the last quarter of the 19th century onwards. It was accompanied by the largely uncontrolled disposal of various forms of mine waste in heaps (tips) on hillslopes near the mines. The Aberfan disaster of October 1966 highlighted the inherent dangers in this practice, with 144 people killed by a rapid flow slide from a Merthyr Vale Colliery spoil heap (Tip 7; Bishop et al., 1969). An initial rotational failure on the front of the tip transformed into a flow slide that travelled downslope at around 10 m/s into the Pantglas school yard. The debris ran out 605 m, building up behind the rear wall of the school, causing it to collapse inwards. The loose waste then filled up classrooms, killing children and teachers (Penman, 2000).

Flow slides develop as a result of the metastable collapse of loose, predominantly cohesionless spoil, resulting in the temporary transfer of part of the total stress onto the pore fluids. This causes a sudden rise in pore pressures and loss in strength. The failed material has a semi-fluid character and can flow at extremely rapid rates, up to 100 m/s (Hutchinson, 1988). The flows can have extensive run-out over relatively gentle slopes, often accompanied by a roaring sound likened to the noise of a jet engine (Siddle et al., 1996). When the material comes to rest, it tends to be relatively dry.

The Aberfan event was not the only flow slide to occur in the coalfield. Between 1898 and 1966, there were a total of 18 events, mainly from tips in the central and eastern parts of the coalfield (Table 6.4) (Siddle et al., 1996, 2000). Most failures involved fresh or recently tipped spoil, highlighting the importance of incremental loading in the initiation of the slides, and the looseness of the spoil in promoting significant run-out. The largest run-out event was at Abergorchi colliery in 1931, 35 years before the Aberfan disaster. The flow slide travelled around 610 m from the initial tip failure, entering the colliery yard and filling the boiler house. This left barely enough steam to raise the 700 miners who were underground at the time. A major disaster was narrowly averted (Siddle et al., 1996). In December 1939, a flow slide from the Cilfynydd colliery tip travelled around 435 m, blocking the main Cardiff–Merthyr road for 5 days and damming the River Taff.

The recurrence interval RI and cumulative annual probability for flow slide events with run-out equalling or exceeding a particular distance have been calculated using the Weibull equation (Table 6.5). Note that the run-out distance is known for only 16 out of the 18 reported flow slides, so that the rank series runs from 1 to 16. In this example, the length of the time series was chosen to be 91 years, 1875–1966. This covers the period from the start of the earliest tipping operations on the failed tip at Pentre mine to the Aberfan disaster (Table 6.4). However, other periods could have been chosen, including between the first and last event in the sequence – that is, 1898 to 1966 (68 years). No events have occurred after 1966, reflecting the subsequently improved tipping practices and the rigorous controls on tip investigation and inspection that followed the disaster.

The results (Table 6.5 and Figure 6.3) suggest that a flow slide running out 605 m (as occurred in the Aberfan disaster) had a 1 in 46 years recurrence interval for the period between 1875 and 1966. The

Estimating the probability of landsliding

Table 6.4 Example 6.5: South Wales coalfield – chronology of flow slide failures at colliery spoil tips

Tip	Date of flow slide	Colliery established	Period of tipping on failed tip	Activity at date of failure	Type of spoil[a]	Run-out: m	Run-out slope: deg
National	1898	1881	?	?	ROM	170	21
Pentre	1909	1864	1875–1908	Disused 1 year	ROM	?	18
Craig-Duffryn	1910	c. 1855	c. 1898–1914	Active	ROM	60	<15
Maerdy	1911	1875	1878–1935	Active	ROM	270	11
Cefn Glas	1925	?	c. 1920–1950	Active?	ROM	190	22
Bedwellty	1926	?	c. 1900–1926	Active	ROM	250	13
Rhondda Main	1928	1912	1912–1924	Disused 4 years	ROM, WD	140	23
Abergorchi	1931	Early 1870s	Pre 1914	?	ROM	610	8
Fforchaman	1935	Pre 1869	1913–1965	Active	ROM	100	10
Cilfynydd	1939	?	1910–1963	Disused 2 months	ROM, WD, BA	495	<17
Glenrhonnda	1943	c. 1910	c. 1920–1966	Active	ROM	280	11
Aberfan	1944	1869	1933–1944	Active	ROM	200	12.5
Nantewlaeth	1947–1960?	1913	c.1937–1960	Active?	ROM, WD	120	8
Fernhill	1960	?	?	Active	ROM, WD	?	15
Aberfan	1963	1869	1958–1966	Active	ROM, T	230	10.5
Mynydd Corrwg Fechan	1963	1905	1958–1970	Active	ROM, WD, T	235	17
Parc	1965	1865	1959–1965	Active	ROM, WD, T	140	<23
Aberfan	1966	1869	1958–1966	Active	ROM, T	605	<12

From Siddle et al. (1996)
[a]ROM, run of mine waste; WD, washery discard; BA, boiler ash; T, tailings

Table 6.5 Example 6.5: South Wales coalfield – flow slide run-out recurrence intervals and cumulative frequencies

Flow slide event	Run-out: m	Rank	RI[a]	Cumulative frequency per year[b]
Craig-Duffryn	60	16	5.8	0.18
Fforchaman	100	15	6.1	0.16
Nantewlaeth	120	14	6.6	0.15
Parc	140	12	7.7	0.13
Rhondda Main	140	12	7.7	0.13
National	170	11	8.4	0.12
Cefn Glas	190	10	9.2	0.11
Aberfan	200	9	10.2	0.10
Aberfan	230	8	11.5	0.09
Mynydd Corrwg Fechan	235	7	13.1	0.08
Bedwellty	250	6	15.3	0.07
Maerdy	270	5	18.4	0.05
Glenrhonnda	280	4	23.0	0.04
Cilfynydd	435	3	30.7	0.03
Aberfan	605	2	46.0	0.02
Abergorchi	610	1	92.0	0.01

[a] $RI = (n+1)/m$, where n is the number of years in the time series (91 years) and m is the rank (1, ..., 16)
[b] Cumulative frequency $= m/(n+1) = 1/RI$

cumulative frequency of this event (a 605 m run-out) being equalled or exceeded was around 0.02 per year. If the time period had been restricted to the dates of the first and last flow slides (68 years), then the recurrence interval and cumulative probability would have been 1 in 35 years and 0.03 per year, respectively.

The measure of magnitude chosen in this example was the run-out distance from the toe of the initial tip failure. Other measures could have been used, including the volume of failed material or the total length of the initial slide and the subsequent run-out. Had the latter measure been used, then the Aberfan event would have had a total distance of 720 m. The recurrence interval and cumulative annual probability for an event of this magnitude would have been the same as for the 605 m run-out event – that is, 1 in 46 and around 0.02.

The magnitude–frequency distribution shown in Figure 6.3 provides an indication of the recurrence interval and annual probability of a 605 m-plus event somewhere in the region (i.e. the colliery spoil tips in the south Wales coalfield). In order to estimate the probability that such an event would have occurred at Aberfan, it is necessary to consider the number of possible sites where such an event might occur – that is, colliery spoil tips. One approach would be to limit the potential sites to those tips in the central and eastern parts of the coalfield where flow slides had been reported – that is, 16 collieries. Thus, the probability of a 605 m plus event at any one of these sites would be

$$P(\geq 605) = 0.02 \times 1/16$$
$$= 0.00125 \text{ (i.e. 1 in 800)}$$

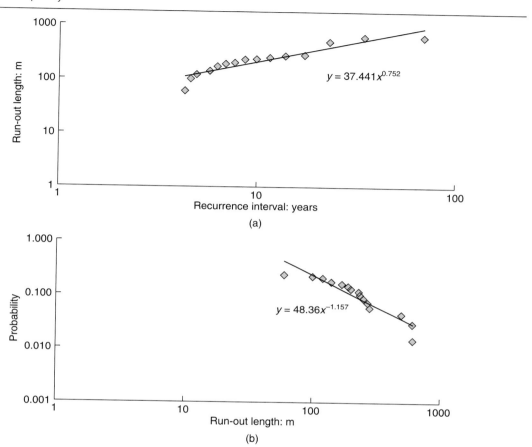

Figure 6.3 Example 6.5: South Wales coalfield flow slides: (a) run-out recurrence interval plot; (b) run-out cumulative frequency

This approach assumes that all the 16 spoil tips are equally likely to generate a large-run-out flow slide. In reality, there are many additional factors that control the potential for flow slides with run-out of 605 m or more, including the nature of the topography, the spoil type, tipping practice, and proximity to springs. Any of these factors might make one particular site more or less likely to fail as a long-run-out event than the region-wide average probability might indicate. This highlights the problem of viewing landslide probability as simply a statistical analysis problem. Local conditions cannot be ignored if realistic site-specific probability estimates are required.

Example 6.6: Submarine landslides in the West Nile Delta, Egypt

The Nile Delta is one of the world's largest deltas, with a submarine fan of about 100 000 km² extending northwards into the Mediterranean Sea, and is a major area of hydrocarbon exploration. The delta extends 20–40 km offshore to a shelf break where the water depth is about 100 m; the maximum water depth beyond the delta slope is about 3000 m in the Herodotus Basin (the eastern Mediterranean Abyssal Plain). At an early stage of gas field development, submarine landslides were identified as posing a potentially serious risk to achieving safe construction and sustainable operation (e.g. Evans et al., 2007; Moore et al., 2007).

A preliminary review of the hazards posed by landslides focused on the record of clay-rich organic mud turbidite deposits in the Herodotus Basin. These deposits are believed to be the product of large-scale landslide activity in the West Nile Delta (e.g. Ducassou et al., 2009). Each turbidite may represent a single massive landslide event or a short-duration period of intense landslide activity across the delta slope. Analyses of deep-sea cores collected on the *Marion Dufresne* Cruise 81 has identified that over 500 km^3 of turbidite sediment was generated from the delta over the period 30 000–7000 BP (Reeder et al., 1998, 2002).

A total of nine West Nile-derived turbidites were identified in the cores (labelled a–o), separated by pelagic and hemipelagic sediments (Table 5.2). The deposits range in size from 0.1 to 190 km^3 of material, and are believed to have been generated by mega-landslide events. The turbidite emplacement dates were determined by carbon-14 (^{14}C) dating of planktonic foraminifera in the horizons above and below the turbidites (Reeder et al., 2002). The raw ^{14}C dates were converted to calendar years by applying a correction factor:

$$\text{calendar years} = 1.24 \times (^{14}\text{C years}) - 440$$

The magnitude–frequency curve for this data set is presented in Figure 6.4. The recurrence interval RI and cumulative annual probability for turbidite events equalling or exceeding a particular volume were calculated using the Weibull equation (Table 6.6). A time period of 30 000 years was chosen for the analysis since this was estimated to be date of the oldest sediments in the sea bed cores – that is, nine events have occurred in 30 000 years. The average annual frequency of major landslide events is nine per 30 000 years, or 0.0003. The results also suggest that an event that could generate deposits with a volume equal or greater than 80 km^3 has a recurrence interval of 1 in 10 000 years and a cumulative frequency of 0.0001 per year.

The analysis highlighted the scale of landslide activity that has taken place within the delta over the last 30 000 years and ensured that submarine landsliding was a prominent item on the project risk register. However, subsequent detailed assessment revealed that the historical record potentially overestimated the landslide threat to any proposed development. This is because the occurrence of past events has been strongly influenced by relative sea level, the sediment load of the Nile and the pore pressure regime within the delta slope sediments (see Figure 5.6).

During the Weichselian glacial maximum (around 25 000–20 000 BP) the Nile Delta was relatively inactive (Ducassou et al., 2009). This inactivity corresponds to a low stand in sea level and a period of arid climate and relatively low sediment discharge from the River Nile, and therefore low sedimentation rates on the delta slope. However, the system was very active during the subsequent period of rapidly rising sea level, between 15 000 and 7000 BP, which was also associated with a wetter continental climate and increased sediment and water discharge from the Nile. Increased sediment deposition on the delta slope would have led to higher pore pressures and relatively frequent large-scale delta slope failure. The Nile system was largely inactive after around 5000 BP. This widespread inactivity is due to retreat of the coastline away from the continental shelf break and to a more arid continental climate and reduced discharge of sediment from the Nile. Since the commissioning of the Aswan High Dam in 1964, the downstream discharge has been reduced to 50% of the 19th-century flows and 64% of the early 20th-century flows.

It follows that the magnitude–frequency curve derived from the 30 000-year turbidite record (see Figure 6.4) covers at least three different time periods, each with its own set of environmental

Figure 6.4 Example 6.5: West Nile Delta: (a) 30 000-year turbidite volume recurrence interval plot; (b) 30 000-year turbidite volume cumulative frequency; (c) Weichselian Termination to Mid Holocene turbidite volume recurrence interval plot

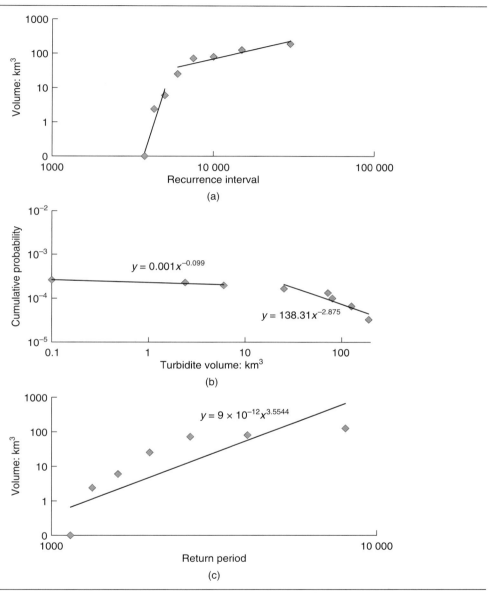

conditions. With the exception of the event at 28 800 BP, large-scale delta slope landslide activity was mainly confined to the period of rapid sedimentation and rising sea levels between 15 000 and 7000 BP. For this one period, the average annual frequency of major landslide events is 8/8000 years, or 0.001 (Table 6.7). During this period, an event that could generate deposits with a volume equal or greater

Table 6.6 Example 6.6: Herodotus Basin – turbidite recurrence intervals and cumulative frequencies

Event	Turbidite volume: km^3	Rank	RI[a]	Cumulative frequency per year[b]
a	0.1	8	3 750	0.00027
f	0.1	8	3 750	0.00027
Debrite	2.4	7	4 286	0.00023
b	6	6	5 000	0.00020
e	25.2	5	6 000	0.00017
g	72	4	7 500	0.00013
c	80	3	10 000	0.00010
d	126	2	15 001	0.00007
o	190	1	30 000	0.00003

[a]$RI = (n + 1)/m$, where n is the number of years in the time series (30 000 years) and m is the rank (1, ..., 8; events a and f are 8th equal)
[b]Cumulative frequency $= m/(n + 1) = 1/RI$

than 80 km^3 would have had a recurrence interval of 1 in 4000 years and a cumulative annual probability of 0.00025 (see Figure 6.4). After this period, there has been no major landslide activity, because of the changing environmental conditions.

In this example, statistical analysis on its own provides a very limited estimate of current landslide probability. Indeed, it is clear that changes in the environmental conditions over the time period covered by the landslide database need to be taken into account. One approach is to adjust (normalise)

Table 6.7 Example 6.6: West Nile Delta – frequency of major landslide events (as indicated by the Herodotus Basin turbidite sequence)

Period	Date[a]	Number of events	Event frequency
1 Mid Holocene to 20th century	7000 BP to pre-Aswan High Dam[b]	0	0
2 Weichselian Termination to Mid Holocene	15 000–7000 BP	8	1/1000 years
3 Weichselian Low Stand	30 000–15 000 BP	1	1/15 000 years

[a]The dates for the time periods are somewhat contentious. The last glacial maximum is currently dated at c. 22 000 BP following recalibration of radiocarbon dates. The sea-level curve of Fairbanks shows the sea level at −121 m at 21 800 BP. By 15 000 BP, the global sea level is thought to have risen to c. −95 m (i.e. a rise of 26 m), and at 7000 BP, it is about −8 m. Some authorities still use 15 000 BP for the Weichselian Termination; however, the sea level had risen significantly by this time
[b]Since the commissioning of the Aswan High Dam in 1964, the downstream discharge has been 50% of 19th-century flows and 64% of early 20th-century flows

the base rate frequency to reflect the conditions over particular time periods. This approach is described in Section 6.8.

6.6. Statistical methods: estimating landslide exceedence probability

The exceedence probability is the probability that a certain event value (e.g. landslide magnitude or number/frequency of events) is going to be exceeded in a given time period. Using the binomial and Poisson models described in Chapter 5, the exceedence probability is the probability of experiencing one or more landslides (n) during time t:

$P(n) = 1 - (1 - P)^t$ (the binomial model)

$P(n) = 1 - e^{-tp}$ (the Poisson model)

As the mean recurrence interval $RI = 1/P$, these equations can be re-written as

$P(n) = 1 - \left(1 - \dfrac{1}{RI}\right)^t$ (the binomial model)

$P(n) = 1 - e^{-t/RI}$ (the Poisson model)

In these statistical models, the historic mean recurrence interval is assumed to be the same as the future mean recurrence interval. The models are very simple to use and only require a database of landslide events that includes the date of occurrence. They can be applied over a region or a subset of the region (e.g. grid cells, municipalities or terrain units), as described below. No knowledge of landslide processes or geological conditions is needed in order to apply the models. This, of course, is a good reason for caution when using the results obtained from this type of approach.

In both of the examples below, a historical landslide inventory was used to calculate the mean recurrence interval within a spatial unit: a 625 m² grid cell in Seattle and variably sized municipalities in Umbria. The focus is on landslide numbers and not landslide types or magnitudes. The hazard, therefore, is assumed to be a function of the rate of landslide occurrence and not the nature and size of the individual events. This is unlikely to be true for most situations.

Example 6.7: City of Seattle, USA

In 1997, the US Geological Survey Landslide Hazards Program, in association with the Urban Geologic and Hydrologic Hazards Initiative, began a project aimed at assessing landslide hazards within the City of Seattle (Coe et al., 2000; Crovelli, 2000). Prior to their studies, landslides had caused widespread damage, especially in the 1996/97, 1997/98 and 1998/99 winter seasons. In the winter of 1996/97, damage to City facilities alone exceeded \$34 million (Coe et al., 2000). Future landslide occurrence was estimated in terms of the probability of having at least one landslide during a specified time period (i.e. the exceedence probability). The analysis was based on an 88.4-year (1909–1997) landslide database that had been compiled for the City of Seattle. The database included location and time of occurrence information, as well as many other landslide characteristics.

The city area was overlain with a grid of 25 m × 25 m (625 m²) cells roughly equivalent in area to an average-sized city lot. A circle covering an area of 40 000 m² (4 ha, roughly equivalent in area to the largest landslides that have occurred in the city) was then placed at the centre of each cell and the number of landslides occurring within the circle were counted. The landslide density, or number of landslides within each 4 ha circle, was then assigned to the 625 m² cell at the centre of each circle.

The historical mean recurrence interval for each cell was calculated as

$$RI = \frac{\text{database time period}}{\text{number of landslides recorded in the cell}}$$

$$= \frac{88.4}{\text{number of landslides recorded in the cell}}$$

The number of landslides recorded in a cell ranged from 1 to 26, resulting in mean recurrence intervals of 88.4 years (88.4/1) to 3.4 years (88.4/26).

The exceedence probability was calculated for each grid cell using the Poisson model:

$$P(n) = 1 - e^{-t/RI}$$

Exceedence probabilities were calculated for cells with historical landslide counts ranging from 1 to 26 for a range of time periods: 1, 5, 10, 25, 50 and 100 years (Table 6.8). For example, cells with ten historical landslides were all assigned the exceedence probability over 1 year the value 0.107, 0.677 over 10 years, 0.997 over 50 years and so on (see Table 6.8). The exceedence probability values for each cell were compiled in map form, producing a landslide hazard map for the city area.

The results would have been exactly the same if the exceedence probability had been calculated from the historical frequency or annual probability (the number of landslides in the cell/88.4) using

$$P(n) = 1 - e^{-tp}$$

As indicated in Table 6.9, the analysis revealed that around 0.8% of the city (174.97 out of 21 544 hectares) has historical landslide frequencies of over 7/88.4 years. For these areas, the annual exceedence probability ranges from 0.05 to 0.25 (i.e. the chance of one landslide occurring in the cell in any given year). The highest landslide counts, lowest mean recurrence intervals and highest exceedence probabilities occur along coastal cliffs developed in glacial deposits along Puget Sound.

Almost identical results could be obtained using the binomial model (Table 6.10). However, as Crovelli (2000) notes, for relatively high probabilities and short periods of time the binomial model significantly over estimates the exceedence probability (see Table 5.4).

Example 6.8: Umbria, Italy

The general approach used in Seattle has been used in Italy to produce regional landslide hazard maps. Guzzetti (2005) used the national AVI landslide database (AVI is an Italian acronym for 'Areas Affected by Landslides and Floods in Italy': *Aree Vulnerate Italiane*) to establish the number of landslides that had occurred in 92 individual municipalities in Umbria over the 85-year period between 1917 and 2001. Mean recurrence intervals were calculated for each municipality:

$$RI = \text{number of landslides in municipality}/85 \text{ years}$$

The Poisson model was then used to estimate exceedence probabilities in all 92 municipalities over a range of time periods (5, 10, 20, 25 50 and 100 years). For a 5-year period, only five municipalities have a 0.9 or larger probability of experiencing at least one event. However, over 100 years, all the municipalities have a 50% or larger probability of experiencing a landslide.

Table 6.8 Example 6.7: City of Seattle – probability of one or more landslides occurring during a particular time period for 4ha cells with historical landslide counts ranging from 1 to 26, using the Poisson model

Number of landslides in a 4 ha cell	Mean RI[a]	Historical frequency[b]	Time period: years[c]					
			1	5	10	25	50	100
1	88.40	0.01	0.011	0.055	0.107	0.246	0.432	0.677
2	44.20	0.02	0.022	0.107	0.202	0.432	0.677	0.896
3	29.47	0.03	0.033	0.156	0.288	0.572	0.817	0.966
4	22.10	0.05	0.044	0.202	0.364	0.677	0.896	0.989
5	17.68	0.06	0.055	0.246	0.432	0.757	0.941	0.997
6	14.73	0.07	0.066	0.288	0.493	0.817	0.966	0.999
7	12.63	0.08	0.076	0.327	0.547	0.862	0.981	1.000
8	11.05	0.09	0.087	0.364	0.595	0.896	0.989	1.000
9	9.82	0.10	0.097	0.399	0.639	0.922	0.994	1.000
10	8.84	0.11	0.107	0.432	0.677	0.941	0.997	1.000
11	8.04	0.12	0.117	0.463	0.712	0.955	0.998	1.000
12	7.37	0.14	0.127	0.493	0.743	0.966	0.999	1.000
13	6.80	0.15	0.137	0.521	0.770	0.975	0.999	1.000
14	6.31	0.16	0.146	0.547	0.795	0.981	1.000	1.000
15	5.89	0.17	0.156	0.572	0.817	0.986	1.000	1.000
16	5.53	0.18	0.166	0.595	0.836	0.989	1.000	1.000
17	5.20	0.19	0.175	0.618	0.854	0.992	1.000	1.000
18	4.91	0.20	0.184	0.639	0.869	0.994	1.000	1.000
19	4.65	0.21	0.193	0.659	0.883	0.995	1.000	1.000
20	4.42	0.23	0.202	0.677	0.896	0.997	1.000	1.000
21	4.21	0.24	0.211	0.695	0.907	0.997	1.000	1.000
22	4.02	0.25	0.220	0.712	0.917	0.998	1.000	1.000
23	3.84	0.26	0.229	0.728	0.926	0.999	1.000	1.000
24	3.68	0.27	0.238	0.743	0.934	0.999	1.000	1.000
25	3.54	0.28	0.246	0.757	0.941	0.999	1.000	1.000
26	3.40	0.29	0.255	0.770	0.947	0.999	1.000	1.000

From Crovelli (2000)
[a] Mean RI = 88.4/number of landslides in cell
[b] Historical frequency = number of landslides in cell/88.4 = 1/RI
[c] The values in columns 4–9 were calculated as $P(n) = 1 - e^{-t/RI}$, where t is the period (1, 5, 10, 25, 50 or 100) and RI is taken from column 2

Landslides are a frequent occurrence in the hills of the 79 km² Collazzone area of Umbria, ranging from degraded deep-seated landslides to shallow slides and flows. Guzzetti et al. (2006) described the development of a landslide inventory through analysis of a sequence of aerial photographs taken between 1941 and 1997. Geomorphological field mapping was undertaken to extend the inventory over the period 1998–2004. A total of 2787 landslides were identified in the 64-year period between 1941 and 2004, resulting in an overall landslide density of 35 landslides per km². The inventory includes 2490 for which the date of occurrence was known. The mapped landslides ranged in size from 51 m² to 1.45 km².

Table 6.9 Example 6.7: City of Seattle – percentage of Seattle (21 544 ha) encompassed by each annual exceedence probability value

Historical landslide count: number of landslides in a 4 ha cell	Annual exceedence probability ($T = 1$ year)	Area: ha	Percentage of entire land area of Seattle
2–6	0.02–0.05	644.65	2.99
7–10	0.05–0.1	119.02	0.55
11–14	0.1–0.15	42.18	0.19
15–18	0.15–0.18	10.12	0.05
19–22	0.18–0.22	2.96	0.01
23–26	0.22–0.25	0.69	0.003

From Coe et al. (2000)

Table 6.10 Example 6.7: City of Seattle – probability of one or more landslides occurring during a particular time period for 4 ha cells with historical landslide counts ranging from 1 to 30, using the binomial model

Number of landslides in a 4 ha cell	Mean RI[a]	Time period: years[b]					
		1	5	10	25	50	100
1	88.40	0.011	0.055	0.108	0.248	0.434	0.679
2	44.20	0.023	0.108	0.205	0.436	0.682	0.899
3	29.47	0.034	0.159	0.292	0.578	0.822	0.968
4	22.10	0.045	0.207	0.371	0.686	0.901	0.990
5	17.68	0.057	0.253	0.441	0.767	0.946	0.997
6	14.73	0.068	0.296	0.505	0.827	0.970	0.999
7	12.63	0.079	0.338	0.562	0.873	0.984	1.000
8	11.05	0.090	0.378	0.613	0.907	0.991	1.000
9	9.82	0.102	0.415	0.658	0.932	0.995	1.000
10	8.84	0.113	0.451	0.699	0.950	0.998	1.000
11	8.04	0.124	0.485	0.735	0.964	0.999	1.000
12	7.37	0.136	0.518	0.768	0.974	0.999	1.000
13	6.80	0.147	0.549	0.796	0.981	1.000	1.000
14	6.31	0.158	0.578	0.822	0.987	1.000	1.000
15	5.89	0.170	0.605	0.844	0.990	1.000	1.000
16	5.53	0.181	0.632	0.864	0.993	1.000	1.000
17	5.20	0.192	0.656	0.882	0.995	1.000	1.000
18	4.91	0.204	0.680	0.897	0.997	1.000	1.000
19	4.65	0.215	0.702	0.911	0.998	1.000	1.000
20	4.42	0.226	0.723	0.923	0.998	1.000	1.000
21	4.21	0.238	0.742	0.934	0.999	1.000	1.000
22	4.02	0.249	0.761	0.943	0.999	1.000	1.000
23	3.84	0.260	0.778	0.951	0.999	1.000	1.000
24	3.68	0.271	0.795	0.958	1.000	1.000	1.000
25	3.54	0.283	0.810	0.964	1.000	1.000	1.000
26	3.40	0.294	0.825	0.969	1.000	1.000	1.000

[a] Mean RI = 88.4/number of landslides in cell
[b] The values in columns 3–8 were calculated as $P(n) = 1 - (1 - 1/RI)^t$, where t is the period (1, 5, 10, 25, 50 or 100) and RI is taken from column 2

Estimating the probability of landsliding

Table 6.11 Example 6.8: Collazzone, Umbria – numbers of terrain units in the three classes of exceedence probabilities

Exceedence probability range	Number of terrain units with the exceedence probability range for different time periods (percentage of total area in parentheses)			
	5	10	25	50
0–0.4	860 (89.2%)	697 (59.1%)	477 (36.0%)	302 (17.3%)
0.4–0.8	34 (10.7%)	197 (40.9%)	291 (34.5%)	279 (25.3%)
0.8–1.0	0 (0%)	0 (0%)	126 (29.4%)	313 (53.9%)

From Guzzetti et al. (2006)
The table indicates that 313 out of 894 terrain units have an exceedence probability of over 50 years in the range 0.8–1.0. These 313 units represent 53.9% of the total area of Collazzone

GIS software was used to subdivide the area into 894 individual terrain units bounded by drainage lines and catchment boundaries. The number of landslides within the 64-year period for which the date was known (i.e. the 2490 subset of the inventory) was determined for each of these units.

Mean recurrence intervals were calculated for each terrain unit as follows:

$$RI = \frac{\text{inventory time period}}{\text{number of landslides recorded in the cell}}$$

$$= \frac{64}{\text{number of landslides recorded in the unit}}$$

The exceedence probability of one or more landslide occurring in a unit was calculated using the Poisson model over 5-, 10-, 25- and 50-year periods:

$$P(n) = 1 - e^{-t/RI}$$

where t is the time period (5, 10, 25 or 50 years).

The results were used to generate an exceedence probability map for the Collazzone area. As indicated in Table 6.11, 29% of the area has an exceedence probability (i.e. the chance of one or more landslides occurring in the unit) of 0.8–1.0 over a 25-year period. This value rises to 54% of the area over a 50-year period.

6.7. Statistical methods: estimating probability of cliff recession through simulation models

In the case of eroding clifflines, the focus is normally concerned with determining how much recession will occur over a particular time period in the future. Effort needs to be directed towards estimating the probability of sequences of events, rather than trying to estimate the likelihood of a single landslide. Cliff recession in over-consolidated materials proceeds primarily via occasional landslide episodes, separated by periods of relative inactivity that may last for more than 100 years on some coastlines (Lee, 1998; Lee and Clark, 2002). This is in stark contrast to the continuous process that has been implicit in many deterministic approaches to predicting cliff recession, such as the extrapolation of historical trends (see Example 9.1). The recession process is complex and far from random. Recession is not an inevitable consequence of the arrival of a storm that removes material from the cliff base or raises groundwater levels within the cliff. In order to fail, the cliff must already be in a state of deteriorating stability, so as to render it prone to the effects of an initiating storm event.

The pattern of past recession events is the result of a unique sequence of wave, tide, weather and environmental conditions. A different set of conditions would have generated a different recession scenario. The inherent randomness in the main causal factors (e.g. wave height and rainstorms) dictates that the future sequence of recession events cannot be expected to be an accurate match with the historical records; there could, however, be a similar average recession rate with contrasting variability between measurements, trends and periodicity. Probabilistic methods offer an improvement on conventional deterministic predictions, because they aim to represent the variability and uncertainty inherent in the recession process (Lee et al., 2001).

The main elements in developing a probabilistic model are as follows (Lee and Clark, 2002; Meadowcroft et al., 1997).

1. Establishment of a cliff behaviour model, with particular emphasis on assessing potential events in terms of size and timing (i.e. recurrence intervals).
2. Assigning probability distributions to represent variability and uncertainty in the key parameters (e.g. event size, event timing and extreme wave heights). Some parameters, such as extreme wave heights, have been extensively studied, and probability distributions for these can be established using standard methods. Other aspects, such as future beach levels, may be established on the basis of historical data, combined, if possible, with the results of modelling. Some factors, however, are more difficult to quantify and may call for a degree of subjective assessment, but this should be guided, wherever possible, by informed arguments about what ranges of values are likely and with what degrees of confidence. Distributions do not have to conform to the standard analytical forms; any probability distribution that can be envisaged can be simulated.
3. Development of a probabilistic prediction framework and selecting a simulation strategy. Simulations may be static, that is, assessing responses at a given point in time, or dynamic that is, simulating a given time period using a time-stepping approach. The static approach is simply a Monte Carlo simulation of the model. There is no attempt to simulate any variation with time, though future prediction can be made by setting, for example, climate parameters to their predicted future values over particular time periods.
The dynamic approach, ideal for long-term prediction, involves repeating many simulations of the required time period in order to establish a histogram of the probability distribution of the given response at a given point in time. The dynamic approach means that events that will occur in the future can be included: as well as random loadings, this could include deterioration of a structure or management intervention.
4. Running repeated simulations. The key requirements here are for a pseudo-random-number generator that produces a stream of values between 0.0 and 1.0 and the inverse probability distribution functions from which values for each of the variables are selected, based on the value of the random variable. Correlated variables require additional functions to ensure that sampled values are correctly correlated. After a large number of simulations, the frequency distributions and correlations of the sampled data should conform to the specified probability distributions, and the result will be a stable frequency distribution, reflecting the variability of the input data and the form of the response function.

Simple response functions and models can be accommodated on a spreadsheet and can be set up and run quickly. More complex models can be built on the basis of existing numerical models, provided that these are not prohibitively slow to run. An advantage of the ever-increasing speed of computers is that multi-simulation techniques can now be used even with relatively complex process–response models, and can include long-term prediction.

Cliff recession data and predictions can be presented in a variety of ways, including probability density functions of the cliff position at a given time or the time required for cliff recession to reach a given point (Figure 6.5). Alternatively, it may be useful to present the results as a hazard zoning based on the cumulative probability distribution of cliff recession over a given time.

Note that the probabilities that demarcate the zones are arbitrary and can be varied to suit the purpose. More detail (i.e. more zones) may be justified in areas with more assets at risk, although choosing too many zones can give a false impression of precision. This form of presentation does not differentiate between different locations within the same zone, although in reality properties at the landward margins of a zone will have lower probabilities of being affected by cliff recession than those at the seaward margins.

Example 6.9: Sussex cliffs, UK

A *two-distribution* model has been developed that assumes that the cliff foot can withstand a given number of storms before the cliff fails (Hall *et al.*, 2002; Lee and Clark, 2002; Lee *et al.*, 2001; Meadowcroft *et al.*, 1997). In this model, a storm that causes undercutting of the cliff foot is defined as a wave height and water level with a certain return period. The return period, together with the number of storms required to initiate failure of the cliff, defines the average time interval between recession events. If a recession event does occur, then a second probability distribution can be used to estimate the likely magnitude of the event – that is, the amount of cliff-top recession. This model therefore has the ability to differentiate between high- and low-sensitivity cliff units by representing the number and magnitude of storm events needed to initiate recession events.

Cliff recession is assumed to proceed by means of a series of discrete landslide events, the size and frequency of which are modelled as random variables. A discrete model for the probabilistic cliff recession X_t during time period t is

$$X_t = \sum_{i=1}^{N} C_i$$

where N is a random variable representing the number of cliff falls that occur in time period t and C_i is the magnitude of the ith recession event. This model can be used to simulate synthetic time series of recession data, which conform statistically to the cliff recession measurements (e.g. Hall *et al.*, 2002; Lee *et al.*, 2001). Three typical realisations of the model are shown in Figure 6.6. The time series are stepped, reflecting the episodic nature of the cliff recession process. Multiple realisations of the simulation are used to build up a probability distribution of cliff recession.

The model is defined by two distributions.

- An *event timing distribution* describes the timing of recession events. The model incorporates physical understanding of the cliff recession process by representing the role that storms play in destabilising cliffs and initiating recession events. Note that in this example it is assumed that

Landslide Risk Assessment

Figure 6.5 Sample results from the two-distribution probabilistic model. (From Lee and Clark, 2002)

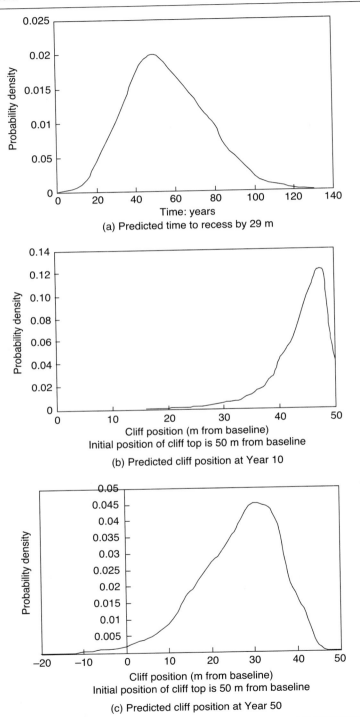

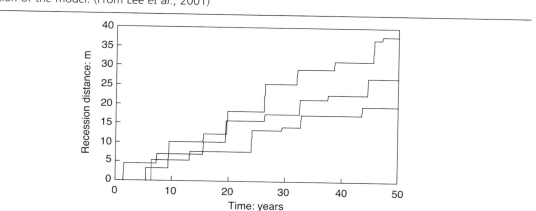

Figure 6.6 Example 6.9: typical realisations of the two-distribution simulation model (see the text for the explanation of the model. (From Lee et al., 2001)

recession is driven by storm events; in other instances, groundwater and other factors will be important. The approach has links to renewal theory (Cox, 1962), inasmuch as the cliff is considered to be progressively weakened by succeeding storms. The arrival of damaging storms is assumed to conform to a Poisson process – that is, successive storms are assumed to be independent incidents with a constant average rate of occurrence. After a number of storms of sufficient severity, a cliff recession event occurs. The time between successive recession events can therefore be described by a gamma distribution. The shape of this distribution is defined by a scaling parameter λ (the reciprocal of the return period of the significant storm event) and a shape parameter k (the number of storms above a certain threshold that cause weakening to the foot of the cliff that is sufficient to trigger failure).

- An *event size distribution* describes the magnitude of recession events in terms of the mean size and variability. The form and parameters of this distribution should reflect the frequency distribution of actual cliff failures and are likely to be site-specific. The model uses a log-normal distribution, following the conclusions of the wave basin tests on a model cliff undertaken by Damgaard and Peet (1999). A log-normal distribution is non-negative, which reflects the non-existence of negative cliff recession events. The probability density rises to a maximum value and then decreases to approach zero as the recession distance becomes large (i.e. very large cliff recession events are extremely unlikely).

The cliff recession model is thus characterised by four parameters: λ and k from the gamma distribution, and the mean μ and variance δ of the log-normal distribution. The model parameters can be estimated by maximum likelihood or Bayesian estimation methods (Hall et al., 2002). The maximum likelihood method is based on optimally fitting the parameters to the available data, while the Bayesian method also makes use of expert knowledge about the size and frequency of landslide events. There is therefore scope to include geomorphological knowledge of event size and timing, which may not necessarily be revealed through examination of the historic data record.

This method has been tested using historical recession data for 20 m-high cliffs in East Sussex, UK, developed in sandstones overlain by Wadhurst Clay, both dating from the lower Cretaceous. The position of the cliff top was obtained from 1:2500-scale topographic maps produced in 1907, 1929, 1936, 1962 and 1991. Cliff-top locations were extracted at eight positions along the 400 m length of

Table 6.12 Example 6.9: historical cliff recession rates (average annual recession rates in m/year)

Location	Sea cliff 1907–1929	Sea cliff 1929–1936	Sea cliff 1936–1962	Sea cliff 1962–91	Sea cliff 1907–1991
1	0.09	1.5	1.31	0	0.57
2	2.09	2.0	0.31	0.06	0.83
3	1.63	0.57	0	0.34	0.57
4	0.91	0.57	0.23	0.31	0.48
5	0.64	0.28	0.31	0.28	0.5
6	0	0.28	0.15	0.34	0.19
7	0.09	0	0.54	0.18	0.24
8	1.27	0	0.54	0	0.26
Mean	0.84	0.65	0.42	0.19	0.45
Standard deviation	0.72	0.68	0.38	0.14	0.20

From Lee and Clark (2002)

coastline. For each period between map dates, the mean recession rate (m/year) was calculated for the eight locations. In addition, overall recession rates from 1907 to 1991 were calculated. For each of the five measurement periods, the standard deviation of recession rate between the different locations was calculated, as well as the mean rate (Table 6.12).

The event timing distribution was chosen using a maximum likelihood parameter estimation model (Hall et al., 2002), with parameters $k = 0.8$ and $\lambda = 0.046$. With more frequent events, the statistical model would not generate sufficient variability as compared with the data. Furthermore, the number of zero recession rates in the data record (Table 6.12) indicated that the characteristic time between recession events was quite long. For example, during the 7-year period from 1929 to 1936, two of the locations showed no recession at all, indicating a significant probability (about 0.25) that the interval between recession rates could be greater than seven years. This type of reasoning was used to constrain the simulation model parameters.

Monte Carlo simulation was used to generate multiple realisations (simulations) of the recession process. Table 6.13 shows results of two simulations from the calibrated model. These were obtained by simulating the time period 1907–1991 and extracting results at the relevant years so that these could be compared directly with the measured values. As this is a sampling approach, different simulations give different results, so the two example simulations shown in Table 6.13 give different individual values. Nevertheless, the general characteristics of the model results are similar to the measured values in Table 6.12.

The statistical model was then used to make probabilistic predictions of

- the time for the cliff to undergo recession by a certain distance, to assess when in the future a hypothetical fixed asset currently 29 m from the cliff top will be lost (Figure 6.5(a))
- the cliff position after 10 and 50 years (Figures 6.5(b) and (c)); the cliff position is measured relative to a fixed baseline 50 m landward of the initial cliff position, so greater than 50 m recession appears as a negative value (i.e. it is landward of the baseline).

Table 6.13 Example 6.9: two-distribution probabilistic model: simulation results (compare the mean and standard deviations with historical data in Table 6.12)

Simulation 1

Location	Number of years and period				
	22 1907–1929	7 1929–1936	26 1936–1962	29 1962–1991	84 1907–1991
1	0.00	0.00	0.18	0.37	0.18
2	0.26	0.36	0.75	0.88	0.63
3	0.85	0.60	0.68	0.36	0.61
4	0.23	0.90	0.20	0.77	0.46
5	0.40	0.43	0.51	0.39	0.44
6	0.05	0.07	0.30	0.40	0.25
7	0.76	0.00	0.70	0.45	0.57
8	1.34	1.64	0.24	0.54	0.75
Mean	0.49	0.50	0.45	0.52	0.49
Standard deviation	0.43	0.52	0.23	0.19	0.18

Simulation 2

Location	Number of years and period				
	22 1907–1929	7 1929–1936	26 1936–1962	29 1962–1991	84 1907–1991
1	0.34	0.60	0.42	0.42	0.42
2	0.58	0.74	0.90	0.69	0.73
3	0.53	0.94	0.38	0.53	0.52
4	0.97	0.00	0.46	0.67	0.63
5	0.50	0.51	0.08	2.53	1.07
6	0.13	0.00	0.28	0.13	0.17
7	0.20	0.39	0.31	0.24	0.27
8	0.54	0.19	0.22	0.97	0.56
Mean	0.47	0.42	0.38	0.78	0.55
Standard deviation	0.24	0.32	0.23	0.71	0.26

From Lee and Clark (2002)

Since these are numerical simulation results, the final distribution is not completely smooth.

Examination of Table 6.12 reveals that the average annual recession rate appears to have been progressively declining during the time period covered by the topographic maps, assuming that these maps are uniformly accurate. This suggests that is there is a trend in the data set that cannot be viewed as a stationary series. This is a common problem for coastal sites, where future conditions are not expected to resemble past conditions, owing to significant natural or human-induced change in the physical processes. In such circumstances, historic records alone cannot be used for recession prediction. However, even under changing conditions, a statistical prediction based on an assumption of stationary long-term average recession rate can usefully provide upper or lower bounds on future recession, as well as past recession rates and rates of change (Hall *et al.*, 2002). Under changing conditions (especially where coast protection schemes have been recently implemented or are planned), deterministic methods combined with engineering judgement have tended to dominate in the past, although process simulation models are becoming available (e.g. Lee *et al.*, 2002).

6.8. Conceptual models: normalisation of base-rate frequency

The *frequentist* approach to estimating landslide probability involves a number of constraining assumptions. The extrapolation of past event frequency into the future is valid as long as the statistical properties that describe the variability within the data series (e.g. mean and variance) have remained constant over time – that is, the data series is stationary. However, this is often a problem for landslide data series, since it is unusual for the environmental controls such as climate and land use to have remained constant over the period of the historical record. In Example 6.6, it was noted that the magnitude–frequency curve for the West Nile Delta that had been derived from the 30 000-year turbidite record actually consisted of at least three different time periods, each characterised by its own distinctive set of environmental conditions and historical frequencies (Table 6.7). The conditions over the next 100 years or so are expected to be controlled by negligible Nile sediment loads (post-Aswan High Dam) and relative sea-level rises of 0.5–1 m. Very slow sedimentation rates are expected, with a very low, dissipating pore pressure regime. As a result, major landslide activity could be expected to be significantly lower than that experienced in the Weichselian Termination to Mid Holocene period. However, this conclusion does not follow from a strict application of relative frequency statistics, since it draws on the use of judgement and knowledge of the controls on landslide activity (Lee, 2007).

Another important assumption is that the occurrence of landslide events across the area is random and not biased towards particular geological or geomorphological settings. In Example 6.3, the annual probability of failure of any one of the eight original intact slopes in South Bay, Scarborough was estimated to be:

annual probability of failure in section $s = 4/(8 \times 256) = 0.00195$

This assumes that the ground conditions at the intact slope sites were uniform around the bay. However, subsequent borehole investigations revealed that this was not the case (High Point Rendel, 1996, 1999). For example, the conditions at the 1993 failure site (Holbeck Cliff) were shown to be significantly different from the adjacent Holbeck Gardens site (an intact cliff; section C in Figure 6.2). Both cliffs are developed in around 25–30 m of glacial till overlying Scalby Formation sandstones and argillaceous beds. However, the glacial tills within the Holbeck Gardens cliffs are dominated by sandy silty clays (soil 1; 87–92%) with only minor layers of sand (soil 2) or stiff plastic silty clay (soil 3) – a marked contrast to the Holbeck Cliffs, which are predominantly stiff plastic silty clay (soil 3; 40–87%) with localised thick lenses of sand (soil 2; 54%).

One approach to integrate additional knowledge or experience into the statistical analysis of landslide inventories is to adjust (normalise) the base-rate frequency to better reflect the conditions experienced in a subset of the sample area or time period or at a similar location outside of the sample area. The approach involves the use of judgement to estimate the degree of departure from the base-rate frequency:

$$P(\text{event}) = \frac{\text{number of recorded events}}{\text{number of years in the record}} \times \text{adjustment factor}$$

In this simple conceptual model, the effects of site-specific variations in controlling factors (e.g. slope angle and aspect, geological setting and ground materials) and preparatory factors (e.g. the presence of tension cracks, stream undercutting or high groundwater levels) can be brought together in one or more adjustment factors.

There can be problems in normalising landslide frequency statistics. This is because every slope is unique but is also a member of the population of slopes from which the inventory was derived. However, there is no 'average' slope in the population against which all the other slopes can be compared. As a result, the significance of particular features can be difficult to gauge. Determining the degree of departure from the base-rate frequency is a matter of judgement.

Example 6.10: Berkely Escarpment, Vancouver, Canada

The Berkely Escarpment in the district of North Vancouver, Canada, is around 60 m high, with slopes ranging from 30° to 45°. The escarpment is developed in extremely dense, glacial tills (the Vashon Drift), overlain by glacio-marine Capilano sediments (stiff, laminated sands, silts and clays), colluvium and up to 4 m of fill. The fill is derived from the glacio-marine sediments, and comprises sand to cobble-size fragments of silt set in a loose to compact fine sand matrix. From around 1950, houses were constructed along the escarpment crest and at the scarp foot. Often, the lots were levelled by pushing or end-tipping local and imported materials over the escarpment crest, so that the uppermost slopes were often formed of made-ground. Retaining walls, many comprising timber cribbing or concrete blocks, were constructed in several locations to increase stability (Porter et al., 2007).

Since 1972, a total of six rapid flows slides have occurred along the escarpment, generated by four major storms (December 1972, December 1979, January 1999 and January 2005). The January 2005 event destroyed two houses at the base of the slope, killing one person and seriously injuring another. Concerns over the potential impact of future landslides prompted the Municipal Council to commission a landslide risk assessment (Porter et al., 2007).

The annual probability of a flow slide was estimated from the historical landslide frequency:

$$P(\text{landslide}) = \frac{\text{number of historical events}}{\text{years in historical record}}$$

$$= \frac{6}{33}$$

$$= 0.18$$

The landslide source areas along the escarpment are typically 15–25 m wide, indicating a total of about 75 potential source areas along the crestline. The average annual probability of a slide from any one of these 75 source areas was calculated as

$$P(\text{landslide, source 1–75}) = P(\text{landslide})/75$$

$$= 0.18/75$$

$$= 0.0024 = 2.4 \times 10^{-3}$$

An adjustment factor was applied to normalise the base-rate annual probability for each potential source area (Table 6.14):

$$P(\text{landslide, source } s) = P(\text{landslide, source 1–75}) \times \text{adjustment factor}$$

Table 6.14 Example 6.10: flow slide probability scores[a]

Slope score	Loose soil score	Water score	Slope deformation score
$<35° = 0.8$ $35–40° = 1.0$ $>40° = 1.25$	Approved mechanical stabilisation at and below crest = 0.35 <1 m deep at crest and <2 m deep below crest = 0.35 <2 m deep at and below crest = 0.5 >2 m deep at or below crest = 1.0 >2 m deep at and below crest = 0.5	All adjacent properties connected to storm sewer and street storm water properly managed = 0.35 Connected to storm sewer but adjacent properties not connected = 0.5; otherwise Run-off from backyard = 0.5 … and half roof = 0.75 … and full roof = 1.0 … and driveway = 1.25 … and street = 2.0	None observed = 0.5 Deformation at or below crest = 1.0 Deformation at and below crest = 1.0

From Porter et al. (2007)
[a]The overall minimum/maximum adjustment score = 0.05–10

The adjustment factor was the product of four separate scores that reflect site-specific slope attributes related to the main landslide controls

- slope steepness
- the presence of thick layers of weak or collapsible soils (e.g. loose fill and colluvium)
- surface and subsurface drainage conditions that promote high groundwater levels, especially during heavy rainfall events
- the presence of deformation at or below the crest

adjustment factor = slope score × loose soil score × water score × slope deformation score

These attributes were selected because they could be assessed through visual inspection and shallow hand-auguring. The weightings were assigned through judgement. However, the conceptual model was calibrated against past slope performance to ensure that the calculated probability of a flow slide somewhere along the escarpment was similar to the historical average.

Example 6.11: Cut-slope failures, Hong Kong

Failure of cut slopes presents a significant hazard in Hong Kong and occasionally leads to fatalities. For example, a 27 m-high cut on the Fei Tsui Road failed in August 1993. The landslide involved an estimated 14 000 m³ of debris, which ran out across the road, killing a young boy and injuring his father, who had been walking along the pavement (GEO, 1996; see Example 10.15).

An approach for assessing the probability of failure of individual cut slopes has been developed that is based on the historical frequency of failure throughout Hong Kong, with appropriate adjustment factors for local site conditions (Fell et al., 1996a; Finlay and Fell, 1995). Systematic recording of landsliding in Hong Kong has been undertaken since 1978, with landslide incident reports prepared for all recorded slides since 1983 (this includes incidents as small as a single boulder or 1 m³ of debris). Table 6.15 summarises the landslide statistics for the period 1984–1993 and provides a performance database for the estimated 20 500 cut slopes in the area. During this 10-year period a total of 2177 cut slope failures were reported. The average historical frequency of failure of 217.7 events/year was used as the basis of generating an estimate of the average annual probability of failure for every one of the entire population of cut slopes:

$$\text{annual probability of failure (Hong Kong-wide)} = \frac{\text{number of historical events}}{\text{total number of cut slopes}}$$

$$= \frac{217.7}{20\,500} = 0.0106$$

This Hong Kong-wide estimate does not differentiate between cut slopes that are more or less likely to fail. However, by using a combination of site-related factors, it is possible to adjust the estimate for a particular cut:

$$\text{probability of failure (cut } c) = \text{probability of failure (Hong Kong-wide)} \times \text{adjustment factor } F$$

The adjustment factors were based on experienced gained from a comprehensive archive of reports and records of landslides dating back to the 1970s and the opinions of professional staff from GEO. If the slope is considered less likely than average to fail, then the adjustment factor F will be less than 1.0, whereas it will be over 1.0 if the slope is considered more likely than average to fail.

The adjustment factor F comprises a *primary site factor F'* and a factor F_e that takes account of the *history of instability* at the site.

The primary site factor F' is the product of five independent factors (Table 6.16):

$$\text{primary site factor } F' = F'_1 \times F'_2 \times F'_3 \times F'_4 \times F'_5$$

The independent factors are as follows.

- *Slope age F'_1*: for those slopes that were constructed before the establishment of geotechnical control procedures in 1977, a factor of 1.25 is used; for post-1977, the factor is 0.25.
- *Geology F'_2*: the key conditions assumed to influence failure are unfavourable discontinuity orientations and the presence of colluvium within the cutting (see Table 6.17(a)).
- *Slope geometry F'_3*: although the potential for failure tends to increase with slope angle, experience has shown that the higher cuts have generally been better designed; therefore the slope geometry factor (Table 6.17(b)) decreases with slope height.
- *Geomorphology F'_4*: the most significant geomorphological condition is considered to be the gradient of the terrain above the cut (Table 6.17(c)).
- *Groundwater F'_5*: Table 6.17(d) presents the groundwater factors, which reflect a combination of evidence of groundwater seepage and the percentage *chunam* cover (a concrete coating sprayed over the cut face). The factor is further adjusted by taking into account drainage conditions and the presence of vegetation upslope (Table 6.17(e)). For example, a factor of 2.0 should be used for a slope with 10% chunam cover and no visible seepage; a further 0.25 and 1.0 should be added if there are grass and leaking pipes upslope, respectively.

Landslide Risk Assessment

Table 6.15 Example 6.11: landslides in Hong Kong recorded by GEO

Year	Cut slopes		Retaining walls		Fill slopes		Natural slopes		Rock falls		Others		Total	
	All slides	Major slides	All slides	Major slides	All slides	Major slides	All slides	Major slides	All slides	Major slides	All slides	Major slides	All slides	Major slides
1984	70	4	11	0	14	1	4	1	7	0	14	0	120	6
1985	145	4	25	3	13	1	10	1	17	0	44	1	254	10
1986	115	6	26	0	18	3	9	2	29	0	36	0	233	11
1987	193	7	27	0	14	1	14	0	28	0	31	1	307	9
1988	73	3	14	0	15	1	7	1	22	0	26	0	157	5
1989	435	39	41	9	22	5	25	1	30	2	67	0	620	56
1990	41	1	5	0	12	2	8	2	9	0	24	1	99	6
1991	49	2	5	0	9	1	9	0	13	0	3	1	88	4
1992	439	14	44	3	55	6	51	2	40	1	12	0	641	26
1993	617	71	25	0	38	10	94	11	41	0	12	1	827	93
Total	2177	151	223	15	210	31	231	21	236	3	269	5	3346	226
Average	217.7	15.1	22.3	1.5	2.1	3.1	23.1	2.1	23.6	0.3	26.9	0.5	334.6	22.6

From Fell et al. (1996a)

Estimating the probability of landsliding

Table 6.16 Example 6.11: listing of factors influencing cut slope failure probabilities

Factor	Maximum value	Minimum value	Independent factor	Factor component	Relative importance (maximum)
F'_1	1.25	0.25	Age	Age	High
F'_2	4	0.9	Geology	Unfavourable joints	Very high
				Recent colluvium	High
F'_3	2	0.1	Slope geometry	Slope angle	Very high
				Slope height	High–average
F'_4	4	0.1	Geomorphology	Angle above slope	Very high
F'_5	4	0.5	Groundwater	Groundwater	Very high
				Percentage chunam cover	Very high
				Drain condition	High–average
				Drain blockage	Very high
				Vegetation upslope	High

From Fell et al. (1996a)

Table 6.17 Example 6.11

(a) Values of geology factor F'_2: primary site factors

Condition	F'_2
Continuous adversely oriented joints, sufficient to cause a landslide of significant magnitude	4
Extensive discontinuities adversely oriented joints, sufficient to cause a landslide of significant magnitude	3
Some discontinuities adversely oriented joints, sufficient to cause a landslide of significant magnitude	2
Recent colluvium of greater than 1 m depth present, sufficient to cause a landslide of significant magnitude	2
Otherwise use	0.9

From Fell et al. (1996a)

(b) Values of slope geometry factor F'_3

Slope height: m	F'_3		
	Slope angle <50°	Slope angle 50–60°	Slope angle >60°
<5	0.7	1.1	2.0
5–10	0.5	1.0	1.8
10–20	0.4	0.85	1.6
>20	0.3	0.7	1.2

From Fell et al. (1996a)

Table 16.7 Continued

(c) Values of geomorphology factor F'_4

Terrain gradient: deg	F'_4
0–5	0.1
5–15	1.0
15–30	1.1
30–40	1.3
40–60	2.0
>60	4.0

From Fell et al. (1996a)

(d) Values of groundwater factor F'_5

Percentage chunam cover	Groundwater factor F'_5		
	Water existing within slope in upper two-thirds of cut height	Water existing within slope in lower third of cut height	No visible seepage
0–25	4.0	3.0	2.0
25–50	3.7	2.7	1.7
50–80	3.4	2.4	1.4
80–100	3.0	2.0	1.0

From Fell et al. (1996a)

(e) Additional adjustment values for groundwater factor F'_5

Factor	Description	Value to adjust F'_5 (add or subtract)
Drain condition and discharge capacity	Poor	Add 0.25
	Fair	No adjustment
	Good	Subtract 0.25
Drain blockage	Yes	Add 0.25
	No	No adjustment
Vegetation upslope	None/grass	Add 0.25
	Shrubs/trees	No adjustment
	Paved	Subtract 0.25
Service pipes upslope	Present, leaking	Add 1.0
	Present, not leaking	Add 0.5
	Present	Add 0.25
	Not present	No adjustment

From Fell et al. (1996a)

Table 6.18 Example 6.11: multiplying factor F_e for evidence of instability and history of instability

		Multiplying factor F_e		
		Major distress, e.g. slumping, large cracks	Some signs of distress, e.g. minor cracking	No evidence of instability
History of instability	Yes	10	3	1.5
	No	6	2	0.5
Limits to probability values	P_{max}	1.0	0.1–1.0	0.1–1.0
	P_{min}	0.1	No limit	No limit

From Fell et al. (1996a)

Finlay and Fell (1995) recommended that the product of these five site factors then be *sense-checked* against previous slope inspections and a number of empirical cut slope assessment methods that have previously been developed in Hong Kong, such as the CHASE approach (Brand and Hudson, 1982) and the Ranking System (Koirala and Watkins, 1988). Adjustments should be made to the primary site factor if there is disagreement between the results.

The previous history of instability at the site, together with observations about the current presence or absence of signs of landslide activity, provides a final adjustment factor F_e (Table 6.18).

For an individual cut slope, therefore, the estimated annual probability of failure is

probability of failure (cut c) = probability of failure (Hong Kong-wide)

× primary site factor F'

× history of instability factor F_e

A worked example for a hypothetical slope is shown in Table 6.19.

Although this method is specific to man-made slopes in Hong Kong, the general approach has broader applicability. Of particular note is the establishment of the probability of failure for an average slope in an area, followed by the use of a series of adjustment factors to estimate the probability of failure of specific slopes relative to the average probability.

6.9. Conceptual models: estimating probability from landslide-triggering events

Most landslides are associated with a particular triggering event, such as a heavy rainstorm or large earthquake. An indication of the probability of landsliding can be obtained through establishing initiating thresholds between parameters, such as rainfall or seismic activity, and landsliding. The most readily defined threshold is one that identifies the minimum conditions (or *envelope*) for landslide activity; above this, the conditions are necessary, *but not always sufficient*, to trigger landslides, and, below this, there is insufficient impetus for failure.

If the frequency of these triggering thresholds can be determined from analysis of climatic (e.g. Schrott and Pasuto, 1999) or earthquake records (e.g. Keefer, 1984), this can be used as a basis for estimating

Landslide Risk Assessment

Table 6.19 Example 6.11: assessment of probability of failure of a hypothetical cut slope

Factor	Site factor	Site comment	Value
Primary site factor F'	Slope age	Pre-GEO slope	1.25
	Geology	Colluvium present	2.0
	Geometry	10–15 m high cut, 50–60° slope	0.85
	Geomorphology	35–40° slope above cut	1.3
	Groundwater	0% chunam and seepage in the lower third of cut	3.0
		Trees upslope	+0.0
		No drainage upslope	+0.0
		No service pipes upslope	+0.0
	Total ($F' = F'_1 \times F'_2 \times F'_3 \times F'_4 \times F'_5$)		8.28
History of instability factor F_e	No history of instability		2
	Signs of minor cracks		
Total adjustment factor $F = F' \times F_e$			16.56
Probability of failure (Hong Kong-wide)			0.0106
Estimated annual probability of cut failure = Hong Kong-wide × F			0.176

Based on Finlay and Fell (1995)

landslide probability. In doing so, however, it is important to recognise that the occurrence of an event that exceeds the triggering threshold may not necessarily lead to slope failure, since some events will be redundant or ineffective because of the recent history of slope development.

In this conceptual model, the probability of a landslide event is the product of the annual probability of the trigger (e.g. earthquake or high groundwater) and the *conditional probabilities* of the subsequent slope response:

$$P(\text{landslide}) = P(\text{trigger}) \times P(\text{landslide}|\text{trigger})$$

Triggering thresholds provide a measure of the average likelihood of landsliding in an area. However, the stability of individual slopes and hence the response to a triggering event will vary from place to place. Some slopes will be very susceptible to a triggering event, whereas others may be able to withstand much higher magnitude events.

In each of the following four examples, the approach relies on the identification of triggering thresholds from historical records. Example 6.12 is a very simple case where only a single triggering event is considered. If the event occurs (the 1 in 1000-year earthquake), then a landslide will occur. Example 6.13 is concerned with the reactivation of a large, deep-seated landslide complex. This type of event is generally associated with prolonged heavy rainfall. In general, the deeper the slide, the longer the period of antecedent heavy rainfall needed to initiate failure. The period may vary from several days (e.g. Reid, 1994) to many months (e.g. Van Asch *et al.*, 1999). In many areas, it may be that the pattern of wet years appears to control the occurrence of landslides (e.g. at the Roughs, Kent, UK; Bromhead *et al.*, 1998).

Example 6.14 considers the impact of particular rainstorms on landslide activity. Whereas weekly or monthly rainfall patterns may be adequate to explain the pattern of deep-seated landslides, it is often the high-intensity events of limited duration (e.g. hours) that are critical for controlling shallower failures. Shallow landslides may either be associated with a critical pore water pressure threshold being exceeded (e.g. Corominas and Moya, 1996; Terlien, 1996; Terlien et al., 1996) or be a result of the increased weight of the saturated soil (e.g. Van Asch et al., 1999). Harp (1997) reported several hundred landslides triggered during and in the days following an exceptional rainstorm in Washington State, USA (over 50 cm in seven days). Casale and Margottini (1995) describe how widespread catastrophic landslide activity in Northern Italy during 1994 was associated with exceptional one- and two-day rainfall totals that exceeded all previous historical maxima.

Example 6.15 considers the combined probability of a landslide triggered by an earthquake from three different sources, in this case separate fault systems in California. The approach combines an assessment of the likelihood of the maximum-magnitude earthquake on each fault over a particular period with the use of expert judgement to estimate the probability of this event actually triggering a landslide at the study site.

Example 6.12: Alborz Mountains, Iran

Earthquake-triggered landslides are a common feature in the Alborz Mountains of northern Iran (e.g. Berberian, 1994). The Manjil earthquake of 1990 ($M_s = 7.7$), for example, triggered more than 120 landslides, including the Fatalak landslide, which completely buried a village with its residents (Haeeri et al., 1993). The major highway that connects Tehran to the Caspian Sea region runs through the mountains and is exposed to rockfalls and rock slides. Indeed, it was closed for 45 days in 1998 because of rockfalls and a major landslide at Emamzadeh Ali. As a result, this site was selected by Mousavi et al. (2011) to determine the landslide threat to the highway and a small local community.

The assessment focused on earthquake-triggered landslides on the 100–150 m high valley-side slopes that are developed in volcanic soils capped by up to 27 m of travertine deposits. A probabilistic seismic hazard assessment was undertaken to define the earthquake magnitude–frequency relationship for the area, using a database of 138 $M_s > 4$ earthquakes between 1900 and 2004.

As described in Chapter 5 (see Table 5.6), the 1 in 1000-year earthquake event has around a 5% probability of exceedence in 50 years. For the Alborz Mountains, this event coincides with a peak ground acceleration of $0.37g$, and is roughly equivalent to an $M_s > 7$ earthquake. This acceleration value was used to model the stability of the slopes under earthquake (i.e. dynamic) conditions using both finite element and conventional stability analysis methods. The results indicated that during an $M_s \geq 7$ earthquake the factor of safety will fall below 1.0, leading to extensive slope failure.

The 1 in 1000-year earthquake has an annual probability of 0.001. As the stability analysis indicated complete slope failure under the modelled earthquake event, it was considered that $P(\text{landslide}|\text{trigger})$ could be assumed to be 1.0. As a result,

$P(\text{landslide}) = P(\text{trigger}) \times P(\text{landslide}|\text{trigger})$

$= 0.001 \times 1.0$

$= 0.001$

This is a very simple example, using just a single earthquake scenario. It assumes that there is no slope response to any earthquake smaller than $M_s = 7$. In other situations, it might be useful to model a series of earthquake events with different magnitudes and frequencies. For smaller events, the landslide response to the triggering event might be more uncertain. As a result, different probability values could be assigned for $P(\text{landslide}|\text{trigger})$ under different earthquake scenarios.

Example 6.13: Ventnor, Isle of Wight, UK

Many pre-existing landslide systems are sensitive to variations in groundwater levels and hence sequences of wet and dry years. An assessment of the climatic influence on landslide activity can therefore be used to assess the probability of reactivation. Lee et al. (1998) describe how a combination of landslide systems mapping, review of historical records and rainfall analysis provided a pragmatic tool for assessing the annual probability of significant ground movement events in different parts of the 12 km-long landslide (the Undercliff) on the south coast of the Isle of Wight, UK.

Landslide activity tends to occur in the winter when rainfall totals are higher and evaporation rates are lower, and consequently much more of the rainfall is effective in raising groundwater levels. The relationship between landslide activity and winter rainfall is not a simple one. Some landslide systems within the Undercliff are more sensitive to rainfall events, while others appear only to show signs of movement during extremely rare conditions. Indeed, some systems only show signs of significant movement (i.e. resulting in the development of tension cracks, subsidence features, etc.) during extreme conditions. An assessment of the probability of significant movement has formed the basis for a pragmatic approach to landslide forecasting by the Isle of Wight Council.

The relationship between landslide reactivation and rainfall was established as follows.

1. *Identification and delimitation of landslide systems.* Detailed geomorphological mapping, at 1:2500 scale, of the Undercliff delimited a series of discrete landslide units within broader landslide systems (Lee and Moore, 1991; Moore et al., 1995).
2. *Analysis of historical records.* Reports of past landslide events were consolidated by a systematic review of available records, including local newspapers (from 1855 to the present day). Over 200 reported incidents have occurred over the last two centuries. Each record was classified according to the nature and scale of event. Events were matched to the relevant landslide system (identified by the geomorphological mapping), revealing marked concentrations of past activity at the extreme eastern and western ends of the Undercliff.
3. *Analysis of rainfall records.* A composite data set was derived from the three rain gauges that have operated within the Undercliff since 1839. The antecedent effective rainfall was calculated for four-month moving periods between August and March (the wettest period of the year), from 1839/40 to 1996 – this having previously been shown to be a sufficiently long time series to allow the identification of the prolonged periods of heavy rainfall that appear to control landslide activity in the Undercliff (Lee and Moore, 1991). Two methods were used to calculate the effective rainfall for the Undercliff:
   - 1839/40–1992: the effective rainfall was calculated using Thornthwaite's formula from monthly data from the Ventnor area and temperature data for Southampton (1855–1989) and Ventnor (1959–92).
   - 1992–1996: an automatic weather station was installed in Ventnor in 1992, and records both daily rainfall and potential evapotranspiration (using the Penman–Montieth method). The monthly effective rainfall was calculated by subtracting potential evapotranspiration from the rainfall total. Although this approach provided similar results to the effective rainfall

derived from the long-term data set, there was not an exact match; it was recognised that this introduced a degree of uncertainty into the subsequent stages of the analysis.

The combined data series was used to calculate the likelihood of different four-month antecedent effective rainfall totals (4AER) occurring within the whole Undercliff in any single year (i.e. the return period) as

$$\text{return period } T_r = \frac{\text{number of years } (156) + 1}{\text{ranking in the sequence}}$$

Figure 6.7 shows the 4AER totals that may be expected to be equalled or exceeded, on average, for particular recurrence intervals.

4   *Assessment of threshold conditions.* This involved relating the historical record for each landslide system to the 4AER data series to identify the minimum return period rainfall that is associated with landslide activity in a particular area. For example, in the westernmost system, Blackgang, significant movements are a frequent occurrence, and the minimum rainfall threshold needed to initiate significant movement appears to have been a 1 in 2-year event.

The 4AER associated with recorded ground movement events in particular areas is indicated in Figure 6.7 to highlight the varying degrees of sensitivity of different parts of the Undercliff. Thus, 4AER totals that may be expected to be equalled or exceeded, on average, every year can lead to ground movement at a number of highly sensitive areas including The Landslip and near Mirables. Elsewhere, on low-sensitivity areas such as Upper Ventnor and The Orchard, near Niton, ground movement has been associated with 4AER totals that have occurred, on average, every 10–30 years. During extreme winter rainfall periods, many more landslide systems may become active, as witnessed in 1960/61 and in January 1994. However, parts of Ventnor and St Lawrence have not displayed increased ground movement over time, even when affected by events with return periods of 100 years or more.

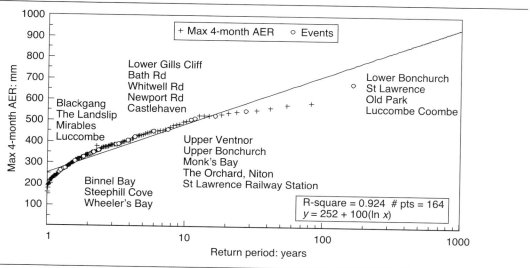

Figure 6.7 Example 6.13: relationship between antecedent effective rainfall and the locations of landslide events in the Isle of Wight Undercliff. (After Halcrow Group, 2003; Lee et al., 1998)

Table 6.20 Example 6.13: indication of the estimated annual probabilities of significant movement in four areas within the Isle of Wight Undercliff

Location	Annual probability of threshold 4AER P(4AER)	Annual probability of threshold 4AER triggering movement P(O|4AER)	Estimated conditional probability of significant movement P_m
Blackgang	0.9	0.1	0.09 (1 in 11)
Luccombe	0.25	0.2	0.05 (1 in 20)
Upper Ventnor	0.02	0.5	0.01 (1 in 100)
St Lawrence	0.005	0.5	0.0025 (1 in 400)

After Lee et al. (1998)

5   *Assessment of the probability of landsliding.* The fact that ground movement does not always occur when the 4AER thresholds shown on Figure 6.7 are exceeded highlights the importance of other factors in controlling landslide activity – that is, preparatory and triggering factors. An assessment was therefore made of the annual probability of a 4AER event of a particular magnitude actually triggering landslide activity. An estimate was made of the number of times that landsliding in a particular system had actually been initiated when 4AER exceeded the threshold value, compared with the number of times that this threshold had been exceeded over the last 150 years – that is, the proportional response.

The conditional annual probability of significant ground movement in a particular landslide system, P_m, was calculated as

$$P_m = P(4AER) \times P(movement|4AER)$$

where $P(4AER)$ is the annual probability of a threshold 4AER being equalled or exceeded in a particular year and $P(movement|4AER)$ is the annual probability of an event given the occurrence of the threshold 4AER being equalled or exceeded.

Table 6.20 provides an indication of the estimated probabilities of significant movement in a number of parts of the Undercliff. For example, in the westernmost system, Blackgang, significant movements are a frequent occurrence, and the minimum rainfall threshold needed to initiate significant movement appears, in the past, to have been a 1 in 1.1-year event. In contrast, significant movements are much less frequent in Ventnor, where the minimum rainfall threshold is a 1 in 50-year event.

Example 6.14: North Island, New Zealand

Landslides triggered by rainfall cause widespread damage throughout New Zealand, with average annual losses up to the mid-1980s estimated to be around US $33 million (Hawley, 1984). The New Zealand Earthquake Commission has paid US $14.8 million for landslide insurance claims since the mid-1970s, at an average of US $0.67 million per year (Glade and Crozier, 1996). Costs of remedial and preventative measures (e.g. soil conservation, erosion control, sustainable land management programmes and education) amounted to US $38.15 million for the period 1990–1995 (Glade and Crozier, 1996).

The probability of landsliding has been established by compiling a database of landslide events and the daily rainfall totals associated with these events (Glade, 1996, 1997, 1998; Glade and Crozier, 1996). At a regional level, the probability of landsliding was estimated as the probability of landslide activity given a triggering event of a particular magnitude

$$P(\text{landslide}) = P(\text{trigger}) \times P(\text{landslide}|\text{trigger})$$

Triggering events were identified through an analysis of daily rainfall data and landslide records (Crozier and Glade, 1999; Glade, 1998). The 24-hour rainfall records from climate stations in Hawke's Bay, Wairarapa and Wellington were compiled and assigned to 20 mm-wide rainfall classes. The 24-hour rainfall totals associated with landslide occurrence were classified in the same way. Thresholds were defined that marked the limits to the relationship between rainfall and landsliding (Figure 6.8). A minimum threshold corresponded with the daily rainfall class below which no landslide activity has been recorded and above which it may occur under certain conditions (i.e. the 24-hour rainfall *sometimes* causes landsliding). The maximum threshold was defined by the 24-hour rainfall class above which landsliding has always occurred (i.e. probability = 1). Table 6.21 summarises these thresholds for each of the three areas.

The probability of these thresholds occurring was defined through analysis of the daily rainfall records. Figure 6.9 presents the return period of the 24-hour rainfall totals for the Wellington area. This indicates that the maximum threshold associated with landsliding (140 mm) has a return period of 20.1 years (probability ≈ 0.05). The minimum threshold (20 mm) has a return period of less than 1 year.

The probability of landslide occurrence was calculated for each 20 mm rainfall class as

$$P(\text{landslides, class } r) = \frac{\text{number of triggering events, class } r}{\text{total number of landslide rainfall events, class } r}$$

In each area, the probability of landsliding increases with increased 24-hour rainfall (Figure 6.8).

Table 6.22 summarises the *conditional probabilities* of landslide activity given the occurrence of a rainfall event of a particular magnitude.

Both the Isle of Wight Undercliff (Example 6.13) and New Zealand studies demonstrate how the analysis of triggering events can help provide a framework for estimating the probability of landsliding at the local and national level, respectively. However, both are products of detailed research programmes that have drawn on lengthy rainfall records and an extensive database of historical landslide incidents. Unfortunately, in many areas, the absence of reliable historical data will present a major constraint on the use of this type of approach.

Example 6.15: San Andreas Fault, California, USA

Earthquakes are a major cause of landslides in neotectonic regions such as California, USA. Kovach (1995) provides the hypothetical example of a geothermal power plant to be located in landslide-prone terrain near San Francisco. The site is threatened by first-time failure triggered by seismic activity at any one of three nearby active strike–slip faults: the Maacama, Healdsburg–Rodgers Creek and San Andreas Faults.

The probability of an earthquake-triggered landslide at the proposed site was estimated as follows.

1. *Determining the maximum-magnitude* earthquake that could occur on each of the three faults. This assessment is based on studies of historical earthquake records and empirical investigations

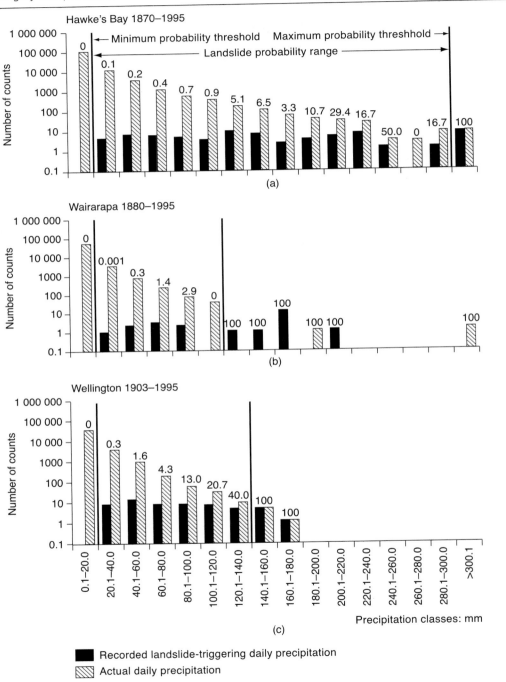

Figure 6.8 Example 6.14: probabilities (%) of landslide occurrence associated with rainfall of a given magnitude: (a) Hawke's Bay; (b) Wairarapa; (c) Wellington (note that a value of 50 means 50% of all measured daily rainfalls in a given category that produced landslides in the past) (after Crozier and Glade, 1999)

Estimating the probability of landsliding

Table 6.21 Example 6.14: daily rainfall probability thresholds associated with landsliding in different areas of New Zealand

Area	Minimum threshold: mm	Maximum threshold: mm
Hawke's Bay	20	>300
Wairarapa	20	>120
Wellington	20	>140

Based on Glade (1998)

that correlate fault length with earthquake magnitude. The larger the fault rupture length, the greater the size of the potential earthquake. The maximum size events were:

Fault:	Maximum-size earthquake:
Maacama Fault	$M = 6.5$
Healdsburg–Rodgers Creek Fault	$M = 7.0$
San Andreas Fault	$M = 8.5$

2 *Establishing the return period for the maximum-magnitude earthquake.* Earthquake recurrence data along individual fault segments can be expressed in the form $\log N = a - bM$, where $N(M)$ is the number of earthquakes with magnitude $\geq M$ and a and b are constants. From this relationship the return periods for the predicted maximum events on the faults were:

Fault:	Maximum-size earthquake:	Return period (annual probability):
Maacama Fault	$M = 6.5$	1 in 4000 (0.00025)
Healdsburg–Rodgers Creek Fault	$M = 7.0$	1 in 380 (0.0026)
San Andreas Fault	$M = 8.5$	1 in 498 (0.0020)

3 *Calculating the probability of the earthquake occurrence over the project design life* (the exceedence probability in 30 years), using the Poisson distribution:

$P(\text{Maacama Fault}) = 1 - e^{-30/4000} = 0.007$

$P(\text{H–R Creek Fault}) = 1 - e^{-30/380} = 0.08$

$P(\text{San Andreas Fault}) = 1 - e^{-30/498} = 0.06$

Figure 6.9 Example 6.14: return periods of daily precipitation in the Wellington region, New Zealand (note that the data set is presented in two parts: <100 mm and >100 mm). (After Crozier and Glade, 1999)

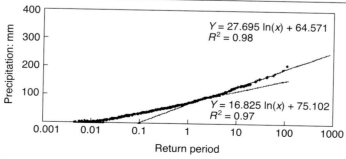

251

Table 6.22 Example 6.14: conditional probabilities of landslide activity triggered by different rainfall events in the Wellington area

Daily rainfall event: mm	Approximate probability of daily rainfall	Probability of landslide response	Conditional probability of landsliding
>140	0.05	1	0.05
120–140	0.1	0.4	0.04
100–120	0.15	0.2	0.03
80–100	0.33	0.13	0.043
60–80	0.66	0.04	0.026
40–60	1	0.016	0.016
20–40	1	0.003	0.003

Based on Crozier and Glade (1999)

4. *Estimating the probability that the maximum-size earthquake would trigger a landslide event.* This judgement was based on a geotechnical appraisal of the site conditions and the level of ground acceleration and shaking that would occur at the site during the earthquake. The probability of landsliding in response to earthquake activity on each of the three faults was judged to be

    Maacama Fault $= 0.2$

    H–R Creek Fault $= 0.03$

    San Andreas Fault $= 0.16$

5. *Calculating the conditional probability of landsliding given a maximum-size earthquake.* The individual landslide probabilities during the 30-year design life of the power plant were calculated as

    $P(\text{landslide; fault } f) = P(\text{earthquake}) \times P(\text{landslide})$

    $P(\text{landslide; Maacama Fault}) = 0.007 \times 0.20 = 0.001$

    $P(\text{landslide; H–R Creek Fault}) = 0.08 \times 0.03 = 0.002$

    $P(\text{landslide; San Andreas Fault}) = 0.06 \times 0.16 = 0.01$

    A landslide could be caused by activity on any of the three faults. These scenarios are not mutually exclusive (see Chapter 5), and hence the overall probability of the landsliding is not simply the sum of the probabilities but is calculated as

    $P = 1 - \{[1 - P(\text{Maacama Fault})] \times [1 - P(\text{H–R Creek Fault})] \times [1 - P(\text{San Andreas Fault})]\}$

    $= 1 - [(1 - 0.001)(1 - 0.002)(1 - 0.01)]$

    $= 0.013$

6.10. Conceptual models: estimating probability through expert judgement

Expert judgement involves the use of experience, expertise and general principles to assign probabilities to landslide scenarios, preferably in an explicit and consistent manner. Such judgements are inevitably

Estimating the probability of landsliding

Table 6.23 Landslide classes for subjective probability assessment

Description	Indicative slope condition	Estimated annual probability	Class name
Landsliding is imminent	Actively unstable	>0.1 (1 in 10)	Frequent
Landsliding should be expected within the design lifetime	Unstable	>0.01 (1 in 100)	Probable
Landsliding is possible within the design lifetime, but not likely	Marginally stable	>0.001 (1 in 1000)	Occasional
Landsliding is highly unlikely, but not impossible, within the design lifetime	Stable	>0.0001 (1 in 10 000)	Remote
Landsliding is extremely unlikely within the design lifetime	Stable	>0.00001 (1 in 100 000)	Improbable
Landsliding will not occur within the design lifetime	Stable	>0.000001 (1 in 1 000 000)	None/negligible

Adapted from AGS (2000) and Hungr (1997)

subjective (see Chapter 5), but, by proposing several possible scenarios followed by the systematic testing and elimination of options as a result of additional investigation and discussion, it is possible to develop reliable estimates.

At the simplest level, probabilities may be rated in one of a number of indicative 'bands' (e.g. Table 6.23; see Chapter 5), drawing on the available knowledge of site conditions. These bands give a broad indication of the judged event probability and should not be viewed as implying a rigorous quantification of the likelihood of slope failure. As the available knowledge increases, so the judgements should become more reliable. Simple field observation methods can also be of value in supporting the judgement of landslide probabilities. For example, Crozier (1984, 1986) developed a ranking system that relates geomorphological descriptions of slopes with increasing probability of landslide occurrence, as shown in Table 6.24. The key factors include

- the presence or absence of instability features at a site, adjacent sites or similar sites (i.e. the same geological and geomorphological setting)
- the presence or absence of visible signs of active or recent movement (Table 6.25)
- surface (landform) evidence, which will also give an indication of the type of movement that could be anticipated at a site – at the simplest level, unfailed portions of unstable slopes may be prone to first-time failure (i.e. rapid, large displacements), whereas failed slopes (slid areas) may be prone to reactivation (i.e. usually slower, less dramatic movements).

A slope model developed from historical evidence, map sources and geomorphological assessment, together with site investigation and monitoring results, can provide a framework for estimating the probability of landsliding. Figure 6.10 presents a flow chart developed by Fell *et al.* (1996b) that uses observational factors to provide an indication of the probability of landsliding. The method

Table 6.24 Landslide probability classification

Class	Description	Estimated annual probability
I	Slopes that show no evidence of previous instability and that by stress analysis, analogy with other slopes or analysis of stability factors are considered to be highly unlikely to develop landslides in the foreseeable future	>0.0001 (1 in 10 000)
II	Slopes that show no evidence of previous landslide activity but that are considered likely to develop landslides in the future. Landslide potential is indicated by stress analysis, analogy with other slopes or analysis of stability factors; several subclasses may be defined	>0.001 (1 in 1000)
III	Slopes with evidence of previous landslide activity but that have not undergone movement in the previous 100 years	>0.01 (1 in 100)
IV	Slopes infrequently subject to new or renewed landslide activity. Triggering of landslides results from events with recurrence intervals of greater than 5 years	>0.1 (1 in 10)
V	Slopes frequently subject to new or renewed landslide activity. Triggering of landslides results from events with recurrence intervals of up to 5 years	>0.2 (1 in 5)
VI	Slopes with active landslides. Material is continually moving and landslide forms are fresh and well defined. Movement may be continuous or seasonal	$\approx$1 (certain)

Modified after Crozier (1984)

involves a structured form of expert judgement, with the probability values based on the authors' experience of conditions on the inter-bedded sedimentary rocks of the Sydney Basin, Australia. It is likely that the approach has generic value and is applicable in areas of similar geology elsewhere, after calibration.

The following examples rely on the judgement of an experienced panel of experts. However, all involve (1) developing a conceptual model of the landslide process, (2) drawing on an understanding of the site conditions, based on geomorphological and subsurface investigations, and (3) focusing awareness on the main controls on landsliding and the historical evidence of landslide behaviour or activity in the area of interest (e.g. past landslide events or the occurrence of triggering events).

Example 6.16: Storfjord, Norway

Landslide-triggered tsunamis are a major hazard along the fjord coast and lake margins of Norway. In the 20th century, a total of 170 people were killed by tsunamis triggered by massive rockslides at Loen in 1905 and 1936, and Tafjord in 1934 (Anda and Blikra, 1998; Bjerrum and Jørstad 1968). The Åknes area in the Storfjord in western Norway is prone to frequent rockslides, typically $0.5 - 5 \times 10^6$ m^3 in size, which present a major threat to communities around the fjord (Lacasse and Nadim, 2011).

The Åknes/Tafjord project involved an assessment of rockslide and tsunami risk, together with the development of an early warning system (e.g. Lacasse et al., 2008). A simple conceptual model was developed, with two tsunami-generating landslide scenarios, namely rapid rockslides of 8×10^6 m^3

Table 6.25 Features associated with active and inactive landslides

Active	Inactive
Scarps, terraces and crevices with sharp edges	Scarps, terraces and crevices with rounded edges
Crevices and depressions without secondary infilling	Crevices and depressions infilled with secondary deposits
Secondary mass movement on scarp faces	No secondary mass movement on scarp faces
Surface of rupture and marginal shear planes show fresh slickensides and striations	Surface of rupture and marginal shear planes show old or no slickensides and striations
Fresh fractured surfaces on blocks	Weathering on fractured surfaces of blocks
	Integrated drainage system
Disarranged drainage systems, many ponds and undrained depressions	Marginal fissures and abandoned levees
Pressure ridges in contact with slide margin	
No soil development on exposed surface of rupture	Soil development on exposed surface of rupture
Presence of fast-growing vegetation species	Presence of slow-growing vegetation species
Distinct vegetation differences 'on' and 'off' slide	No distinction between vegetation 'on' and 'off' slide
Tilted trees with no new vertical growth	Tilted trees with new vertical growth above inclined trunk
No new supportive, secondary tissue on trunks	New supportive, secondary tissue on trunks

From Crozier (1984)

and 35×10^6 m^3. Tsunami run-up values were estimated for locations around the Storfjord (Table 6.26) (e.g. Eidsvig et al. 2008). The results indicate that a 35×10^6 m^3 rockslide could trigger a tsunami with a run-up of up to 35 m at Hellesylt.

The probability of landsliding and tsunami propagation was estimated through the use of expert judgement. Event trees were constructed to provide a logical framework for the assessment. Three potential landslide triggers were considered: earthquake, high porewater pressures, and creep and weakening of the shear surface. Event tree nodes were established for

- whether the slide mass remains largely intact or breaks up into smaller pieces
- the landslide volume (using mutually exclusive classes: $>35 \times 10^6$ m^3 or $5 \times 10^6 - 35 \times 10^6$ m^3 for massive slides and $0.5 \times 10^6 - 5 \times 10^6$ m^3 or $<0.5 \times 10^6$ m^3 for broken-up slides)
- the resulting tsunami run-up (using mutually exclusive run-up height classes: ≤ 5 m, 5–20 m or >20 m).

The event tree analysis was undertaken over several days at a workshop in October 2007, where the participants included engineers, scientists and stakeholders (e.g. local politicians, journalists, the

Landslide Risk Assessment

Figure 6.10 Flow chart for estimating the probability of landsliding: soil slides. (After Fell et al., 1996b)

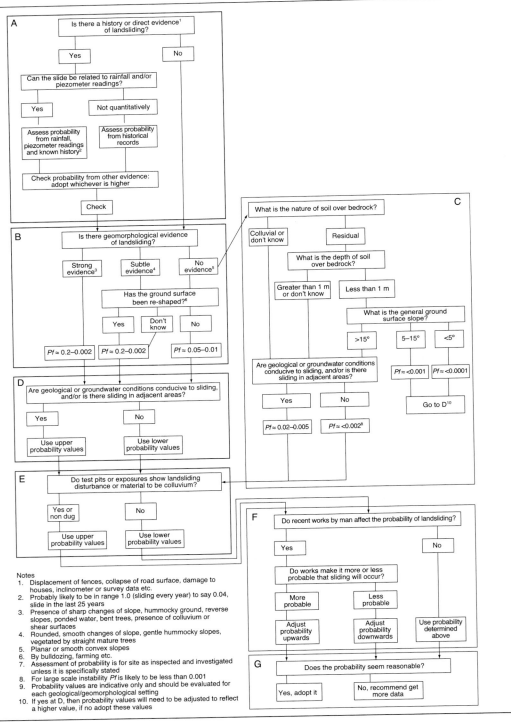

Table 6.26 Example 6.16: Åknes rockslides – estimated tsunami run-up heights

Location	Run-up height: m	
	Rockslide 8×10^6 m^3	Rockslide 35×10^6 m^3
Hellesylt	8–10	25–35
Geiranger	8–15	20–40
Stranda	1–3	3–6
Fjøra	1–2	5–7
Tafjord	3–5	12–18

From Lacasse and Nadim (2011)

emergency services and planners; see Lacasse and Nadim, 2011). Consensus judgements were established for the probability of a rockslide and the values at each of the nodes.

An example of the event tree for tsunami propagation is presented in Figure 6.11 (note that Lacasse et al (2008) state that the probability values are given to illustrate the assessment process and are not estimates for the Åknes slope). Assuming that the probability of a rockslide is 0.001, then the conditional probability model for a run-up $R > 20$ m caused by a 35×10^6 m^3 massive slide is

$$P(R > 20 \text{ m}, 35 \times 10^6 \text{ m}^3 \text{ slide}) = P(\text{landslide}) \times P(\text{massive event}|\text{landslide})$$

$$\times P(35 \times 10^6 \text{ m}^3 \text{ volume}|\text{massive event})$$

$$\times P(R > 20 \text{ m}|35 \times 10^6 \text{ m}^3 \text{ volume})$$

$$= 0.001 \times 0.8 \times 0.1 \times 0.9$$

$$= 7.2 \times 10^{-5}$$

Run-up >20 m can also be caused by a massive 5×10^6 m^3 slide ($0.001 \times 0.8 \times 0.9 \times 0.3 = 2.2 \times 10^{-4}$) and a broken slide of 0.5×10^6 m^3 ($0.001 \times 0.2 \times 0.5 \times 0.001 = 1 \times 10^{-7}$). The total probability of a run-up >20 m is the sum of these three scenarios (2.9×10^{-4}).

The early warning and emergency preparedness system was implemented in early 2008, based on the monitoring of extensometers installed on the rock slopes to measure crack opening. An emergency preparedness centre in Stranda is in operation continuously to coordinate the response to changes in rockslide displacement velocities.

Example 6.17: Caucasus Pipeline, Georgia

An existing pipeline passes through a number of landslide-prone terrains in the Caucasus. Although the majority of landslides in these areas are of some antiquity, they remain prone to reactivation during periods of heavy rainfall or by earthquakes. It was recognised that landslide activity could present a threat to pipeline integrity, since ground displacements could result in pipe rupture. As part of a broader assessment of the risks to the pipeline, a preliminary review of the likelihood of

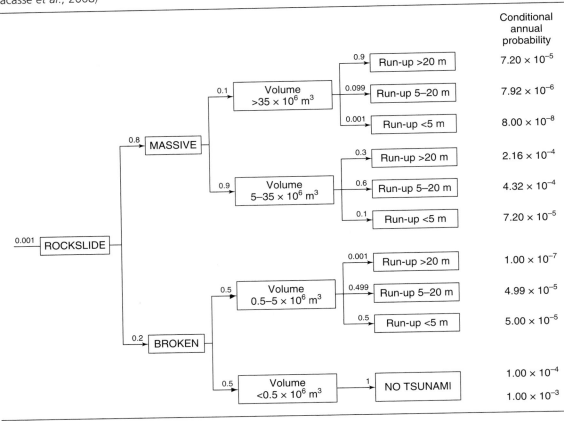

Figure 6.11 Example 6.16: event tree developed to model landslide-triggered tsunami hazards in Storfjord. (After Lacasse et al., 2008)

pipeline rupture due to landsliding was undertaken by an expert panel comprising local geologists and two international landslide specialists.

Before the expert panel visit, the local geologists had concentrated on the identification of all recognisable landslides along the route. An aerial photograph interpretation and subsequent fieldwork programme were directed towards the identification of pipe rupture scenarios that provided a framework for assessing the probability of landsliding. In the absence of historical records of movement and a detailed understanding of landslide behaviour, the review had to rely on expert judgements to estimate the probability of landslide-related pipe ruptures. In doing so, the expert panel considered

- the potential for a particular type of landslide to generate a pipeline failure scenario
- the annual probability of landslide triggering events (as described earlier for earthquakes and heavy rainfall)
- the expected frequency of pipeline failure events at a particular site, given the occurrence of a landslide-triggering event, (i.e. not all landslide movements will lead to pipeline failure).

The assessment involved the following.

1. Developing a broad understanding of the causes and mechanisms of the local landsliding.
2. Developing an appreciation of the stability (i.e. activity state) of the landslide systems that could threaten the pipeline. Landslide activity was assessed from field evidence, based on the following categories: dormant/inactive, relatively inactive (no signs of contemporary movement), relatively active (localised signs of contemporary movement), periodically active (likely to be active in most wet years) and seasonally active (likely to be active in most wet seasons). Judgements about the expected *landslide behaviour* were made from the field assessment of landslide form and the materials likely to be involved (Table 6.27).
3. Identification, through group discussion involving the whole field party, of plausible sequences of events (scenarios) that could lead to landslide activity and subsequent pipeline failure. The field judgements on landslide activity and behaviour were used to consider the following range of pipeline rupture scenarios at each site:
   - *displacement* – pipe rupture as a result of differential horizontal and/or vertical movement of the landslide mass
   - *spanning* – pipe failure as a result of removal of support along a significant length of pipeline due to landslide movement.
4. Estimating the likelihood of these scenarios occurring using an event tree approach. This approach considers the probability (i.e. chance) of a sequence of events progressing, for example, from an initial trigger (an *initiating event* such as an earthquake) to the progressive reactivation of successive parts of the pre-existing landslide systems. The chance of each stage of a scenario occurring was determined through group discussion based on the broad appreciation of landslide behaviour and stability – that is, the probabilities assigned to each scenario were the agreed 'judgements' of a team of experts.

Estimates of the likelihood of a triggering event were based on the historical frequency of earthquake-triggered landslide activity (around 2.5 events/10 years) and rainfall-triggered landslide activity (around 2 events/10 years) (Kuloshvili and Maisuradze, 2000; Tatashidze *et al.*, 2000). These two figures can be combined to provide an estimate of the annual probability of a landslide triggering event of the order of 0.45 (4.5 events every 10 years). However, this relates to a country-wide likelihood

Table 6.27 Example 6.17: classification of landslide behaviour types

Mechanism (flow)	Type of behaviour (fluid-type movement)
Sliding	Plastic deformation Plastic/block deformation Slab/block deformation Brittle failure
Heave	Plastic deformation Plastic/block deformation Slab/block deformation Brittle failure
Fall	Brittle failure Creep and slow settlement

Landslide Risk Assessment

Figure 6.12 Example 6.17: framework for estimating landslide probabilities – pipeline crosses a recorded landslide

Location	Landslide type	Activity	Backscar position	Initiating event: Annual prob. triggering event	Response: Annual prob. significant landslide movement	Outcome: Annual prob. pipe damaging event	Conditional annual prob. pipe damaging event	Estimated frequency pipe damaging event (1 in n years)
Pipe passes behind landslide backscar	Shallow slides (slab slides and debris slides)	Active	Backscar within 10 m of pipe	0.1	0.1	0.001	0.000 01	100 000
			Backscar within 10–25 m of pipe	0.1	0.01	0.001	0.000 001	1 000 000
			Backscar 25–50 m from pipe	0.1	0	0.001	0	0
		Relatively active	Backscar within 10 m of pipe	0.1	0.01	0.001	0.000 001	1 000 000
			Backscar within 10–25 m of pipe	0.1	0	0.001	0	0
			Backscar 25–50 m from pipe	0.1	0	0.001	0	0
		Dormant	Backscar within 10 m of pipe	0.1	0.001	0.001	0.000 000 1	10 000 000
			Backscar within 10–25 m of pipe	0.1	0	0.001	0	0
			Backscar 25–50 m from pipe	0.1	0	0.001	0	0
	Mudslide systems and shallow rotational slides	Active	Backscar within 10 m of pipe	0.1	0.1	0.01	0.0001	10 000
			Backscar within 10–25 m of pipe	0.1	0.01	0.01	0.000 01	100 000
			Backscar 25–50 m from pipe	0.1	0	0.01	0	0
		Relatively active	Backscar within 10 m of pipe	0.1	0.01	0.01	0.000 01	100 000
			Backscar within 10–25 m of pipe	0.1	0	0.01	0	0
			Backscar 25–50 m from pipe	0.1	0	0.01	0	0
		Dormant	Backscar within 10 m of pipe	0.1	0.001	0.01	0.000 001	1 000 000
			Backscar within 10–25 m of pipe	0.1	0	0.01	0	0
			Backscar 25–50 m from pipe	0.1	0	0.01	0	0

of such an event occurring. It was felt that, on average, only 1 in 4 of these events were likely to act as potential landslide triggers along the pipeline route, since it only passes through part of the country and avoids the more seismically active, high-rainfall areas. Thus, the expected annual probability of a triggering event was judged to be 0.1 (1 in 10).

Two generic models were developed specifically for the study (Figures 6.12 and 6.13) to provide a framework for the systematic estimation of the annual probability of a landslide response to a triggering event (not all potential triggers will actually initiate movement of a specific landslide) and subsequent damaging outcomes (not all landslide movements will cause pipeline failure). These models address two key situations.

Estimating the probability of landsliding

Figure 6.13 Example 6.17: framework for estimating landslide probabilities – pipeline passes upslope of a recorded landslide

Pipe crosses landslide			Estimated factor of safety	Initiating event — Annual prob. triggering event	Response — Annual prob. significant landslide movement	Outcome — Annual prob. pipe damaging event	Conditional annual prob. pipe damaging event	Estimated frequency pipe damaging event (1 in n years)
	Shallow slides (slab slides and debris slides)	Active	0.9–1.1	0.1	1	0.01	0.001	1000
		Relatively active	1.1–1.25	0.1	0.1	0.01	0.0001	10 000
		Dormant	>1.25	0.1	0.01	0.01	0.000 01	100 000
	Mudslide systems and shallow rotational slides	Active	0.9–1.1	0.1	0.2	0.1	0.002	500
		Relatively active	1.1–1.25	0.1	0.1	0.1	0.001	1000
		Dormant	>1.25	0.1	0.01	0.1	0.0001	10 000
	Deep-seated landslides (rotational and compound)	Active	0.9–1.1	0.1	1	1	0.1	10
		Relatively active	1.1–1.25	0.1	0.5	1	0.05	20
		Dormant	>1.25	0.1	0.1	1	0.01	100

- *Where the pipe crosses a recorded landslide*, the potential for pipeline failure is a function of the likelihood of landslide reactivation resulting in significant lateral displacement within pipe depth. It was assumed that this varies with landslide type (i.e. behaviour mechanism) and activity state – parameters that were estimated in the field. An active landslide (i.e. one with a low factor of safety) will be more sensitive to a potential triggering event (either heavy rainfall or an earthquake), and will have a higher annual probability of significant movement than a dormant one with a higher factor of safety. For example, an active mudslide may respond to every triggering event (i.e. a probability of response of 1.0), whereas a dormant mudslide might only be reactivated by less frequent, larger triggers (say 1 in 100-year events, i.e. a probability of response of 0.01).

  The landslide type and the behaviour mechanism are important in determining whether landslide movement results in pipeline damage. For example, shallow slab slides involving plastic deformation of the displaced materials are less likely to cause pipeline failure than large deep-seated landslides characterised by brittle/slab styles of movement (estimates of the probability of pipeline damage, given the occurrence of significant movement, were 0.01 and 1 for these two slide types, respectively).

- *Where the pipeline passes behind (i.e. upslope of) a recorded landslide*, the chance of pipeline failure is a function of the potential for a first-time failure of the ground upslope of the landslide

Table 6.28 Example 6.17: indicative probability bands used in the review of the existing chemical products pipeline

Estimated annual probability of a pipe-fracturing event	Likelihood of pipeline damage and annual chance of a fracturing event
0	None
>0.000001	Extremely improbable (1 in 1 000 000)
>0.00001	Improbable (1 in 100 000)
>0.0001	Remote (1 in 10 000)
>0.001	Possible (1 in 1000)
>0.002	Occasional (1 in 500)
>0.01	Probable (1 in 100)
>0.1	Frequent (1 in 10)
1	Certain (1 in 1)

backscar that would undermine the pipe. This potential reflects a combination of landslide type and activity state, together with the distance between the pipeline and the current position of the landslide backscar. As a general guide. it was considered extremely unlikely that any of the landslide situations encountered had the potential to generate more than 25 m of backscar retreat in a single event, or experience an average annual backscar recession rate in excess of 0.5 m/year over an extended period.

The *conditional probability* for each scenario was calculated as

$$P(\text{scenario}) = P(\text{initiating event}) \times P(\text{response}) \times P(\text{outcome})$$

The structured use of event trees enabled the conditional annual probability of a landslide-related pipeline failure to be rated in one of a number of 'bands' based on a nine-point scale (Table 6.28).

Example 6.18: Lyme Regis, UK

The coastal slopes forming the eastern fringes of Lyme Regis, Dorset, UK, are covered by active landslide systems, which form the seaward part of a larger area of dormant landsliding that extends inland for around 0.5 km. Since the 1960s, these landslides have displayed progressive reactivation, and future movements present a significant hazard to the local community (e.g. Hutchinson, 1962; Lee, 1992). As a consequence, an expert panel was convened in 1998 with the task of establishing possible landslide reactivation scenarios in order to provide a framework for assessing landslide risk, and for testing the economic viability of different coast protection and landslide management options (Lee *et al.*, 2000, 2001).

The expert panel consisted of representatives from the local authority (West Dorset District Council), a consultant and two landslide specialists. The range of landslide reactivation scenarios it identified were based on the understanding of the causes and mechanisms of landslide behaviour developed during an extensive, ongoing programme of ground investigations, undertaken as part of the Lyme Regis environmental improvements scheme (Clark *et al.*, 2000; Fort *et al.*, 2000; Sellwood *et al.*, 2000). Of particular importance were the likely reactivation sequences, an in-depth appreciation of the stability of the landslide systems and the interrelationships between adjacent landslide units.

In general, each scenario involved the progressive expansion inland of the zone of active instability as pre-existing landslide units suffer toe-unloading in turn, due to the movement of downslope (seaward) landslide units (which had provided passive support to the upslope units); each phase of reactivation is promoted by the occurrence of high groundwater levels. While these sequences of events might, or might not, be expected to occur at some time within the timescale under consideration (i.e. the next 50 years), the precise timing of the initiating events and subsequent responses will be controlled by the almost-random combination of potential initiating events and prevailing antecedent conditions. In addition, the coastal slope conditions are progressively deteriorating, with the chance of failure expected to increase over time, owing to a combination of the decline in structural integrity of the existing seawalls and the future increases in both storminess and winter rainfall totals predicted as a result of climate change (see Chapter 12).

For each landslide system along the coastal frontage, a series of event trees (reflecting different initiating events and associated estimates of scenario probabilities) were established as follows.

1. *Identification and characterisation of landslide systems.* Detailed geomorphological mapping of the coastal slopes delimited a series of discrete landslide units within broader landslide systems. Within each system, a complex arrangement of individual landslide units was recognised, reflecting the wide variety of landslide types and processes present. These systems and units formed the framework for understanding the contemporary ground behaviour of the Lyme Regis area and were progressively refined as a result of a detailed ground investigation (Sellwood et al., 2000).

2. *Identification of landslide reactivation scenarios.* Combining an understanding of the form and character of the landslide systems (i.e. aerial photograph interpretation, surface mapping and subsurface geotechnical data) with monitoring data (ground movements and piezometric levels; Fort et al., 2000), together with analysis of past events and building damage, led to the development of a range of credible reactivation scenarios at each site. Each scenario was developed from an initiating event (i.e. seawall failure or wet-years sequences), with subsequent responses and outcomes identified as the destabilising effects of the initiating event were transmitted inland and upslope through the adjacent landslide units. It should be noted that all of these scenarios are 'do nothing' scenarios, in the sense that nothing is considered to have been done to prevent an initiating event or to control the subsequent responses – that is, landslide movements would be allowed to develop unchecked.

3. *Development of event trees.* Each scenario comprised an initiating event followed by a response (i.e. movement of parts of the landslide system: response 1). In turn, this response might act as an initiating event for a second movement (response 2) and so on. Ultimately, the combination of initiating event and resultant responses would lead to particular outcomes (i.e. impacts of coastal slope assets: scenario outcome elements S1, S2, ...). Each sequence of initiating event–response–outcome was simplified to a series of simple event trees (Figure 6.14), with responses to a previous event either occurring or not occurring (i.e. yes/no options).

4. *Estimation of the annual probability of initiating events.* This involved the estimation of the likelihood of seawall failure and wet-years sequences in each year for future years 1–50. The annual probability of seawall failure was assessed for individual sections of seawall by the local authority, West Dorset District Council, who also estimated an expected annual rate of increase in the chance of failure to reflect the gradual deterioration of the structures (under a 'do nothing' scenario).

   Analysis of rainfall records for the period 1868–1998 indicated that there were eight wet-years sequences of 3–6 years' duration in 130 years, suggesting an annual probability of around 1 in 16 (0.06). The frequency of these sequences (and possibly the duration) appeared to have

Figure 6.14 Example 6.18: Langmoor Gardens, Lyme Regis, UK: event trees. (From Lee et al., 2000)

S1 Localised damage to property, seawalls, services etc.
S2 Extensive loss of amenity gardens, sea front property, seawalls, services etc.
S3 Loss of up to 20 m of cliff top land, including gardens, tennis courts, access lane to Gardens, property

A. Initiating event = seawall failure (leading to toe failure, landslide reactivation and rear cliff failure)

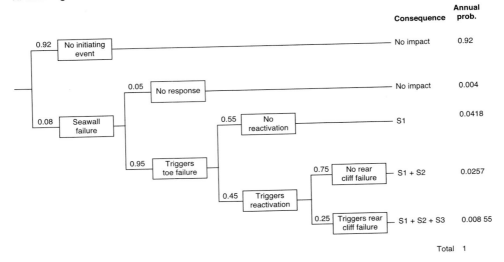

B. Initiating event = high groundwater levels

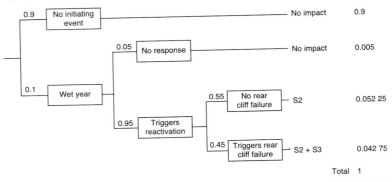

increased over the last three decades of the period, suggesting a current annual probability of around 0.1 (1 in 10); this figure was used in the analysis.

5 *Estimation of the probability of responses.* Probabilities were assigned to each of the event tree 'branches/forks' at each site, mindful that the sum of probabilities at each fork must equal 1.0. This was achieved by first obtaining the 'expert judgement' of the individual project team members and then holding discussions to reach consensus on the 'best-guess' figures.
The approaches used to develop convergence or agreed consensus in this study were *open forum* and a *Delphi panel* (Roberds, 1990) (see Chapter 5): Each individual in the project team was provided with the same set of background information and was asked to conduct and document (in writing) a self-assessment. These assessments were then consolidated to identify areas of disagreement where further discussion was required in an open workshop meeting. Typically,

the individual assessments tended to converge after discussion. Such iterations were continued until consensus was achieved.

It was found through discussions that the most acceptable approach to identifying probabilities at each branching of the event tree, or 'fork', was to identify a time period over which the team believed there was a 95% chance of the 'failure' route being realised. This was used to identify a corresponding annual probability that would deliver this cumulative probability over the agreed time period (this assumed a binomial distribution of events). The estimated annual probability, cumulative probability and the time by which an event is almost certain to have occurred, are related by

$$\text{probability of occurrence in } x \text{ years} = 1 - (1 - \text{annual probability})^x$$

6 *Calculation of conditional probabilities for each scenario.* For year 1 (the initiating event and response occurred in year 1), the conditional probability associated with each 'branch' of an event tree (i.e. a unique sequential combination of scenario outcome elements, e.g. S1, S2, S3, ...) was calculated as

$$\text{scenario probability} = P(\text{initiating event}) \times P(\text{response 1}) \times P(\text{response 2}) \times P(\text{response } n)$$

For subsequent years (the initiating event and response occurred in the same year), the calculation is essentially the same as the above, with the exception that the annual probability of the initiating event was changing over time (e.g. the estimated probability of seawall failure increased at 5% per year). In addition, in calculating the probability of a combination of scenario elements occurring in year 2, it was necessary to take into account the possibility that the scenario actually occurred in year 1 and hence could not occur in year 2. Thus, the annual probability for year 2 (and subsequent years) was modified as follows:

probability of failure in year i = annual probability of failure in year i

$\times$ [probability failure did not occur in year $(i-2)$

$-$ probability failure occurred in year $(i-1)$]

However, a response might be delayed or lagged, so as to occur in any year after an initiating event – that is, if the initiating event occurred in year 1, then the response need not be in year 1, but could be in year 2 or any other year up to year 50. Thus, calculating the combined probability of a response occurring in a particular year was more complex. For example, the probability of an initiating event leading to a response occurring in year 4 involved the combination of four possibilities: P(seawall failure in year 1 and the response three years later), P(breach in year 2 and response two years later), P(breach in year 3 and response one year later) and P(breach in year 4 and response nil years later). For the probability of the response in year 50, there would be 50 combinations of probabilities.

The analysis necessitated the development of a sequence of related worksheets for each landslide system. Each worksheet comprises a 50×50 matrix of probabilities derived from multiplying P(Initiating event) by P(response) for all possible combinations of timings. Figure 6.15 presents an annotated worksheet illustrating how the analysis was built up. The example produces the probability of response 1 following the occurrence of an initiating event. The results from this sheet (the total probability column) then form the input data (along with the probability distribution for response 2) to the next sheet, and so on.

The basic principles of this method for assessing landslide risks in situations where there is no historical precedent are relatively straightforward. The aim was to assess the annual likelihood of a landslide

Landslide Risk Assessment

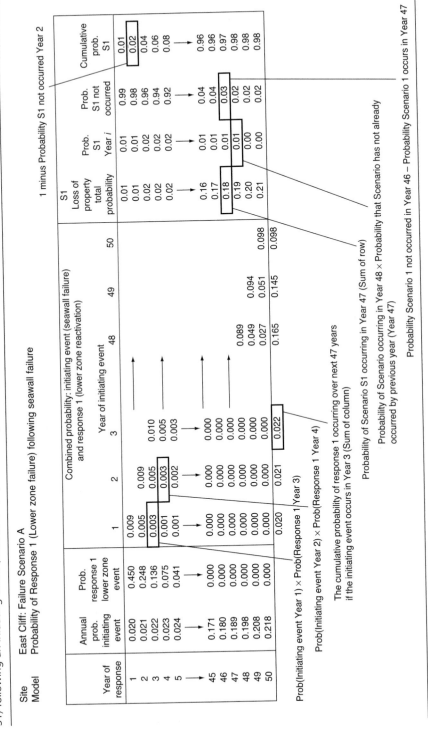

Figure 6.15 Example 6.18: annotated example of the worksheet used to define the conditional probability of an outcome (scenario outcome element S1) following an initiating event (seawall failure) and subsequent response (lower-zone landslide reactivation). (After Lee et al., 2000)

event (here, a reactivation) and how this likelihood may change through time. In view of the long time-scales, complex processes involved and sparse data, it was inevitable that expert judgement would play an important role.

6.11. Reliability methods: estimating probability through use of probabilistic stability analysis

Stability analysis provides a quantitative measure of the stability of a slope or pre-existing landslide (e.g. Bromhead, 1986; Duncan, 1996a,b; Graham, 1984; Nash, 1987). The slope form and materials are modelled theoretically, together with the loadings on the slope. A failure criterion is then introduced. The analysis indicates whether the failure criterion is reached and allows a comparison between the modelled conditions and those under which the slope would just fail. The results are generally presented as a factor of safety F, given by

$$F = \frac{\text{shear strength}}{\text{shear stress}}$$

The factor of safety is best viewed as the numerical answer to the question 'By what factor would the strength have to be reduced to bring the slope to failure by sliding along a particular potential slip surface?' (Duncan, 1996a).

The approach is deterministic, since single values of shear strength, pore water pressure and loading are the input parameters to the model. However, there can be significant uncertainties associated with the modelling, not least because of the variability of the soil and the possible presence of geotechnical 'anomalies', such as thin weak layers and discontinuities that may have a major influence on stability.

Stability analysis can be used to generate estimates of the probability that the factor of safety will be less than 1, based on many simulations using variable parameter values. In probabilistic stability analysis, the probability of failure is defined with respect to a performance function G(x) that represents slope stability as a function of a combination of slope parameters, each with a continuous probability distribution (i.e. random variables; see Chapter 5).

Example 6.19: Whitby, UK

The coastal cliffs at Whitby, UK, are developed in a highly variable sequence of over-consolidated sandy clay tills. Part of the cliffline, West Cliff, has been protected by a seawall and stabilised since the late 1920s, with the most recent works undertaken between 1988 and 1990 (see Clark and Guest, 1991). Concerns have been raised subsequently about the potential for a deep-seated failure of the stabilised slope that would threaten the integrity of the seawall. Indeed, a major breach in the seawall occurred in about 1962 as a result of deep-seated landsliding.

An assessment of the probability of deep-seated landsliding was undertaken during a recent review of the cliff management strategy. This was based on the use of stability analysis and the calculation of a reliability index for the slopes. It involved the following stages.

1 *Stability analysis.* A slope stability model was developed using Bishop's simplified method for circular failures (Bishop, 1955). The range of input parameters (unit weight, effective cohesion and angle of friction) were derived from previous studies (e.g. Clark and Guest, 1991) and more recent site investigation results (High Point Rendel, 2003).

A number of iterations (50) of the stability model were carried out, using pre-selected combinations of the input parameters. The model runs included
- a 'best estimate', which was assumed to correspond to the mean slope conditions
- a lower-bound estimate, based on an assumed combination of input parameters that were believed to correspond to the 'worst case' scenario
- an upper-bound estimate, based on an assumed combination of input parameters that were believed to correspond to the 'best case' scenario.

The deep-seated slip circle with the lowest factor of safety was considered to represent the stability of the slope for each model run. The results of the stability modelling indicated a mean factor of safety F of 1.21 (i.e. the mean value of all the lowest factors of safety), with a standard deviation of F of 0.08.

2 *Calculation of a reliability index.* The reliability index β was calculated from

$$\beta = \frac{\text{mean } F - 1}{\text{standard deviation of } F}$$
$$= \frac{0.21}{0.08}$$
$$= 2.62$$

3 Estimation of the probability of failure. The probability of failure was approximated by the cumulative standard normal distribution Φ evaluated at $-\beta$ (see Table 5.11). This indicated a probability of 0.004332 (i.e. 4.3×10^{-3} or 1 in 230.8).

Example 6.20: Mad Dog and Atlantis, Gulf of Mexico, USA

The Mad Dog and Atlantis oil prospects in the Gulf of Mexico are located in 2200–3000 m of water at the 100–235 m-high Sigsbee Escarpment, which has formed in clay and sand layers at the edge of an extensive salt sheet (Jeanjean et al., 2003; Orange et al., 2003). Owing to the disposition of the developable reserves, it was found necessary to locate the production wells and associated facilities as close as possible to the escarpment, either at the crest (Mad Dog) or at the foot (Atlantis). However, there is widespread evidence for numerous major rotational failures and debris flow events during the relatively recent geological past. For example, at Atlantis, there have been numerous deep-seated failures of the steep scarp slopes, which have transformed into debris flows. Some of the larger debris flows identified would have run out over 7 km at inferred speeds up to 100 km/h (Niedeoroda et al., 2003). Entrained within the flows were intact blocks of sediments comparable in size to the Houston Astrodome.

The presence of these landslide hazards led to considerable concern about whether it was possible to safely develop the prospects. As a result, the field development was supported by an integrated programme of studies directed towards developing an understanding of the level of risk associated with landslides, for a range of field layouts. This programme included the use of both deterministic and probabilistic stability analysis (PSA) to estimate the annual probability of deep-seated scarp slope failure (Nadim, 2006; Nadim and Locat, 2005; Nadim et al., 2003). For both analyses, a simple two-wedge model (i.e. a sliding and collapsing block; Figure 6.16) was used. The deterministic analysis of a critical slope section ('Slump E') predicted a factor of safety of 1.54.

The PSA was undertaken using a first-order reliability model (FORM). The following parameters were considered to be random variables and represented by assumed probability distributions in the analysis.

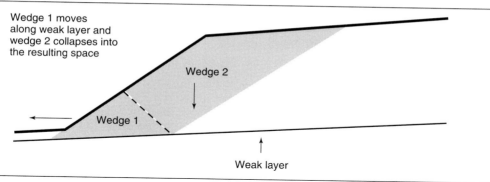

Figure 6.16 Example 6.20: the simple two-wedge model (i.e. a sliding and collapsing block) used in the stability modelling. (After Nadim et al., 2003)

- The submerged unit weight of each clay and sand layer.
- The undrained shear strength parameters in each layers.
- The removed overburden in each layer.
- The removed overburden at the toe of the scarp.
- The shear strength anisotropy.

In addition, a random variable ε was used to account for the uncertainty in the modelling; the assigned values (standard deviation = 5%) were determined by expert judgement and on the basis of previous experience. For Atlantis Slump E, the PSA gave a reliability index of 3.34 and a failure probability of 4.2×10^{-4} for the critical failure mechanism. This corresponds to a median factor of safety of 1.48.

However, the computed probabilities do not relate to a particular time period – that is, they are 'timeless'. Nadim et al. (2003) suggested that the reliability method could provide an estimate of the 'static' probability that the escarpment would fail immediately if it were 'constructed' today (e.g. as a man-made embankment). However, the scarp at slump E was probably created by a deep-seated failure around 60 000 years ago. As it is still standing, the probability of failure was believed by Nadim et al. (2003) to be zero, provided that no triggering events occur or that the slope conditions do not unexpectedly deteriorate. Detailed regional and site-specific studies concluded that seismicity, wave action and other potential processes were not credible triggering mechanisms under the current environmental conditions. This suggested that the only mechanisms by which the slope would fail in the future was the accumulation of sediment, loading the slope and changing the slope geometry. Assuming a future sedimentation rate the same as the current rate (12 cm/1000 years), over the next 10 000 years, the probability of failure would decline to around 4.85×10^{-4} (the factor of safety declines at a rate of 0.0008/1000 years).

Nadim et al. (2003) suggested two approaches that could be used to yield an annual probability of deep-seated failure at slump E.

- The model results represented the cumulative probability of failure since the last event. Using relatively simple statistical models (e.g. the binomial model), the annual probability of failure could be derived it was known when the last event occurred (around 60 000 years ago). This approach yielded an annual probability of deep-seated failure at slump E in the range 1×10^{-8} to 1×10^{-6}.

- In the second approach, the probability that the factor of safety would fall below 1.0 during the next year was calculated. This model assumed that failures were not 'triggered' as such, but related to a progressive decline in the factor of safety due to sedimentation at the rate of 0.0008/1000 years (as modelled above). This approach yielded an estimate for the annual probability of failure in the range 1×10^{-9} to 1×10^{-7}.

The results suggested that relatively high 'static' failure probabilities generated from the PSA translated into a very low annual probability of deep-seated failure at Atlantis, thereby providing the project with a level of assurance about the safe development of the site. However, despite the appearance of being a rigorous, objective method, the PSA was actually heavily reliant on judgement. As is the case in many landslide studies, the PSA was based on limited shear strength data and a large number of assumptions about the distributions of the random variables. Among the most critical of these were the assumptions made in establishing a probability distribution for shear strength and associated model uncertainties. As Lacasse and Nadim (2006) noted, computed failure probabilities are difficult to interpret because of the uncertainties arising from lack of knowledge of soil conditions.

The estimation of an annual probability from the calculated 'timeless' probabilities is sensitive to the assumptions that were made.

- The cumulative probability approach relied on accurate dating of the last event: orders of magnitude errors in this date produced orders of magnitude differences in the computed annual probability of failure. In addition, the model assumed that the factor of safety remained stationary over the period since the last event – this would potentially be a significant over-simplification at many sites, considering the various time-dependent processes that might be at work.
- The progressive sedimentation and declining factor of safety approach was sensitive to the known and estimated sedimentation rates and to the modelled impact that this deposition had on slope stability.

Example 6.21: Lushan, Taiwan
Lushan is a popular hot-spring mountain resort in Nantou County, Taiwan, where the average annual rainfall is around 2400 mm (Lee and Chi, 2011). Significant creep movement developed on the mountain slopes in 2006, during Typhoon Winnie, resulting in large cracks to roadways and severe damage to buildings. Investigations were able to demonstrate a direct link between rainfall, groundwater levels and landslide activity on the slopes, which are developed in Miocene slates and weathered slates and sandy slates (Huang et al., 2008).

The probability of landsliding was estimated as part of a landslide risk assessment, using a simple conditional probability model:

$$P(\text{landslide}) = P(\text{trigger}) \times P(\text{landslide}|\text{trigger})$$

The analysis that supported this assessment involved an interesting combination of the development of rainfall–groundwater scenarios and probabilistic stability analysis.

Rainfall and the resulting groundwater levels were considered to be the landslide trigger. Three groundwater level scenarios were defined by a combination of expert judgement and rainfall analysis (Lee et al., 2009).

Estimating the probability of landsliding

Table 6.29 Example 6.21: Lushan, Taiwan – shear strength parameters for probabilistic slope stability analysis

Unit	C: kPa		ϕ: deg	
	Mean	Coefficient of variation	Mean	Coefficient of variation
Slate	5	0.4	28	0.16
Sandy slate	10	0.4	35	0.2
Weathered slate	250	0.5	35	0.5

From Lee and Chi (2011)

- The normal condition of rainfall, which produces a 'low' groundwater table (25–40 m below ground level). This was estimated to have an annual probability of occurrence of 0.795.
- The scenario of a daily rainfall of 400 mm, which produces 'medium' groundwater table levels (10–15 m below ground level). Analysis of rainfall records indicated that this scenario has had an annual probability of 0.2 (Huang et al., 2008).
- Torrential rainfall, which produces surface flows across the entire area and causes a 'high' groundwater table. The probability of this scenario was estimated to be 0.005.

These scenarios were used as separate deterministic inputs into the probability stability analysis. Five slope representative profiles at Lushan were defined for analysis, using the Chen and Morgenstern (1983) method of slices. The mean and coefficient of variation (i.e. the standard deviation divided by the mean) of the shear strength parameters used in the analysis are shown in Table 6.29.

The probabilistic stability model generated values for the reliability index β from the factor of safety F for each of the three groundwater scenarios:

$$\beta = \frac{\text{mean } F - 1}{\text{standard deviation of } F}$$

The probability of landsliding for each scenario was approximated by the cumulative standard normal distribution Φ evaluated at $-\beta$ (see Chapter 5). The results for two of the five profiles are shown in Table 6.30. It is interesting to note that for both profiles the highest annual probability of landsliding,

Table 6.30 Example 6.21: Lushan, Taiwan – conditional probability of landsliding at two slope profiles

Profile	Groundwater scenario	P (trigger)	P (landslide|trigger)	P (landslide)	Factor of safety
1	Low	0.795	0.17	0.135	1.19
	Medium	0.2	0.42	0.084	1.11
	High	0.005	0.589	0.003	0.92
	Total	0.222			
2	Low	0.795	0.2405	0.191	1.18
	Medium	0.2	0.4879	0.098	1.09
	High	0.005	0.7002	0.004	0.98
	Total	0.292			

From Lee and Chi (2011)

P(landslide), is associated with the low-groundwater-level scenario, despite the higher factors of safety under these conditions. This is because the triggering event is more likely than with the other two scenarios. The total annual probability of landsliding at each profile was calculated as the sum of the annual probabilities for each of the three groundwater scenarios (low, medium and high).

The probability model was also used to demonstrate the potential effectiveness of a slope drainage and stabilisation system. Lee and Chi (2011) suggested that, with the scheme in place, the annual probabilities of the medium- and high-groundwater scenarios could be reduced to near zero. It follows that the total probability of landslide occurrence would be equal to the occurrence probability under the low-groundwater scenario – that is, 0.135 and 0.191 for profiles 1 and 2 respectively.

6.12. Estimating probability: precision or pragmatism?

This chapter has presented a wide range of approaches that can be used to estimate the probability of landsliding, all of which rely on either the historical record of landsliding or an understanding of slope conditions (i.e. geomorphology, geology or geotechnics), or both in combination. None of these approaches can, however, be *guaranteed* to provide a reliable estimate, for reasons including the following.

- Incomplete or inaccurate data sets can mean that the historical frequency of recorded landsliding is an imprecise guide to the future probability of failure. The record may be too short to contain evidence of high-magnitude low-frequency events or may yield a frequency that does not conform to the long-term probability.
- Empirical relationships between landslide activity and the occurrence of triggering events assume that the critical landslide threshold conditions have remained constant over the period of the historical record (i.e. the data series is *stationary*). However, the fundamental nature of the relationship between the occurrence of triggering events and landsliding may have changed over the period in question, because of land use change or progressive weathering of the slope – that is, the data series is *non-stationary*. In addition, future environmental conditions and slope responses could be significantly different to those in the past because of the combined effects of human activity and predicted climate or environmental change.
- The use of expert judgement can lead to estimates that are heavily biased by the experience and personality of the experts. This may lead to a wide range of estimates, even when using exactly the same basic information.
- The estimates derived from probabilistic stability analysis and the use of reliability indices can be very sensitive to the amount of geotechnical information available about the slope conditions.

It is important to appreciate, therefore, that *estimates* of the probability of landsliding can only be estimates. The commonly expressed desire for increasing precision in the estimation process needs to be tempered by a degree of pragmatism that reflects the reality of the situation and the limitations of available information. For most projects, estimates will generally need to be *fit for purpose* (i.e. support the development of a risk assessment and guide decision-making; see Chapter 2) rather than the product of a lengthy academic research programme.

Perhaps of greater concern to those seeking precision when estimating the probability of landsliding is the fact that slope behaviour does not completely conform to a probabilistic model. Indeed, natural systems can be classified in terms of their degree of complexity and randomness of *behaviour* (Weinberg, 1975).

- Highly organised simple systems that can be modelled and analysed using deterministic mathematical functions (e.g. the prediction of tides or the movement of the planets).
- Unorganised complex systems that exhibit a high degree of random behaviour and hence can be modelled by stochastic methods (e.g. the probability of waves of particular heights or flood discharges).
- Organised complex systems, where the behaviour involves both deterministic and random components.

Slopes can be viewed as organised complex systems (e.g. Lee, 2003). The response of a slope to triggering events of a particular size is controlled by the antecedent conditions. Landsliding may not conform particularly well to either the deterministic or the random models. Landslide events are not independent, but can be influenced by the size and location of previous events. In other words, landsliding is a process with a *memory* (insofar as the current and future behaviour is influenced by the effects of past events on the system). Indeed, the chaotic nature of the short-term preparatory and triggering factors (e.g. Essex *et al.*, 1987), and of geomorphological systems in general (e.g. Hallet, 1987), suggests that there may well be a limit to the predictability of the landsliding process.

REFERENCES

Anda E and Blikra L (1998) *Rock-avalanche hazard in Møre og Romsdal, Western Norway. 25 Years of Snow Avalanche Research*. Norwegian Geotechnical Institute, Oslo. Publication No. 203.

AGS (Australian Geomechanics Society) (2000) Landslide risk management concepts and guidelines. *Australian Geomechanics* **35**: 49–52.

Berberian M (1994) Natural hazards and the first earthquake catalogue of Iran. *Historical hazard in Iran prior to 1900*. International Institute of Earthquake Engineering and Seismology (IIEES), Tehran, vol.1, pp. 560–603.

Bishop AW (1955) The use of the slip circle in the stability analysis of slopes. *Géotechnique* **5**: 7–17.

Bishop AW, Hutchinson JN, Penman ADM and Evans HE (1969) *Geotechnical Investigation into the Causes and Circumstances of the Disaster of 21 October 1966*. HMSO, London.

Bjerrum L and Jørstad F (1968) *Stability of Rock Slopes in Norway*. Norwegian Geotechnical Institute, Oslo. Publication No. 79.

Brand EW and Hudson RR (1982) CHASE – an empirical approach to the design of cut slopes in Hong Kong soils. *Proceedings of the 7th Southeast Asian Geotechnical Conference*, Hong Kong Vol. 1, pp. 1–16.

Bromhead EN (1986) *The Stability of Slopes*. Surrey University Press, London.

Bromhead EN, Hopper AC and Ibsen ML (1998) Landslides in the Lower Greensand escarpment in south Kent. *Bulletin of Engineering Geology and the Environment* **57**: 131–144.

Brunsden D and Lee EM (2004) Behaviour of coastal landslide systems: an inter-disciplinary view. *Zeitschrift fur Geomorphologie* **134**: 1–112.

Bunce C, Cruden DM and Morgenstern NR (1997) Assessment of the hazard from rock fall on a highway. *Canadian Geotechnical Journal* **34**: 344–356.

Casale R and Margottini C (eds) (1995) *Meteorological Events and Natural Disasters: An Appraisal of the Piedmont (North Italy) Case History of 4–6 November 1994 by a CEC Field Mission*. European Commission, Brussels.

Chaytor JD, ten Brink US, Solow AR and Andrews BD (2009) Size distribution of submarine landslides along the U.S. Atlantic margin and its implication to tsunami hazards. *Marine Geology* **264**: 16–27.

Chen AY and Morgenstern NR (1983) Extension to the generalized method of slices for stability analysis. *Canadian Geotechnical Journal* **20(1)**: 104–119.

Clark AR and Guest S (1991) The Whitby cliff stabilisation and coast protection scheme. In *Slope Stability Engineering: Developments and Applications* (Chandler RJ (ed.)). Thomas Telford, London, pp. 283–290.

Clark AR and Guest S (1994) The design and construction of the Holbeck Hall landslide coast protection and cliff stabilisation emergency works. *Proceedings of the 29th MAFF Conference of River and Coastal Engineers*, Loughborough, 3.3.1–3.3.6.

Clark AR, Fort DS and Davis GM (2000) The strategy, management and investigation of coastal landslides at Lyme Regis, Dorset, UK. In *Landslides: In Research, Theory and Practice* (Bromhead ES, Dixon N and Ibsen ML (eds)). Thomas Telford, London, pp. 279–286.

Clements M (1994) The Scarborough experience – Holbeck landslide, 3–4 June 1993. *Proceedings of the Institution of Civil Engineers, Municipal Engineers* **103**: 63–70.

Coe JA, Michael JA, Crovelli RA and Savage WZ (2000) *Preliminary Map Showing Landslide Densities, Mean Recurrence Intervals, and Exceedance Probabilities as Determined from Historic Records, Seattle, Washington*. US Geological Survey, Reston, VA. Open File Report 00-303.

Corominas J and Moya J (1996) Historical landslides in the Eastern Pyrenees and their relation to rainy events. In *Landslides* (Chacon J, Irigaray C and Fernandez T (eds)). Balkema, Rotterdam, pp. 125–132.

Cox DR (1962) *Renewal Theory*. Methuen, London.

Crovelli RA (2000) *Probability Models for Estimation of Number and Costs of Landslides*. US Geological Survey, Reston, VA. Open File Report 00-249.

Crozier MJ (1986) *Landslides: Causes, Consequences and Environment*. Croom Helm, London.

Crozier MJ and Glade T (1999) Frequency and magnitude of landsliding: fundamental research issues. *Zeitschrift fur Geomorphologie NF Suppl.Bd* **115**: 141–155.

Crozier MJ and Preston NJ (1998) Modelling changes in terrain resistance as a component of landform evolution in unstable hill country. In *Lecture Notes in Earth Science* (Hergarten S and Neugebauer HJ (eds)) **78**: 267–284.

Damgaard JS and Peet AH (1999) Recession of coastal soft cliffs due to waves and currents: experiments. *Proceedings of Coastal Sediments '99: 4th International Symposium on Coastal Engineering and Science of Coastal Sediment Processes*, Long Island, NY, pp. 1181–1191.

Dimmock P, Mackenzie B and Mills A (2012) Probabilistic slope stability analysis in the west Nile delta, offshore Egypt. Offshore Site Investigation and Geotechnics Integrated Geotechnologies – Present and Future. *Proceedings of the 7th International Conference 12–14 September 2012*. Royal Geographical Society, London, UK.

Ducassou E, Migeon S, Mulder T, Murat A, Capotondis L, Bernasconi SM and Mascle J (2009) Evolution of the Nile deep-sea turbidite system during the Late Quaternary: influence of climate change on fan sedimentation. *Sedimentology* **56**: 2061–2090.

Duncan JM (1996a) Soil slope stability analysis. In *Landslides: Investigation and Mitigation* (Turner AK and Schuster RL (eds)). National Academy Press, Washington, DC, pp. 337–371. Transportation Research Board Special Report 247.

Duncan JM (1996b) State-of-the-art: limit equilibrium and finite-element analysis of slopes. *Journal of Geotechnical Engineering* **122(7)**: 557–596.

Eidsvig UK, Lacasse S, Nadim F and Høeg K (2008) *Event Tree Analysis of Hazard and Risk Associated with the Åknes Rockslide*. ICG/NGI, Oslo. Report 20071653-1.

Essex C, Lookman T and Nererberg MRH (1987) The climate attractor over short time scales. *Nature* **326**: 64–66.

Evans SG, Guthrie RH, Roberts NJ and Bishop NF (2007) The disastrous 17 February 2006 rockslide – debris avalanche on Leyte Island, Philippines: a catastrophic landslide in tropical mountain terrain. *Natural Hazards and Earth Systems Science* **7**: 89–101.

Evans TG, Usher N and Moore R (2007) Management of geotechnical and geohazard risks in the West Nile Delta. *Proceedings of the 6th International Offshore Site Investigation and Geotechnics Conference: Confronting New Challenges and Sharing Knowledge*. Society for Underwater Technology, London.

Fell R, Finlay P and Mostyn G (1996a) Framework for assessing the probability of sliding of cut slopes. In *Landslides* (Senneset K (ed.)). Balkema. Rotterdam, pp. 201–208.

Fell R, Walker BF and Finlay PJ (1996b) Estimating the probability of landsliding. *Proceedings of the 7th Australia/New Zealand Conference on Geomechanics, Adelaide*. Institution of Engineers Australia, Canberra, pp. 304–311.

Finlay PJ and Fell R (1995) *A Study of Landslide Risk Assessment in Hong Kong*. Geotechnical Engineering Office, Hong Kong.

Fort D S, Clark AR and Savage DT (2000) Instrumentation and monitoring of the coastal landslides at Lyme Regis, Dorset, UK. In *Landslides: In Research, Theory and Practice* (Bromhead EN, Dixon N and Ibsen ML (eds)). Thomas Telford, London, pp. 573–578.

Geological Engineering Office (GEO) (1996) *Report on the Fei Tsui Road Landslide of 13 August 1995*. Geotechnical Engineering Office, Hong Kong.

Glade T (1996) The temporal and spatial occurrence of landslide-triggering rainstorms in New Zealand. *Beitrage zur Physiogeographie – Festschrift für Dietrich Barsch* **104**: 237–250.

Glade T (1997) The temporal and spatial occurrence of rainstorm-triggered landslides in New Zealand. PhD thesis, Department of Geography, Victoria University of Wellington.

Glade T (1998) Establishing the frequency and magnitude of landslide-triggering rainstorm events in New Zealand. *Environmental Geology* **55(2)**: 160–174.

Glade T and Crozier MJ (1996) Towards a national landslide information base for New Zealand. *New Zealand Geographer* **52**: 29–40.

Graham J (1984) Methods of stability analysis. In *Slope Instability* (Brunsden D and Prior DB (eds)). Wiley, Chichester, pp. 171–216.

Guthrie RH (2009) The occurrence and behavior of rainfall-triggered landslides in coastal British Columbia. PhD Thesis, University of Waterloo, Ontario.

Guthrie RH and Evans SG (2004a) Analysis of landslide frequencies and characteristics in a natural system, coastal British Columbia. *Earth Surface Processes and Landforms* **29**: 1321–1339.

Guthrie RH and Evans SG (2004b) Magnitude and frequency of landslides triggered by a storm event. Loughborough Inlet, British Columbia. *Natural Hazards and Earth System Sciences* **4**: 475–483.

Guzzetti F (2005) Landslide hazard and risk assessment. Dissertation, University of Bonn.

Guzzetti F, Reichenbach P, Ardizzone F, Cardinali M and Galli M (2005) Probabilistic landslide hazard assessment at the basin scale. *Geomorphology* **72**: 272–299.

Guzzetti F, Galli M, Reichenbach P, Ardizzone F and Cardinali M (2006) Landslide hazard assessment in the Collazzone area, Umbria, Central Italy. *Natural Hazards and Earth System Sciences* **6**: 115–131.

Haeeri SM, Sttari S and Sammiee A (1993) Instabilities due to Manjil earthquake of 1990. *8th International Seminar of Earthquake Prediction*, Tehran, pp. 185–193, 230.

Halcrow Group (2003) *Coastal Instability Risk: Ventnor Undercliff, Isle of Wight*. Isle of Wight Council, Newport.

Hall JW, Meadowcroft IC, Lee EM and van Gelder PHAJM (2002) Stochastic simulation of episodic soft coastal cliff recession. *Coastal Engineering* **46**: 159–174.

Hallet B (1987) On geomorphic patterns with a focus on stone circles viewed as a free-convection phenomenon. In *Irreversible Phenomena and Dynamic Systems Analysis in Geosciences* (Nicolis C and Nicolis G (eds)). NATO ASI, Brussels, pp. 533–553.

Harp EL (1997) *Landslides and Landslide Hazards in Washington State Due to February 5–9 1996 Storm*. US Geological Survey, Reston, VA. Administrative Report.

Hawley JG (1984) Slope instability in New Zealand. In *Natural Hazards in New Zealand* (Speden IG and Crozier MJ (eds)). UNESCO, Wellington, pp. 88–133.

High Point Rendel (1996) *Holbeck Gardens Coast Protection and Slope Stabilisation*. Report to Scarborough Borough Council, Scarborough.

High Point Rendel (1999) *Holbeck-Scalby Mills Coastal Defence Strategy*. Report to Scarborough Borough Council, Scarborough.

High Point Rendel (2003) *Whitby Coastal Defence Strategy*. Scarborough Borough Council, Scarborough.

Huang JC, Chang YT, Yen CY, Huang SC and Huang JK (2008) The engineering geology investigation on the sliding slope above Lushan hot spring area – a case study. *Sino-Geotechnics* **117**: 71–80 (in Chinese).

Hungr O (1997) Some methods of landslide hazard intensity mapping. In *Landslide Risk Assessment* (Cruden D and Fell R (eds)). Balkema, Rotterdam, pp. 215–226.

Hutchinson JN (1962) *Report on Visit to Landslide at Lyme Regis, Dorset*. Building Research Station Note c890.

Hutchinson JN (1988) General report: Morphological and geotechnical parameters of landslides in relation to geology and hydrogeology. In *Landslides* (Bonnard C (ed.)). Balkema, Rotterdam, pp. 3–35.

Jeanjean P, Hill A and Taylor S (2003) The challenges of confidently siting facilities along the Sigsbee Escarpment in the Southern Green Canyon area of the Gulf of Mexico. Framework of integrated studies. *Proceedings of the Offshore Technology Conference, Houston, TX*, Paper 15156.

Jones DKC (1992) Landslide hazard assessment in the context of development. In *Geohazards: Natural and Man-made* (McCall GJH, Laming DJC and Scott SC (eds)). Chapman and Hall, London, pp. 117–141.

Keefer DK (1984) Landslides caused by earthquakes. *Geological Society of America Bulletin* **95**: 406–421.

Keystone (2009) Keystone XL Project pipeline risk assessment and environmental consequence analysis. Appendix P: Analysis of incident frequencies and spill volumes for environmental consequence estimation for the Keystone XL Project. Available from: http://keystonepipeline-xl.state.gov/documents/organization/182239.pdf (accessed 2013).

Koirala NP and Watkins AT (1988) Bulk appraisal of slopes in Hong Kong. In *Landslides* (Bonnard C (ed.)). Balkema, Rotterdam, 2, pp. 1181–1186.

Kovach RL (1995) *Earth's Fury: An Introduction to Natural Hazards and Disasters*. Prentice-Hall, Englewood Cliffs.

Kuloshvili SI and Maisuradze GM (2000) Geological-geomorphological aspects of landsliding in Georgia (Central and Western Caucasus). In *Landslides: In Research, Theory and Practice* (Bromhead EN, Dixon N and Ibsen ML (eds)). Thomas Telford, London, pp. 861–866.

Lacasse S and Nadim F (2006) Hazard and risk assessment of slopes. *International Conference on Slopes, Malaysia*, 3–34.

Lacasse S and Nadim F (2011) Learning to Live with Geohazards: From Research to Practice. *ASCE GSP 224. Proceedings, Geotechnical Risk Assessment and Management*, pp. 64–116.

Lacasse S, Eidsvik U, Nadim F, Høeg K and Blikra LH (2008) Event Tree analysis of Åknes rock slide hazard. *IV Geohazards Québec, 4th Canadian Conference on Geohazards*, pp. 551–557.

Lee EM (1992) Urban landslides: impact and management. In *The Coastal Landforms of West Dorset* (Allison R (ed.)). Geologists Association Guide No. 47, pp. 80–93.

Lee EM (1998) Problems associated with the prediction of cliff recession rates for coastal defence. In *Coastal Defence and Earth Science Conservation* (Hooke JM (ed.)). Geological Society Publishing, Bath, pp. 46–57.

Lee EM (1999) Coastal Planning and Management: The impact of the 1993 Holbeck Hall landslide, Scarborough. *East Midlands Geographer* **21**: 78–91.

Lee EM (2003) *Coastal Change and Cliff Instability: Development of a Framework for Risk Assessment and Management*. Unpublished PhD thesis, University of Newcastle upon Tyne.

Lee EM (2007) Landslide risk assessment: the challenge of estimating landslide probability. In *Engineering Geology in Geotechnical Risk Management*. Hong Kong Regional Group of the Geological Society of London, pp. 13–31.

Lee EM (2009) Landslide risk assessment: the challenge of estimating the probability of landsliding. *Quarterly Journal of Engineering Geology and Hydrogeology* **42**: 445–458.

Lee EM and Charman JH (2005) Geohazards and risk assessment for pipeline route selection. In *Terrain and Geohazard Challenges Facing Onshore Oil and Gas Pipelines* (Sweeney M (ed.)). Thomas Telford, London, pp. 95–116.

Lee EM and Clark AR (2000) The use of archive records in landslide risk assessment: historical landslide events on the Scarborough coast, UK. In *Landslides: In Research, Theory and Practice* (Bromhead EN, Dixon N and Ibsen ML (eds)). Thomas Telford, London, pp. 904–910.

Lee EM and Clark AR (2002) *Investigation and Management of Soft Rock Cliffs*. Thomas Telford, London.

Lee EM and Moore R (1991) *Coastal Landslip Potential Assessment: Isle of Wight Undercliff, Ventnor*. Department of the Environment, London.

Lee EM and Moore R (2007) Ventnor Undercliff: development of landslide scenarios and quantitative risk assessment. In *Landslides and Climate Change: Challenges and Solutions* (McInnes R, Jakeways J, Fairbank H and Mathie E (eds.)). Balkema, pp. 323–334.

Lee EM and Sellwood M (2002) An approach to assessing risk on a protected coastal cliff: Whitby, UK. In *Instability – Planning and Management* (McInnes RG and Jakeways J (eds)). Thomas Telford, London, pp. 617–624.

Lee EM, Clark AR and Guest S (1998a) An assessment of coastal landslide risk, Scarborough, UK. In *Engineering Geology: The View from the Pacific Rim* (Moore D and Hungr O (eds)). Balkema, Rotterdam, pp. 1787–1794.

Lee EM, Moore R and McInnes RG (1998b) Assessment of the probability of landslide reactivation: Isle of Wight Undercliff, UK. In *Engineering Geology: The View from the Pacific Rim* (Moore D and Hungr O (eds)). Balkema, Rotterdam, pp. 1315–1321.

Lee EM, Brunsden D and Sellwood M (2000) Quantitative risk assessment of coastal landslide problems, Lyme Regis, UK. In Landslides: In *Research, Theory and Practice* (Bromhead EN, Dixon N and Ibsen ML (eds)). Thomas Telford, London, pp. 899–904.

Lee EM, Hall JW and Meadowcroft IC (2001) Coastal cliff recession: the use of probabilistic prediction methods. *Geomorphology* **40**: 253–269.

Lee EM, Meadowcroft IC, Hall JW and Walkden MJ (2002) Coastal landslide activity: a probabilistic simulation model. *Bulletin of Engineering Geology and the Environment* **61**: 347–355.

Lee YF and Chi YY (2011) Rainfall-induced landslide risk at Lushan, Taiwan. *Engineering Geology* **123**: 113–121.

Lee YF, Lee DH and Chi YY (2009) A probabilistic method for evaluating slope failure by reliability index. *Journal of the Chinese Institute of Civil and Hydraulic Engineering* **21(1)**: 91–103 (in Chinese).

Meadowcroft I, Brampton A and Hall J (1997) Risk in an uncertain world – finding practical solutions. *Proceedings of the MAFF Conference of River and Coastal Engineers,* H.2.1–H.2.11. MAFF Publications, London.

Moore R, Lee EM and Clark AR (1995) *The Undercliff of the Isle of Wight: a review of ground behaviour*. South Wight Borough Council, Sandown.

Moore R, Usher N and Evans T (2007) Integrated Multidisciplinary Assessment and Mitigation of West Nile Delta Geohazards. *Proceedings of the 6th International Offshore Site Investigation and Geotechnics Conference: Confronting New Challenges and Sharing Knowledge*, London.

Mousavi SM, Omidvar B, Ghazban F and Feyzi R (2011) Quantitative risk analysis for earthquake-induced landslides – Emamzadeh Ali, Iran. *Engineering Geology*, pp. 191–203.

Nadim F (2006) Challenges to geo-scientists in risk assessment for sub-marine slides. *Norwegian Journal of Geology* **86**: 351–362.

Nadim F and Locat J (2005) Risk assessment for submarine slides. In *Landslide Risk Management* (Hungr O, Fell R, Couture R and Eberhardt E (eds)). Balkema, Rotterdam, pp. 321–333.

Nadim F, Krunic D and Jeanjean P (2003) Probabilistic slope stability analyses of the Sigsbee Escarpment. *Proceedings of the Offshore Technology Conference*, Houston, TX, Paper 15203.

Nadim F, Einstein H and Roberds W (2005) Probabilistic stability analysis for individual slopes in soil and rock. In *Landslide Risk Management* (Hungr O, Fell R, Couture R and Eberhardt E (eds)). Balkema, Rotterdam, pp. 63–98.

Nash D (1987) A comparative review of limit equilibrium methods of stability analysis. In *Slope Stability* (Anderson MG and Richards KS (eds)). Wiley, Chichester, pp. 11–75.

Niedoroda A, Reed C, Hatchett L, Young A and Kasch V (2003) Analysis of past and future debris flows and turbidity currents generated by slope failures along the Sigsbee Escarpment. *Proceedings, Offshore Technology Conference, Houston, TX*, Paper 15162.

Orange D, Angell M, Brand J, Thompson J, Buddin T, Williams M, Hart B and Berger B (2003) Shallow geological and salt tectonic setting of the Mad Dog and Atlantis field: relationship between salt, faults, and seafloor geomorphology. *Proceedings of the Offshore Technology Conference*, Houston, TX, Paper 15157.

Penman ADM (2000) The Aberfan flow slide, Taff Valley. In *Landslides and Landslide Management in South Wales* (Siddle HJ, Bromhead EN and Bassett MG (eds)). National Museums and Galleries of Wales, Cardiff, pp. 62–68. Geological Series No. 18.

PHSMSA (2010) PHMSA pipeline incident statistics. Website: http://primis.phmsa.dot.gov/comm/reports/safety/PSI.html (accessed 2013).

Porter M, Jacob M, Savigny KW, Fougere S and Morgenstern N (2007) Risk management for urban flow slides in North Vancouver, Canada. In *OttawaGeo 2007: Proceedings of 60th Canadian Geotechnical Conference*, Ottawa.

Radbruch-Hall DH, Colton RB, Davies WE, Lucchitta I, Skipp BA and Varnes DJ (1982) *Landslide Overview Map of the Conterminous United States*. US Geological Survey, Reston, VA. Professional Paper 1183.

Reeder MS, Rothwell RG, Stow DAV, Kahler G and Kenyon NH (1998) Turbidite flux, architecture and chemostratigraphy of the Herodotus Basin, Levantine Sea, south-eastern Mediterranean. In *Geological Processes on Continental Margins: Sedimentation, Mass Wasting and Stability* (Stoker MS, Evans D and Cramp D (eds)). Geological Society, London, pp. 19–41. Special Publication 129.

Reeder MS, Stow DAV and Rothwell RG (2002) Late Quaternary turbidite input into the eastern Mediterranean basin: new radiocarbon constraints on climate and sea-level control. In *Sediment Flux to Basins: Causes, Controls and Consequences* (Jones SJ and Frostick LE (eds)). Geological Society, London, pp. 267–278. Special Publication 191.

Reid ME (1994) A pore pressure diffusion model for estimating landslide inducing rainfall. *Journal of Geology* **102**: 709–717.

Rendel Geotechnics (1994) *Preliminary Study of the Coastline of the Urban Areas within Scarborough Borough: Scarborough Urban Area*. Scarborough Borough Council, Scarborough.

Roberds WL (1990) Methods for developing defensible subjective probability assessments. *Transportation Research Record* **1288**: 183–190.

Schofield J (1787) *An Historical and Descriptive Guide to Scarborough and its Environs*. Blanchard, York.

Schrott L and Pasuto A (eds) (1999) Temporal stability and activity of landslides in Europe with respect to climate change (TESLEC). *Geomorphology*, Special Issue 30, Nos 1–2.

Schuster RL (ed.) (1991) *The March 5, 1987, Ecuador Earthquakes – Mass Wasting and Socioeconomic Effects*. National Academy Press, Washington, DC.

Sellwood M, Davis GM, Brunsden D and Moore R (2000) Ground models for the coastal landslides at Lyme Regis, Dorset, UK. In *Landslides: in Research, Theory and Practice* (Bromhead EN, Dixon N and Ibsen ML (eds)). Thomas Telford, London, pp. 1361–1366.

Siddle HJ, Wright MD and Hutchinson JN (1996) Rapid failures of colliery spoil heaps in the South Wales Coalfield. *Quarterly Journal of Engineering Geology* **29**: 103–132.

Siddle HJ, Wright MD and Hutchinson JN (2000) Rapid failures of spoil heaps in the South Wales Coalfield. In *Landslides and Landslide Management in South Wales* (Siddle HJ, Bromhead EN and Bassett MG (eds)). National Museums and Galleries of Wales, Cardiff, pp. 32–35.

Sweeney M, Gasca AH, Garcia-Lopez M and Palmer AC (2005) Pipelines and landslides in rugged terrain: a database, historic risks and pipeline vulnerability. In *Terrain and Geohazard Challenges Facing Onshore Oil and Gas Pipelines* (Sweeney M (ed.)). Thomas Telford, London, pp. 641–660.

Tatashidze ZK, Tsereteli ED and Khazaradze RD (2000) Principal hazard factors and mechanisms causing landslides (Georgia as an example). In *Landslides: in Research, Theory and Practice* (Bromhead EN, Dixon N and Ibsen ML (eds)). Thomas Telford, London, pp. 1449–1452.

ten Brink US, Giest EL and Andrews BD (2006) Size distribution of submarine landslides and its implication to tsunami hazard in Puerto Rico. *Geophysics Research Letters* **33**: L11307.

Terlien MJM (1996) *Modelling Spatial and Temporal Variations in Rainfall-triggered Landslides*. International Institute for Aerospace and Earth Sciences (ITC), Enschede. Publication 32.

Terlien MJM, De Louw PGB, Van Asch TWJ and Hetterschijt RAA (1996) The assessment and modelling of hydrological failure conditions of landslides in the Puriscal area (Costa Rica) and the Manizalis Region (Colombia). In *Advances in Hillslope Processes* (Anderson MG and Brooks SM (eds)). Wiley, Chichester, pp. 837–855.

Van Asch TWJ, Buma J and Van Beek LPH (1999) A view on some hydrological triggering systems in landslides. *Geomorphology* **30**: 25–32.

Voight B, Glicken H, Jandra RL and Douglass PM (1981) Catastrophic rockslide avalanche of May 18. In *The 1980 eruptions of Mount St. Helens, Washington* (Lipman PW and Mullineaux DR (eds)). US Geological Survey, Reston, VA, pp. 347–378. Professional Paper 1250.

Weinberg GM (1975) *An Introduction to General Systems Thinking*. Wiley, New York.

Whittaker M (1984) *The Book of Scarborough Spaw*. Barracuda Press, Birmingham.

Chapter 7
Exposure

7.1. Introduction

Will the landslide adversely affect something of value? The concept of *exposure* provides an important link between hazard assessment (threats) and consequence assessment (e.g. fatalities or costs). As outlined in Chapter 1, it generally encompasses two separate issues: the chance that a landslide will actually impact people and the things valued by people (the 'elements at risk') and exactly how many people or other assets are likely to be present at that particular moment in time. In this way, exposure helps to determine the transformation of a landslide from simply being a physical phenomenon to a hazard event with the potential to cause harm. The stage in the event sequence where the occurrence of a landslide actually attains the ability to cause harm has come to be known as the 'top event' (see Chapter 2, Figure 2.5). In other words, it is the point where an 'event sequence' is transformed into an 'accident sequence' (see Figure 2.4). Should the landslide miss all the assets or occur at a time when the impact zone (i.e. danger zone) is unoccupied, then there will be no 'top event'. However, this simple *hit or miss model* of top events needs to be extended to include those situations where the actual landslide might cause negligible or limited impact (an initial miss in the first case), yet might still lead to significant adverse consequences.

Common examples of 'top events' include the following.

- A landslide hits a building and causes structural damage and subsequent collapse. This can lead to injuries or fatalities if the building was occupied at the time.
- A landslide hits a person or vehicle travelling through or stationary (albeit temporarily) in the impact zone. This may result directly in injuries or fatalities. In other cases, the landslide causes no direct harm, but the moving debris causes a vehicle to swerve or brake suddenly and thereby cause a crash.
- Landslide debris is deposited along a transport route and forms an obstacle for traffic. For example, rockfall debris on a railway track may lead to a derailment if a train fails to stop in time and collides with it. Alternatively, a vehicle may swerve to avoid landslide debris on the road and crash into a tree or on-coming traffic.

Losses or damage occur when the location of elements at risk intersects with the locations of landslides – that is, they are 'hit' by landsliding or are affected as a consequence of landsliding. For example, a shelter at the base of coastal cliff is only exposed to that proportion of rockfalls from the cliffline that occur above that particular site; the events elsewhere along the cliffline are 'misses' or even 'near misses'. The proportion of events in an area that could intersect the location of a particular asset is the *spatial probability*, and can be defined by a simple conditional probability model that describes the chance of being in the 'wrong place' (i.e. in the danger zone) when a landslide occurs:

$$P(\text{wrong place}) = P(\text{spatial}|\text{landslide})$$

Exposure also displays marked variations at short to very short timescales due to the dynamic characteristics of a population and the mobile nature of certain assets. This is termed *temporal probability* (Australian Geomechanics Society, 2000, 2007; Morgan *et al.*, 1992) or *temporal vulnerability*, and can be defined as being *present at the 'wrong time'* – that is, when the landslide occurs:

$$P(\text{wrong time}) = P(\text{temporal}|\text{landslide})$$

The overall probability of a 'hit' can be defined as being both in the 'wrong place' and at the 'wrong time' when the landslide occurs:

$$P(\text{hit}) = P(\text{landslide}) \times P(\text{spatial}|\text{landslide}) \times P(\text{temporal}|\text{landslide})$$

When considering exposure, it is important to distinguish between fixed and mobile assets. Buildings or pipelines are *permanent* or *stationary* assets that could be damaged irrespective of the timing of the landslide event ($P(\text{wrong time}) = 1$). As these assets are always within the zone of potential landslide impact, the adverse consequences for a particular magnitude of event can be assumed to remain constant. However, there will be uncertainty regarding the location of the landslide event and whether it will impact (i.e. 'hit') or miss a particular asset.

For *temporary* or *non-stationary* assets the degree of risk can vary with the timing of the event, be it night or day, a weekday or a weekend, the tourist season or off-season. The consequences will reflect the chance of the event occurring at a time when mobile assets are either within the zone of impact (e.g. a train passing through a rock cutting at the time of a rockfall) or at a relatively high level of concentration (i.e. occupancy rates for housing are higher at night). Temporary exposure can be represented by a factor on the scale from 0 (never present) to 1 (permanently present fixed asset).

As there is uncertainty regarding exactly where or when a landslide will occur, it is difficult to reliably predict how many people or what value of temporary assets will be present to be impacted by the event (i.e. being in the 'wrong place' at the 'wrong time').

In addition, there are many different ways of looking at the exposure of people to landslide hazards, including

- a specific individual person for a one-off visit or repeated visits (e.g. daily, weekly) to the landslide area
- a hypothetical individual who stays within the landslide area for 100% of the time period of interest (this forms the basis for the estimation of location-specific individual risk: see Chapter 10)
- a specific group of people visiting the landslide area on a single occasion (e.g. a geological field trip) or on a regular basis (e.g. a maintenance crew inspecting road-side drains)
- an equivalent population of individuals who are theoretically present in the landslide area for 100% of the time period of interest.

7.2. Spatial probability: stationary assets

The probability of a landslide hitting a permanent stationary asset depends on the relative locations of the asset, the path the landslide is likely to travel from its source and the travel distance.

- For assets located on a pre-existing landslide or in the source area of a first-time failure, the probability of being in the 'wrong place' (P(spatial|landslide)) when reactivation or movement occurs is 1.0.
- For assets located below a landslide-prone slope or unstable cliff there will be uncertainty about the exact position of the source areas/paths of future landslide events. The spatial probability will depend on the relationship between relative widths of the potential source area (e.g. the length of a cliffline), the landslide event and the asset. The probability of being in the 'wrong place' (P(spatial|landslide)) when the landslide occurs will vary between >0 and 1.0.
- For assets located along a potential landslide pathway, such as on a channelised debris flow fan at the mouth of a stream catchment, there will be uncertainty about whether the debris flow travels far enough to reach the asset and whether the landslide path intersects the position of the asset. The spatial probability will depend on the travel distance of the landslide, as well as the relative widths of the debris fan, the landslide event and the asset. The probability of being in the 'wrong place' (P(spatial|landslide)) when the landslide occurs will vary between 0 and 1.0.

Example 7.1

A 500 m-long (alongshore) section of the seaside promenade at Brighton, UK, is located directly beneath a high chalk cliff that is regularly affected by small rockfalls (average 1 m wide) that have occurred at a frequency of 1 every 10 years (P(landslide) = 0.1). The proportion of the promenade that is affected by a fall (i.e. the spatial probability) is

$$P(\text{spatial}) = \frac{\text{size of rockfall}}{\text{length of promenade}}$$

$$= 1/500$$

$$= 0.002$$

A 3 m-wide bench has been placed on the promenade to allow visitors to rest and admire the view out to sea. The spatial probability of the bench being in the rockfall path is

$$P(\text{spatial}) = \frac{\text{length of bench}}{\text{length of promenade}}$$

$$= 3/500$$

$$= 0.006$$

The probability of a person being hit by a rockfall is considered in Example 7.7.

Example 7.2

One of the main threats to offshore gas development projects in the West Nile Delta are slab slides with a shallow shear surface (Dimmock et al., 2012; Moore et al., 2007). Geophysical data indicate that these failures appear to have occurred randomly over the gently sloping (generally <5°) delta slope that lies seaward of the continental shelf, in over 1500 m of water. Individual failures vary in size, but are typically around 500 m long and 400 m wide. Within a 5000 km^2 (50 km × 100 km) section of the delta slope, the historical frequency of slab slide events has been estimated to be 1 in 100 years (P(landslide) = 0.01).

The spatial probability of a production well located in this section of the delta slope being in the path of slab slide is

$$P(\text{spatial}) = \frac{\text{landslide width}}{\text{delta slope width}} \times \frac{\text{landslide length}}{\text{delta slope length}}$$

$$= 400/50\,000 \times 500/100\,000$$

$$= 0.00004$$

This model can be simplified to

$$P(\text{spatial}) = \frac{\text{landslide area}}{\text{delta slope area}}$$

The probability of the well being in the wrong place and 'hit' when a slab slide occurs is

$$P(\text{hit}) = P(\text{landslide}) \times P(\text{spatial}|\text{landslide})$$

$$= 0.01 \times 0.00004 = 0.0000004 \text{ or } 4 \times 10^{-7}$$

Example 7.3

A road-side restaurant (25 m wide) has been built at the base of a landslide-prone hillslope in the Black Sea Mountains of Turkey, similar to that demolished in the Çatak disaster of 1988 (Jones et al., 1989). The debris slides are typically 100 m wide, and could occur anywhere within the 1 km-wide landslide-prone section. Historical evidence indicates that landslide frequency is around 1 event every 25 years.

Any 'hit' could cause structural damage and lead to building collapse. The probability that a landslide will hit any part of the restaurant needs to take into account the width of the building and the position of the landslide centrelines that would 'hit' at least part of the building – that is, half the landslide width either side of the restaurant (Figure 7.1). Consideration of centrelines is also important because 100 m-wide landslides cannot have centrelines within 50 m of each end of the landslide-prone section of slope (see denominators below).

$$P(\text{spatial}) = \frac{\text{restaurant width} + (0.5 \times \text{landslide width} \times 2)}{\text{width of landslide prone section} - (0.5 \times \text{landslide width} \times 2)}$$

$$= \frac{25 + (50 \times 2)}{1000 - (50 \times 2)}$$

$$= 0.139$$

The spatial probability that the restaurant receives a 'direct hit' from a landslide (i.e. the wole building impacted)

$$P(\text{spatial}) = \frac{\text{restaurant width} + [((0.5 \times \text{landslide width}) - \text{restaurant width}) \times 2]}{\text{width of landslide prone section} - (0.5 \times \text{landslide width} \times 2)}$$

$$= \frac{25 + 50}{900}$$

$$= 0.083$$

Figure 7.1 Example 7.3: landslide intersection with a restaurant

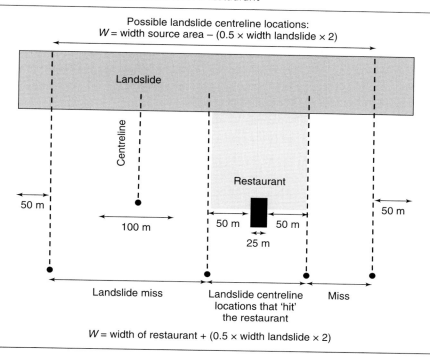

Example 7.4

A sub-sea oil-well centre (10 ha; 500 m wide and 200 m long) comprising a cluster of ten wells, flowlines and pipeline manifolds is to be located close to the base of a steep escarpment in the Gulf of Mexico. Geophysical data indicates that the 10 km-long escarpment has had a long history of mudflow activity, with events occurring on average every 25 years over the last 10 000 years ($P = 0.04$). Over this period, the individual mudflows have been around 100 m wide and the run-out has varied from 100 m from the scarp foot to a maximum distance of 1 km. A run-out probability distribution was developed from the analysis of over 1000 events: for example, the probability of a run-out of between 400 m and 500 m was 0.1.

The probability that a mudflow will hit a well centre sited 450 m from the escarpment foot is

$$P(\text{spatial}) = \frac{\text{well centre width} + (0.5 \times \text{mudflow width} \times 2)}{\text{escarpment width} - (0.5 \times \text{mudflow width} \times 2)} \times P(\text{run-out of 450 m})$$

$$= \frac{500 + 100}{10\,000 - 100} \times 0.1$$

$$= 0.006$$

The probability of the well being in the 'wrong place' and 'hit' when a mudflow occurs is

$$P(\text{hit}) = P(\text{landslide}) \times P(\text{spatial}|\text{landslide})$$

$$= 0.04 \times 0.006$$

$$= 0.00024 \text{ or } 2.4 \times 10^{-4}$$

Example 7.5

An electricity pylon with a base width of 15 m is located at the foot of a relatively steep 350 m-wide slope (Roberds, 2005). The slope is prone to debris slides that typically have a 20 m-wide debris trail, but variable run-out distances. Mapping of the downslope extent of previous events indicates that the run-out angle for the debris slides (i.e. the angle from the horizontal between the top of the slide to the distal end of the run-out) follows a normal distribution with a mean of 20° and a standard deviation of 5°. The upslope edge of the pylon corresponds with a 15° run-out angle.

The probability that the run-out angle exceeds 15° can be calculated using the NORMDIST function in Excel: $= \text{NORMDIST}(15, 20, 5, \text{TRUE}) = 0.158$.

The probability that a mudflow will hit a pylon sited 450 m from the escarpment foot is

$$P(\text{spatial}) = \frac{\text{pylon width} + (0.5 \times \text{slide width} \times 2)}{\text{slope width} - (0.5 \times \text{slide width} \times 2)} \times P(\text{run-out angle} > 15°)$$

$$= \frac{15 + 20}{350 - 20} \times 0.158$$

$$= 0.0167$$

7.3. Spatial and temporal probability: non-stationary assets

The probability of a landslide hitting a non-stationary asset depends on the amount of time that the asset spends in the danger zone: the longer the time, the higher the temporal probability of a landslide impact. Assuming that the landslide event is instantaneous, the temporal probability depends on the time spent in the danger zone and the number of people/assets that might be exposed over a particular time period (e.g. a day or year).

- Temporal probability ($P(\text{temporal})$) is 1.0 for stationary assets, as they are present in the danger zone for 100% of the time.
- For a non-stationary asset that is present in the danger zone for <100% of the time (e.g. a car parked below a cliff), there will be uncertainty about coinciding with the timing of the landslide event. $P(\text{temporal}|\text{landslide})$ will vary between >0 and <1.0 depending on the time spent in the danger zone. The temporal probability is calculated as the time in the danger zone expressed as a proportion of a year.
- For a non-stationary asset passing through a danger zone (e.g. across a pre-existing landslide or below an unstable cliff), the time exposed will depend on the speed of the person or vehicle (i.e. the distance travelled per hour) and the width of the danger zone, usually expressed as a proportion of a year.

The temporal and spatial probabilities are often combined for vehicle impacts, considering factors such as vehicle and landslide size, as well as journey times through (i.e. presence within) the danger zone.

The time spent within the danger zone is usually expressed as a proportion of a year, simply because landslide probability is usually expressed as the annual probability. In those instances where there are multiple events in a year, then it is necessary to consider each event as a separate statistical trial and model this scenario using the binomial distribution (see Chapter 5):

$$P(\text{hit}) = 1 - \{1 - [P(\text{spatial}) \times P(\text{temporal})]\}^{N \text{ events}}$$

Example 7.6
A house is sited downslope of a road constructed partly across uncompacted fill (Fell et al., 2005). Because of the loose, saturated nature of the fill, it is possible for rapid debris flows to reach the house. Four people live in the house: one is there 20 hours a day, seven days a week. The temporal probability for this person is

$P(\text{temporal}) = \text{time in house per year}/(24 \times 365)$

$= 7300/8760$

$= 0.83$

The other three people are present for 12 hours a day, 2 days a week, and the temporal probability for each individual is

$P(\text{temporal}) = \text{time in house per year}/(24 \times 365)$

$= (12 \times 2 \times 52)/(24 \times 365)$

$= 0.14$

Example 7.7
With reference to Example 7.1, the proportion of time that an individual walking at 2.5 km per hour spends on the 500 m-long section of the seaside promenade at Brighton, UK is

$P(\text{temporal}) = \dfrac{\text{length of promenade}}{\text{speed}(\text{m/hour})}$

$= 500/2500$

$= 0.2 \text{ hours} = 0.2/(365 \times 24)$ of a year

$= 0.0000228$ or 2.28×10^{-5}

The annual probability of the walker being hit by a rockfall (annual $P = 0.1$) is

$P(\text{hit}) = P(\text{landslide}) \times P(\text{spatial}) \times P(\text{temporal})$

$= 0.1 \times 0.002 \times 0.0000228$

$= 0.0000000046$ or 4.6×10^{-9}

If a promenade user interrupts the walk by sitting on the 3 m-wide bench admiring the view out to sea for an hour, the temporal probability of being hit by a rockfall when sat on the bench would be

$P(\text{temporal}) = \text{seating time}/(24 \times 365)$

$= 1/(24 \times 365) = 1/8760 = 0.000114$ or 1.14×10^{-4}

The chance of the seated person being hit by a rockfall ($P = 0.1$) while seated on the bench would be

$P(\text{hit}) = 0.1 \times 0.002 \times (1.14 \times 10^{-4})$

$= 0.0000000228$ or 2.28×10^{-8}

The overall probability of the person being hit whilst on the promenade needs to take account of the time spent both walking and sitting on the bench:

$$P(\text{hit}) = P(\text{walking hit}) + P(\text{seated hit})$$
$$= (4.6 \times 10^{-9}) + (2.28 \times 10^{-8})$$
$$= 2.73 \times 10^{-8}$$

Example 7.8

An old mine spoil tip lies on a hillside, immediately upslope of a road and an isolated farmhouse (Fell et al., 2005). There are concerns about the stability of the tip, which could fail and generate large flow slides that might reach the assets at risk. The house is occupied by four people for 10 hours a day, 325 days a year, with a temporal probability for each individual of

$$P(\text{temporal}) = (\text{hours in house}/24 \text{ hours}) \times (\text{days in house}/365 \text{ days})$$
$$= (10/24) \times (325/365)$$
$$= 0.37$$

Vehicles on the road are exposed to flow slides over a 100 m-long section. Given that the average vehicle speed through this section is 30 km/hour, then the temporal probability for a single journey is

$$P(\text{temporal}) = \frac{\text{length of section (m)}}{\text{speed (m/hour)} \times 365 \times 24}$$
$$= 100/(30\,000 \times 8760)$$
$$= 3.8 \times 10^{-7}$$

If the vehicle makes 250 journeys along this section of road throughout the year, then the temporal probability can be estimated as

$$P(\text{temporal, 250 journeys}) = 3.8 \times 10^{-7} \times 250 = 9.5 \times 10^{-5}$$

However, an alternative approach would be to treat each journey as a separate trial, and model the temporal probability using the binomial distribution (see Chapter 5):

$$P(\text{temporal, 250 journeys}) = 1 - [1 - P(\text{temporal})]^{N \text{ trials}}$$
$$= 1 - (1 - 3.8 \times 10^{-7})^{250}$$
$$= 9.5 \times 10^{-5}$$

Example 7.9

A 1350 m section of the Lawrence Hargrave Drive, south of Sydney, Australia, was constructed through steep cliffs that rise to a height of some 300 m above the road (Wilson et al., 2005). The road has had a history of severe rockfall and debris flow problems (Hendrickx et al., 2011). The exposure (i.e. temporal probability) of a single vehicle travelling through the landslide prone area at

60 km/hour would be

$$P(\text{temporal}) = \frac{\text{journey time through exposed area}}{24 \times 365 \text{ hours}}$$

$$= \frac{\text{distance/speed}}{8760}$$

$$= 0.0225/8760$$

$$= 0.00000256 \text{ or } 2.56 \times 10^{-6}$$

For a 5 m-long vehicle, the spatial probability is

$$P(\text{spatial}) = \text{length of car/length of section}$$

$$= 5/1350$$

$$= 0.003704$$

Given the occurrence of a landslide event ($P = 1$), the probability of an impact over the course of a year is

$$P(\text{hit}) = P(\text{landslide}) \times P(\text{spatial}) \times P(\text{temporal})$$

$$= 1 \times 0.003704 \times 0.00000256$$

$$= 9.513 \times 10^{-9}$$

This can also be calculated by combining the spatial and temporal components of the exposure:

$$P(\text{hit}) = 1 \times \frac{\text{number of vehicles per year} \times \text{vehicle length}}{24 \times 365 \times \text{vehicle speed (m/hour)}}$$

$$= 1 \times (1 \times 5)/(24 \times 365 \times 60\,000)$$

$$= 9.513 \times 10^{-9}$$

If the number of vehicles driving through the landslide area is 1000 per day (all of the same size and with the same journey time), then

$$P(\text{hit}) = P(\text{landslide}) \times P(\text{spatial, temporal})$$

$$= 1 \times (365 \times 1000 \times 5)/(24 \times 365 \times 60\,000)$$

$$= 0.00347$$

If there are, on average, 100 landslide events per year, then $P(\text{hit})$ can be estimated using the binomial model (see Chapter 5):

$$P(\text{hit}) = 1 - \{1 - [P(\text{spatial}) \times P(\text{temporal})]\}^{100}$$

$$= 1 - (1 - 0.00347)^{100}$$

$$= 0.29$$

Example 7.10

Construction of a two-lane road to a privately owned ski resort involved a series of 50 rock cuts that are prone to small rockfalls, ranging in size from 0.5–1 m diameter (Fell et al., 2005). The average rockfall frequency is 100 events per year (0.27/day), spread evenly between the cuttings (i.e. 2/year for each cutting). Rockfall simulation modelling indicates that 60% of rocks fall on the inside lane, with 10% onto the outside lane. The average frequency of falls onto the inside lane is 0.16/day (0.6 × 0.27) and 0.027/day for the outside lane (0.1 × 0.27).

On average, 2000 vehicles (average length 6 m) use the road each day, travelling at 60 km/hour. The daily probability of a vehicle occupying a section of the road where a rockfall occurs (i.e. the spatial, temporal probability) is

$$P(\text{spatial, temporal}) = \frac{\text{number of vehicles/day} \times \text{length of vehicle} \times \text{number of lanes affected}}{24 \times \text{speed (m/hour)}}$$

$$= \frac{2000 \times 6 \times 1}{24 \times 60\,000}$$

$$= 0.0083$$

For the inside lane the daily probability of a hit can be calculated using the binomial model:

$$P(\text{hit, inside lane}) = 1 - [1 - P(\text{spatial, temporal})]^{0.16}$$

$$= 1 - (1 - 0.0083)^{0.16}$$

$$= 0.0013$$

Example 7.11

In the previous examples, the scenario has involved a landslide impacting a non-stationary asset, and the exposure of multiple vehicles passing through a landslide-prone area was calculated by simply treating each journey as being a repeat of a single journey:

$$P(\text{hit}) = P(\text{landslide}) \times P(\text{spatial}) \times P(\text{temporal}) \times \text{number of vehicles/year}$$

However, there are instances where the interaction between vehicles in a stream of traffic is important in determining the landslide risk. For example, the two-lane Gardesana highway in northern Italy lies at the foot of a steep, 1000 m-high rock slope. The road serves an important tourist resort, Lago di Garda. It was closed in 2001, following a landslide that resulted in a fatality (Roberds, 2005).

A number of scenarios were considered to estimate the risk to the road users, including

- scenario 1 – a landslide directly impacting a vehicle
- scenario 2 – a follow-on accident, after a vehicle has been impacted by a landslide
- scenario 3 – a vehicle hitting landslide debris that has accumulated on the road.

The probability that a landslide will impact a non-stationary vehicle was defined as

$$P(\text{scenario 1}) = 1 - \frac{[V_S - (D_L + V_W) \times (V_V/D_V)] - D_W}{V_S + V_L}$$

where V_S is the average vehicle spacing, calculated as (vehicle speed/vehicle frequency) – vehicle length (for a vehicle frequency of 1000/hour travelling at an average speed of 50 km/hour, the average spacing is 40 m: (50 000/1000) − 10); V_V is the vehicle speed (50 km/hour); V_L is the vehicle length (10 m); V_W is the vehicle width (3 m); M_L is the volume of landslide debris on the road; D_L is the length (up slope) of landslide debris, calculated as $M_L^{0.33}$; D_W is the width of landslide debris along the road, calculated as $M_L^{0.33}$; and D_V is the landslide velocity (100 km/hour).

For a landslide volume of 10 m³, the probability of intersection with a passing vehicle is 0.33.

The probability of a follow-on accident (scenario 2) is primarily a function of the traffic characteristics. If the impacted vehicle stops and blocks the road, then the driver of the vehicle behind will try to stop. The vehicle behind will either hit the impacted vehicle or swerve off the road if the stopping distance is greater than the vehicle spacing.

Roberds (2005) suggests that the actual stopping distance can be expressed as a log-normal distribution with a specified mean (assumed to be the nominal stopping distance) and a coefficient of variation (the standard deviation/the mean). The probability of there being insufficient stopping distance (i.e. the probability of the crash) is

$$P(\text{scenario 2}) = 1 - \phi\{\ln(V_x) - \ln(mV_D)/\ln[\text{COV}(V_D) \times mV_D]\}$$

where V_x is the available stopping distance; mV_D is the nominal stopping distance; $\text{COV}(V_D)$ is the coefficient of variation of the stopping distance; and $\phi\{\ldots\}$ is the standard normal cumulative distribution (using Excel this is set up using the function NORMSDIST).

For a nominal stopping distance of 40 m (due to the speed, reaction time and vehicle breaking) with a COV = 0.1, and the available stopping distance of 60 m, then the probability of the crash is

$$P(\text{scenario 2}) = 1 - \phi\{\ln(V_x) - \ln(mV_D)/\ln[\text{COV}(V_D) \times mV_D]\}$$

$$= 1 - \phi\{0.292\}$$

$$= 1 - 0.615$$

$$= 0.385$$

The probability of a vehicle crashing into the landslide debris on the road (scenario 3) is similar to the previous case, except that the debris might not be in the driver's lane. $P(\text{lane})$ is the probability that the debris will be in the driver's lane. In this example, it is assumed to be 0.5 if the debris is <10 m³ and 1.0 if the debris is ≥10 m³. For an event <10 m³:

$$P(\text{scenario 3}) = P(\text{lane}) \times (1 - \phi\{\ln(V_x) - \ln(mV_D)/\ln[\text{COV}(V_D) \times mV_D]\})$$

$$= 0.5 \times (1 - \phi\{0.292\})$$

$$= 0.5 \times (1 - 0.615)$$

$$= 0.192$$

Further scenarios involving a variety of follow-on accidents are presented in Roberds (2005).

7.4. Occupancy and population models

There can be considerable uncertainty over the number of people present when a landslide hits a building or non-stationary assets such as a group of people in open space or a vehicle. This uncertainty is usually addressed by developing a model of how the population in the danger zone varies over time, between day and night or from season to season (a population or occupancy model). This is used to determine a hypothetical average value for the exposed population that represents the most likely number of people present when a random landslide event occurs. One approach for estimating the average daily exposure for a residential population is to calculate variations in the actual number of people within each different type of building at different times of the day using a population model, and then sum the values for the area as a whole. Although the actual numbers present when the event occurs may be lower or higher than the average value, these scenarios will be less likely.

Example 7.12

An apartment block in Hong Kong has a total population of 1000, but the actual population varies between 95% at night (some people are always away from home or on night shift) to around 25% in the middle of the afternoon (Table 7.1).

The average daily population at risk can be estimated as follows:

$$\begin{aligned}\text{Average population} &= \text{total population} \times [(6/24 \times 0.95) + (3/24 \times 0.75) \\ &\quad + (3/24 \times 0.5) + (3/24 \times 0.5) + (3/24 \times 0.25) \\ &\quad + (3/24 \times 0.75) + (3/24 \times 0.9)] \\ &= \text{total population} \times 0.69 \\ &= 690\end{aligned}$$

The daily variation in the number of people within the block is equivalent to a fixed population of 690 exposed for 100% of the time, which is 690 people with an exposure of 1. Alternatively, it can be expressed as 1000 people (maximum possible) with an exposure of 0.69. If this estimation is considered too crude because of the scale of the daily variation, then grouping the data results in four separate values:

1000 people × 0.9375 for 37.5% of the time (21 hours to 06 hours)

1000 people × 0.75 for 25% of the time (06–09 hours and 18–21 hours)

Table 7.1 Example 7.12: apartment block – population variation

Time of day	Residential population: % of total population
00–06 hours	0.95
06–09 hours	0.75
09–12 hours	0.5
12–15 hours	0.5
15–18 hours	0.25
18–21 hours	0.75
21–24 hours	0.90

1000 people × 0.5 for 25% of the time (09 hours to 15 hours)

1000 people × 0.25 for 12.5% of the time (15–18 hours)

The statistics can be of value in terms of developing 'least impact' and 'worst case' scenarios.

Example 7.13

Climate change and relative sea-level rise over the next century are expected to result in an increase in the landslide risk within the Isle of Wight Undercliff, UK (see Example 6.13). This concern led to a re-evaluation of the potential benefits of major landslide stabilisation works in the Ventnor area. A risk assessment was undertaken to determine whether this is likely to be a cost-effective approach to managing the landslide risks (Lee and Moore, 2007). This involved the assessment of both economic losses and societal risk.

In order to determine the societal risk, a population model was developed to determine the proportion of the overall population that was exposed to landsliding for 100% of the time. The estimate of the exposed daily population considered three separate elements.

1. *Residential population* – it was assumed that the average occupancy for each of the 332 properties within the landslide area under consideration was ten, reflecting the numbers of flats, apartments and hotels, yielding an overall population of 3320. This is not, however, a permanent population, and has to be adjusted to take account of variable occupancy during the night and day (each taken as 12 hours' duration). The night-time and daytime occupancy factors were assumed to be 1 and 0.5, respectively:

    average population = (3320 × night occupancy × 50% of the time)

    + (3320 × day occupancy × 50% of the time)

    This gives an overall exposed daily population of 3320 × 0.75 = 2490.

2. *Tourism* – it is assumed that the influx of tourists increases the residential population by 100% during the 6 months from May to October (50% of the year), inclusive. The exposed residential population, therefore, was calculated as

    average population (winter) = 2490 × 1 = 2490

    average population (summer) = 2490 × 1.5 = 3735

    average population = (average population (winter) × 0.5)

    + (average population (summer) × 0.5)

    = 1245 + 1876.5

    = 3112.5

3. *Traffic* – figures provided by the local council indicate a traffic flow into the town of around 4700 vehicles per day. Assuming an average vehicle occupancy of 1.71, this yields a transient traffic population of 8037 per day. The exposure of this population is limited to the time spent driving through the landslide as part of a journey. Assuming average journey speeds of 10 km/hour through the town and a journey length over the landslide of 1.5 km give an

exposure time of 0.15 hour/vehicle (equivalent to 0.006 day/vehicle). The exposure of a single occupant is this figure multiplied by the average vehicle occupancy, giving a daily exposure factor of 0.0107 for each of the 8037 travellers. The traffic flow is, therefore, the equivalent of about 86 people (0.0107×8037) exposed to the landslide for 100% of the time.

The total equivalent population exposed for 100% of the time is

exposed population = exposed residential and tourist population

+ exposed traffic exposed population

$= 3198.5$

7.5. Reducing exposure through landslide detection

In many instances, it will be possible to detect landslide movement, and move people out of harm's way, significantly reducing the spatial/temporal probability. Landslide monitoring programmes can be developed to collect, record and analyse landslide movements and provide the technical framework for an early warning system. These systems mitigate risk by reducing the exposure. The systems are designed to issue alerts or warnings early enough to give sufficient lead time to implement actions to protect persons and/or property.

An example is the Hong Kong Landslip Warning System, operated jointly by the Geotechnical Engineering Office (GEO) and the Hong Kong Observatory to alert the public to landslide danger during periods of heavy rainfall (GEO, 2009). This system is based on rainfall data from 110 automatic weather stations supported by radar and satellite images to monitor changes and movement of storm cells. When defined thresholds are exceeded, the public is alerted by national and local media, such as radio and television. Pedestrians are advised to avoid walking or standing close to steep slopes, and motorists to avoid driving in hilly areas or on mountain roads, especially at locations where signs indicating dangerous hillsides or retaining walls have been erected. When the landslide threat is viewed as becoming serious, the public is advised to stay in a safe shelter or at home (Cheung et al., 2006).

In other situations, it may simply be that someone in the path of a relatively slow-moving landslide might see the debris approaching and run out of the way. As Roberds (2005) notes, they must be aware of the landslide threat, detect the movement and then be mobile enough to get away in time.

The uncertainty about whether a landslide movement will be detected and avoided can be addressed in a variety of ways. Roberds (2005) suggested that the basic risk model can be expanded to include additional factors that reflect the probability that the landslide is detected and/or the person is mobile enough to get away if movement is detected:

$P(\text{hit}) = P(\text{landslide}) \times P(\text{spatial}|\text{landslide})$

$\times P(\text{temporal}|\text{landslide}) \times P(\text{detected}|\text{landslide}) \times P(\text{avoided}|\text{detected})$

These factors will vary between 0 and 1.0, depending on how far away the landslide was detected, the speed of movement of the landslide, and the mobility of the person. With regard to the general warning issued by an individual, the risk model is extended further to include the effectiveness of the warning system and the subsequent response.

An alternative approach would be to include the detection and avoidance conditions as nodes within an event tree (see Chapter 5), or to develop a series of population models that reflect different detection scenarios (detected and complete avoidance, detection and partial evacuation, etc.).

Example 7.14

The Lei Yue Mun squatter villages in Hong Kong are located beneath abandoned 20–40 m-high quarry faces. A number of landslides occurred in a major rainstorm in August 1995, resulting in damage to the squatter houses. Fortunately, loss of life was narrowly avoided. Following this incident, a risk assessment was undertaken to assist in decision-making with regard to rehousing the squatters (Ho et al., 2000).

The squatter villages were divided into a series of 20 m × 20 m blocks. Site surveys were undertaken to develop a population model for each block. Event trees were generated for each of the reference blocks, which traced the different credible scenarios, considering the landslide type, the timing of failure, the responses to the GEO landslip warning (see above), the effectiveness of emergency response and the nature of secondary hazards. An extract from one of the event trees (a daytime minor debris slide) is reproduced in Figure 7.2. In this example, it was assumed that the probability of the warning not

Figure 7.2 Example 7.14: sample of the event tree developed for the Lei Yue Mun squatter camp. (After Ho et al., 2000)

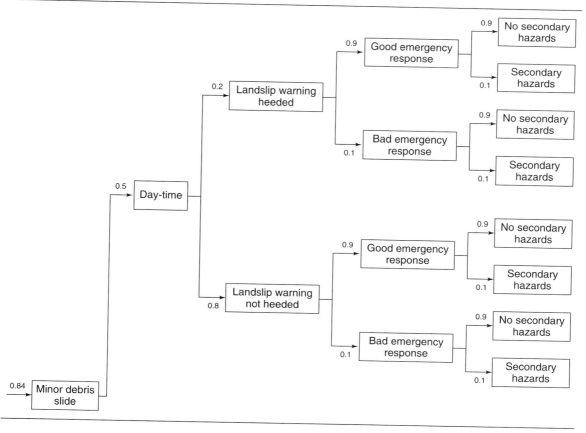

being heeded was 0.8, followed by a 0.9 probability of a good emergency response. The model predicts no fatalities, but a combination of minor and major injuries.

REFERENCES

Australian Geomechanics Society (AGS) (2000) Landslide risk management concepts and guidelines. *Australian Geomechanics* **35**: 49–52.

AGS (2007) Practice note guidelines for landslide risk management. *Australian Geomechanics* **42(1)**: 63–114.

Cheung SPY, Wong MC and Yeung LHY (2006) Application of rainstorm Nowcast to real-time warning of landslide hazards in Hong Kong. *WMO PWS Workshop on Warnings of Real-Time Hazards by Using Nowcasting Technology*, Sydney.

Dimmock P, Mackenzie B and Mills AJ (2012) Probabilistic slope stability analysis in the west Nile delta, offshore Egypt. Offshore Site Investigation and Geotechnics Integrated Geotechnologies – Present and Future. *Proceedings of the 7th International Conference 12–14 September 2012*. Royal Geographical Society, London, UK.

Fell R, Ho KKS, Lacasse S and Leroi E (2005) A framework for landslide risk assessment and management. In *Landslide Risk Management* (Hungr O, Fell R, Couture R, and Eberhardt E (eds)). Balkema, Rotterdam, pp. 3–26.

Geotechnical Engineering Office (GEO) (2009) *Landslip Warning System*. Standards and Testing Division, GEO, Hong Kong. Information Note 02/2009.

Hendrickx M, Wilson RA, Moon AT, Stewart I and Flentje P (2011) Slope hazard assessment on a coast road in New South Wales, Australia. *Australian Geomechanics Journal* **46(2)**: 95–108.

Ho K, Leroi E and Roberds B (2000) Quantitative risk assessment: application, myths and future direction. *Proceedings of the Geo-Eng Conference*, Melbourne, Publication 1, pp. 269–312.

Jones DKC, Lee EM, Hearn G and Genc S (1989) The Catak landslide disaster, Trabzon Province, Turkey. *Terra Nova* **1**: 84–90.

Lee EM and Moore R (2007) Ventnor Undercliff: development of landslide scenarios and quantitative risk assessment. In *Landslides and Climate Change: Challenges and Solutions* (McInnes R, Jakeways J, Fairbank H and Mathie E (eds)). Balkema, Rotterdam, pp. 323–334.

Moore R, Usher N and Evans T (2007) Integrated multidisciplinary assessment and mitigation of West Nile Delta geohazards. *Proceedings of the 6th International Offshore Site Investigation and Geotechnics Conference: Confronting New Challenges and Sharing Knowledge*.

Morgan GC, Rawlings GE and Sobkowicz JC (1992) Evaluating total risk to communities from large debris flows. *Geotechnique and Natural Hazards: A Symposium Sponsored by the Vancouver Geotechnical Society and the Canadian Geotechnical Society*. BiTech Publishers, Vancouver, pp. 225–236.

Roberds W (2005) Estimating temporal and spatial variability and vulnerability. In *Landslide Risk Management* (Hungr O, Fell R, Couture R and Eberhardt E (eds)). Balkema, Rotterdam, pp. 129–158.

Wilson RA, Moon AT, Hendrickx M and Stewart IE (2005) Application of quantitative risk assessment to the Lawrence Hargrave Drive Project, New South Wales, Australia. In *Landslide Risk Management* (Hungr O, Fell R, Couture R and Eberhardt E (eds)). Balkema, Rotterdam, pp. 589–598.

Chapter 8
Vulnerability

8.1. Introduction: multiple meanings of vulnerability

Will a landslide of a specific size actually damage something of value? Vulnerability, simply stated, is the 'potential for something to suffer harm from a human perspective' (see Chapter 1), and forms a crucially important component of consequence assessment, by providing a measure of the propensity of the various elements at risk to suffer harm when a landslide event occurs (e.g. Alexander, 1989, 2005; Douglas, 2007). People's lives and health can be directly affected by the destructive effects of a landslide. Their incomes and livelihood can be adversely affected because of the destruction of houses, factories, crops, livestock, equipment, artefacts and the infrastructure that they depend on. Major events can even have severe effects on sizeable private companies due to adverse impacts on their operations, assets and share price, while national governments of small countries can suffer the economic consequences of reduced gross domestic product.

The concept of vulnerability, as developed within the social sciences, and is of crucial importance for two reasons.

1. Its very existence imparts the attribute of 'hazard' onto physical phenomena (i.e. landslides), due to the fact that 'hazard only exists because of the presence of vulnerability'.
2. It provides the means of linking the assessed levels of hazard and exposure, thereby enabling the determination of levels of loss and the generation of quantitative assessments of risk.

As it is considered to be a condition that exists independent of landslide hazard, vulnerability addresses the scale of adverse impact that can be expected to be produced by a landslide of a particular magnitude. This can be expressed in a variety of ways, including

- the probability that a person hit by the landslide will be killed or injured
- the number of people within an exposed group or population that are likely to be killed or injured by a landslide event of a particular magnitude
- the probability that a building or structure, such as a road, railway or pipeline, will suffer a particular level of damage (destruction, severe damage, partial damage, superficial damage, etc.)
- the proportion of the building stock within the threatened area that is likely to be destroyed or damaged by a landslide.

There is an extensive literature on vulnerability, although relatively little is specifically focused on landsliding. Three main categories have been identified.

- Physical – the vulnerability of buildings, structures, infrastructure, etc.
- Societal – the vulnerability of socio-politico-economic systems, including industrial production, agricultural production, assets such as housing and other forms of revenue generation.
- Social – vulnerability of livelihoods, community resilience and coping mechanisms.

Research has tended to focus on physical or technical vulnerability, involving modelling the structural capacity of buildings or structures to withstand geohazard impacts (e.g. earthquake engineering, earthquake-proofing, etc.). However, over the last two decades or so, the mounting disaster casualty figures in parts of the developing world has resulted in much work on 'social' or 'community' vulnerability (e.g. Anderson, 2000; Cannon, 2000; Cutter, 1996; Cutter et al., 2003; IFRC, 2008; UN, 2011; Wisner et al., 2004; World Bank, 2010). The result has been a bewildering multiplicity of meanings and uses of the term 'vulnerability'; indeed, Cutter (1996) revealed the existence of 18 definitions in the literature. Wisner et al. (2004) react by going so far as to restrict use of the term 'vulnerability' to people and 'not to buildings (susceptible, unsafe), economies (fragile), nor unstable slopes (hazardous) or regions of the earth's surface (hazard-prone)'. To them, the proper definition of vulnerability is 'the characteristics of a person or group and their situation that influence their capacity to anticipate, cope with, resist and recover from the impact of a natural hazard'.

As the primary requirement of landslide risk assessment is to establish the likely general nature of impacts resulting from a landslide of a particular magnitude, it is perfectly acceptable to take the broader view that uses 'vulnerability' to describe levels of impact sustained by a wide range of objects, artefacts and activities. This is consistent with the UN Office for Disaster Risk Reduction (2009) view that vulnerability is 'the characteristics and circumstances of a community, system or asset that makes it susceptible to the damaging effect of a hazard'.

It is important, therefore, not to confine consideration to buildings and infrastructure, but to address how individuals and groups within society are affected by and respond to the impact of hazards such as landsliding. This is a combination of societal and social (community?) vulnerability, and involves defining the coping capacity of the society, and of the individuals within the society, both during and following a landslide impact. The main focus is on the underlying structural factors that reduce the capacity of human societies to cope with a range of hazards (well portrayed in the *pressure and release model* of Wisner et al. (2004)), including both sensitivity and adaptive capacity (e.g. Adger, 2006).

Resilience is somewhat the antithesis of vulnerability, and therefore helps to defines it. According to Timmerman (1981), 'vulnerability is the degree to which a system, or a part of a system, may react adversely to the occurrence of a hazardous event. The degree and quality of that adverse reaction are partly conditioned by the system's resilience, the measure of a system's, or part of a system's, capacity to absorb and recover from the occurrence of a hazardous event'.

Resilience is very important, and has been well redefined by the UN Office for Disaster Risk Reduction (2009) as 'the ability of a system, community or society exposed to hazards to resist, absorb, accommodate to and recover from the effects of a hazard in a timely and efficient manner'. It is now recognised that an appreciation of the interaction between social/societal vulnerability and resilience is central to the understanding as to why a hazard impact can become a disaster (Buckle, 2006; Tapsell et al., 2010). However, predicting, assessing and quantifying these aspects as part of a landslide risk assessment is hugely problematic, and is discouraged, except at the most general level (see Section 8.5 on social and societal vulnerability).

Landsliding clearly has the potential to cause environmental damage (e.g. land degradation and loss of soil resources) and ecosystem change (e.g. loss/gain of habitat and species). *Environmental vulnerability* is used to describe the extent to which environmental quality and amenity can be adversely affected by landsliding. Measuring such impacts in terms of loss of amenity is extremely difficult, and is usually done qualitatively, although economic valuation techniques can be used (see Bateman *et al.*, 2002; DTLR, 2002; Pearce *et al.*, 2006).

Ecological vulnerability (or fragility) is the extent to which fauna and flora can be adversely affected by a change in environmental conditions brought on slowly, as in the case of climate change, or more suddenly, as in the case of a pollution episode or impact of a geohazard such as a landslide. When dealing with non-human organisms in the environment, it is usual to refer to disturbing influences (stressors) acting on spatially identifiable organisational units (ecosystems).

8.2. Physical vulnerability

The International Society of Soil Mechanics and Ground Engineering (ISSMGE) defines vulnerability as 'the degree of loss to a given element or set of elements within the areas affected by the hazard' (ISSMGE, 2004). From this perspective, the concept of vulnerability addresses the physical impact of the landslide event on an asset at risk (i.e. physical vulnerability).

The vulnerability (V) of property can be envisaged as the level of potential damage, or degree of loss, of a particular asset (expressed on a scale of 0 to 1) subjected to a landslide impact or 'hit' of a given intensity (Fell, 1994). It can be represented by simple conditional probability models:

$$\text{consequences (property)} = P(\text{damage}|\text{hit}) \times \text{value of elements at risk}$$

The vulnerability of people is usually expressed as the probability (between 0 and 1) that a person is killed if they are 'hit' by a landslide of a given intensity:

$$\text{consequences (people)} = P(\text{fatality}|\text{hit})$$

Vulnerability to landsliding is dependent on the nature and intensity of the mechanical stresses generated by the actual failure (differential ground movement, subsidence, heave, loading, etc.) and the capacity of the 'elements at risk' to withstand an impact of a given degree of severity. In practice, the concept of physical vulnerability has proved difficult to apply (e.g. van Westen *et al.*, 2006). This is because very little is known about the actual vulnerability of structures or people to landslides of different types and intensity (e.g. Amatruda *et al.*, 2004).

For each element at risk, the vulnerability level varies depending on the types and magnitudes of the different types of landslides that it is exposed to (e.g. Zêzere *et al.*, 2007). Fragility curves that relate a measure of landslide intensity to damage are rarely available (e.g. Glade, 2003; Roberds, 2005). 'Crash-test dummies' have not been subjected to repeated landslide impacts to establish the probability of a fatal injury for events of different sizes, hardness and strength, impact angles, etc. Although the interaction between rockfalls and structures is well documented (e.g. Volkwein *et al.*, 2011), there have been no systematic trials to determine, for example, the level of building damage associated with landslide events of differing magnitude or velocity. Available databases of landslide damage are often incomplete or biased. Only events that have caused substantial damages tend to be recorded, and precise information on the type, characteristics and damages due to landsliding are often absent (e.g. ENSURE, 2009).

Unfortunately, there is no presently available single methodology for defining the vulnerability of elements at risk to different types and intensities of landslides or to individual/combinations of landslide processes (Glade and Crozier 2005). As a result, the vulnerability values adopted in virtually all consequence assessments are generally subjective, based on expert judgement (e.g. Dai *et al.*, 2002).

To compound the problem, these judgements tend to be made by landslide specialists, rather than structural engineers or medical doctors experienced in landslide impact injuries (contrast the situation in the case of earthquakes). Many assessments draw upon pioneering vulnerability assessments undertaken in places such as Hong Kong (e.g. Finlay et al., 1999), Australia (e.g. Fell and Hartford, 1997) and Switzerland (e.g. Romang et al., 2003). However, the findings of these studies (analogues) are not particularly transferable because of the variations that exist between countries regarding the dominant landslide types and velocities, and the durability/robustness of the elements at risk (e.g. ENSURE, 2009).

8.3. Physical vulnerability of buildings and infrastructure

The main controls that determine the susceptibility of buildings and infrastructure to damage include the quality of the building materials used, structural design, workmanship and maintenance history (e.g. Amatruda et al., 2004; Papathoma-Köhle et al., 2007). For example, a post-event survey after a major landslide in Italy revealed that 60% of the destroyed buildings were made of stone and 40% of masonry; none of them were constructed of concrete (EPFL, 2002).

Alexander (2002) recognised the following significant factors for buildings.

- Construction type – load-bearing walls (mudbrick, random rubble, dressed stone, bonded brick, etc.).
- Nature of vertical and horizontal load-bearing members (e.g. brick wall, steel beam).
- Size of building, number of floors, number of wings, square metres of space occupied or cubic metres of capacity.
- Regularity of plan and elevation.
- Degree to which construction materials and methods are mixed in the building, or, alternatively, whether there is a single building technique and set of materials.
- Age category of the building – this can be generalised according to the dominant building material of the period (usually this involves categories such as pre-1900, 1900–1940s, 1950s–1965, 1966 onwards).
- The building's state of maintenance (excellent, good, mediocre, bad).

Damage to structures and infrastructures can be classified as

- *superficial* (cosmetic or minor damage), where the functionality of buildings and infrastructure is not compromised, and the damage can be repaired quickly and at low cost
- *functional* or *operational* (serious), where the functionality of structures or infrastructure is compromised, and the repairs take time and significant expenditure
- *structural* or *severe to total*, where buildings or transportation routes are severely damaged or destroyed, necessitating extensive and extremely costly demolition and reconstruction.

Examples of structural damage classification schemes for buildings are presented as Tables 8.1 and 8.2.

Estimates of vulnerability can be based on the inferred relationship between the intensity and type of the expected landslide (see Chapter 3), and the likely damage the landslide would cause (see Table 8.2). Factors that influence the impact of a landslide on buildings include (Australian Geomechanics Society (AGS), 2000, 2007)

- the volume and velocity of the slide
- the position of the structure relative to the slide
- the magnitude of displacements, and relative displacements within the landslide.

Table 8.1 Structural damage classification

Grade	Description
None	The building has sustained no significant damage
Slight	There is non-structural damage, but not to the extent that the cost of repairing it will represent a significant proportion of the building's value
Moderate	There is significant non-structural damage, and light-to-medium structural damage. The stability and functionality of the building are not compromised, although it may need to be evacuated to facilitate repairs. Buttressing and ties can be used for short-term stability
Serious	The building has sustained major non-structural damage and very significant structural damage. Evacuation is warranted in the interests of personal safety, but repair is possible, although it may be costly and complex
Very serious	The building has sustained major structural damage and is unsafe for all forms of use. It must be evacuated immediately and either demolished or substantially buttressed to prevent it from collapsing
Partial collapse	Portions of the building have fallen down. Usually these will be cornices, angles, parts of the roof or suspended structures such as staircases. Reconstruction will be expensive and technically demanding; demolition of the remaining parts will be the preferred option
Total collapse	The site will have to be cleared of rubble. A few very important buildings may need to be reconstructed (usually for cultural reasons) even though they have collapsed totally, but most will not be rebuilt

From Alexander (2002)

Table 8.3 presents a simple classification, developed by Cardinali *et al.* (2002), of the expected damage to buildings and roads by landslides of different types (rockfall, debris flow or slide) and intensities (slight, medium, high or very high). However, such intensity scales are of limited use unless properly adjusted to local conditions in terms of the range of building types present, local building materials, indigenous building practices, the quality of workmanship, building styles, the nature of building codes and the level of their enforcement, and economic conditions in the area under consideration.

Using the three-grade division described earlier (Alexander, 2002), it is possible to develop a simple vulnerability index (on a scale of 0 to 1) by estimating the relative proportion of the overall value of assets that would be lost in an event that causes each of these three grades of damage (superficial, functional and structural). For example, the InterRisk Assess project in the Swabian Alb, south of Tubingen and Reutlingen, Germany, established five levels of vulnerability associated with landslide damage to buildings (Table 8.4; Blöchl and Braun, 2005). For example, if a building is predicted to have cracks in the walls as a result of landslide movement, then it is assumed that the loss will be the equivalent of 20–30% of the asset value. In this context, the concept of vulnerability is used to provide a link between the intensity of building damage and the resulting losses:

loss = (proportion of building value|damage intensity) × building value

Table 8.2 Landslide damage intensity scale

Grade	Description of damage
0	*None*: building is intact
1	*Negligible*: hairline cracks in walls or structural members; no distortion of structure or detachment of external architectural details
2	*Light*: building continues to be habitable; repair not urgent. Settlement of foundations, distortion of structure and inclination of walls are not sufficient to compromise overall stability
3	*Moderate*: walls out of perpendicular by 1–2°, or substantial cracking has occurred to structural members, or foundations have settled during differential subsidence of at least 15 cm; building requires evacuation and rapid attention to ensure its continued life
4	*Serious*: walls out of perpendicular by several degrees; open cracks in walls; fracture of structural members; fragmentation of masonry; differential settlement of at least 25 cm compromises foundations; floors may be inclined by 1–2°, or ruined by soil heave; internal partition walls will need to be replaced; door and window frames too distorted to use; occupants must be evacuated and major repairs carried out
5	*Very serious*: walls out of plumb by 5–6°, structure grossly distorted and differential settlement will have seriously cracked floors and walls or caused major rotation or slewing of the building (wooden buildings may have detached completely from their foundations). Partition walls and brick infill will have at least partly collapsed; outhouses, porches and patios may have been damaged more seriously than the principal structure itself. Occupants will need to be rehoused on a long-term basis, and rehabilitation of the building itself will not be feasible
6	*Partial collapse*: requires immediate evacuation of the occupants and cordoning off the site to prevent accidents with falling masonry
7	*Total collapse*: requires clearance of the site

After Alexander (1989)

In recent years, efforts have been made to improve the way in which vulnerability is estimated and accommodated within landslide risk assessment (e.g. Duzgun and Lacasse, 2005; Uzielli *et al.*, 2008). Li *et al.* (2010), for example, have described physical vulnerability in terms of the landslide intensity (see Chapter 3) and the resistance of the element at risk:

$$\text{vulnerability} = \text{landslide intensity}/\text{resistance}$$

In this context, resistance (R) reflects the ability of exposed elements to withstand damage when impacted by a landslide event of a particular intensity (I). The resistance is considered to be a function of the foundation depth, structure type, maintenance state and building height. The intensity is defined in terms of

- the dynamic intensity of the landslide, related to the landslide velocity
- the geometric intensity, which is defined in terms of deformation within the body of the landslide or deposition depth downslope of the landslide.

Table 8.3 Vulnerability of a range of assets to landslide events of different intensity

Landslide intensity		Buildings	Main roads	Secondary roads	Minor roads	Buried pipelines	Railway lines
Light	Rockfall	C[a]	C	C	C	C	C
	Debris flow	C	C	F	F	C	C
	Slide	C	C	F	S	F–S	C
Medium	Rockfall	F	F	F	F	C	F
	Debris flow	F	F	F	F	C	F
	Slide	F	F	S	S	S	F
High	Rockfall	S	S	S	S	C	S
	Debris flow	S	S	S	S	C	S
	Slide	S	S	S	S	S	S
Very high	Rockfall	S	S	S	S	C	S
	Debris flow	S	S	S	S	C	S
	Slide	S	S	S	S	S	S

Based on Cardinali et al. (2002)
[a] C, superficial/cosmetic damage; F, functional damage; S, structural damage

Intensity and resistance are defined such that when the intensity is equal to half of the resistance of an element, the expected value of vulnerability (V) would be 0.5:

when $I/R < 0.5$: $\quad V = 2I^2/R^2$

when $I/R = 0.5\text{--}1$: $\quad V = 1 - \dfrac{2(R-I)^2}{R^2}$

when $I/R > 1$: $\quad V = 1$

Table 8.4 Vulnerability of buildings according to the type of damage through landslides

Damage intensity	Proportion of building value (vulnerability: 0–1)
Slight non-structural damage, stability not affected, furnishings or fittings damaged	0.01–0.1
Cracks in the walls, stability not affected, reparation not urgent	0.2–0.3
Strong deformation, huge holes in walls, cracks in supporting structures, stability affected, doors and windows unusable, evacuation necessary	0.4–0.6
Structural breaks, partly destructed, evacuation necessary, reconstruction of destroyed parts	0.7–0.8
Partly or totally destroyed, evacuation necessary, complete reconstruction	0.9–1

From Blöchl and Braun (2005)

Figure 8.1 Theoretical relationships between landslide intensity (*I*), resistance (*R*) and vulnerability (*V*). (From Li et al., 2010)

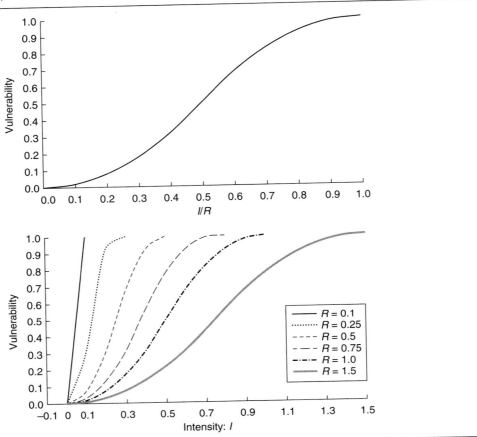

The resulting relationship between I/R and vulnerability V is shown in Figure 8.1. The same intensity may result in different levels of damage to the exposed elements, depending on the resistance: low-resistance buildings may be destroyed, whereas others may only be slightly damaged. This type of model represents the current research frontier. Further calibration against landslide damage data would be required before similar models become a reliable alternative to the expert-judgement-based methods that are currently used in most assessments.

Example 8.1

In Iceland, two catastrophic snow avalanches killed 34 people in 1995. As a consequence, in 2000 the Icelandic Ministry of the Environment issued a new regulation on hazard zoning with respect to snow avalanches and landslides (Regulation No. 505/2000), which aims to prevent people living or working within the areas most at risk. Bell and Glade (2004) describe the development of a risk assessment for Bildudalur village in western Iceland, where the processes posing a threat to the population are mainly debris flows, rockfalls, snow avalanches, slush flows and floods.

Table 8.5 Example 8.1: Bildudalur village risk assessment – vulnerability values for buildings and roads

Event	Low-magnitude event		Medium-magnitude event		High-magnitude event	
	Buildings	Roads	Buildings	Roads	Buildings	Roads
Debris flow	0.1	0.2	0.2	0.4	0.5	0.6
Rockfall	0.1	0.1	0.3	0.2	0.5	0.4

From Bell and Glade (2004)

Most of the houses in the village are fairly weak timber or concrete constructions, with relatively large windows built facing towards the mountainside. Vulnerability values were developed for debris flows and rockfalls of different magnitude and return period (Table 8.5). These values were estimated by expert judgement, taking account of buildings styles and experience of snow avalanches. In this example, vulnerability is defined as the proportion of the asset value that is lost if a building or road is hit by a landslide of a particular magnitude:

loss = asset value × (proportion of asset value|landslide magnitude)

Four classes of asset value were defined: very low (0–36€/m: 16 buildings and the power line), low (>36–480€/m: roads, infrastructure and buildings), medium (>480–960€/m), high (>960–1440€/m) and very high (>1440€/m).

For example, if a very high-value building (100 m^2; value = 144 000€) is hit by a high-magnitude debris flow, then the losses would be

144 000 × 0.5 = 72 000€

Example 8.2

In Umbria, Italy, landslides range from rockfalls and rapidly moving debris flows in mountain areas to slow-moving complex failures in the hilly parts of the region (e.g. Guzzetti et al., 2003). A catalogue of historical landslides covering the period 1900–2001 lists 169 sites where buildings and other structures have been damaged by landslides, together with 514 sites where roads and railways were damaged (Salvati et al., 2006). In the catalogue, landslide damage was classified qualitatively as 'light', where damage was aesthetic; 'severe', where the functionality of the building or transport route was compromised; and 'total', where a building was destroyed or the use of a road or railway was interrupted (Galli and Guzzetti, 2007). Along the transport network, about 34% of the damage was classified as light, 62% as severe and 4% as total. For the built-up areas, 26% of the damage was classified as light, 63% as severe and 11% as total.

Galli and Guzzetti, (2007) analysed this historical database to determine the vulnerability of buildings and roads to slow moving landslides. They recognised that the proportion of direct damage depends largely on the area of the landslide. However, there is not a linear relationship between landslide area and damage. They analysed the damage information for 103 landslides that occurred between January 1982 and December 2005. For each landslide, the proportion of the damage caused to each element at risk was ranked on a scale from 0 (no damage) to 1 (complete loss). Vulnerability was

Figure 8.2 Example 8.2: Umbria, Italy – proportion of landslide damage to buildings as a function of landslide area (the grey zone highlights the range of damage values between the minimum and maximum vulnerability thresholds). (After Galli and Guzzetti, 2007)

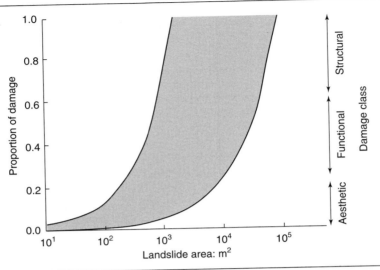

defined as the proportion of the asset value that is lost if the building or road is hit by a landslide of a particular size (i.e. area):

loss = (proportion of asset value|landslide area) × asset value

Landslides affecting buildings ranged in size from 258 m² to 165 237 m². The smallest landslide (a soil slide 13 m wide by 20 m long) produced the least ascertained damage ($V = 0.05$), and the two largest landslides (the Monteverde landslide of 22 December 1982, area = 165 237 m², and the Valderchia landslide of 6 January 1997, area = 73 222 m²) produced the most severe damage ($V = 1.0$).

Plots of landslide area (A) against damage revealed a considerable scatter of data (Figure 8.2). However, the plots do indicate minimum and maximum vulnerability thresholds for the different elements at risk. These thresholds were described using power law functions (Table 8.6; see Chapter 5). The results were used to develop a landslide vulnerability map of the Collazzone area in central Umbria. This involved linking landslide area to the established vulnerability thresholds for each of the 2490 landslides in the landslide inventory (recorded between 1937 and 2004) and mapping the

Table 8.6 Example 8.2: Umbria, Italy – minimum and maximum vulnerability values for buildings and major roads

	Minimum vulnerability	Maximum vulnerability	Range of landslide area
Buildings	$V = 0.0006A^{0.62}$	$V = 0.0045A^{0.7}$	250–20 000 m²
Major roads	$V = 0.0013A^{0.62}$	$V = 0.0050A^{0.85}$	50–40 000 m²

From Galli and Guzzetti (2007)

distribution of the minimum and the maximum expected vulnerability to landslides. Given the nature of landslide events in the Collazone area, it was found that all the slope failures affecting buildings are expected to cause impact in the range severe structural damage to total destruction ($V > 0.8$).

Example 8.3
In Castries, St Lucia, low-quality housing has been built on steep hillslopes prone to shallow debris slides and debris flows. Many of the buildings are wooden or concrete block constructions on concrete piles or cut and fill terraces. As part of a study of the benefits of local-scale landslide mitigation measures, Holcombe et al., (2012) developed a vulnerability model to take account of the variety of landslide hazards that these hillslope communities are exposed to. Vulnerability was defined as the proportion of the asset value lost in a landslide of a particular magnitude/intensity:

loss = asset value × (proportion of asset value|landslide scenario)

Four vulnerability scenarios were developed to represent the range of damage that could arise to the housing in different parts of a shallow landslide.

- Undercutting of <20% of the house at the landslide margins (crest and side-slopes) or within landslide body, where the shear surface is deeper than the foundations, leading to minor structural damage, but not loss of possessions (vulnerability = 0.5).
- Undercutting of 20–100% of the house at the landslide margins (crest and side-slopes) or within landslide body, where the shear surface is deeper than the foundations, leading to structural damage, but not loss of possessions (vulnerability = 1.0).
- Deposition of run-out material at a depth less than half the height of the house, leading to minor structural damage and loss of possessions due to flooding of property (vulnerability = 0.5).
- Deposition of run-out material at a depth greater than half the height of the house, leading to structural damage and loss of possessions due to collapse of the building and/or flooding (vulnerability = 1.0).

8.4. Human vulnerability

The physical juxtaposition of people and a landslide does not necessarily imply that there will be human casualties. The behaviour of the landslide is important in terms of its unexpectedness, suddenness, speed, ground disturbance characteristics and violence. In some circumstances, the landslide may be the main cause of human losses, but often it is the result of secondary events such as the collapse of buildings, transport accidents, fires and local flooding. How people are likely to be affected is also important in determining deaths and injuries. Are people going to be hit by missiles, engulfed by torrents of debris, buried alive or trapped within collapsed or partially collapsed structures? Will they be crushed, battered, suffocated, burned, drowned or suffer hypothermia, and will the effects of shock and resulting depression also take their toll?

The AGS (2000, 2007a) has identified a number of factors that determine *human vulnerability* in the context of landsliding and rockfalls.

- Volume of the slide or fall.
- Type of slide, mechanism of slide initiation and velocity of sliding.
- Depth of the slide.
- Whether the debris buries the person(s).

Table 8.7 Vulnerability (probability of a fatality) factors for open hillside landslides in Hong Kong

Landslide volume: m^3	Vulnerability (indoor population)	Vulnerability (outdoor population)
50	0.0002	0.03
100	0.006	0.054
500	0.011	0.078
1000	0.026	0.11
2500	0.04	0.15
5000	0.17	0.48

From DNV Technica (1996) and Halcrow Asia Partnership (1999)

- Whether the person(s) is in the open or enclosed in a vehicle or building.
- Whether the vehicle or building collapses when impacted by debris.
- The type of collapse if the vehicle or building collapses.

While these are fine for risk assessment purposes, the assumption made is that all humans are equally vulnerable, which is untrue. Age, gender and disability appear significant factors. The very young and the very old are often most at risk, especially if rescue and medical aid is delayed, and old people frequently fail to live long after their lives and homes have been severely impacted by a hazard event. Work on earthquakes has revealed that survivors aged over 60 and females are most likely to suffer severe physical injuries, and that females suffer most from post-event psychiatric problems (Chen et al., 2001; Peek-Asa et al., 2002). Disability hinders both evacuation and rescue.

Table 8.7 presents the vulnerability factors developed as part of a research programme into the application of risk assessment methods in areas exposed to natural terrain hazards in Hong Kong (Halcrow Asia Partnership, 1999). The factors reflect the probability of the death of a person located in an area affected by landslide events of different magnitude, and take account of whether the person is inside or outside an affected building.

Even small slides and single boulders can kill. The ability to escape from a landslide is partly related to the velocity of movement. Following a review of 27 fatal landslide incidents in Hong Kong, Finlay et al. (1999) found that a person within a run-out zone is very vulnerable in the event of complete or substantial burial by debris, or the collapse of a building (Table 8.8). If the person is buried by debris, death is most likely to result from asphyxia rather than crushing or impact. If the person is not buried, injuries are much more likely than death. For people located on the landslide, the vulnerability is likely to be lower because of a much longer warning time available at slide initiation (precursors), which provides the opportunity of escape before the slide gathers momentum. People located below a slow-moving landslide will also have lower vulnerabilities due to the increased warning time. However, research on earthquake casualties (Coburn and Spence, 1992) has shown that building design and building materials have an important influence on survival rates, and this should be taken into consideration in the case of landsliding.

Cruden (1997) has suggested that a ratio of three injured for every death might be a useful starting point for estimating the fatality rate associated with landslides. This may be satisfactory for sluggish and intermediate slides, but for major, extremely rapid landslides and falls the rate of fatalities to survival injuries is likely to be much higher, as in the case of volcanic eruption phenomena, which

Table 8.8 Summary of vulnerability factors for people affected by landsliding in Hong Kong

Case	Range of data	Recommended value	Comments
Person in open space, if struck by a rockfall	0.1–0.7	0.5	May be injured but unlikely to cause death
Person in open space, if buried by debris	0.8–1.0	1.0	Death by asphyxia almost certain
Person in open space, if not buried	0.1–0.5	0.1	High chance of survival
Person in vehicle, if vehicle is buried/crushed	0.9–1.0	1.0	Death is almost certain
Person in vehicle, if vehicle is damaged only	0–0.3	0.3	High chance of survival
Person in a building, if the building collapses	0.9–1.0	1.0	Death is almost certain
Person in a building, if the building is inundated with debris and the person buried	0.8–1.0	1.0	Death is highly likely
Person in a building, if the debris strikes the building only	0–0.1	0.05	Very high chance of survival

From Australian Geomechanics Society (2000) and Finlay et al. (1999)

'leave few walking wounded; a sharp dividing line separates the quick and the dead' (Baxter, 1990). For example, the 1985 Armero mudflows in Colombia killed 21 000 people, and only 65 were rescued from the debris (e.g. Voight, 1990).

The potential for persons to be impacted by a landslide is a function of a large number of factors, some involving chance (i.e. temporal exposure), some relating to the robustness and resilience of the people who are being threatened, some relating to the social conditions prevailing, and some relating to the nature and violence of the ground movements and subsequent hazards. Clearly, age, health, fitness and awareness are all important. Similarly, the stage in the evolution of a landslide at which a person or people become involved can also determine the outcome, as does the speed and efficiency of search and rescue operations. Finally, chance plays a role, not merely in involvement but also in survival or the limiting of injury. There are numerous tales of miraculous survivals, including that of a man who survived the Sale or Salashan, China, loess flowslide of March 1983 (Jones, 1992; Sassa, 1992) by clinging to a lone tree that floated upright on a small raft of ground for 700 m surrounded by 3 Mm^3 of flowing loess that obliterated three villages, killing over 200 people.

Example 8.4
The Lawrence Hargrave Drive, south of Sydney, Australia, was constructed through steep cliffs that rise to a height of some 300 m above the road (Wilson et al., 2005; see Example 7.9). The cliffs are prone to rockfalls, debris slides and debris flows. The road was closed by the Ministry of Roads in August 2003 for safety reasons. A quantitative risk assessment was undertaken to demonstrate that remedial measures could reduce the risk to acceptable levels (Wilson et al., 2005).

Estimates of vulnerability were developed by the risk assessment team through a workshop, drawing on knowledge and experience from other projects and the guidance presented by Finlay et al. (1999)

Table 8.9 Example 8.4: Lawrence Hargrave Drive, Australia – vulnerability values for various landslide scenarios

Volume of landslide debris crossing road: m³	Rockfalls		Debris flows	
	Hits car	Car hits debris	Hits car	Car hits debris
0.03	0.05	0.006	NA	NA
0.3	0.1	0.002	NA	NA
3	0.3	0.03	0.001	NA
30	0.7	0.03	0.01	0.001
300	1	0.03	0.1	0.003
3000	1	0.03	1	0.003

From Wilson et al. (2005)

and the AGS (2000). Vulnerability was defined as the probability of death given an impact from a landslide of a particular magnitude:

$$P(\text{fatality}) = P(\text{death}|\text{landslide magnitude})$$

Two scenarios were considered.

- The vulnerability of persons in a vehicle directly hit by a landslide.
- The vulnerability of persons in a vehicle that runs into landslide debris on the road. It was assumed that the landslide reaches the road 40 m or less in front of the vehicle and that the driver does not have time to avoid a collision. This distance was based on a 2 second response time at a travel speed of 60 km/hour, plus an allowance for breaking and avoidance (to avoid death but not necessarily car damage).

Table 8.9 presents a selection of the vulnerability values used in the risk assessment.

Example 8.5

In the winter of 2000/2001 there was a series of cliff collapses behind Brighton Marina, UK, including a major collapse in April 2001 that led to the closure of the Asda supermarket at the foot of Black Rock. As a result, a series of works were designed to enable the cliffs to be made sufficiently safe to reopen the promenade between Brighton Marina and Saltdene (the Undercliff Walk) and to safeguard the cliff-top A259 and marina access roads. The rockfall and rock slide risks to promenade users were estimated in order to provide the local authority with a baseline against which risk reduction measures could be compared (High Point Rendel, 2005).

A site-specific vulnerability model was developed that provides estimates of the chance of death for different landslide hazards along the cliffline:

$$P(\text{fatality}) = P(\text{death}|\text{landslide scenario})$$

The main factors influencing the vulnerability were considered to include

- the nature of the landslide event – that is, the size/volume and velocity of movement

- the event pathway from detachment to collision with an individual – this can be either direct (i.e. no dissipation of kinetic energy) or may involve several stages at which kinetic energy is lost (e.g. a bouncing boulder)
- the nature of the collision with the individual, e.g. burial by landslide debris or a blow to the head or the lower body.

The model is presented in Table 8.10, and is the product of the expert judgement of the project team, guided by published case studies (e.g. AGS, 2000; Finlay *et al.*, 1999) and an assessment undertaken for the Sandown to Shanklin cliffs, Isle of Wight (High Point Rendel, 2004).

8.5. Societal and social vulnerability

Societal vulnerability is concerned with broader issues of the potential effects on the economy, society and government at the scale of the area, region or state. Major hazard events can cause adverse consequences with national economies, as has been well described by Benson and Clay (2004) with respect to earthquakes. This vulnerability *of economic systems* is the extent to which economic activity is reduced as a consequence of hazard impacts of differing severity. Damage or destruction of factories, businesses and sites of production, loss of cash crops, the dislocation of transport routes, disruption to power and water supplies, and the reduction in the available workforce are some of the factors that lead to economic consequences over short to long timescales. In the case of major disasters, such consequences are increasingly measured in terms of changes in gross domestic product (GDP). Where the costs of destruction and damage also represent a significant proportion of GDP, then the financial burden of reconstruction will also have an adverse effect on economic performance.

Landslides rarely achieve such impacts, although they are important contributors in the case of landslide-generating events such as tropical revolving storms and earthquakes. More usually, the adverse effects of landsliding exacerbate the effects of economic marginalisation. Economic marginality describes the situation where the returns of an economic activity barely exceed the costs. Destruction, damage and disruption due to landsliding can lead to both costs and loss of output, which can combine to render an activity uncompetitive or even determine that the activity should not recommence (i.e. no repair, reconstruction or restoration), all of which will have knock-on effects for the local economy. Thus, economic consequences may increase with time after an event. Risk assessments may need to identify whether such economic vulnerability exists and the scale and duration of the consequences.

Social vulnerability arises from a mix of physical and socio-economic conditions: poverty, gender, class, ethnicity, etc. (Bankoff *et al.*, 2004; Birkmann, 2006). The frailties of people, together with any limitations in terms of lack of protection, survival capabilities or the ability to recover, owe much to prevailing socio-political conditions, or what Hewitt (1997) terms 'the making of vulnerability by human activity'. This notion applies equally to the creation of conditions prior to a hazard impact as to what happens during and immediately after an event. The potential for adverse consequences associated with many follow-on hazards (e.g. looting, disease), as well as some secondary hazards (e.g. fire), is largely a function of management failure. Similarly, incompetent search and rescue operations after an event, including failure to mobilise adequate medical services, can result in increased losses due to what Alexander (2000) calls *secondary vulnerability*.

Attempting to apply the social science-based concept of vulnerability to people and societies in a quantitative fashion turns out to be far more problematic than in the case of physical structures and infrastructure. Impacts on humans are not simply confined to the death and injury of individuals, but include long-term and more subtle effects, including loss of well-being. Thus, it is not merely a

Table 8.10 Example 8.5: the Undercliff Walk, Brighton – vulnerability factors for different hazard events, expressed as the probability of a fatality given that a person has been hit by the event

Hazard	Event size (volume): m^3	Typical event sequence	Vulnerability (chance of fatality)	Comment
H1: gravel and cobbles	0.0001	Fall detaches from the cliff face and hits sloping apron above the crest of the blockwork wall at the foot of the cliff and bounces across the promenade	0.000001	Bouncing gravels and cobbles extremely unlikely to cause more than non-priority *serious* injury
		Fall detaches from the cliff face and hits the projecting bed on the cliff face and bounces outwards, landing on the promenade before bouncing	0.00001	Bouncing gravels and cobbles extremely unlikely to cause more than priority *serious* injury
H2: small boulders	0.0025	Fall detaches from the cliff face and hits the sloping apron above the crest of the blockwork wall at the foot of the cliff and bounces across the promenade.	0.0001	Bouncing small boulder unlikely to cause more than priority *serious* injury
		Fall detaches from the cliff face and hits the projecting bed on the cliff face and bounces outwards, landing on the promenade before bouncing	0.001	Bouncing small boulders unlikely to cause more than priority *serious* injury
H3: small-mass failure	0.25–1	Failure detaches and slides rapidly down the cliff face, crossing the sloping apron above the crest of the blockwork wall at the foot of the cliff, and accumulates on the promenade	0.01	Priority and non-priority *serious* injury expected; remote possibility of death
H4: medium-mass failure	1–5	Failure detaches and slides rapidly down the cliff face, crossing the sloping apron above the crest of the blockwork wall at the foot of the cliff, and accumulates on the promenade	0.1	Priority and non-priority *serious* injury expected, possibility of death if person knocked over and hit on head
H5: large-mass failure	5–100	Failure detaches and slides rapidly down the cliff face, crossing the sloping apron above the crest of the blockwork wall at the foot of the cliff, and accumulates on the promenade. The promenade fills up with debris over a 25 m length, some debris spills over the wave wall into the Asda facilities	0.95	*Instantaneous or subsequent death* almost certain
H6: very large-mass failure	>100	Failure detaches and slides extremely rapidly down the cliff face, crossing the sloping apron above the crest of the blockwork wall at the foot of the cliff, and accumulates on the promenade. The promenade fills up with debris over a 30 m length. Debris flows over the wave wall into the Asda facilities	1.0	*Instantaneous death* certain

High Point Rendel (2005)

Figure 8.3 A framework for social vulnerability assessment. (From Parker and Tapsell, 2009)

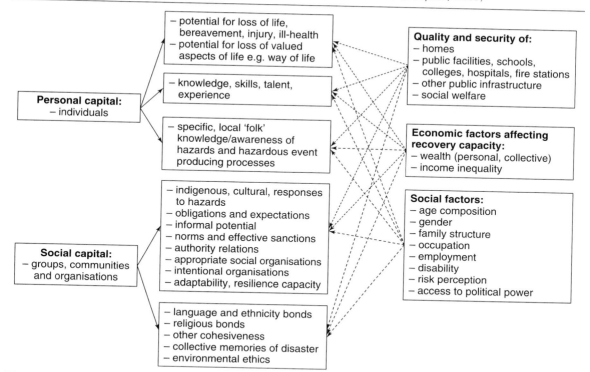

case of measuring the robustness of humans in the context of the physical threats they may face, because individuals form parts of groups and the impact on individuals, such as the death or rendering disabled of the main family income earner, can have profound and life-changing consequences for others. Such impact was graphically displayed in the Aberfan landslide disaster that occurred at 9.15 am on 21 October 1966, when a flow-slide of colliery waste killed 144 people, including 116 children, resulting in deep social and psychological scars on the community that took decades to heal. Quantifying these costs in hindsight is difficult enough, attempting to predict them in the future is problematic in the extreme, and must surely lie beyond the scope of most landslide risk assessments (see the last paragraph of this section).

Social and *societal vulnerability* are clearly closely interrelated, and form parts of a complex concept that is still being refined in an active and often confusing literature (e.g. Tapsell *et al.*, 2010). A framework for societal vulnerability is presented in Figure 8.3, central to which are the relationships between 'personal' and 'social' capital with a variety of security, economic and social factors (Table 8.11; Parker and Tapsell, 2009).

Vulnerability also determines the variable impact of landslides on different sections of a society (i.e. poorer neighbourhoods and cheap or shanty housing tend to be more vulnerable) and the ability of a community as a whole to withstand a landslide impact and recover (e.g. Alexander, 2005). *Societal vulnerability* includes three interrelated factors (Blaikie *et al.*, 1994; Cannon, 1993; Galderisi *et al.*, 2010).

Table 8.11 Key components of societal vulnerability

Component of societal vulnerability	Description
Security	Issues of safety and longer-term stability: these factors incorporate the 'physical' impact of an event on the natural and built environment where people are located. Also considered is the ability for key institutions to respond and manage the event effectively to cause minimal disruption to exposed communities
Economic	Levels of vulnerability are highly dependent upon the economic status of individuals, communities and nations. Economic factors exert a profound influence upon social vulnerability, to the extent that the two can be difficult to untangle and thus we may see references to 'socio-economic vulnerability'. It is not the lack of wealth directly that makes an individual or community socially vulnerable; it is the provision and access to resources that 'money can buy' that is of interest. The economic vitality of an area in general has been shown to influence quality of life: conditions prior to a hazard event (e.g. out-migration, economic recession) are likely to continue post-hazard event (Cutter and Emrich, 2006)
Social	The characteristics of the at-risk individual or community that alter the degree of susceptibility and sensitivity to hazard impact. These may include demographic characteristics such as age, gender, family structure, health and disability, occupation and employment, as well as access to political power

From Parker and Tapsell (2009)

- *Resilience*, or the ability to maintain a system and to recover after impact, is used in the context of communities as well as of the economic or wealth generating systems within an area. An economy dependent on a limited number of transportation routes or a single industry may be highly vulnerable to landsliding.
- *Robustness*, or the ability to respond to a spectrum of uncertainties or actual physical threats, is applied to organisations, communities and individuals within a group, reflecting social, health and economic strength. Insurance, for example, can make individuals more able to recover from the losses incurred in a landslide.
- *Preparedness*, which can reflect both the protection provided to a group and their willingness to act on their own behalf.

To many social scientists and planners, the vulnerability of groups within a society may be of greater importance in determining the impact of a landslide than its intensity. A highly vulnerable group, such as a squatter camp on the unstable slopes at the margins of a city, may often be badly affected by large, fast-moving landslides, whereas wealthy, low-vulnerability groups might experience minimal impacts from similar events. Thus, numbers of people displaying particular levels of vulnerability, with respect to landslides of a particular intensity, will provide a measure of the potential impact.

Social vulnerability is most apparent after a hazard event has occurred, when different patterns of loss, suffering and recovery are observed across the population (Cutter *et al.*, 2003). While all people living in landslide hazard areas are vulnerable, the social impacts of hazards often fall disproportionately on the most vulnerable people in society – the poor, minorities, children, the elderly and the disabled. These groups are often the least prepared for an emergency, have the fewest resources with which to prepare for a hazard, tend to live in the highest-risk locations in substandard housing, and lack knowledge or the social and political connections necessary to take advantage of resources that would speed their recovery (e.g. Alexander, 2005; National Research Council, 2006).

Thus, to estimate social and societal vulnerability for the purposes of a risk assessment it is necessary to attempt to obtain some or all of the following information.

- Scenarios of landslide development in terms of physical characteristics and intensity.
- Estimations of patterns and intensities of damage/destruction to buildings and infrastructure.
- Estimations of the likelihood, nature and severity of secondary hazards.
- Estimations of the number of people expected to be within the area affected by primary and secondary hazards, after consideration of timing and the possibility of warning/evacuation.
- Estimation of relevant characteristics of people within danger area (age/sex, etc.).
- Estimation of survival rates based on the anticipated nature of destruction and the likely efficiency of emergency response.

Oven (2009) described an investigation into social vulnerability to landsliding in the Upper Bhote Koshi Valley, Nepal, based on household surveys and interviews. The background to the research was a dramatic increase in landslide fatalities over the period 1978–2005, attributed by Petley *et al.* (2007) to the rural road-building programme. Although the older housing is sited along stable ridge crests away from landslide-prone areas on the lower valley slopes adjacent to the valley floor, the construction of the Arniko Highway in 1967 has been accompanied by the characteristic growth of roadside housing. The new properties have been built at the bottom of steep, unstable slopes and sited on colluvial and alluvial deposits commonly adjacent to stream channels. They are occupied mainly by local migrants from high-caste, low-caste and hill tribe backgrounds who were attracted by the economic opportunities associated with the road. These households were both relatively rich and relatively poor. As a result, there was no apparent correlation between caste/ethnic group and poverty level and the increase in vulnerability. However, the vulnerability to landslide events was seen to vary between households, reflecting a household's resilience and coping capacity, which appears to be linked to wealth.

8.6. Societal and social vulnerability in landslide risk assessment

At this point it is time to take stock. Landslide risk assessment is a process that starts with the landslide and ends with the costs within society. It begins in the realms of science, dealing with the probabilities of physical phenomena, and ends within the social sciences, attempting to assess the possibilities of complex webs of consequences, many of which are unquantifiable in terms of cost (e.g. 'fates worse than death'). Under these circumstances, it is important to question exactly how far, and in what detail, the potential accident sequences need to be traced as part of a risk assessment.

Vulnerability and risk assessments tend to be focused at a range of levels, from the global to the local, and the approaches and techniques vary accordingly. As most instances of widespread landsliding tend to be subsumed under the main primary hazard, such as earthquakes or tropical revolving storms, the majority of detailed landslide risk assessments tend to be location-specific, albeit in some instances extending over significant areas. The purpose is to identify and quantify the likely scales of impact, together with the possibilities of catastrophic scenarios. Thus, although there has been some progress in the development of quantitative assessments of societal/social vulnerability (Hill and Cutter, 2001; UN Development Programme, 2003), these have mainly been at larger spatial scales (e.g. the UN Disaster Risk Index (HDI), Peduzzi et al., 2009) or focused on climate/environmental change (IISD et al., 2009; van Aalst et al., 2007; Willows and Connell, 2003).

While most agree that vulnerability assessments at local scales are the ideal, it is a hugely difficult task, and there are major difficulties in obtaining sufficient socio-economic data at this level. Perhaps the best to date is the 'Index of Social Vulnerability' (SoVI) developed by Cutter et al. (2003) for use at the county level in the USA. The study of resilience suffers from the same problems (see Cutter et al., 2008). These difficulties have led to debates as to the correct balance that needs to be achieved between quantitative and qualitative data and even whether it is actually possible to quantify vulnerability within social systems (Wisner et al., 2004).

Clearly, this is a difficult field. As a result, any landslide risk assessment that identifies the possibility of serious social/community consequences needs to proceed with caution. The advice is to keep it simple, using expert input, unless the possibility of significant levels of impact requires deeper investigation, in which case engage suitably qualified social scientists.

REFERENCES

Adger WN (2006) Vulnerability. *Global Environmental Change* **16**: 268–281.
Australian Geomechanics Society (AGS) (2000) Landslide risk management concepts and guidelines. *Australian Geomechanics* **35**: 49–52.
AGS (2007) Practice note guidelines for landslide risk management. *Australian Geomechanics* **42(1)**: 63–114.
Alexander DE (1989) Urban landslides. *Progress in Physical Geography* **13(2)**: 157–191.
Alexander DE (2000) *Confronting Catastrophe*. Terra, Harpenden.
Alexander DE (2002) *Principles of Emergency Planning and Management*. Terra, Harpenden.
Alexander DE (2005) Vulnerability to landslides, In *Landslide Hazard and Risk* (Glade T, Anderson MG, Crozier MJ (eds)). Wiley, Chichester, pp. 175–198.
Amatruda G, Bonnard Ch, Castelli M, Forlati F, Giacomelli L, Morelli M, Paro L, Piana F, Pirulli M, Polino R, Prat P, Ramasco M, Scavia C, Bellardone G, Campus S, Durville J-L, Poisel R, Preh A, Roth W and Tentschert EH (2004) A key approach: the IMIRILAND project method. In *Identification and Mitigation of Large Landslide Risks in Europe: Advances in Risk Assessment* (Bonnard Ch, Forlati F and Scavia C (eds)). Balkema, Rotterdam, pp. 13–44.
Anderson MB (2000) Vulnerability to disaster and sustainable development. In *Storms* (Pielke RA and Pielke RA (eds)). Routledge, London, vol. 1, pp. 11–25.
Bankoff G, Frerks G and Hilhorst D (2004) *Mapping Vulnerability: Disasters, Development and People*. Earthscan, London.
Bateman, I, Carson, RT, Day B, Hanemann WM, Hanley N, Hett T, Jones-Lee M, Loomes G, Mourato S, Ozdemiroglu E and Pearce DW (2002) *Economic Valuation with Stated Preference Techniques: A Manual*. Edward Elgar, Cheltenham.
Baxter PJ (1990) Medical effects of volcanic eruptions. *Bulletin Volcanologique* **52**: 532–544.
Benson C and Clay EJ (2004) *Understanding the Economic and Financial Impacts of Natural Disasters*. The World Bank, Geneva.
Bell R and Glade T (2004) Quantitative risk analysis for landslides – examples from Bildudalur, NW-Iceland. *Natural Hazards and Earth System Sciences* **4**: 117–131.
Birkmann J (ed.) (2006) *Measuring Vulnerability to Natural Hazards: Towards Disaster Resilient Societies*. United Nations University Press, Tokyo.

Blaikie P, Cannon T, Davis I and Wisner B (1994) *At Risk: Natural Hazards, Peoples' Vulnerability and Disasters*. Routledge, London.

Blöchl A and Braun B (2005) Economic assessment of landslide risks in the Swabian Alb, German – research framework and first results of homeowners' and experts' surveys. *Natural Hazards and Earth System Sciences* **5**: 389–396.

Buckle P (2006) Assessing social resilience. In *Disaster Resilience: An integrated approach* (Paton D and Johnston DM (eds)). Charles C. Thomas, Springfield, IL, pp. 88–103.

Cannon T (1993) A hazard need not a disaster make: vulnerability and the causes of 'natural' disasters. In *Natural Disasters: Protecting Vulnerable Communities* (Merriman PA and Browitt CWA (eds)). Thomas Telford, London, pp. 92–105.

Cannon T (2000) Vulnerability analysis and disasters. In *Floods* (Parker DJ (ed.)). Routledge, London, pp. 45–55.

Cardinali M, Reichenbach P, Guzzetti F, Ardizzone F, Antonini G, Galli M, Cacciano M, Castellani M and Salvati P (2002) A geomorphological approach to the estimation of landslide hazards and risks in Umbria, Central Italy. *Natural Hazards and Earth System Sciences* **2**: 57–72.

Chen CC, Yeh TL, Yang YK et al. (2001) Psychiatric morbidity and post-traumatic symptoms among survivors in the early stage following the 1999 earthquake in Taiwan. *Psychiatry Research* **105**: 13–22.

Coburn A and Spence R (1992) *Earthquake Protection*. Wiley, Chichester.

Cruden DM (1997) Estimating the risks from landslides using historical data. In *Landslide Risk Assessment* (Cruden D and Fell R (eds)). Balkema, Rotterdam, pp. 177–184.

Cutter SL (1996) Vulnerability to environmental hazards. *Progress in Human Geography* **20(4)**: 529–539.

Cutter SL, Boruff BJ and Shirley WL (2003) Social vulnerability to environmental hazards. *Social Science Quarterly* **84**: 242–261.

Cutter SL, Barnes L, Berry M et al. (2008) A place-based model for understanding community resilience to natural disasters. *Global Environmental Change* **18**: 598–606.

Cutter SL and Emrich CT (2006) Moral hazard, social catastrophe: the changing face of vulnerability along coasts. *The Annals of the American Academy of Political and Social Science* **604**: 102–111.

Dai FC, Lee CF and Ngai YY (2002) Landslide risk assessment and management: an overview. *Engineering Geology* **64(1)**: 65–87.

DNV Technica (1996) *Quantitative Landslip Risk Assessment of Pre-GCO Man-made Slopes and Retaining Walls*. Geotechnical Engineering Office, Hong Kong.

Douglas J (2007) Physical vulnerability modelling in natural hazard risk assessment. *Natural Hazards and Earth System Sciences* **7**: 283–288.

Duzgun HSB and Lacasse S (2005) Vulnerability and acceptable risk in integrated risk assessment framework. In *Landslide Risk Management* (Hungr O, Fell R, Couture R and Eberhardt E (eds)). Balkema, Rotterdam, pp. 505–515.

DTLR (Department for Transport, Local Government and the Regions) (2002) *Economic Valuation with Stated Preference Techniques: A Summary Guide. DTLR Appraisal Guidance*. DTLR, London.

ENSURE (2009) *Methodologies to Assess Vulnerability of Structural Systems. Enhancing Resilience of Communities and Territories Facing Natural and Na-tech Hazards*. ENSURE Project (Enhancing resilience of communities and territories facing natural and na-tech hazards), EC, Brussels. Contract 212045.

EPFL (Ecole Polytechnique de Lausanne) (2002) Relevant criteria to assess vulnerability and risk. Unpublished deliverable (D16) of project IMIRILAND: Impact of Large Landslides in the Mountain Environment, 2002. EPFL, Lausanne.

Fell R (1994) Landslide risk assessment and acceptable risk. *Canadian Geotechnical Journal* **31**: 261–272.

Fell R and Hartford D (1997) Landslide risk management. In *Landslide Risk Assessment* (Cruden D and Fell R (eds)). Balkema, Rotterdam, pp. 51–108.

Finlay PJ, Mostyn GR and Fell R (1999) Landslides: prediction of travel distance and guidelines for vulnerability of persons. *Proceedings of the 8th Australia/New Zealand Conference on Geomechanics, Hobart*. Australian Geomechanics Society, vol. 1, pp. 105–113.

Galderisi A, Ceudech A, Ferrara FF and Profice AS (2010) *Integration of Different Vulnerabilities vs. Natural and Na-tech Hazards*. ENSURE Project, EC, Brussels. Deliverable 2.2.

Galli M and Guzzetti F (2007) Landslide vulnerability criteria: a case study from Umbria, central Italy. *Environmental Management* **40**: 649–664.

Glade T (2003) Vulnerability assessment in landslide risk analysis. *Die Erde* **134**: 123–146.

Glade T and Crozier MJ (2005) The nature of landslide hazard and impact. In *Landslide Hazard and Risk* (Glade T, Anderson MG and Crozier MJ (eds)). Wiley, Chichester, pp. 43–74.

Guzzetti F, Reichenbach P, Cardinali M, Ardizzone F and Galli M (2003) Impact of landslides in the Umbria Region, central Italy. *Natural Hazards Earth System Science* **3**: 469–486.

Halcrow Asia Partnership (1999) *Stage 2 Detailed Site-specific QRA Report*. Geotechnical Engineering Office, Hong Kong.

Hewitt K (1997) *Regions of Risk: A Geographical Introduction to Disasters*. Addison Wesley Longman, London.

High Point Rendel (2004) *Shanklin – Sandown Cliff Stability Survey Quantitative Risk Assessment*. Isle of Wight Council, Newport.

High Point Rendel (2005) *Black Rock Stage 2 Cliff Stabilisation, Brighton Marina Quantitative Risk Assessment: Black Rock*. Brighton and Hove City Council, Brighton.

Hill A and Cutter SL (2001) Methods of determining disaster proneness. In *American Hazardscapes The Regionalisation of Disasters and Disasters* (Cutter SL (ed.)). Joseph Henry Press, Washington, DC.

Holcombe E, Smith S, Wright E and Anderson MG (2012) An integrated approach for evaluating the effectiveness of landslide risk reduction in unplanned communities in the Caribbean. *Natural Hazards* **61**: 351–385.

IFRC (2008) *Vulnerability and Capacity Assessment – Guidelines*. International Federation of Red Cross and Red Crescent Societies, Geneva.

IISD et al. (2009) *CRiSTAL: Community-based Risk Screening Tool- Adaptation and Livelihoods. User Manual v4.0*. IISD, Geneva.

ISSMGE (2004) *Risk Assessment – Glossary of Terms*. TC32, Technical Committee on Risk Assessment and Management Glossary of Risk Assessment Terms, ISSMGE, London.

Jones DKC (1992) Landslide hazard assessment in the context of development. In *Geohazards: Natural and Man-made* (McCall GJH, Laming DJC and Scott SC). Chapman and Hall, London, pp. 117–141.

Li Z, Nadim F, Huang H, Uzielli M and Lacasse S (2010) Quantitative vulnerability estimation for scenario-based landslide hazards. *Landslides* **7**: 125–134.

National Research Council (2006) *Facing Hazards and Disasters: Understanding Human Dimensions. Committee on Disaster Reduction in the Social Sciences: Future Challenges and Opportunities*. National Academy Press, Washington, DC.

Oven KJ (2009) Landscape, livelihoods and risk: community vulnerability to landslides in Nepal. PhD thesis, Durham University.

Parker D and Tapsell S (2009) *Relations between Different Types of Social and Economic Vulnerability*. ENSURE Project, EC, Brussels. Deliverable 2.1.

Papathoma-Köhle M, Neuhauser B, Ratzinger K, Wenzel H and Dominey-Howes D (2007) Elements at risk as a framework for assessing the vulnerability of communities to landslides. *Natural Hazards and Earth Systems Science* **7**: 765–779.

Pearce D, Atkinson G and Mourato S (2006) *Cost–benefit Analysis and the Environment: Recent Developments*. Organisation for Economic Co-operation and Development, Paris.

Peduzzi P, Dao H, Herold C and Mouton F (2009) Assessing global exposure and vulnerability towards natural hazards: the disaster risk index. *Natural Hazards and Earth System Sciences* **9**: 1149–1159.

Peek-Asa C, Ramirez M, Shoaf K and Seligson H (2002) Population-based case-control study of injury factors in the Northridge earthquake. *Annals of Epidemiology* **12**: 525–6.

Petley DN, Hearn GJ, Hart A et al. (2007) Trends in landslide occurrence in Nepal. *Natural Hazards* **43**: 23–34.

Roberds W (2005) Estimating temporal and spatial variability and vulnerability. In *Landslide Risk Management* (Hungr O, Fell R, Couture R and Eberhardt E (eds)). Balkema, Rotterdam, pp. 129–158.

Romang H, Kienholz H, Kimmerle R and Böll A (2003) Control structures, vulnerability, cost-effectiveness – a contribution to the management of risks from debris torrents. In *Debris-flow Hazards Mitigation: Mechanics, Prediction, and Assessment* (Rickenmann D and Chen C (eds)). Millpress, Rotterdam, pp. 1303–1313.

Salvati P, Bianchi C and Guzzetti F (2006) *Landslide and Flood Historical Catalogue in Umbria*. CNR IRPI and Fondazione Cassa di Risparmio di Perugia, Perugia (in Italian).

Sassa K (1992) Landslide volume – apparent friction relationship in the case of rapid loading on alluvial deposits. *Landslide News* **6**: 16–19.

Tapsell S, McCarthy S, Faulkner H and Alexander M (2010) *Social Vulnerability and Natural Hazards*. Flood Hazard Research Centre, Middlesex University, London. CapHaz-Net WP4 Report.

Timmerman P (1981) *Vulnerability, Resilience and Collapse of Society*. Institute for Environmental Studies, Toronto.

UN (2011) *Global Assessment Report on Disaster Risk Reduction*. United Nations, New York. http://www.preventionweb.net/english/hyogo/gar/2011/en/home/download.html.

UN Development Programme (2003) *World Vulnerability Report*. United Nations, New York.

UN Office for Disaster Risk Reduction (2009) UNISDR terminology on disaster risk reduction. http:/www.unisdr.org/we/inform/terminology#letter-d (accessed 14/9/2013).

Uzielli M, Nadim F, Lacasse S and Kaynia AM (2008) A conceptual framework for quantitative estimation of physical vulnerability to landslides. *Engineering Geology* **102**: 251–256.

Van Aalst MK, Helmer M, de Jong C, Monasso F, van Sluis E and Suarez P (2007) *Red Cross/Red Crescent Climate Guide*. Red Cross/Red Crescent Climate Centre, The Hague.

Van Westen CJ, Van Asch TWJ and Soeters R (2006) Landslide hazard and risk zonation – why is it still so difficult? *Bulletin of Engineering Geology and Environment* **65(2)**: 167–184.

Voight B (1990) The 1985 Nevado del Ruiz volcano catastrophe – anatomy and retrospection. *Journal of Volcanology and Geothermal Research* **42**: 151–188.

Volkwein A, Schellenberg K, Labiouse V, Agliardi F, Berger F Bourrier F, Dorren LKA, Gerber W and Jaboyedoff M (2011) Rockfall characterisation and structural protection – a review. *Natural Hazards and Earth Systety Sciences* **11**: 2617–2651.

Willows RI and Connell RK (eds) (2003) *Climate Adaptation: Risk, Uncertainty and Decision-Making*. UKCIP, Oxford.

Wilson RA, Moon AT, Hendrickx M and Stewart IE (2005) Application of quantitative risk assessment to the Lawrence Hargrave Drive Project, New South Wales, Australia. In *Landslide Risk Management* (Hungr O, Fell R, Couture R and Eberhardt E (eds)). Balkema, Rotterdam, pp. 589–598.

Wisner B, Blaikie P, Cannon T and Davis I (2004) *At Risk: Natural Hazards, People's Vulnerability and Disasters*. Routledge, London.

World Bank (2010) Reducing human vulnerability: helping people help themselves. *World Development Report 2010*. World Bank, Geneva, pp. 87–123.

Zêzere JL, Garcia RAC, Oliveira SC and Reis E (2007) Probabilistic landslide risk analysis considering direct costs in the area north of Lisbon (Portugal). *Geomorphology* **94**: 467–495.

Chapter 9
Estimating the consequences

9.1. Introduction

What would happen if a landslide occurred? This question is crucial to the determination of risk. However, to answer it requires that some form of *consequence assessment* be undertaken to estimate the likely damages and losses that could be expected. These losses or detriment can be extremely diverse in nature, and involve loss of life, injury and impairment of persons; destruction of property, resources and heritage; disruption of activities, and denial of supplies and services; and cultural, spiritual and ethical violations (Hewitt, 1997).

Only a proportion of landslides cause detriment (see Figure 3.1). Of those that do, most cause temporary, short-lived losses, but some can have long-term effects. Usually the losses are confined to a limited and readily definable area and are, therefore, relatively easy to identify, but sometimes the adverse consequences spread across neighbouring areas like ripples on a pond, becoming ever more difficult to distinguish with distance and the passage of time. In some instances the losses are easily quantified, such as repairing a building or replacing a cow, but often they include adverse consequences that are extremely difficult to value in monetary terms, such as the psychological effects experienced by survivors of a destructive landslide (see Lacey, 1972).

It is possible to distinguish a range of landslide events based on the scale and complexity of the adverse consequences that they generate, including the following.

- *Simple events*, which cause detriment as a direct consequence of a single event or repeated movement at a location (e.g. a rotational failure or mudslide). For example, on 24 January 2012, a landslide in the Southern Highlands of Papua New Guinea (PNG) buried the villages of Tumbi and Tumbiago. The PNG Red Cross estimated that between 25 and 60 people were killed, although no official death toll exists.
- *Compound events*, which cause detriment when a triggering event produces a cascade of different types of landslide (falls, flows and slides) that, if topographic conditions and material availability allow, can achieve great size, speed and violence (e.g. the 1970 Huascarán rock/ice avalanche destroyed the town of Yungay, Peru, killing between 15 000 and 20 000 people; Plafker and Ericksen, 1978).
- *Multiple events*, involving widespread landslide activity and considerable detriment. This type of event is often associated with earthquakes, extensive and intense rainstorms, volcanic eruptions and forest clearance by humans. For example, when Hurricane Mitch came virtually to a halt over Central America for six days in October 1998 and dropped 2 m of rain over mountainous terrain, more than 9000 landslides were triggered in Guatemala alone (Bucknam, 2001); the Northridge (California) earthquake of 17 January 1994 triggered more than 11 000 failures (Harp and Jibson, 1995, 1996); and over 4000 landslides were generated

in Umbria (Italy) by snowmelt following a sudden change in temperature on 1 January 1997 (Cardinali et al., 2002).
- *Complex events*, in which a significant proportion of the overall damage is the product of the generation of *secondary geohazards*, such as floods, tsunamis or volcanic eruptions (see Chapter 3). For example, collapse of the northern flank of Mount St Helens, Washington State, USA, on 18 May 1980, was triggered by an earthquake ($M = 5.1$). The huge landslide exposed the gas-rich magma that had been intruded into the volcano during the preceding months. Decompression caused the magma to explode and disintegrate into a cloud of gas and volcanic debris that raced downslope at 300–500 km/hour. This flow, together with the blast from the air pushed ahead of it, laid waste some 600 km^2 of land within 2 minutes (Lipman and Mullineaux, 1981).

Landslide magnitude does not, in itself, determine the potential for loss: some major failures do not produce adverse consequences. The diverse nature and scale of adverse consequences is a reflection of the interaction of two sets of variables that change over space and time: the huge variety of landslide events in terms of character, magnitude, timing and location (*hazard potential*), interacting with the distribution of humans and the objects, activities and environments valued by humans that are, to varying degrees, susceptibility to impact by landsliding (vulnerability; see Chapter 8) and therefore capable of suffering harm (*exposure potential*).

If the hazard and exposure potential are plotted as the axes of a diagram (Figure 9.1), then four categories of landslide can be identified.

- *Category A* landslides, due to their small size or location, result in little or no impact from a human perspective and present low risk or no risk.
- *Category B* landslides can be of any size, and result in significant impacts and detriment.

Figure 9.1 The hypothetical division of contemporary landslides into four groups with reference to hazard significance, based on the relationship between hazard potential and exposure potential (see the text for discussion). The shaded box represents the envelope of scenario values that might be produced as part of a risk assessment of a specific site

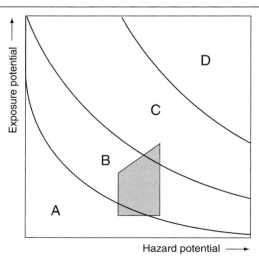

- *Category C* landslides can result in major impacts, some of which are termed 'disasters' by the media.
- *Category D* landslides result in severe impacts, sometimes resulting in zones of total destruction; these are the events that tend to be referred to as major disasters, mega-disasters, calamities or catastrophes by the media, as well as in some sections of the academic literature.

It should be noted that although *catastrophe potential* tends to be associated with high magnitude–low frequency (rare) landslide events, it can also be achieved by smaller landslides if they affect particularly vulnerable locations or sites, or through *accidents of timing*.

The shaded area in Figure 9.1 illustrates the range of possible situations with respect to a hypothetical landslide risk assessment, and shows that although most of the scenarios are likely to point to variations within category B, the *best case scenario* will indicate an outcome within category A, while the *worst case* will show a much higher level of detriment associated with category C. This variation often has as much to do with the precise location and timing of a landslide as with changes in landslide magnitude.

The approach to estimating the consequences of an event has to be somewhat different in each of these cases. However, the basic principle of consequence assessment is the development of *consequence models* or *scenarios,* which attempt to define and quantify the possible adverse outcomes. In each case, the key questions that will need to be considered include the following.

- What is there to be impacted?
- How, and to what extent, will it be impacted?
- What is the likelihood that the landslide event will generate significant losses through the development of *secondary hazards* or *follow-on hazards*, such as accidents and fires?
- How can the adverse consequences be valued?

Clearly, this stage of the landslide risk assessment process sees the focus shift from the nature and probability of the landslide threat to assessing the 'harm' that varying scales of landsliding could generate; in other words, to assessing what features valued by humans ('elements at risk') could be adversely impacted, the scale of impact and the costs that would flow from the disruption.

There is a need to develop a structured approach to assessing adverse consequences that is supported by the historical record, but based on an awareness of the key factors that determine landslide damage. These factors are

- the nature and scale of likely ground movements (*ground behaviour*; see Chapter 3)
- the nature and value of the *assets* or *elements at risk* within the threatened area
- the extent to which the *elements at risk* are likely to be *exposed* to the hazard(s) generated by the predicted ground movements (see Chapter 7)
- the *vulnerability* of the exposed *elements at risk* to the hazard(s) generated by predicted ground movements (see Chapter 8).

Timing is crucial, as the actual losses incurred in a particular event can vary greatly depending on unique combinations of circumstances. For example, it is estimated that 66 people were killed in 1988 by a landslide along the Trabzon–Erzurum highway in Çatak, Turkey (Jones et al., 1989). A small landslide blocked the road at midnight on the wet night of 22/23 June and led to the delayed travellers congregating in a road-side coffee house. A grader was despatched to clear the debris, but operations were delayed because of the difficulty of night-time working in heavy rainfall. Eight hours later a very much larger slide occurred on the same slope, burying the coffee house and killing all inside. Similar-sized landslide events elsewhere along this stretch of road would not have caused so many fatalities.

Consequences are the result of particular unique combinations of physical and social conditions. In Çatak, a different set of conditions could have generated a different landslide damage scenario. The inherent randomness in the main factors involved in such disasters (the build-up of people in a particular location, night-time, etc.) dictates that future landslides cannot be expected to cause the same levels of adverse consequences, even if they are of the same magnitude.

Although the damage information contained in the historic record provides an important guide as to the potential adverse consequences of future landslides, despite inevitable incompleteness and bias, it cannot properly indicate the true range of possibilities. The past, therefore, is not a complete guide to the future because circumstances change with respect to both the 'physical' and 'human' or 'socio-economic' environments. New hazards are identified, new vulnerabilities develop, and new accident sequences emerge.

9.2. Elements at risk

The term *asset* is used to describe all objects or qualities valued by humans. All assets occurring in an area that could be adversely affected by a hazard, such as a landslide, are known as *elements at risk*. These elements are extremely diverse in nature and are usually divided into the following major groupings.

- *People*. At the simplest level, this is the number of people present in the area likely to suffer impact. More detailed analysis could use age–sex distributions and an indication of the state of health (e.g. good health, poor health, infirm), as these aspects do influence death rates and the nature and severity of injuries within a population (see Chapter 8). Detriment is usually expressed in terms of loss of life or injury, although the longer-term consequences of physical impairment, psychological effects and ill-health should also be considered.
- *Buildings, structures, services and infrastructure*. The value of these physical assets can usually be determined from real estate agents or local authority tax bands for housing stock and from the owners/operators for services and infrastructure. Damage can be total or repairable (i.e. partial loss).
- *Property*. This includes the contents of houses, businesses and retailers, machinery, vehicles, domesticated animals and personal property. Information on values may be obtained from trade organisations and the insurance industry.
- *Activities*. This group includes all activities, whether for financial gain or pleasure. The main components are business, commerce, retailing, entertainment, transportation, agriculture, manufacturing and industry, mineral extraction and recreation. The losses incurred through landsliding are equivalent to the temporary or permanent disruption of these activities, and are, in the main, expressed in terms of loss of revenue.
- *Environment*, including flora, fauna, environmental quality and amenity.

9.3. Using the historical record

Finding out how landslides occurred in similar situations elsewhere, how they impacted on the societies concerned and how the costs were calculated are all important ingredients in the development of realistic consequence assessments. Clearly, the use of local examples is preferable. Should examples be from abroad, then cultural and societal differences (e.g. building materials and design) have to be taken into account when using such *analogues*. It is also necessary to note that the character and behaviour of landsliding varies across the globe due to differences in the tectonic setting, geological make-up, geomorphological history and prevailing climatic regime, which means that the applicability of potential landslide analogues must also be ascertained.

The historical record provides a vital source of data regarding the adverse consequences associated with past landslide events in general, which provide guidance as to what *could* happen and the magnitude–frequency characteristics of losses (see Chapter 3). It also contains descriptions and studies of what *actually* occurred in specific events, together with estimations of loss, which can be used as analogues for predicting impacts in the future. However, it is generally recognised that global data on landslide impacts is both incomplete and unreliable, especially prior to the mid-20th century, when many significant events failed to be recorded because of remoteness or for political reasons.

The communicability of human loss figures, together with their availability relatively quickly after a hazard impact, has made them a favourite with the media. As a consequence, death tolls have become 'the currency of disaster', often determining the threshold at which serious impacts should be called 'disaster' (i.e. 10 deaths, 20, 25, 50, 100, etc.), and are even employed by some to indicate the relative and absolute significance of disasters. This focus does, however, ensure that records of numbers killed are likely to be the best kept statistics of landslide impacts.

Multiple events can occasionally cause 100 000 deaths, with the figure possibly exceeding 500 000 under quite exceptional circumstances (e.g. the Shansi earthquake, 1556, when loess failures entombed huge numbers of people). However, death tolls attributable to the landsliding caused by such events are more usually limited due to uncertainty as to the precise attribution of death. For example, an intense storm on 14–16 December 1999 caused numerous catastrophic landslides and flooding along a 40 km coastal strip north of Caracas from La Guaira to Naiguita, Venezuela (Wieczorek *et al.*, 2001). It was estimated that around 30 000 people died (USAID, 2000), but precisely how many were killed by landsliding is unknown.

Landslide death tolls tend to be less than 15 000, with relatively frequent records in the range 100–1000. The maximum death toll in a single landslide event is around 20 000, as in the cases of the catastrophic rock/ice avalanche at Huascarán, Peru (1970; Plafker and Ericksen, 1978) and the volcanic mudflow that destroyed Armero, Colombia, during the eruption of the Nevado del Ruiz volcano (1985; Herd *et al.*, 1986). More usually, the numbers killed by individual landslides, or the floods created by the failure of landslide dams, are less than 1000, and the frequency appears to increase dramatically with diminishing casualty figures.

Because of the poor quality of the historical record, there remains uncertainty as to the actual numbers of people killed by landsliding per year, the magnitude–frequency distribution of landslide-generated death tolls and the regional patterns of casualty figures. Early disaster statistics produced by the Red Cross indicated 3006 deaths due to landslide disasters over the period 1900–1976 (Crozier, 1986), or a trivial 40 deaths per annum out of a combined total of 60 000 deaths a year due to natural disasters. These data are now seen to be gross underestimations.

Since 1988, the World Health Organization's Collaborating Centre for Research on the Epidemiology of Disasters (CRED; e.g. Sapir and Misson, 1992) has been maintaining an emergency events database (EM-DAT: http://www.emdat.be/database). The database contains records of the occurrence and effects of over 18 000 disasters in the world from 1900 to the present. Between 2004 and 2010 there were 130 recorded landslides that have caused loss of life (a minimum of ten fatalities), resulting in 7431 fatalities. This suggests an average rate of 1062 fatalities per year over the 7-year period.

The Durham Fatal Landslide Database (DFLD; http://www.landslidecentre.org/database.htm) has been compiled from internet searches and corroborated with analysis from government statistics, aid agency reports, and research papers (e.g. Petley *et al.*, 2005). Excluding earthquake-triggered landslides, during the 7-year period between 1 January 2004 and 31 December 2010 there were 2620 landslide incidents that led to fatalities (Petley, 2012). A total of 32 322 people were reported as having been killed in these incidents, over four times more fatalities than indicated by the EM-DAT database. The main event clusters of these non-seismic incidents occur around the margins of populated mountainous regions, especially in Asia.

The DFLD average figure of 4617 deaths/year for the period 2004–2010 (Petley, 2012) would rise dramatically if/when the data set becomes truly global, and the effects of individual impacts within landslide generating events, such as earthquakes, floods and tropical revolving storms, are included. And then there are the countless number of unreported instances where individuals or small groups have been killed by landslides, especially in remote areas. The 'true' average figure for landslide-induced deaths is probably >10 000/year. Irrespective of the actual figure and the spatial variation of landslide intensity, the conclusion has to be that deaths from landsliding are not unusual and need to be considered carefully in the majority of landslide risk assessments.

Among the best-known high-fatality landslide events are the following.

- The Vaiont disaster of 9 October 1963, in Italy, when a 250 Mm^3 rockslide entered the reservoir, sending a wave around 100 m high over the crest of the concrete dam. The flood wave destroyed five downstream villages and killed around 1900 people (Hendron and Patton, 1985; Kiersch, 1964).
- The Aberfan disaster of October 1966 in south Wales. Parts of a 67 m-high colliery spoil tip collapsed and flowed downslope into the village below. Twenty houses and a primary school were overwhelmed, and 144 persons died, including 116 out of 250 pupils in the school (Bishop *et al.*, 1969; Miller, 1974). The disaster led to stricter control over the disposal of mine waste in Britain, under the Mines and Quarries (Tips) Acts of 1969 and 1971.
- Over 250 people were killed by landslides in Hong Kong during a period of heavy rainfall in June 1972. Most of the deaths were as a result of two major landslides (Government of Hong Kong, 1972a,b). At Sau Mau Ping Resettlement Estate, Kowloon, a 40 m-high embankment collapsed, and the resulting flowslide left 71 persons dead and 60 injured. A 12-storey building collapsed under the impact of an extremely rapid flowslide from the slopes above Po Shan Road, in the Hong Kong mid-levels, killing 67 persons and injuring a further 20. The public outcry from these events, and the 18 deaths caused by subsequent flowslides on the Sau Mau Ping Resettlement Estate in 1976, led to the establishment of a Geotechnical Control Office (now named the Geotechnical Engineering Office, GEO) in 1977. This is a central policing body to regulate the planning, investigation, design, construction, monitoring and maintenance of man-made slopes (Chan, 2000; Malone, 1998).
- In October 1985, tropical storm Isabel dropped over 560 mm of rainfall within a 24-hour period on the island of Puerto Rico. The rainfall triggered the Mameyes rock slide, which destroyed 120 homes and killed an estimated 129 people, though only 39 bodies were recovered (Jibson, 1992). This death toll is the largest from a single landslide in North America.

- On 30 July 1997 a landslide occurred in the Thredbo ski village, Australia. The landslide caused a section of the Alpine Way, the major road in the area, to collapse. The Carinya Lodge, directly under the Alpine Way, moved down the slope about 100 m before hitting the Bimbadeen Lodge. Eighteen people were killed, with only one survivor emerging from the rubble (Hand, 2000).
- On 5 May 1998, Sarno and neighbouring villages in the Campania region of Italy were devastated by a series of mudslides and debris flows (Guzzetti, 2000). Around 160 people died, prompting the development of new legislation on landslide risk assessment procedures in Italy (*Gazzetta Ufficiale della Repubblica Italiana*, 1998).
- Over 380 people were killed by a landslide on the slopes of Mount Elgon, Uganda, in March 2010 (Gill, 2011). A number of school pupils had taken shelter from the heavy rainfall in a health centre, which was buried beneath landslide debris.

In addition to these high-fatality events, there are numerous incidents that cause fewer deaths. The following examples can only begin to show something of the diversity of the unfortunate circumstances.

- In July 1989 a large rockfall (1400 m^3) collapsed onto a rock shed protecting traffic on National Highway Route 35 in Echizen-cho, Japan. The crest of the shed was displaced, crushing a micro-bus that had been passing at that moment; all 15 passengers were killed (Yoshimatsu, 1999).
- In April 1988 a 17 m-high railway embankment at Coledale, a suburb of Wollongong, Australia, failed during heavy and prolonged rainfall. The mudslide enveloped a house situated about 20 m from the base of the embankment, trapping four occupants. Two were rescued from the debris, but a mother and her 2-year-old son were killed (Davies and Christie, 1996).
- A rockfall from a cutting on British Columbia Highway 99 in 1982 fell on a vehicle that was stopped in traffic, killing a woman and disabling her father (Bunce *et al.*, 1997; Cory and Sopinka, 1989).
- 15 people were killed in May 2005 when the village of Kuzulu, in Sivas Province, Turkey, was buried by a debris avalanche (12.5 Mm^3) that had travelled over 2 km down the valley at around 15 m/s (e.g. Gokceoglu *et al.*, 2005; Ulusay *et al.*, 2007).
- On 10 January 2005 a landslide struck the community of La Conchita in Ventura County, California, destroying 13 houses and killing ten people (Jibson, 2006). The event involved the remobilisation of part of a landslide that dated from 1995. The 2005 landslide occurred at the end of a 15-day period that produced record and near-record amounts of rainfall in many areas of southern California.
- Five people were killed on the island of St Lucia, Eastern Caribbean, when their house was destroyed by a major debris avalanche triggered by the category 2 hurricane Tomas, which hit the island in October–November 2010 (ECLAC, 2011).

Most events like these would only have been recorded locally prior to the relatively recent establishment of disaster databases (e.g. EM-DAT) and the Durham Fatal Landslide Database and, therefore, their existence will only be counted if researched either locally or at a national level. When this is done, the numbers of landslide-induced deaths rise dramatically. For example, Ikeya (1976) found that in Japan over the five years (1967–1972), no fewer than 662 people had been killed by debris flows and a further 682 by landslides, which gives a death rate of 269 per year. Similarly, research in Peru (Taype, 1979) identified 34 975 deaths in 50 years due to 168 slides, 37 mudflows and 3734 huaycos (debris torrents) – that is, 700 deaths per year, of which an unspecified proportion should be attributed to flooding. In Italy, it has been estimated that more than 5600 people were killed by landslides in the 20 century (Guzzetti, 2000).

The uncertainty regarding death-tolls is compounded further in the case of physical injuries. Red Cross/Red Crescent statistics indicate that for the period 1973–1997, 267 people a year were injured in landslide disasters (International Federation of the Red Cross and Red Crescent Societies, 1999). However, the actual number is undoubtedly many times greater, especially as the average annual death toll for the same period is quoted as 790 per year in the same publication.

The fact that widespread landsliding is often caused by other, more conspicuous, geohazard events such as tropical revolving storms, volcanic eruptions or earthquakes (*landslide-generating events*) means that the losses caused by landsliding are often attributed to these other hazards (Alexander, 2005; Jones, 1995). This is especially the case with reference to seismic activity, where the huge losses of life and destruction associated with the Shansi (1556), Kansu (1920), Armenian (1988), Tajikistan (1988) and Iranian (1990) earthquakes were, to a large extent, the result of landsliding (see US Geological Survey, 1989). This further reduces the perceived significance of landsliding as the cause of detriment, despite the work that has been undertaken to show just how much landsliding can be generated by seismic activity (Keefer, 1984, 2002). For example, the 1950 Assam earthquake ($M_s = 8.6$) is thought to have caused the displacement of a total of 50 000 Mm^3 of material over an area of 15 000 km^2 (Kingdon-Ward, 1952, 1955), while the 1999 Chi-Chi earthquake in Taiwan ($M_w = 7.5$) triggered over 20 000 landslides (Lin et al., 2003; Wang et al., 2003). The Guatemala City earthquake of 1976 ($M_w = 7.5$) caused 10 000 landslides over 15 000 m^3, and 11 over 100 000 m^3 (Harp et al., 1981).

Similarly, the spatial (geographical) record is extremely variable and generally of medium to poor quality. While records kept in developing countries may be adequate, especially in the case of landslides that caused major impacts or were conspicuous/dramatic, no such records exist for extensive tracts of the developing world, although attempts have been made to address this problem as part of the UN International Decade for Natural Disaster Reduction and subsequently. Dependence on the media as a source of information has tended to introduce significant bias in terms of emphasis on events that were of interest to readers or viewers (i.e. *close to home*), conspicuous, dramatic, rapid on-set, had *human interest* and involved human deaths. These influences were clearly identified in a study of American TV coverage of natural disasters (Adams, 1986), which revealed a Western prioritisation dominating over severity, so that, in terms of newsworthiness, the death of one Western European equalled three Eastern Europeans, equalled nine Latin Americans, equalled 11 Middle Easterners, equalled 12 Asians. It is clear that only the undertaking of detailed, systematic studies of landsliding will reveal both the geographical patterns of landslide hazard and the relative significance of past events.

There undoubtedly exist additional accounts of landslides for most areas held in files, archives, books, reports, newspapers and journals, which can be searched to improve the record, as was attempted in the case of the UK Government sponsored 'Review of Research on Landsliding in Great Britain' (see Jones and Lee, 1994). However, the historical record in the public domain provides a biased and incomplete picture of how landslides cause detriment.

Nevertheless, there now exist numerous well-documented case histories of the impact of major geohazard events, such as catastrophic landslides, which reveal that an enormous variety of adverse consequences or detriment can be produced. People are killed and injured, buildings and property are damaged or destroyed, services are cut-off, 'normal life' is disrupted, and the survivors may experience hunger, exposure, illness and despair. In every major event, there are innumerable individual accident sequences (see Chapter 2; Figure 2.4) that result in a cascade of detriment that may last for years and which can be almost impossible to fully unravel after an event, let alone predict in advance of some unspecified time in the future. Thus, consequence assessments have, through necessity, to be increasingly pragmatic and broad-brush with the increasing extent of the envisaged impact. For small sites and well-defined events, however, more specific detail can be modelled.

Finally, it has to be noted that the range of reported impacts does not cover all possibilities for the future, not simply because the record is incomplete but because evolutionary changes in society and technology have, and undoubtedly will, result in new forms and combinations of impacts. Thus, the historical record provides an important guide but must not be a straitjacket; consequence assessments need to consider new possibilities and, in some cases, to 'think the unthinkable'.

9.4. Categories of adverse consequences

The adverse consequences of landsliding can be extremely varied in terms of nature and timing. In order to develop consequence models and scenarios, this complexity needs to be distilled into a simple and clearly structured framework in order to facilitate prediction. The simplest division is into *direct effects* and *indirect effects*. *Direct effects* are the first-order consequences that are intimately associated with an event, or arise as a direct consequence of it (e.g. destruction), while *indirect effects* emerge later, such as mental illness, longer-term economic problems or relocation costs (Smith, 2001).

Direct and indirect effects consist of both *gains* and *losses*, and can be further divided into *tangible* and *intangible* categories, depending on whether or not it is possible to assign generally accepted monetary values to the losses/gains (e.g. van Westen *et al.*, 2006). However, it is unsatisfactory to totally exclude consideration of certain relevant consequences simply because they cannot be quantified: they should be indicated as possibilities within a qualitative addition to the core quantitative assessment. Building on the previous discussion of vulnerability (see Chapter 8), the main categories are

- direct impact on people (loss of life or injury)
- direct economic costs arising from direct impact on buildings, structures and infra-structure
- other economic costs that can be estimated in advance, including indirect economic costs, possessions, etc.
- intangible losses, including direct and indirect economic, social and societal losses that cannot be predicted in advance.

Hindsight reviews of the impact of environmental hazards have resulted in the use of the terms 'direct economic losses', 'indirect economic losses' and 'intangibles'. Examples will be given in the following sections to give an idea of the scale and complexity of the adverse consequences that can be generated by landsliding. In these instances, careful analysis 'after the dust has settled' has revealed statistics of destruction and estimations of cost, although the latter could remain uncertain for years as the longer-term consequences are played out.

The terms *direct economic losses* and *indirect economic losses* are generally applied to those losses capable of being given monetary values because of the existence of a market, with all other losses classified as *intangibles*. *Direct economic losses* arise principally from the physical impact of a landslide on property, buildings, structures, services and infrastructure.

Indirect economic losses or *indirect costs* are those that subsequently arise as a consequence of the destruction, damage and disruption caused by a primary hazard (i.e. the landslide itself), secondary hazards or follow-on hazards. The distinction between 'direct' and 'indirect' can be somewhat arbitrary and a matter of subjective preference. A further focus of uncertainty concerns whether the impact of secondary hazards (e.g. fires and floods) and follow-on hazards (e.g. looting) count as direct or indirect losses.

Intangible losses are the vague and diffuse consequences that arise from an event and which cannot easily be valued in economic terms (e.g. changes in environmental quality, ecosystem degradation). From a risk assessment perspective, many of the adverse outcomes will be difficult to impossible to predict, let alone quantify, and so the *intangible* category becomes significantly bigger.

There are strong grounds for including the physical impacts of secondary hazards, including fires and directly attributable accidents, within direct economic losses because they arise primarily as a consequence of ground movement. However, it is often difficult to distinguish where one set of destructive forces ceases and another begins. For example, if a landslide descends into a lake or dams a valley, then it is illogical for the costs of inundation and downstream flooding to be excluded from the losses attributable to the landslide (e.g. the Vaiont dam disaster of 1963). The same logic is applicable in the case of tsunamis, even though the costs may be borne by other societies. Similarly, if a landslide directly causes an accident, then the costs of the accident should be attributed to the landslide. Thus, care should be taken not to confuse *secondary impacts* and *indirect effects* with indirect losses.

An alternative approach advocated by Hewitt (1997) is to subdivide losses into *primary*, *secondary* and *tertiary damages*, with secondary damages largely the product of secondary hazards but including fire, and tertiary damages arising from the impairment of general functions such as disease, delayed economic effects and forced out-migration. This is a useful framework for use in hindsight studies concerned with assessing the effects of past major impact events, where the passage of time allows the identification of longer-term political, economic, social and environmental consequences. However, it is of limited value for determining adverse consequences in advance as part of a risk assessment, where a simpler and more pragmatic approach is required.

9.5. Loss of life and injury

One of the main priorities of any risk assessment is the identification of the possibility of human loss of life or physical injury (e.g. Fell *et al.*, 2005). This is not merely a response to an increasingly litigious society but reflects a combination of other factors: the human-centred nature of risk assessment, the fact that humans generally tend to value human life above all other things; and because casualty figures often convey a sense of loss far more effectively than does any other form of impact statistic and, as a consequence, can generate highly adverse publicity.

Clearly the level of detail that can be undertaken in a landslide risk assessment will very much depend on the nature and scope of the assessment and the complexity of the consequence scenarios in terms of adverse effects on humans. However, some estimates of casualty figures and costs should be feasible. These estimates need to take into account the following.

- *Will humans be impacted in the open?* In which case the characteristics of the landslide event will be important. According to Alexander (2002) the three primary causes of death (physiological pathologies) associated with landslides are general trauma, crush injuries and suffocation, with drowning a further possibility. Clearly, the relative significance of these will vary with the type, material properties, volume and velocity of the anticipated landslides. For example, Sanchez *et al.* (2009) report that 90% of the 43 people killed when over 250 landslides and debris flows were generated in Micronesia following a tropical storm died from suffocation, while Guzzetti (2000) claimed that 80% of landslide deaths in Italy are due to fast moving slides.

- *Are humans likely to be trapped within buildings and affected by building collapse?* In which case both the characteristics of the landslide event and the design and robustness of the buildings will influence the outcome. Should the building disintegrate on impact, then the consequences for those inside will be the same as above. However, if the building collapses or partially collapses, then fractured bones, concussion and loss of blood need to be added to the list. The literature on earthquake impact is a useful guide to what can happen, how building design affects survival rates and the speed with which people die in collapsed buildings ('fade away times'; e.g. Coburn and Spence, 1992).
- *The age, sex and health of the population at risk,* because agility/mobility determines the ability to avoid moving material. Also, it is the young and old that tend to have the highest mortality rates in hazard events (e.g. Pradhan *et al.*, 2007), and women that experience the highest levels of post-traumatic stress (e.g. Peek-Asa *et al.*, 2002).

The term 'injury' covers a wide spectrum of differing levels of harm in terms of severity, costs involved in recovery and the extent to which full recovery can be achieved. The simple division of injuries into 'light', 'moderate', 'severe' and 'critical' is a useful first approximation. Alternatively, the triage injury categories can be used

- dead or unsaveable
- life-threatening injuries requiring immediate medical attention and hospitalisation
- injury requiring hospital treatment
- light injury not requiring hospitalisation ('the walking wounded'), including lacerations, cuts and bruising.

Alternatively, a scheme based on the degree of incapacitation might also be adopted

- injury causing major incapacitation
- injury causing minor incapacitation
- injury not affecting performance.

It is unfortunate, therefore, that despite much research on measures of injury severity (see Olser, 1993), there is still no internationally agreed standard injury classification and no agreed basis for data collection. Thus, existing statistics must still be viewed as of dubious quality, especially as more trivial injuries tend to be under-recorded. Also, some schemes fail properly to address the differing types of physical injury, or the extent to which pre- impact normal life can be resumed due to individuals being crippled, blinded, deafened or experiencing other forms of long-term physical or psychological disability.

It follows from the above that casualty estimation is extremely problematic. Although there has been much work in the context of earthquake impacts, there is relatively little information available for other geohazards, such as landslides. Estimation must, therefore, be based on the development of scenarios, using, for example, the following categories of impact on humans (Alexander, 2002).

- No injury.
- Slight injury. Minor medical attention will solve the problem, and transport to hospital is not required; medical assistance is not urgently required.
- Serious injury that does not require immediate priority treatment. The patient will be taken to hospital but will not be among the first to be treated. Injuries are not life-threatening and will not lead to a worsening of the patient's condition if they must wait to receive treatment.

- Serious injury that requires priority treatment in order to produce some significant improvement in the patient's long-term prognosis, or simply to avoid deterioration in the patient's condition.
- Instantaneous death.
- Subsequent death due to stress-induced heart attacks, catastrophic deterioration of pre-existing medical conditions or complications arising from initially non-fatal injuries, including general decline due to lack of resilience (a feature of older people), increased vulnerability to infectious diseases and suicide due to depression.

In most risk assessments, fatalities will be expressed in terms of the annual probability of death (e.g. individual risk) or potential loss of life (i.e. societal risk). However, in some cases it may be necessary to apply monetary values to death and injuries. This is generally undertaken to support policy analysis rather than to determine levels of compensation (the distinction is important).

The valuation of deaths and injuries is very contentious, largely due to confusion as to what is being valued. In the case of injuries, it is possible to estimate the costs of rescue operations, of emergency treatment and of hospitalisation for differing categories of injury (i.e. light, moderate, severe, etc.). This will produce average costs per type of injury, which are important, but fails to value the incapacitation and discomfort to the person concerned. Rather more difficult is the problem of identifying long-term costs in terms of loss of function (i.e. blindness) and loss of earnings. Research can produce figures for these, based on certain assumptions, but it is a time-consuming and expensive exercise, so for most landslide risk assessments these items will be considered to be *intangibles*.

The same is generally true of *death*, although there have been efforts to place a value on saving/extending a human life in an abstract sense. The basis of this work is well described in Mooney (1977), Jones-Lee (1989), Marin (1992), Pearce *et al.* (1995, 2006) and Organisation for Economic Co-operation and Development (OECD) (2011), and focuses on establishing the *value of a statistical life* (VOSL or VSL), a notion that a significant proportion of the population finds morally repugnant and ethically unacceptable, largely because of the confusion between VOSL and an actual person. VOSL is used to estimate the benefits of reducing the risk of death (Viscusi, 2003) or the financial value of reducing the average number of deaths by one. Put another way, 'VOSL is most appropriately measured by estimating how much society is willing to pay to reduce the risk of death' (Australian Government, 2008).

Pearce *et al.* (1995) describe the approach as follows:

> Attributing a monetary value to a 'statistical life' is controversial and raises a number of difficult theoretical and ethical issues. It is important to understand that what is valued is a change in the risk of death, not human life itself. In other words, the issue is how a person's welfare is affected by an increased mortality risk, not what his or her life is worth. If 100 000 people were exposed to an annual mortality risk of 1 : 100 000, there will, statistically, be one death incidence per year. Removing the risk would thus save one 'statistical life'. It is this statistical life that has an economic value. It would make no sense to ask an individual how much he or she is willing to pay to avoid certain death. Nor is that the context of social decision making. But it can make sense to ask what individuals are willing to pay to reduce the risk of death or what they are willing to accept to tolerate an increased risk of death.
>
> The reality is that safety is not 'beyond price'. If it were, most of the world's wealth would be spent trying to save lives by reducing accidents and preventing disease. Risks are taken every day, both by individuals and by governments in choosing their social and economic expenditures, some of which are specifically directed at protecting and extending human life. For example, if a government introduces a programme of inoculation for childhood diseases that costs $10 000 000 per year and saves an average of 80 lives per year, a statistical life is implicitly valued at $125 000 at a minimum.

Several methods have been applied in attempts to calculate VOSL, all of which suffer from problems. The *prescriptive* or *normative* approach attempts to set VOSL in terms of what life ought to be worth, and tends to be dismissed as lacking a proper economic basis. By contrast, the *'descriptive'* approach essentially involves establishing how much people are actually willing to spend to reduce the risk of death. Two approaches have mainly been used (Pearce *et al.*, 1995).

- *The human capital approach.* This involves treating an individual as an economic agent capable of producing an output that can be valued in monetary terms. The value of this output, less any consumption that the individual would have made, is assumed to occur when he or she is killed, for example by a landslide. The approach tends to produce extremely low values for those with low earnings, discriminating against the poor.
- *The willingness-to-pay/willingness-to-accept method* (WTP/WTA). This involves valuing a statistical life on the basis of what individuals are willing to pay or accept for changes in the level of risk. Such values can be derived from contingent valuation, where individuals are asked directly how much they would be willing to pay to reduce risks. Other measures include finding out how much people are spending on safety and disease-preventing measures, or by how much wages differ between safe and risky jobs (the hedonic approach). For example, suppose 100 000 workers are paid an additional $15 each to tolerate an increased risk of mortality of 1/100 000. The increased risk will result in one statistical life lost, valued at $15 × 100 000 = $1 500 000.

The WTP approach is the more generally favoured, involving as it does both asking people about their views of risk (*expressed preference*) and studying responses to risk (*revealed preference*). However, other techniques are also used (see OECD, 2011; Pearce *et al.*, 2006), and the resulting calculations reveal very variable values for VOSL depending on the approach adopted, the assumptions made, the techniques used, the types of risk addressed and the context (e.g. health, transport, natural hazards; Table 9.1, Bellavance *et al.*, 2009). Work in the early 1990s suggested that VOSL for the UK should be set at £2–3 million at 1990 prices (Marin, 1992), but other work produced VOSLs for developed countries in the range US $1.8–$9 million, with a best guess average estimate of US $3.5 million. However, estimates for other countries were significantly less at US $300 000 for Russia, and only US $150 000 for China, India and Africa, which provoked outrage about discrimination in certain quarters. A global average of US $1 million was proposed by Pearce *et al.* (1995).

Table 9.1 Table of VOSL values compiled by Bellavance *et al.* (2009) from analysis of 40 studies spread across different countries (values are US dollars (2000))

	Number of studies	Average value/median	Standard deviation
USA	16	$6.27 million/$4.65 million	$4.97 million
Canada	7	$9.16 million/$4.05 million	$10.39 million
UK	3	$17.00 million/$14.18 million	$12.59 million
Australia	2	$11.17 million/$11.17 million	$9.63 million
South Korea	1	$1.55 million	–
India	1	$16.07 million	–
Japan	1	$12.81 million	–
Taiwan	1	$1.20 million	–

These debates and differences are of little concern, because what is required for landslide risk assessments is a generally agreed VOSL for use in a national context for the purpose of policy appraisal. In the case of the UK, Marin (1992) suggested £2–3 million at 1990 prices, but the UK Department for Transport (DfT) subsequently fixed on £1.2 million, while the UK Health and Safety Executive uses a benchmark figure of £1 million at 2001 prices for the value of preventing a fatality (VPF) in cost–benefit analyses (Health and Safety Executive, 2001). Other countries use different figures: for example, the Australian Government (2008) recommends A \$3.5 million at 2007 prices, based on Abelson (2007). The most recent review (OECD, 2011) shows that the UK DfT currently uses £1.638 million (at 2007 prices) to estimate the 'social costs of preventing a fatality', the US Environmental Protection Agency uses US \$7.5 million (2007), but other agencies use different values, while the EU uses values in the range €1–2 million.

With respect to valuing injuries, one approach is to use the *earnings foregone* approach. For example, Alexander (2002) quotes figures for the mid-1990s from the USA based on *earnings foregone* calculations, which indicate that if death results in an average value of US \$2.2 million, then equivalent values for 'moderate' and 'slight' injuries are US \$5000 and US \$200, respectively. However, greater emphasis has come to be placed on a product of VOSL studies, the related concept of a the value of a statistical life year (VSLY or VOLY; Pearce *et al.*, 2006). VOLY can be calculated in several ways (Pearce *et al.*, 2006), the easiest being:

$$\text{VOLY} = \frac{\text{VOSL}}{\text{life expectancy of the person} - \text{present age of the person}}$$

with the remaining years discounted (see Chapter 10). The simplest approach is to value injuries and disabilities by taking the applicable VOSL (interpreted as the value of a year of life free of injury or disability) and multiplying it by a weighting determined by the severity of the injury (see Mathers *et al.*, 1999). Thus, the Australian Government (2008) quotes an amputated foot as $0.3 \times \text{VOLY} = 0.3 \times \text{A} \$151\,000 = \text{A} \$45\,300$ per year. The UK Department for Environment, Food and Rural Affairs currently uses a VOLY value of £29 000 at 2004 prices (OECD, 2011).

9.6. Direct impact on buildings, structures and infra-structure (direct economic costs/losses)

Estimating and predicting economic losses is both complex and problematic. Quite naturally there is a tendency to restrict direct losses so as to make them easier to compute. This can be achieved by restricting direct losses to physical damage and destruction, thereby excluding costs of transport disruption, pollution, etc. Alternatively, temporal limits can be placed on direct costs, so that only those actually caused during the period of ground motion are considered, thereby neatly side-stepping the problems posed by fires, floods and tsunamis. Lastly, spatial limits can be placed on direct losses, so that only those costs caused by the landslide within its limits are considered, as is exemplified by Schuster and Fleming's (1986) definition of direct losses as 'costs of replacement, repair, or maintenance due to damage to installations or property within the boundaries of the responsible landslide'. According to this definition, direct losses are only those losses that are caused by the ground disruption that occurs during landsliding and the physical contact of displaced materials with properties and their contents.

The potential for landslides to cause direct economic damage is enormous because human constructions and wealth are concentrated at, or very near to, the Earth's surface. Thus movements of the ground itself, or of material over the surface of the ground, has the potential to inflict serious impact on fixed assets that may either be shallowly buried or located on the surface, or raised above the surface but founded in the ground. Adverse impacts may range in scale from minor cracks in walls, buildings and roads due to limited displacements, via the destruction of individual buildings or even groups of buildings, to scenes of total devastation which, in the case of the Huascarán catastrophic failure of 1970, resulted in the total destruction of the town of Yungay and the death of most of its inhabitants (Plafker and Ericksen, 1978).

Data on the global effects of landsliding in terms of direct economic costs is recognised to be of very poor quality. However, the widespread occurrence of landsliding suggests that it must be significant, and this can be supported by the work of Alfors et al. (1973) in California, USA, which concluded that landsliding was ranked second to earthquakes in terms of loss potential, accounting for 25.7% of *projected losses* from geohazards over the period 1970–2000.

Among the more dramatic impacts in terms of direct losses have been the following.

- Mudflows that accompanied the 1980 eruption of Mount St Helens, Washington State, USA, and damaged or destroyed over 200 buildings and 44 bridges, buried 27 km of railway and more than 200 km of roads, badly damaged three logging camps, and disabled several community water supply and sewage disposal systems (Schuster, 1983). Landslide damages associated with the eruption are estimated to have been $500 million.
- The Ancona landslide, Italy (Marche region), occurred on 13 December 1982. It involved the movement of 342 ha of urban and suburban land, damage to two hospitals and the Faculty of Medicine at Ancona University, damage to, or complete destruction of, 280 buildings containing 865 apartments, displacement of the main railway and coastal road for more than 2.5 km, one (indirect) death, and the evacuation of 3661 people (Catenacci, 1992; Crescenti, 1986). The economic loss was estimated at US $700 million (Alexander, 1989).
- Heavy rainfall associated with a strong El Niño in the winter and spring of 1998 caused over $158 million in landslide damage in the San Francisco Bay region, USA. This led to ten counties being declared eligible for disaster assistance from the Federal Emergency Management Agency (Godt and Savage, 1999).
- The widespread landslide damage in the Basilicata region of southern Italy where it is estimated that 18.5% of the land area is affected by landsliding, including 1800 deep-seated landslides (Regione Basilicata, 1987). Damaged buildings, unsafe bridges and heavily disturbed roads are ubiquitous. As the majority of the 131 towns (communes) in the region are built on hill tops, it is not surprising to learn that no less than 115 were sufficiently endangered by the progressive upslope development of landsliding as to require either extensive engineered structural measures (consolidation), abandonment in favour of relocation to a new, safer location (transferral), or a combination of the two (Fulton et al., 1987; Jones, 1992). The abandonment of the village of Craco is a particularly well-documented example (Del Prete and Petley, 1982; Jones, 1992), but is not particularly unusual, for there are numerous examples of abandoned villages in the area due to a combination of earthquake shaking and slope failure. Indeed, in the adjacent region of Calabria the cost of damage to roads, railways, aqueducts and houses was estimated at US $200 million for 1972–1973 alone (Carrara and Merenda, 1976), and over the centuries nearly 100 villages have had to be abandoned due to landsliding, involving the displacement of nearly 200 000 people.
- A major landslide occurred at the town of Saint-Jean-Vianney, Quebec on May 4, 1971. An estimated 6.9×10^8 m^3 of quick clay material liquefied and travelled down the Rivière aux Vases at a speed of around 25 km/hour (Tavenas et al., 1971). The landslide destroyed 41 homes and

killed 31 people. The site was subsequently declared unsafe for habitation, and over the next 6 months the survivors were resettled at Arvida. In 1989, it was found that the nearby town of Lemieux was at risk from a similar quick clay failure, and, after consultations with the township, the provincial Ministry of Natural Resources and the local residents, it was decided to relocate the residents to a safer area. Over the next 2 years, the residents were relocated to existing nearby communities at the provincial government's expense. On 20 June 1993, 2 years after the town was officially abandoned, the site was destroyed by a major retrogressive landslide (McIntyre, 2005).

These events are merely extreme examples of what can be a common problem in urban areas located in mountainous or hilly terrain. Typical problems include the costs of clearing up landslide debris from roads and pavements, rebuilding or repairing damaged property, and replacing or repairing lengths of road, sewers, water pipes and fractured gas mains. For example, Murray (2000) reports that in New Zealand there were 800 insurance claims for damage caused by landslides during 1997–1998, at a cost of \$4.7 million.

Significant direct losses can also arise as a result of cliff recession, especially on the coast. It is well known that the Holderness coast of England has retreated by around 2 km over the last 1000 years, and in the process destroyed at least 26 villages listed in the Domesday survey of 1086; 75 Mm^3 of land has been eroded over a 100-year period (Pethick, 1996; Valentin, 1954). Perhaps the most famous example of the effects of rapid recession on the English coast is at Dunwich, Suffolk, where much of the town was lost over the last millennium (e.g. Bacon and Bacon, 1988). Gardner (1754) records that by 1328 the port was virtually useless, and that 400 houses, together with windmills, churches, shops and many other buildings were lost in one night in 1347.

9.7. Other economic costs that may be estimated in advance

There are two categories to be considered here. The first is that identifying buildings, structures and infra-structure as a separate category (see above) leaves the contents of buildings, equipment, machinery, items of personal property, works of art, ornamental vegetation, crops and domesticated animals, etc., as an area for consideration. Sometimes it may be possible to quantify such costs, especially where destruction is inevitable, but this level of analysis may be appropriate only in the case of very localised, time-limited risk assessments involving, for example, a single or small group of buildings.

The second involves *indirect economic losses* or *indirect costs*, which are those costs that subsequently arise as a consequence of the destruction and damage caused by a primary hazard (i.e. the landslide itself), secondary hazards or follow-on hazards. They include the costs due to the long-term disruption to transport, loss of production of agricultural products, manufactured goods and minerals, loss of businesses and retail income, costs of cleaning up, resulting water pollution or contaminated land, the various costs incurred during the recovery process such as unemployment, and the extra costs resulting from increased illness, etc. The effects of post-event civil disorder such as rioting, looting and murder might be included in indirect costs.

Examples of indirect loss include the following:

- An earthquake in 1987 triggered landslides in the Andes of north-eastern Ecuador, which resulted in the destruction or local severance of nearly 70 km of the Trans Ecuadorian oil pipeline and the only highway from Quito to the eastern rain forests and oil fields (Nieto and

Schuster, 1999). Economic losses were estimated at $1 billion. Oil exports were disrupted for almost 6 months, reducing the government's income by 35% (Stalin Benitez, 1989).

- Landslides near Lake Tahoe, USA, led to the closure of the US-50 highway. The direct loss associated with highway repairs was $3.6 million (Walkinshaw, 1992). However, the road was closed for 2.5 months, causing access disruption and loss of tourist revenues of $70 million (*San Francisco Chronicle*, 1983).
- The Thistle landslide, Utah, in 1983 severed the Denver and Rio Grande Western Railroad main line. The landslide closed the main railroad for 3 months, and Routes 6 and 89 for 8 months, during which time the communities of eastern and south-eastern Utah were cut off from the rest of the state. The operations of coal mines, uranium mines, turkey farms, animal feed companies, gypsum mines, and cement and clay factories were all severely impacted. At least two trucking firms and one oil-producing firm suspended or ceased operations. Some people who lived and worked on opposite sides of the landslide area suddenly had commutes exceeding 160 km. The direct cost of the landslide was estimated at $200 million (equivalent to $467 million in 2013). However, some estimates of the total cost reached as high as $400 million (equivalent to $933 million in 2013). The railroad company estimated the slide cost them $1 million for each day that the tracks were out-of-service ($80 million in total lost revenue; University of Utah, 1984). This figure included $19 million in payments to the Union Pacific company for the use of their lines.
- The Fraser River salmon fisheries, Canada, were severely affected by a small rockfall that occurred during construction of the Canadian National Railway in 1914. The fall prevented migrating salmon reaching their spawning grounds. Between 1914 and 1978, the losses to both the sockeye and pink salmon fishery were of the order of $2600 million (International Pacific Salmon Fisheries Commission, 1980).

Indirect costs should also include the costs incurred due to the threat of hazard. These include the costs of planning for and implementing emergency action and evacuation, including the provision of temporary shelter, food and medical services to those who may have had to be moved; relocation costs arising from the need to permanently rehouse people in safer areas; the extra costs of increased geotechnical investigation, monitoring and mitigation measures required as a consequence of the recognition of increased risk following a hazard impact; additional costs to future construction arising from the development and enforcement of new building regulations and codes; the costs of increased insurance premiums arising from the recognition of landsliding as a problem in an area; and any fall in property prices as a consequence of an event or the implementation of a new zoning scheme, etc.

A complete assessment of indirect losses is difficult to achieve, even in hindsight assessments, because many elements of cost may still be incurred a considerable time after the 'hazard event' and result from consequences not specifically restricted to the immediate area physically impacted by the event. They are, therefore, extremely difficult to impossible to predict in advance for the purposes of risk assessment. Often the best that can be expected is a list of broader-scale adverse consequences that could arise under certain sets of conditions. Only in those situations where the risk assessment is site specific should some assessment of direct and indirect costs be attempted.

9.8. Intangible losses

Intangible losses are the vague and diffuse consequences that arise from an event and which cannot easily be valued in economic terms, together with impacts on elements that can be measured but not valued. They include effects on the environment, nature conservation, ecosystem health, amenity, local culture, heritage, aspects of the local economy, recreation and peoples' health, as well as the attitudes, behaviour and sense of well-being of people. Possible examples are landslide scars disfiguring a famous view, or landsliding causing the destruction of an important historical monument, both of which could result in 'costs' in terms of distress and reduced pleasure, as well as adverse effects on the local economy in terms of reduced tourism. The effects of post-traumatic stress on the populations affected can also be significant.

Landsliding can cause significant ecological damage. Changes in terrain, vertical displacement of ground, alterations to slope aspect and drainage, burial of existing ground, changes in river flow, inundation behind landslide dams, catastrophic floods and increased sediment loads in rivers represent some of the more obvious sources of ecological stress. Alteration or destruction of habitat can have profound consequences.

The following three examples illustrate the problematic nature of intangible losses and indicate how extremely difficult it is to incorporate them within risk assessments.

- A major landslide occurred during the construction of a tailings dam to store waste from the Ok Tedi copper mine, Papua New Guinea, in 1984 (e.g. Griffiths *et al.*, 2004). The landslide led to the abandonment of the dam site and the dumping of 80 000 tonnes of mine waste containing lead, cadmium, zinc and copper directly into the river. According to the Australian Conservation Foundation, nearly 70 km of the Ok Tedi River became 'almost biologically dead', with a further 130 km 'severely degraded', so that fish stocks declined by 90%. The chief executive of Broken Hill Properties, the leader of the Ok Tedi Mining Ltd consortium, described the problems that followed the landslide as an 'environmental abyss'. Broken Hill Properties agreed to a $430 million write-down of the value of the asset and to hand its stake to the Papua New Guinea Government.
- In the aftermath of the Aberfan disaster of October 1966, Gaynor Lacey, a consultant psychiatrist at the Merthyr Tydfil Child Guidance Centre, saw 56 children who had developed behavioural problems since the landslide (Lacey, 1972). The most common problems were sleeping difficulties, nervousness, insecurity, enuresis and unwillingness to go to school. Many children had lost all their friends, so there was little point in going out, and once having stayed in for a lengthy period, it was difficult to start going out again. The educational development of many of the children was delayed until they began to overcome the trauma.
- In recent years, concerns have been raised that the ancient Incan stronghold of Machu Picchu in Peru is threatened by active rockslides (e.g. Carreno and Bonnard, 1999; Sassa *et al.*, 2009). The citadel was used by the Incas as a refuge from the Spanish conquistadores in the 16th century and was only rediscovered by Hiram Bingham in 1911. The ruins have been declared a UNESCO World Heritage Site and are widely regarded as a priceless monument.

Intangible costs or losses may be produced as both direct and indirect consequences of landsliding. Because they cannot easily be valued in economic terms, their existence results in the production of impact studies and risk assessments that consist of an unfortunate mixture of statistics (e.g. so many dead and injured), monetary costs and value judgements (e.g. reduced amenity value), which makes comparisons of risks an extremely difficult and subjective process. As a result, there has been much effort in developing approaches and techniques designed to allow monetary values to be assigned to different types of loss, as has been the case in the value of human life. Many of these are described in Bateman *et al.* (2002) and Department for Transport, Local Government and the Regions (2002). The valuing of intangibles, especially within the environment, is considered in more detail in Chapter 12.

9.9. Uncertain consequences
Quantitative consequence assessment is challenged by three sets of uncertainty.

- Uncertainty as to what could happen in terms of the combination of hazard and exposure potential, both now and increasingly into the future.
- Uncertainty regarding the variation in scale of the adverse consequences that could be produced.
- The difficulty of placing meaningful estimates of value on many of the possible adverse consequences.

Many of the challenges relate to *data uncertainty*, which can arise because of measurement errors (random and systematic) or incomplete data. As a result, the conditions will need to be inferred (e.g. interpolated, extrapolated or analytically derived) from other information. However, there may also be imperfect understanding regarding the processes involved or the applicability of transferring knowledge from one site to another.

Then there are so-called *environmental (real world) uncertainties*. For example, it may not be possible to predict how the future choices by governments, businesses or individuals will affect the socio-economic and physical environments within which landslide hazards operate. There is little prospect of reliably predicting what these choices will be (see Chapter 12).

Nevertheless, decisions will need to be made despite the uncertainties. As a result, risk assessments will often consist of a mixture of statistics, monetary costs and value judgements. This can make comparisons of risks an extremely difficult and subjective process. Hence, the objective must be to apply economic valuation wherever possible, and most especially in the case of well-defined, small-scale (site-specific) risk assessments.

9.10. Consequence models
The impact of a landslide is controlled by ground behaviour (e.g. the landslide intensity), the exposure of the elements at risk and their vulnerability to damage. Combining these factors enables landslide consequence models to be developed that reflect the damage signature of particular landslide scenarios. In each of the examples presented in Table 9.2, both the exposure (permanent and/or temporary) and vulnerability factors are used to establish the predicted damages compared with a total loss event in which the assets at risk would be completely lost. Thus, for a landslide of a particular intensity

$$\text{risk} = P(\text{event}) \times \text{adverse consequence}$$
$$= P(\text{event}) \times (\text{total loss} \times \text{exposure} \times \text{vulnerability})$$

For example, a convention centre (valued at $10 million; element 1) is located at the foot of an escarpment prone to debris slides ($P(\text{event}) = 0.01$). The spatial probability of the centre being in the path of a landslide was calculated to be 0.04. The level of damage resulting from the centre being hit by a typical debris slide was estimated to be severe (i.e. vulnerability $= 0.9$). The annual risk, expressed in monetary terms, is

$$\text{risk (element 1)} = 0.01 \times (\$10 \text{ million} \times 0.04 \times 0.9)$$
$$= \$3600$$

Table 9.2 Examples of simple landslide consequence models

Landslide scenario	Consequence factor	Consequence model	Comment
Cliff recession, ongoing cliff retreat causing progressive loss of cliff top property	Ground behaviour	Loss of cliff top land	Damage is the *total loss* of property in year of land loss
	Assets at risk	Properties set back from the cliff edge, at varying distances	
	Exposure	Fixed assets, permanent exposure. No loss of life as population is evacuated before land loss	
	Vulnerability	Total loss of assets as they are declared unsafe before they drop off the cliff edge	
First-time landslide, buried pipeline crossing slope with potential for landsliding	Ground behaviour	Horizontal and vertical displacements and deformation	Uncertain outcome, damage is dependent on the vulnerability of the pipe to different intensity events
	Assets at risk	Buried pipeline runs through the potential landslide area	Damage = *total Loss* × *vulnerability factor*
	Exposure	Fixed asset, permanent exposure	
	Vulnerability	Potential for pipe rupture depends on the intensity of movement	
Landslide reactivation, periodic slow ground movement threatens properties on the unstable slope	Ground behaviour	Horizontal and vertical displacements and deformation	Uncertain outcome, damage is dependent on the location of the property and its vulnerability to different intensity events
	Assets at risk	Properties located on landslide blocks, at varying distances from block boundaries	Damage = $\sum$ *total loss* × *vulnerability factor* for each property
		No loss of life expected because of very slow speed of movement	Note vulnerability factor varies with location
	Exposure	Fixed assets, exposure variable dependent on location relative to landslide block boundaries	
	Vulnerability	Potential for property damage depends on the intensity of movement	

Estimating the consequences

Scenario	Component	Description	Consequences
Debris slide or rockfall, failure of a cliff threatens road users at the cliff foot	Ground behaviour	Detachment and boulder fall with instantaneous impact. Debris slide run-out and instantaneous impact	Uncertain outcome, damage is dependent on the exposure of the population and its vulnerability to different intensity events. $Damage = \sum total\ loss \times exposure \times vulnerability\ factor$ for each event size
	Assets at risk	Vehicles using the mountain road	
	Exposure	Temporary exposure, dependent on number of users per hour and vehicle speed	
	Vulnerability	Potential for loss of life depends on boulder and/or slide and/or intensity	
Rockfall blocks roads and causes temporary road closure and traffic diversion	Ground behaviour	Detachment and boulder fall with instantaneous impact	Damage (and additional transportation costs and opportunity costs) dependent on period of road closure
	Assets at risk	Vehicles using road	
	Exposure	Dependent on number of vehicles and occupants per hour and vehicle speed	
	Vulnerability	Vehicles and occupants travelling will be delayed	
Debris slide or rockfall, failure of a road cutting threatens road users and roadside property	Ground behaviour	Detachment and boulder fall with instantaneous impact. Debris slide run-out and instantaneous impact	Uncertain outcome. Loss of life is dependent on the exposure of the population and its vulnerability to different intensity events. $Damage = \sum total\ loss \times exposure \times vulnerability\ factor$ for each event size. Property damage is dependent on the location of the property and its vulnerability to different intensity events. $Damage = \sum total\ loss \times exposure \times vulnerability\ factor$ for each event size $\times$ exposure to that event size for every property. Note that exposure varies with location because of the variations in the landslide footprint with differing intensities of event
	Assets at risk	Vehicles driving along the road, properties at varying distances from the base of the cutting	
	Exposure	Road users: temporary exposure, dependent on number of vehicles per hour and speed. Property: fixed assets, but exposure varies with location and event intensity	
	Vulnerability	Potential for loss of life depends on boulder and/or slide size and/or intensity. Potential for property damage depends on the intensity of movement	

In most cases it will be necessary to carry out this exercise for each of the individual elements at risk, or even each individual property. For example, if a hotel (element 2, valued at $5 million, with the same spatial probability and vulnerability = 0.8) is adjacent to the convention centre (element 1):

risk (element 2) = P(event) × adverse consequence (element 2)

$\qquad$ = P(event) × (total loss × exposure × vulnerability)

$\qquad$ = 0.01 × ($5 million × 0.04 × 0.8) = $1600

risk = P(event) × $\sum$ adverse consequences (elements 1 to n)

$\qquad$ = 0.01 × (360 000 + 160 000)

$\qquad$ = $5200

A slightly different approach has been adopted by Wong *et al.* (1997) in which the adverse consequence scenarios are considered in relation to the adverse consequences expected from a benchmark, or reference, landslide of known dimensions and impact. The approach was developed for estimating the potential loss of life in Hong Kong, where the reference slide was taken to be a 10 m-wide, 50 m^3-volume, shallow hillside failure. The number of fatalities expected when such a reference slide affects various land use categories in differing locations is shown in Table 9.3. The loss of life associated with a potential landslide is then scaled up or down from the reference slide. The scaling factor used was based on the width of the landslide relative to the reference slide:

$$\text{scale factor} = \frac{\text{potential slide width}}{\text{reference slide width}}$$

A vulnerability factor is used to relate the loss of life associated with the reference landslide to that expected with the potential slide. This vulnerability factor is influenced by a number of variables, including the nature, proximity and spatial distribution of the facilities (assets) and the debris mobility (see Example 9.7, below).

Using the reference landslide approach, the risk is calculated as

risk = P(event) × adverse consequence

$\qquad$ = P(event) × (reference slide expected fatalities × scale factor × vulnerability)

Note that variable exposure is not specifically addressed in this reference landslide approach. However, the probability of a person being caught in the path of the debris (i.e. being in the 'wrong place at the wrong time') can be incorporated into the model, as follows:

risk = P(event) × (reference slide expected fatalities × scale factor × exposure × vulnerability)

This would involve the development of a population model in order to estimate the average number of people expected to be within the danger zone during a given time period (see Chapter 7).

The reference landslide approach has been used by ERM-Hong Kong (1999) in order to assess expected fatalities along selected roads in Hong Kong, including Castle Peak Road (a two-lane highway: see Example 10.16). This highway falls into category 2B ('road with heavy vehicular or pedestrian traffic density') in Table 9.3, for which the reference landslide would lead to one fatality.

Table 9.3 Fatalities associated with a reference landslide (10 m wide, 50 m³ volume) in Hong Kong

Group number	Facilities at risk	Expected number of fatalities
1	A. Buildings with a high density of occupation or heavily used residential building, commercial office, store and shop, hotel, factory, school, power station, ambulance depot, market, hospital/polyclinic/clinic, welfare centre	3
	B. Others: ■ bus shelter, railway platform and other sheltered public waiting areas ■ cottage, licensed and squatter area	3
2	A. Buildings with a low density of occupation or lightly used built-up area (e.g. indoor car park, building within barracks, abattoir, incinerator, indoor games sports hall, sewage treatment plant, refuse transfer station, church, temple, monastery, civic centre, manned substation)	2
	B. Others: ■ road with heavy vehicular or pedestrian traffic density ■ major infrastructure facility (e.g. railway, tramway, flyover, subway, tunnel portal, service reservoir)	1
3	Roads and open space: ■ densely used open space and public waiting area (e.g. densely used playground, open car park, densely used sitting-out area, horticultural garden) ■ quarry ■ road with moderate vehicular or pedestrian traffic density	0.25
4	Roads and open space: ■ lightly used open-air recreation area (e.g. district open space, lightly used playground, cemetery, columbarium) ■ non-dangerous goods storage site ■ road with low vehicular or pedestrian traffic density	0.03
5	Roads and open space: ■ remote area (e.g. country park, undeveloped green belt, abandoned quarry) ■ road with very low vehicular or pedestrian traffic density	0.001

From Wong *et al.* (1997)

Notes: (1) to account for different types of building structure with different detailing of window and other perforations, etc., a multiple fatality factor 1–5 is appropriate for group 1A, to account for the possibility that some incidents may result in a disproportionately larger number of fatalities than that envisaged; (2) for incidents that involve the collapse of a building, it is assumed that the expected number of fatalities is 100

Table 9.4 presents the vulnerability factors for roads at the toe of cut slopes (slope angle = 60°). These factors are expressed in terms of the run-out angle (reach angle; see Figure 3.5) of debris for a range of landslide volumes and slope heights.

The reach angle was calculated as follows ():

reach angle = $\tan^{-1}$(slope height/{distance from slope toe + [slope height/tan(slope angle)]})

Table 9.4 Vulnerability factors for roads at the toes of cut slopes

Failure volume: m³	Run-out of debris: degrees								
	20–25	25–30	30–35	35–40	40–45	45–50	50–55	55–60	>60
<20			0.0015	0.0065	0.019	0.042	0.072	0.095	0.1
20–50			0.03	0.1	0.23	0.37	0.47	0.5	0.5
50–500		0.0015	0.078	0.26	0.48	0.63	0.69	0.7	0.7
500–2000		0.01	0.15	0.48	0.83	0.95	0.95	0.95	0.95
>2000	0.01	0.15	0.48	0.83	0.95	0.95	0.95	0.95	0.95

From ERM-Hong Kong (1999)

For a given reach angle, vulnerability factors have been estimated considering the different lane sections separately and then averaged. For example, considering a failure volume of <20 m³ and a slope height of <10 m

- *Lane 1* (closest to the cut slope). The lane lies between 1.5 m and 4.5 m from the base of the cut slope, and would be covered by debris with a reach angle of over 50° (Table 9.5), indicating a vulnerability factor of 0.072 (Table 9.4).
- *Lane 2* (furthest from the cut slope). The lane extends for between 4.5 m and 8 m from the base of the slope and would be covered by debris with a reach angle of around 40°, indicating a vulnerability factor of 0.019.

The average vulnerability factor is 0.0455 – that is, (0.072 + 0.019)/2. Table 9.6 presents the scale factors that relate the expected fatalities in the reference landslide (see Table 9.3) to the actual failures. For a failure involving less than 20 m³ of material, a scale factor of 0.4 would be used. Thus, for this example, the adverse consequences, expressed as the probability of a fatality, are

adverse consequence = (reference slide expected fatalities × scale factor × vulnerability factor)

$$= 1 \times 0.4 \times 0.0455 = 0.0182$$

Table 9.5 Landslide reach angles (degrees) that would cover different road lanes, for different cut slope heights

Slope height: m	Lane 1 (1.5 m from slope base)	Lane 2 (4.5 m from slope base)	Lane 3 (8 m from slope base)	Lane 4 (12.5 m from slope base)
<10	52.1	40.3	31.3	24.0
10–20	55.9	48.7	42.0	35.3
>20	57.9	54.0	49.8	45.2

From ERM-Hong Kong (1999)

Example 9.1

A small coastal community is located above an eroding cliffline that has been retreating at an average annual rate of 1.25 m/year. Continued recession over the next 25 years will threaten a number of

Estimating the consequences

Table 9.6 Scale factors used to modify number of fatalities expected from a reference landslide (see Table 9.3)

Failure volume: m^3	Average width	Scale factor
<20	4	0.4
20–50	7	0.7
50–500	15	1.5
500–2000	20	2
>2000	25	2.5

From ERM-Hong Kong (1999)

properties and important services, including a gas main and sewer (i.e. exposure = 1.0). These losses will be irreversible – that is, once the land has been eroded, or the cliff top is considered too close for the house to be used safely, then the land or house cannot be regained (i.e. vulnerability = 1.0). The potential for loss of life is considered negligible because the local authority will require occupants to move out of potentially unsafe houses before they become too close to the cliff edge.

The average erosion rate was used to project erosion contours for 25 years into the future in order to determine the expected year of loss of particular houses (Figure 9.2). In addition, a narrow strip of agricultural land on the cliff top adjacent to the properties is also expected to be lost each year. Property prices and agricultural land prices were obtained from a local real estate agent, whereas the value of the services and infrastructure were provided by the operators.

The consequences of continued cliff recession were calculated for each year, from year 0 to year 25, as follows:

consequences (year t) = total loss × exposure × vulnerability

= market value (property and land lost) × 1 × 1

The overall losses arising from destruction within a 25-year period are the sum of the consequences for each year, and have a current market value of £6.25 million (Table 9.7).

Example 9.2

An oil pipeline has been built across the dissected hills of Hong Kong, buried within the colluvium and residual soils that mantle most slopes. The hill-slopes are susceptible to shallow (<3 m deep) debris slides or debris avalanches. Concerns have been expressed about the risk within a 5 km length of the route, where it crosses a series of deeply dissected catchments that have a history of widespread slope movements.

The 5 km length of route through landslide prone terrain was divided into sections, based on bedrock geology and slope class. The Hong Kong Natural Terrain Landslide Inventory (NTLI; Evans and King, 1998; Evans et al., 1997; King, 1997, 1999) statistics were used to generate landslide densities for each section. It should be noted that the figures used were for *recent slides* (i.e. occurred within the last 50 years), as the distribution of these slides is believed to be a good indication of landslide density over a 50-year period (Evans and King, 1998). The frequency per year was calculated by dividing the landslide densities by 50.

Figure 9.2 Example 9.1: erosion contours for modelling consequences of cliff recession. (After Penning-Rowsell et al., 1992)

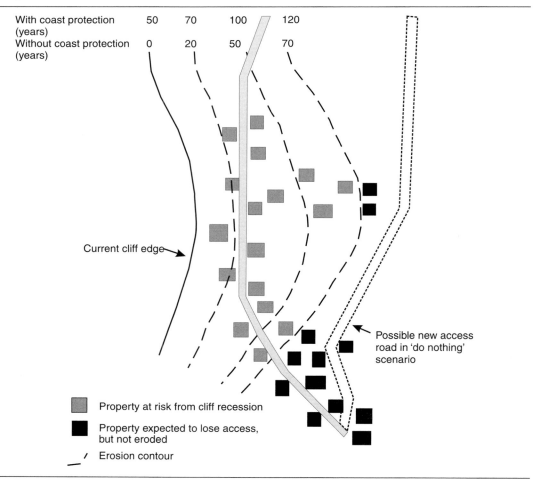

The landslide densities generated from the NTLI statistics were expressed as landslides/km² and not per unit length of pipeline. Although the pipeline right-of-way is around 30 m wide, it can be impacted by landslides occurring on the adjacent hill-slopes, both upslope and downslope. It was assumed, therefore, that the route and that portion of the adjacent hillsides that might generate impact on the pipeline could be considered to be a 250 m-wide corridor. Thus, the NTLI landslide densities were adjusted by a factor of 0.25 to produce landslide frequencies per kilometre of route corridor, which could then be used to yield annual probabilities (Table 9.8).

The pipeline will only be vulnerable to a proportion of all the landslides that affect the right-of-way. From industry experience, it was assumed that only slides >20 m wide *could* damage the pipe. The NTLI statistics revealed that 15% of all recorded slides are greater than 20 m wide (Evans et al., 1997). As these >20 m-wide slides could have a range of intensities and impacts on the pipeline, it

Estimating the consequences

Table 9.7 Example 9.1: consequence model

Year	Property lost due to recession	Market value: £ thousands	Agricultural land: ha/year	Market value: £ thousands
0			0.125	0.625
1			0.125	0.625
2			0.125	0.625
3			0.125	0.625
4			0.125	0.625
5			0.125	0.625
6			0.125	0.625
7	Trunk gas main	450	0.125	0.625
8	1–5 Acacia Avenue	350	0.125	0.625
9			0.125	0.625
10	Cafe	120	0.125	0.625
11	Sunnyview and Dunswimmin	180	0.125	0.625
12	Sewage Pump Station	2300	0.125	0.625
13	Hightrees House	1500	0.125	0.625
14			0.125	0.625
15			0.125	0.625
16			0.125	0.625
17	2–8 Acacia Avene	400	0.125	0.625
18	The Saltings	76	0.125	0.625
19			0.125	0.625
20			0.125	0.625
21	14–20 Rocco Boulevard	850	0.125	0.625
22			0.125	0.625
23			0.125	0.625
24			0.125	0.625
25			0.125	0.625
Total		6226	3.25	16.25

After Ministry of Agriculture, Fisheries and Food (1993)
Note: loss (year t) = property (market value) + land (market value)

was assumed that only 25% *would* lead to rupture of the pipe. Thus, a *pipeline vulnerability factor* of 0.0375 was used in the analysis:

pipeline vulnerability factor $= 0.15 \times 0.25 = 0.0375$

No exposure factor was used in the analysis, as the pipeline is a permanent asset (i.e. the exposure factor would be 1.0).

The consequences of a pipe rupture include the environmental damage and clean-up costs associated with an oil spill, estimated to be, on average, $5 million per event. In addition, supply interruptions lasting from days to weeks can be expected, depending on the scale and location of the pipe-rupturing landslide event. The average length of disruption is 10 days. As the pipe carries 40 000 barrels of oil per

Table 9.8 Example 9.2: consequence model

Section	Landslide-prone length: km	Slope class	Geology	Recent landslide density/km² per 50 years	Recent landslide density: normalised for right of way	Potential natural terrain landslides/year	Vulnerability factor	Frequency of rupture/year	Potential losses: $ million	Risk: $ million
1	1.5	35–40	GC	10.04	3.77	0.0753	0.0375	0.028	15	0.042
2	0.4	30–35	GF	22.8	2.28	0.0456	0.0375	0.0017	15	0.026
3	0.7	30–35	FT	11.63	2.04	0.0407	0.0375	0.0015	15	0.023
4	0.1	35–40	GF	22.7	0.57	0.0114	0.0375	0.0004	15	0.006
5	1.3	25–30	CBS	17.46	5.67	0.1135	0.0375	0.0043	15	0.064
6	0.5	25–30	CS	19.81	2.48	0.495	0.0375	0.0019	15	0.028
7	0.5	20–25	CS	14.04	1.76	0.0351	0.0375	0.0013	15	0.020
Total	5					0.37		0.014		0.209

Geology types: GC, coarse-grained granite; GF, fine-grained granite; FT, fine ash tuff; CBS, conglomerate and breccia; CS, siltstone and sandstone. Landslide density statistics are derived from the Hong Kong NTLI (Evans and King, 1998)
Frequency of rupture/year = potential natural terrain landslides/year × vulnerability factor
Risk = frequency of rupture/year × potential losses

day (estimated value of $1 million), a disruption of operation for 10 days would lead to a loss of $10 million in business.

For each particular section of route, the risk was calculated as follows:

$$\text{risk (section 1)} = P(\text{event}) \times \text{adverse consequences}$$
$$= P(\text{event}) \times (\text{total loss} \times \text{vulnerability factor})$$
$$= P(\text{event}) \times (\$15 \text{ million} \times 0.0375)$$

The overall risk along the 5 km of landslide susceptible route was the sum of the risks for each individual section (see Table 9.8):

$$\text{risk (sections 1 to } n) = \sum P(\text{event}) \times \text{adverse consequences (sections 1 to } n)$$

This worked out at $0.209 million per annum, with around 50% of the total risk produced within Sections 1 and 5. However, these sections actually comprise 56% of the total route, which implies that other sections of the pipeline route have higher levels of risk. Dividing the risk in each section by the length of each section reveals that the highest levels of risk per unit length of pipeline appear to occur in the relatively short lengths of Sections 2 and 4.

Example 9.3

The small town of Runswick Bay, on the North Yorkshire coast, UK, has grown up on a pre-existing landslide complex developed in glacial till. In addition to almost continuous creep, high to very high intensity, very slow movements occur during periods of heavy winter rainfall. The majority of the building stock is brick-built and of Victorian age. Many buildings have been damaged by the cumulative effects of ground movement (e.g. Rozier and Reeves, 1979). The extent of the damage was determined by a systematic survey of contemporary damage to roads and structures. This survey involved the assessment of the type and magnitude of damage, using a classification based on guidelines provided by the Building Research Establishment (1981) and Alexander (2002; see Table 8.1).

In order to provide an indication of the risk, the potential damages were estimated by an expert panel (see Chapter 5) for two landslide reactivation scenarios.

- Scenario 1: an episode of significant ground movement in response to the exceedance of a winter rainfall threshold level (annual probability estimated as 0.04).
- Scenario 2: an episode of major ground movement in response to the exceedance of an extreme winter rainfall threshold level (annual probability estimated as 0.01).

The effects of movements in terms of the damage to the properties (i.e. separate buildings) within the landslide were estimated on a landslide block by landslide block basis. For each block, it was assumed that the level of damage will vary across the unit depending on the precise location. The following method was used to accommodate the uncertainty in damage levels/locations.

- The number of properties within each block and the average property value (total value divided by number of properties) was calculated.
- The proportion of the block surface that would be affected by ground movements of differing magnitude (i.e. *the hazard factor*) was established for both scenarios. Damage was found to be most commonly located at the boundaries between individual landslide blocks, where it is

possible to recognise narrow bands of severe hazard due to differential movement. Indeed, the degree of hazard can vary dramatically within as little as a metre due to the surface exposure of inter-block shear surfaces. Thus, while one property may be severely damaged by differential movement, an adjacent property may be largely unaffected.

- The pattern of hazard was related to the distribution of properties, to reveal the number of properties likely to be subjected to differing levels of ground movement (i.e. *the exposure factor*). These proportions were established through expert panel discussion, based on historical evidence.
- The proportion of each block affected by differing magnitudes of movement in each scenario was then converted into six damage levels for the housing located on these areas (e.g. 'unaffected' to 'write-off'), each of which was assumed to be equivalent to specific percentages of the average property value. This *vulnerability factor* was derived through expert panel discussion.

The method can be illustrated with reference to the predicted ground movement damage for a single landslide block (termed block 1). There are 100 properties located on this block, with an average property value of £100 000 (the values were obtained from a local real estate agent).

A range of *exposure factors* were developed to estimate the likely impact of ground movements associated with each scenario on the properties located on the block. The results are shown in Table 9.9. In scenario 1, for example, it is assumed that 50% of all the properties would be unaffected, with 25% affected by negligible damage, 15% by moderate damage and 5% by both serious and severe damage. By contrast, in scenario 2 some 80% of houses would be affected by serious or worse damage, with 25% considered to be 'write-offs'.

The economic implications of the variation in impact between landslide scenarios are reflected in different levels of loss – that is, the *vulnerability* of the assets vary with the scale and intensity of the movement associated with each scenario. Vulnerability is defined as the level of potential damage, or degree of loss, of a particular asset (expressed on a scale of 0 to 1) subjected to a landslide of a given severity. Estimates of vulnerability are based on an inferred relationship between the severity of ground movement and the proportion of the property value that would have to be spent in undertaking repairs. For example, moderate damage is expected to result in losses equivalent to 10% of the property value, whereas write-off would result in 100% loss (see Table 9.9).

Thus, the risk in block 1 being associated with scenario 1 (an episode of significant ground movement) was calculated as follows (see Table 9.9):

risk (block 1) = P(event) × damage to property within block 1

The value of damage in scenario 1 was estimated for each damage class, as follows:

damage (class d) (block 1) = number of properties (block 1) × proportion affected (scenario 1)

× average value × vulnerability factor (damage class d)

Taking severe damage as an example:

severe damage (block 1) = number of properties × proportion affected

× average value × vulnerability factor

= 100 × 0.05 × 100 × 0.5 = £250 000

Estimating the consequences

Table 9.9 Example 9.3: consequence model for block 1 (containing 100 properties with an average value of £100 000)

Scenario	Annual probability		Unaffected	Negligible–slight damage	Moderate damage	Serious damage	Severe damage	Write-off	Total: £ thousands	Risk: £ thousands
1. Episode of significant movement	0.04	Proportion of properties affected	0.50	0.25	0.15	0.05	0.05	0.00		
		Vulnerability factor	0	0.01	0.1	0.25	0.5	1		
		Number of properties	50	25	15	5	5	0		
		Total property value: £ thousands	5000	2500	1500	500	500	0		
		Damage: £ thousands	0	25	150	125	250	0	550	22
2. Episode of widespread movement	0.01	Proportion of properties affected	0.00	0.10	0.10	0.25	0.30	0.25		
		Vulnerability factor	0	0.01	0.1	0.25	0.5	1		
		Number of properties	0.0	10	10	25	30	25		
		Total property value: £ thousands	0	1000	1000	2500	3000	2500		
		Damage: £ thousands	0	10	100	625	1500	2500	4735	47.35

Damage = number of properties (100) × proportion affected × average value (£100 000) × vulnerability factor
Total damage (scenario s) = $\sum$ damage (damage class unaffected to write-off)
Risk (scenario s) = scenario probability × total damage

The overall damages for scenario 1 were calculated as

damage (block 1) = $\sum$ damage (unaffected to write-off) = £550 000

The exercise was repeated for scenario 2 (see Table 9.9), yielding an overall damage total of £4 735 000.

The risk determined for block 1, based on these two scenarios, is as follows:

risk (scenario 1) = P(event) × damage within block 1

$= 0.04 \times 550\,000$

$= £22\,000$ per year

risk (scenario 2) = P(event) × damage within block 1

$= 0.01 \times 4\,735\,000$

$= £47\,350$ per year

Example 9.4
In Example 7.8, the exposure of a vehicle crossing the 100 m-wide potential flow path of large flow slides from an old mine spoil heap was calculated to be

$$P(\text{temporal}) = \frac{\text{width of section}}{\text{speed (m/hour)} \times 365 \times 24}$$

$= 100/(30\,000 \times 365 \times 24)$

$= 3.8 \times 10^{-7}$

As the vehicle makes 250 journeys along this section of road over a year:

$P(\text{temporal, 250 journeys}) = 3.8 \times 10^{-7} \times 250 = 9.5 \times 10^{-5}$

Based on the high velocity of the flow slide events, it is estimated that the vulnerability of people within the vehicle would be 0.9 (see Table 8.8).

If the vehicle occupied by one person was hit by a flow slide (annual probability = 0.005), then the consequences, expressed as the chance of a fatality, can be calculated as

$P(\text{fatality}) = P(\text{flow slide}) \times P(\text{temporal}) \times \text{vulnerability}$

$= 0.005 \times (9.5 \times 10^{-5}) \times 0.9$

$= 4.285 \times 10^{-7}$

Example 9.5
A mountain road in British Columbia runs below a high cliff that is regularly affected by rockfalls. In the past, rockfalls have occurred at a frequency of five per year, blocking the road and presenting a threat to people using the road.

Estimating the consequences

The probability of a fatal accident to an occupant of a vehicle driving along the road is

risk = P(rockfall) × exposure factor × vulnerability factor

The *exposure factor* is the chance of the vehicle occupying the same space as the rockfall pathway.

Assuming that 100 vehicles per hour (each 5 m long) drive along the road at an average speed of 50 km/hour, then the proportion of the road that is *instantaneously occupied* by a vehicle is

$$P(\text{temporal}) = \frac{\text{number of vehicles per hour} \times \text{average vehicle length(m)}}{\text{speed(m/hour)}}$$

$$= \frac{100 \times 5}{50\,000}$$

$$= 0.01$$

The probability of a rockfall hitting a moving vehicle was modelled using the binomial distribution, and based on the number of falls per year (each fall is a separate trial in the binomial model; see Chapter 5) and the chance of a particular outcome (hit or no hit, depending on the exposure):

$$P(\text{vehicle impact}) = 1 - [1 - P(\text{temporal})]^{\text{number of falls per year}}$$

$$= 1 - (1 - 0.01)^5$$

$$= 4.9 \times 10^{-2}$$

If it is assumed that the traffic is uniformly distributed in time and space throughout the year (i.e. there is no temporal variability in exposure), then the probability of the vehicle occupying the same space as the rockfall pathway, P(spatial), is 1.0.

The annual probability of a rockfall hitting a moving vehicle is

$$P(\text{accident}) = P(\text{vehicle impact}) \times P(\text{spatial})$$

$$= 4.9 \times 10^{-2} \times 1$$

Given that a rockfall has hit the vehicle, a vulnerability factor was used to estimate the probability of loss of life in the incident. The factor is based on the fact that only 25% of a vehicle length is occupied by passengers and that only 1 out of every 3 rockfall impacts are severe enough to cause death. Thus, the vulnerability factor is $0.25 \times 0.33 = 0.0825$.

The annual probability of a rockfall hitting a moving vehicle and causing loss of life is:

$$P(\text{fatal accident}) = P(\text{vehicle impact}) \times P(\text{spatial}) \times \text{vulnerability factor}$$

$$= 4.9 \times 10^{-2} \times 1 \times 0.0825$$

$$= 4.04 \times 10^{-3}$$

Example 9.6
A mountain road between two small towns in the Scottish Highlands is regularly blocked by rockfalls, and the traffic is forced to follow a lengthy diversion around the affected area. The consequences of

temporary road closure can be measured in terms of the *additional transportation costs* that result from the diversion (i.e. additional fuel, oil and depreciation costs incurred in travelling further or at an additional speed) and the *opportunity costs* caused by the delay (i.e. time wasted because of the additional journey time). Note that road closure generates disruption not only on the blocked road but also on the diversion routes because the traffic already using these routes will be slowed down by the diverted traffic.

In the consequence model developed for this example, the *exposure factor* is represented by the average number of vehicles and their occupants using the road. The *vulnerability* of the traffic to the disruption is assumed to be 1.0 – that is, if a road user wants to travel between the two towns on the day of a rockfall, then they will incur the additional costs. Among the factors that influence the scale of traffic disruption costs are

- the traffic flow
- the traffic flow as a proportion of the route capacity
- the rockfall frequency
- the time period(duration) of road closure
- the length of diversions
- the road types along the diversion route(s).

Three rockfall scenarios were identified by an expert panel review (see Chapter 5).

- Scenario 1: a small rockfall (<100 m^3) that results in the road being closed for 1 day, during which the debris is removed by the road maintenance crews (estimated annual probability of 0.3).
- Scenario 2: a relatively small rockfall (<1000 m^3) that results in the road being closed for 5 days because of the need for repairs to the road surface (estimated annual probability of 0.05).
- Scenario 3: a large rockfall ($>10\,000$ m^3) that results in the road being closed for 50 days, during which time emergency works are undertaken to stabilise the unstable rockfall backscar and repair the road surface (estimated annual probability of 0.01).

The effects of the traffic disruption associated with each of the three scenarios were calculated as the additional transportation costs (resource costs) and opportunity costs (delay costs) arising from the road closure, using the method presented in Parker *et al.* (1987). The losses include the following.

- *The marginal transportation costs*, which are a function of vehicle type and speed:

    marginal transportation costs $= a + b/v + cv^2$

    where a, b and c are coefficients given in Table 9.10, and v is the speed in km/hour.

Table 9.10 Example 9.6: parameters used to calculate vehicle operating costs

Vehicle type	Parameter (fuel and non-fuel combined)		
	a	b	c
Cars	3.87	32.47	0.0000430
Light goods vehicle	4.76	62.83	0.0000577

From Highways Agency (1997)

Estimating the consequences

Table 9.11 Example 9.6: vehicle occupancy and resource values of time

Vehicle type	Total occupancy	Value of time: pence/hour
Cars	1.65	673.6
Light goods vehicle	1.47	1166.7
Average road vehicle	1.71	784.4

From Highways Agency (1997)

- *Opportunity costs* caused by the delay. A measure of the impact of the delay can be obtained from values assigned to a traveller's time (Table 9.11).

Traffic disruption was calculated in the following steps, for both a *without landslide* and a *with landslide* case.

1. Construction of a road network diagram (Figure 9.3) that incorporates roads onto which traffic will be diverted when the main route is blocked. The traffic using the main route A–B–C is diverted onto the secondary route D–E–F when the road is blocked by rockfalls along section A. Each section was assigned a road class (Table 9.12).
2. Establishing the average traffic flow (vehicles per hour) along each road section from local authority statistics, based on traffic count data.
3. The traffic flow along each road section was converted to passenger car units (PCUs), to provide an indication of the number of users (i.e. drivers and passengers) and allow for the mix

Figure 9.3 Example 9.6: road network diagram

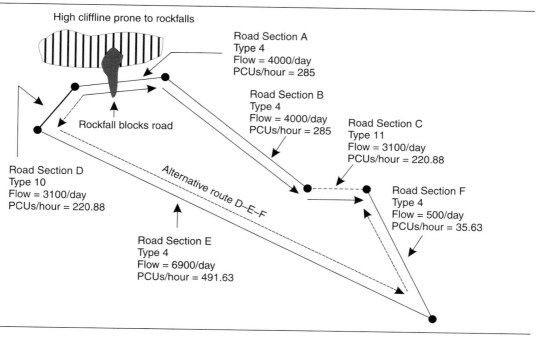

355

Landslide Risk Assessment

Table 9.12 Example 9.6: speed flow relationships for roads

Road type	Class	Free flow speed: km/hour	Free flow limit: PCUs/hour/lane	Limiting capacity: PCUs/hour/lane	Speed at free flow limit: km/hour
4	Single two-lane carriageway	63	400	1400	55
10	Urban two-lane carriageway	35	350	600	25

From Parker et al. (1987)

in the flow (i.e. cars, light vehicles, buses, HGVs, etc.). In this example, the traffic flow was multiplied by 1.71, the occupancy of an average road vehicle (see Table 9.11).

4 The average speed along each section was calculated from Table 9.12, which relates the free-flow speed to road type.
For a class 4 road (single two-lane carriageway) the traffic speed is 63 km/hour up to a flow of 400 PCUs. At the limiting capacity (LC) the speed of flow is 55 km/hour. Between the free flow limit (FFL) and the limiting capacity (1400 PCUs), the speed falls as the flow increases. At 600 PCUs per hour:

$$\text{speed} = \text{speed FFL} - \left((\text{speed FF} - \text{LC}) \times \frac{\text{PCUs} - \text{PCUs FFL}}{1000}\right)$$

$$= 63 - \left(8 \times \frac{600 - 400}{1000}\right)$$

$$= 61.4 \text{ km/hour}$$

5 The resource cost per hour was calculated from the marginal transportation cost ($a + b/v + cv^2$: see above), multiplied by the number of vehicles and the section length:

resource cost = marginal cost × number of vehicles × section length

Values for the coefficients a, b and c were obtained from Table 9.10; for simplicity it has been assumed that the traffic flow is limited to cars ($a = 3.87$, $b = 32.47$, $c = 0.000043$).

6 The delay cost was calculated from the value of time figures presented in Table 9.11, adjusted for the road conditions:

delay cost = value of time × PCUs/hour × section length/travel speed

An average figure of 784.4 pence/hour was used for the value of time.

7 The costs of disruption (expressed as a cost per hour) is the difference between the combined resource cost and delay cost for the *with landslide* and *without landslide* cases:

disruption cost = with landslide (resource cost + delay cost)

− without landslide (resource cost + delay cost)

Table 9.13 Example 9.6: traffic disruption calculations

	Length: km	Type	Flow/day	PCUs/ hour	Traffic speed: km/hr	Resource costs: pence/hour	Delay costs: pence/hour	Total costs: pence/hour
Without landslide (road section)								
A	3.2	4	4000.00	285.00	63.00	4156.04	11355.12	15511.17
B	5.5	4	4000.00	285.00	63.00	7143.20	19516.62	26659.82
C	1.8	11	3100.00	220.88	25.00	2066.07	12474.31	14540.38
Total	10.5					13365.31	43346.05	56711.37
D	2.5	10	3100.00	220.88	35.00	2678.88	12375.31	15054.19
E	13.1	4	6900.00	491.63	63.00	29348.80	80186.69	109535.49
F	4.4	4	500.00	35.63	63.00	714.32	1951.66	2665.98
Total	20					32742.00	94513.66	127255.66
Overall total	30.5					46107.31	137859.71	183967.03
With landslide (road section)								
D	2.5	10	6200.00	441.75	35.00	5357.75	24750.62	30108.37
E	13.1	4	10900.00	776.63	59.99	46463.86	133034.14	179498.00
F	4.4	4	4500.00	320.63	63.00	6428.88	17564.96	23993.84
Total	20					58250.49	175349.72	233600.21

Disruption costs = with landslide total costs − without landslide total costs = 233 600.21 − 183 967.02 = 49 633.18 pence/hour = £11 911.96 per day. Resource costs = $(a + b/v + cv^2)$ × section length × PCUs/hour. Delay cost = value of time (784.4 pence) × PCUs/hour × section length/traffic speed. PCUs/hour = Flow/hour × 1.71 (i.e. the average vehicle occupancy – see Table 9.11)

The results of the analysis are presented in Table 9.13, which indicates a potential loss of £11 900 for each day the road is blocked.

The estimated total disruption costs associated with each of the three scenarios were calculated as

total disruption cost = daily disruption cost × length of delay

The disruption cost associated with each scenario is

disruption cost (scenario 1) = 11.9 × 1 = £11 900

disruption cost (scenario 2) = 11.9 × 5 = £59 500

disruption cost (scenario 3) = 11.9 × 50 = £595 000

Example 9.7

A road and building located at the base of a cut slope in Hong Kong are threatened by a potential failure (500–2000 m³ volume) and run-out of the debris (Wong et al., 1997). The risk is, in part, a function of the travel distance of the debris. However, for a given slope type, landslides of similar mechanism can produce a range of run-out distances because of variations in slope conditions (Wong and Ho, 1996). In order to evaluate the risk, it was necessary to develop a frequency distribution for the potential run-out distances.

Table 9.14 Example 9.7: vulnerability matrix factors

Frequency of occurrence of landslides with different travel angles	Likely probability of death for different shadow angle ranges							
	>60°	55–60°	50–55°	45–50°	40–45°	35–40°	30–35°	25–30°
Scenario 1: $\alpha = 27.5° \pm 2.5°$ – probability = 0.05	0.95 (0.95)	0.95 (0.95)	0.95 (0.95)	0.95 (0.95)	0.95 (0.95)	0.60 (0.95)	0.20 (0.60)	0.05 (0.20)
Scenario 2: $\alpha = 32.5° \pm 2.5°$ – probability = 0.60	0.95 (0.95)	0.95 (0.95)	0.95 (0.95)	0.95 (0.95)	0.60 (0.95)	0.20 (0.60)	0.05 (0.20)	
Scenario 3: $\alpha = 37.5° \pm 2.5°$ – probability = 0.35	0.95 (0.95)	0.95 (0.95)	0.95 (0.95)	0.60 (0.95)	0.20 (0.60)	0.05 (0.20)		
Vulnerability factor calculated	0.95 (0.95)	0.95 (0.95)	0.95 (0.95)	0.83 (0.95)	0.48 (0.83)	0.17 (0.48)	0.04 (0.15)	0.0025 (0.01)

From Wong et al. (1997)
The upper figure in each cell is the vulnerability for a person within a building. The figure in parentheses is the vulnerability for a person on a road

In this example, the run-out distance was modelled in terms of the reach angle (α), defined as the inclination of the line joining the far end of the debris apron to the slope crest (see Figure 3.5). Based on historical events, the reach angle was assumed to range between 25° and 40°, with the following probability distribution:

- scenario 1: $\alpha = 27.5 \pm 2.5°$; probability = 0.05
- scenario 2: $\alpha = 32.5 \pm 2.5°$; probability = 0.60
- scenario 3: $\alpha = 37.5 \pm 2.5°$; probability = 0.35.

The location of the assets at risk is represented by a 'shadow angle', defined as the angle of a line that joins the asset to the slope crest.

Wong et al. (1997) used expert judgement to develop a vulnerability matrix that defines the probability of loss of life for a range of scenarios involving failures with different travel distances and facilities or assets located at different distances from the slope (i.e. shadow angles; Table 9.14).

The risk, expressed as the probability of loss of life, is calculated for each scenario, as follows:

risk (scenario 1) = P(event) × vulnerability factor

overall risk = $\sum P$(event) × vulnerability factor (scenarios 1 to 3)

For a person travelling along a particular lane of the road (shadow angle of 35–40°), the corresponding probability of death would be (from Table 9.14)

risk (scenario 1) = P(event) × vulnerability factor

$$= 0.05 \times 0.95 = 0.0475$$

$$\text{risk (scenario 2)} = P(\text{event}) \times \text{vulnerability factor}$$
$$= 0.6 \times 0.6 = 0.36$$
$$\text{risk (scenario 3)} = P(\text{event}) \times \text{vulnerability factor}$$
$$= 0.35 \times 0.2 = 0.07$$

Should a landslide occur, the overall probability of death for a person at that location would be 0.48.

For a person within a building, again with a shadow angle of 35–40°, the risk of death would be (from Table 9.14)

$$\text{risk (overall)} = (0.05 \times 0.6) + (0.6 \times 0.2) + (0.35 \times 0.05) = 0.17$$

The difference in the level of risk reflects the different degrees of protection afforded to the people in the potential run-out area. In this example, the vulnerability of a person within a building with reference to loss of life is assumed to be over 50% less than a person on a road at the same location.

Note that *variable exposure* is not addressed in this example. However, the probability of a road user being caught in the path of the debris (i.e. in the 'wrong place at the wrong time') can be estimated using the approach described in Example 9.5.

9.11. Multiple-outcome consequence models

The previous examples have been based on a simple, deterministic view of the adverse consequences resulting from a landslide event. If a landslide of a particular magnitude or travel distance occurs, then a particular set of adverse consequences will result. In reality, the precise consequences can often reflect an almost unique and, perhaps, unexpected combination of circumstances that arise at the moment the event occurs and during its aftermath.

Take, for example, a small rockfall from a cutting that lands on the tracks of a mainline railway.

- *The event occurs at night (i.e. 5.00 am) when the line is not in operation.* It is spotted by the track maintenance crew who alert the train operators and arrange for a clean-up team to remove the debris. The impact of the event is restricted to the costs of the clean-up operations, any minor repairs and the temporary delay to the commencement of services.
- *The event occurs at 8.00 am immediately prior to a packed commuter train entering the cutting.* The train hits the debris and is derailed, forcing it into the path of an on-coming goods train including wagons containing hazardous chemicals. The goods train is also derailed, and wagons run down an embankment, spilling chemicals into the river, which is important for producing salmon. The impacts of the event include the loss of life in the commuter train, the driver of the goods train, the trauma affecting many of the survivors and rescue teams, the destruction and damage to railway property, lengthy transport delays, serious contamination of the river with medium-term ecological consequences, including loss of salmon stocks leading to a reduction in the tourist income to the local economy, followed by the bankruptcy of a number of local hotels and businesses.

Chance, therefore, is a major determinant of adverse consequences, as is clearly illustrated by the following report of an actual railway accident in 1883:

This singular accident took place at Vroig cutting on the Cambrian railway on the evening of New Year's Day. The scene of the occurrence was a point between Llyngwril and Barmouth, where the rails skirt, at a considerable eminence from the water, the shore of Cardigan Bay. About eight feet above the railway line in the cliff side is the turnpike road, which a retaining wall protects. This wall with a portion of the road gave way, and fell on the railway. The 5.30 train from Machynlleth to Pwllheli was advancing at its ordinary speed when the engine dashed into the obstruction. The engine and tender rolled over the precipice to the sea-shore, a distance of about fifty feet. The engine-driver and the stoker were instantly killed, their bodies being shockingly mutilated on the jagged rocks. The four carriages and van, which with the engine and tender made up the train, did not go over the precipice. The first carriage turned over on its side, and lay partly overhanging the cliff, the coupling between it and the tender having fortunately broken. The second carriage turned on its side among the rubbish, while the remaining two did not leave the rails. The extent of the disaster was lessened by a second landslip, which took place as the train arrived at the spot, and this prevented the carriages from following the engine and tender by partly burying them. Only a few passengers were in the train. Captain Pryce of Cyffeonydd, Welshpool, Vice-Chairman of the Cambrian Railway, was in the overturned carriage, but he, as well as the other passengers, marvellously escaped without injury. (*The Graphic*, 13 January 1883)

In order to include a range of possible outcomes in a risk assessment it is necessary to develop *multiple consequence models* (consequence scenarios) and evaluate their likelihood. Thus, the risk associated with a particular outcome associated with a particular landslide (i.e. the event) can be represented by

$$\text{risk} = P(\text{event}) \times P(\text{scenario 1}) \times \text{adverse consequences (scenario 1)}$$

The risk associated with all possible outcomes becomes

$$\text{total risk} = P(\text{event}) \times \sum [P(\text{scenario 1 to } n) \times \text{adverse consequences (scenario 1 to } n)]$$

The probabilistic view of consequences is particularly useful for the *back-analysis* of a particular consequence sequence after it has happened. For example, it may be useful to demonstrate that, in the case of the simple example used earlier, 'it was a 10 million to 1 chance that the two trains collided following the landslide'.

This type of problem lends itself to the use of *event trees* (see Chapter 5) to establish the range of outcomes that could be generated by a particular initiating event. Expert judgement is required to identify all the significant consequence scenario components and then develop estimates at each node along the event tree.

The event tree approach uses *forward logic* in that it starts with a single landslide event and traces the possible consequences of the event. An alternative approach is to identify a particular consequence scenario (e.g. the two trains colliding) and use so-called *backward logic* to establish how the accident could happen – that is, the causal relationships that lead to the accident or *top event*. The combinations of factors needed to generate a particular 'top event' can be identified through the use of *fault trees* (see Chapter 5).

Example 9.8

The Lei Yue Mun squatter villages, Hong Kong, are situated at the foot of a 20–40 m-high slope, standing at 65–80°, created by the quarrying of granular fill in the early 20th century (Ho et al., 2000). The natural terrain above the quarried face rises some 200 m above the squatter huts. Both the abandoned quarry face and the hillside have a history of instability. A number of large landslide events occurred during or shortly after a major rainstorm in August 1995, causing severe damage to the squatter dwellings. Loss of life was narrowly avoided by evacuation during the storm.

Table 9.15 Example 9.8: angles of friction for debris slides of different volumes

Debris slide	Small	Medium	Large	Very large	Extremely large
Volume: m^3	<50	50–500	500–5000	5000–50 000	>50 000
Apparent angle of friction: degrees	47	43	35	32	30

After Smallwood et al. (1997)

An event tree approach was used to model the consequences of landslide activity at the site, as part of a risk assessment to support decision-making about the need for re-housing of the squatters (Smallwood et al., 1997).

The approach adopted involved the following steps.

1. Estimating the frequency of landsliding at the site, from a review of aerial photographs covering a 55-year period (1940–1995). A total of 115 events were identified.
2. Estimating the potential run-out of landslides of different magnitudes using the travel distance model of Wong and Ho (1996), who determined an *apparent friction angle* based on a line projected from the failure scar crest to the distal end of the debris (i.e. the reach angle; see Figure 3.5). Apparent angles of friction used at the site are presented in Table 9.15.
3. Classifying the hazard posed by landslide events with different run-out potential into three levels or groupings – major, intermediate and minor – based on judgement. The hazard associated with an event of a particular volume reduces with increasing distance away from the slope foot. With reference to Figure 9.4, small slides presented only what were considered to be minor hazards to buildings and population close to the slope foot (zones A and B), whereas large slides presented a major hazard in these zones, an intermediate hazard in zone C and a minor hazard in zone D.
4. Subdividing the squatter village area into 149 separate 20 m × 20 m grid cells (reference blocks). The number of people and the temporal presence in each block (i.e. the exposure) were determined from a population survey.
5. Establishing an event tree for each reference block. Individual event trees considered landslides from the three hazard groupings (minor, intermediate and major).

Figure 9.5 presents an extract of an event tree developed for one of the reference blocks, and illustrates the multiple consequence scenarios associated with a *minor* debris slide. Each branch of the event tree represents a distinct consequence scenario, with the consequences (i.e. fatalities, minor and major injuries) being determined by the unique combinations of

- the proximity of the affected buildings and population (i.e. the hazard grouping of the landslide event)
- the temporal exposure of the population (i.e. whether it was day or night when the event occurred)
- whether landslide warnings were likely to be heeded or not
- the efficiency of the response of the emergency services
- the occurrence of secondary hazards, such as fire.

Figure 9.4 Example 9.8: landslide hazard groupings for landslide events of different sizes and assets at different distances from the slope base. (After Smallwood *et al.*, 1997)

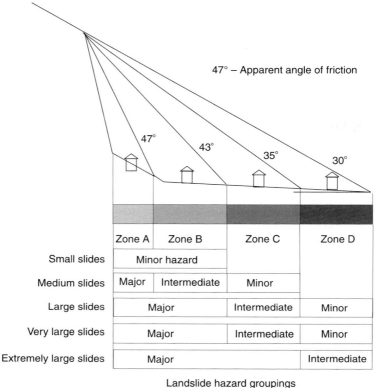

Rule sets, based on expert judgement, were used to estimate the number of minor injuries, major injuries and fatalities resulting from events from each of the three hazard groupings. An *equivalent fatality* statistic was determined as

equivalent fatality = number fatalities + (0.1 × number major injuries)
+ (0.005 × number minor injuries)

The results were used to support an assessment of individual and societal risk (see Chapter 10).

Example 9.9

A small industrial estate has been located in a former chalk quarry. The estate comprises a combination of light structures, used by small firms making electronic components, car repair businesses and service companies, and adjacent parking lots. Following the severe wet winter of 2000–2001, there was significant rockfall activity, raising concerns about the risks to the businesses and, in particular, their staff and customers.

A series of event trees were developed to model the multiple consequence scenarios associated with rockfalls from the quarry face. Individual trees were developed for rockfall events of particular magnitude and frequency. Figure 9.6 presents one of the event trees, illustrating the consequences

Estimating the consequences

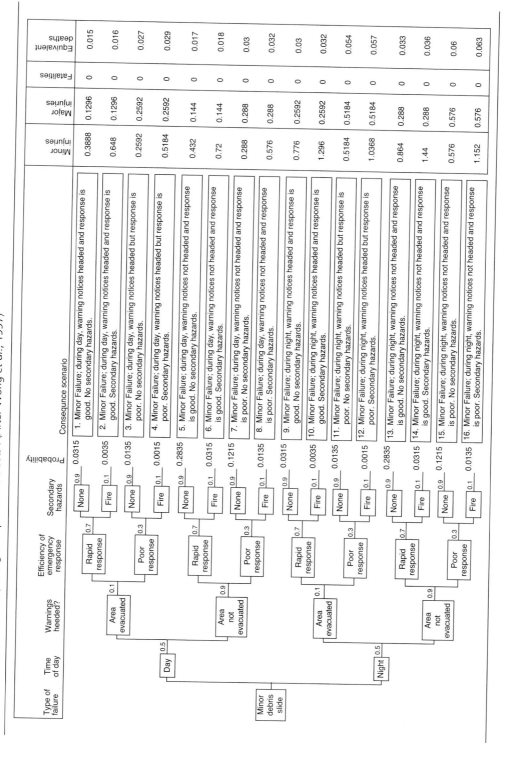

Figure 9.5 Example 9.8: event tree showing the multiple consequences scenarios associated with a minor debris slide. Note that numbers of minor and major injuries were derived by using site-specific 'rule sets'. (After Wong et al., 1997)

Landslide Risk Assessment

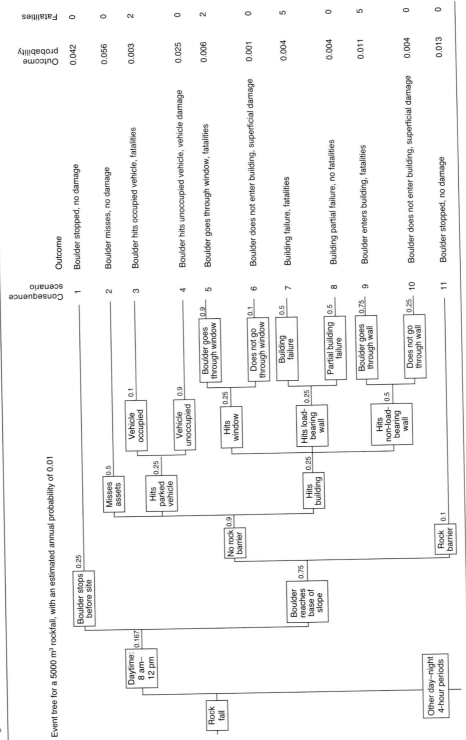

Figure 9.6 Example 9.9: event tree showing the multiple consequence scenarios associated with a small rock fall

of a fall (5000 m³) with an annual probability of 0.01 (established from historical records and expert judgement). The complexity of the tree reflects the need to take account of the following.

- The influence of natural obstacles, such as trees or boulders at the base of the quarry face, which would prevent the rockfall reaching the boundary of the developed area.
- The presence or absence of man-made protective barriers (part of the estate is protected by a wire rock fence).
- The assets at risk on the quarry floor, including open space, the parking lot and light structures.
- The timing of the event. A simple population model was developed to account for the variable occupancy of the site at different times of the day. The day was split into six 4-hour periods, each of which had a different average occupancy value for the different assets at risk.
- The chance that the rockfall would miss all the assets and come to rest without causing any damage or injury.
- The likelihood that the vehicles in the parking lot would be occupied at the time of the rockfall event.
- The possibility that a boulder would pass through a window or the walls of a structure, together with the potential for partial or complete building failure.

Each branch of the event tree represents a different consequence scenario. The consequences were determined by an expert panel (see Chapter 5) who used their judgement and experience of rockfall incidents elsewhere to assign expected fatality statistics to each of the consequence scenarios, taking account of the population model.

The overall risk at the site, expressed in terms of potential for fatalities, was calculated as the product of the probability of the rockfall event and the sum of the fatalities associated with each of the consequence scenarios.

As a range of rockfall event magnitudes/probabilities were considered, the overall risk at the site was calculated as

$$\text{total risk} = \sum (\text{risk, events 1 to } r)$$

Example 9.10

A young child was killed by a gas explosion in a house built on a pre-existing landslide. The explosion took place when an automatic timer switch controlling a washing machine in the basement caused a spark. A gas pipeline had failed as a result of a landslide reactivation event, and the escaping gas had built up in the basement. Figure 9.7 presents a fault tree developed to trace the causal factors involved in the accident and estimate the probability of this particular consequence scenario. Note that it is one of numerous scenarios that could have arisen following the landslide reactivation event, the overwhelming majority of which would not have resulted in a fatality.

The key stages in the accident sequence are as follows.

- The child being present when the gas explodes. The child was playing with a wooden train set that had been laid out on the floor of the basement. The basement was used by the child every day in the morning (8 am to 12 noon) for around 1 hour, so the probability that it was occupied

Landslide Risk Assessment

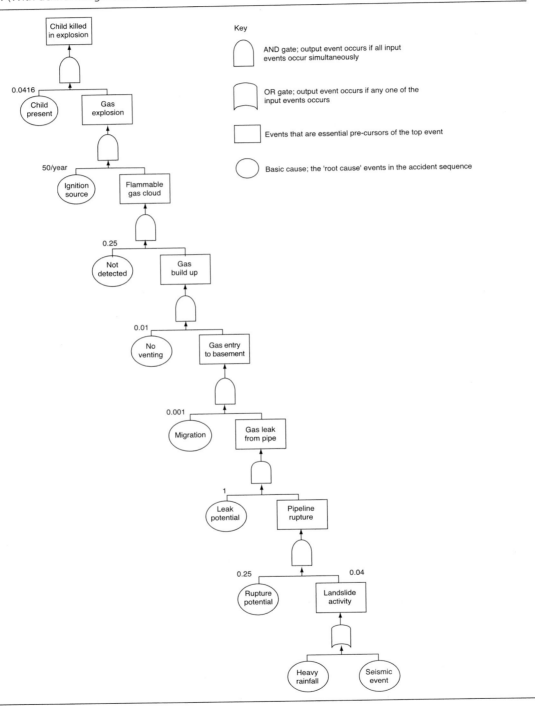

Figure 9.7 Example 9.10: fault tree showing the sequence of events that led to the child being killed in the gas explosion. (With acknowledgements to O'Riordan and Milloy, 1995)

at the time of the explosion was estimated to be

$$P(\text{occupied}) = P(\text{morning explosion}) \times P(\text{child present})$$
$$= 4/24 \times 1/4$$
$$= 0.0416$$

It was assumed that if the child was caught in the explosion the likelihood of death was very high (probability = 1.0).
- A spark in the automatic switch igniting the gas cloud. The switch was set to come on once a week (except when the family was away for 2 weeks on their annual holidays), and, hence, the frequency of the ignition source was 50 times a year.
- The gas build up not being detected as there were no gas detectors in the basement. It was assumed that although the child was not familiar with the smell of leaking gas, the parents might have raised the alarm. Hence, the probability that the gas was not detected was assumed to be 0.25.
- The gas building up in the basement. There was no mechanical ventilation in the basement, but the doors do not provide a perfect seal. The cloud built up to explosive limits overnight. However, the build-up might have been partly dissipated when the basement door was opened in the morning. The probability that opening of the basement doors had failed to prevent the build-up of the gas was judged to be 0.01.
- The gas entering the basement. The concrete slab on the basement floor had a number of cracks, probably as a result of poor workmanship or the effects of previous episodes of ground movement. The probability that the gas would have been able to migrate from the rupture point and enter the basement through the cracks was judged to be 0.001.
- The gas leaking from the pipe. If the pipe had ruptured, then, provided the pipe was carrying gas, there would be a leak. The probability of there being gas in the pipe at the time of rupture was assumed to be 1.0.
- The pipe rupturing. The probability that the landslide movements would have been of sufficient intensity to cause rupture of the pipe was judged to be 0.25 (i.e. 75% of reactivation events would be of insufficient intensity to cause pipe rupture).

From the fault tree analysis, the probability of the fatal accident scenario was estimated to be

$$P(\text{consequence scenario}) = 0.0416 \times 50 \times 0.25 \times 0.01 \times 0.001 \times 1 \times 0.25$$
$$= 0.0000013 \text{ or } 1.3 \times 10^{-6}$$

Given that the annual probability of landslide reactivation was estimated to be 0.04 (based on historical records of landslide activity in the area), the overall probability of the accident was

$$P(\text{fatal accident}) = P(\text{event}) \times P(\text{consequence scenario})$$
$$= 0.04 \times 0.0000013$$
$$= 0.000000052 \text{ or } 5.2 \times 10^{-8}$$

9.12. Complex outcomes and uncertain futures

This chapter has emphasised that the potential for adverse consequences is the product of the specific hazard, exposure and vulnerability conditions at a particular location at a specific time. Uncertainties regarding the magnitude, character and timing of a future landslide event are compounded by uncertainties as to exactly what will suffer impact and how severe that impact will be. Slight variations in the magnitude or timing of a slope failure can result in quite dramatic variations in the level of detriment, especially where human lives are concerned. As a result, there might be need for an enormous number of possible consequence models or scenarios, in order to define and quantify the possible outcomes, especially for compound, multiple or complex events.

A balance needs to be found between the effort and resources involved in evaluating all possible outcomes and the generalisations involved in attempting to characterise a 'best estimate' scenario. On the one hand, it may be the scale of a 'worst case' scenario (e.g. the train collision described earlier in this chapter) that drives decision-making, on the other hand the analysis of such extreme or catastrophic scenarios may divert attention away from more likely but less dramatic consequences (e.g. the impact of recurrent building damage caused by ground movement on the local economy). Carefully defined scoping can be useful in limiting any consequence assessment to the particular requirements of a specific risk assessment, but, even so, problems will occur in determining which of the full range of adverse consequences to focus on.

There is, of course, a danger that limiting the scope of a risk assessment could increase the likelihood that 'game-changing' catastrophic consequences are overlooked. In many assessments, the estimated consequences are restricted to loss of life, effects on buildings and direct economic losses, with the indirect and intangible losses considered to be of minor importance. However, in some cases it is these 'knock-on' effects that can transform the impact of the initial event from a predictable and manageable situation into a disaster. This is especially so in poorer, highly vulnerable societies. As Alexander (2005) noted, although very serious landslide hazards threaten densely populated areas in Hong Kong and Japan, the devastation and loss of life is generally less than would be the case for similar hazards in Venezuela, Brazil and Nicaragua, due to differences in vulnerability. It is also the case that landslide hazards can cause severe disruption in complex economic systems where there can be global dependencies on the uninterrupted supply of services. For example, the Baku–Tbilisi–Ceyhan (BTC) pipeline transports 1 million barrels of crude oil per day from the Caspian Sea, through the Lesser Caucasus and Anatolian Mountains to the Turkish Mediterranean coast, from where the oil is shipped via tankers to European markets. Landslides risk assessment played an important role in defining the route through the mountainous sections (e.g. Lotter *et al.*, 2005; Shilston *et al.*, 2005). As the BTC pipeline carries 1% of the global oil supply, any disruption to operations could have a significant impact on oil prices.

It is important to appreciate that whichever scenarios are considered, they will have a short life-span in terms of their validity, because the nature and value of the assets at risk are almost always continuously changing. While it is difficult enough to foretell patterns of human activity in the short term and even to obtain agreement as to the values that should be placed on different artefacts and environmental resources, it becomes especially hard to assess what the situation will be like at some future date. Exponents of *futurology* attempt to foretell the nature of society and how it will operate at varying times into the future, but hindsight reviews indicate that such 'predictions' have not been particularly successful to date (e.g. why are we not all dressed in aluminium foil suits and travelling in hover-cars?).

Population growth, economic growth, environmental change and the progressive spread of human activity and infrastructure into increasingly remote and environmentally hostile areas all indicate that the potential for landslide generated losses will grow. In addition, increasing standards of living will result in rising property values and the increasingly widespread ownership of expensive household goods such as televisions, washing machines and fridge-freezers, all of which are vulnerable to damage. Suleman *et al.* (1988) suggest that the *damage potential* with reference to floods rose by over 50% for short-duration flood events and 100% for long-duration flood events between 1977 and 1987; a similar trend can be envisaged for landslide events.

Ideally, those scenarios considering the possibility of events in the more distant future could attempt to include development outcomes, changes in population distribution, numbers and concentrations, and changes in the values of assets, although all will become more uncertain with increasing time into the future. In reality, however, most consequence assessments use the current distribution and value of assets as a baseline and then project this into the future, thereby controlling the speculation.

REFERENCES

Abelson P (2007) *Establishing a Monetary Value for Lives Saved: Issues and Controversies.* Department of Finance and Deregulation, Canberra. WP 2008-2.

Adams WC (1986) Whose lives count? TV coverage of natural disasters. *Journal of Communication* **36(2)**: 113–122.

Alexander DE (1989) Urban landslides. *Progress in Physical Geography* **13(2)**: 157–191.

Alexander DE (2002) *Principles of Emergency Planning and Management.* Terra, Harpenden.

Alexander DE (2005) Vulnerability to landslides. In *Landslide Hazard and Risk* (Glade T, Anderson MG, Crozier MJ (eds)). Wiley, Chichester, pp. 175–198.

Alfors JT, Burnett JL and Gay TE (1973) Urban geology masterplan for California. *California Division of Mines and Geology Bulletin* 198, 1–109.

Australian Government (2008) *Best Practice Regulation Guidance Note: Value of Statistical Life.* Office of Best Practice Regulation, Canberra.

Bacon J and Bacon S (1988) *Dunwich Suffolk.* Segment Publications, Colchester.

Bateman I, Carson RT, Day B, Hanemann WM, Hanley N, Hett T, Jones-Lee M, Loomes G, Mourato S, Ozdemiroglu E and Pearce DW (2002) *Economic Valuation with Stated Preference Techniques: A Manual.* Edward Elgar, Cheltenham.

Bellavance F, Dionne G and Lebeau M (2009) The value of a statistical life: a meta-analysis with a mixed effects regression model. *Journal of Health Economics* **28(2)**: 444–64.

Bishop AW, Hutchinson JN, Penman ADM and Evans HE (1969) *Geotechnical Investigation into the Causes and Circumstances of the Disaster of 21 October 1966. A Selection of Technical Reports Submitted to the Aberfan Tribunal 1–80.* Welsh Office, HMSO, London.

Bucknam RC (2001) *Landslides Triggered by Hurricane Mitch in Guatemala: Inventory and Discussion.* US Geological Survey, Renton, VA.

Building Research Establishment (1981) *Damage in Low-rise Buildings.* BRE, Watford. Digest 251.

Bunce C, Cruden DM and Morgenstern NR (1997) Assessment of the hazard from rock fall on a highway. *Canadian Geotechnical Journal* **34**: 344–356.

Cardinali M, Reichenbach P, Guzzetti F, Ardizzone F, Antonini G, Galli M, Cacciano M, Castellani M and Salvati P (2002) A geomorphological approach to the estimation of landslide hazards and risks in Umbria, Central Italy. *Natural Hazards and Earth System Sciences* **2**: 57–72.

Carrara A and Merenda L (1976) Landslide inventory in northern Calabria, Southern Italy. *Bulletin of the Geological Society of America* **87**: 1153–1162.

Carreno R and Bonnard C (1999) Rock slide at Macchupicchu, Peru. In *Landslides of the World* (Sassa K (ed.)). Kyoto University Press, Kyoto, pp. 323–326.

Catenacci V (1992) Il dissesto geologico e geoambientale in Italia dal dopoguerra al (1990) Memorie Descrittive della Carta Geolog-ica d'Italia. *Servizio Geologico Nazionale* **47**: 301 (in Italian).

Chan RKS (2000) Hong Kong slope safety management system. *Proceedings of the Symposium on Slope Hazards and Their Prevention*, The Jockey Club Research and Information Centre for Landslip Prevention and Land Development, The University of Hong Kong, Hong Kong.

Coburn A and Spence R (1992) *Earthquake Protection*. Wiley, Chichester.

Cory J and Sopinka J (1989) John Just versus Her Majesty The Queen in right of the Province of British Columbia. *Supreme Court Report* **2**: 1228–1258.

Crescenti U (1986) La grande frana di Ancona del 13 dicembre 1982. *Studi Geologici Camerti, Special Issue, Camerino* **1** (in Italian).

Crozier MJ (1986) *Landslides: Causes, Consequences and Environment*. Croom Helm, London.

Davies WN and Christie HD (1996) The Coledale mudslide, New South Wales, Australia – a lesson for geotechnical engineers. In *Landslides* (Senneset K (ed.)). Balkema, Rotterdam, pp. 701–706.

Del Prete M and Petley DJ (1982) Case history of the main landslide at Craco, Basilicata, South Italy. *Geologia Applicato e Idrogeolia*. **17**: 291–304.

Department for Transport, Local Government and the Regions (2002) *Economic Valuation with Stated Preference Techniques: A Summary Guide. DTLR Appraisal Guidance*. Department for Transport, Local Government and the Regions, London.

Economic Commission for Latin America and the Caribbean (ECLAC) (2011) Saint Lucia: Macro socio-economic and environmental assessment of the damage and losses caused by Hurricane Tomas, a geo-environmental disaster. ECLAC 7 February 2011.

ERM-Hong Kong (1999) *Slope Failures along BRIL Roads: Quantitative Risk Assessment and Ranking*. Geotechnical Engineering Office, Hong Kong. GEO Report No. 81.

Evans NC and King JP (1998) *The Natural Terrain Landslide Study: Debris Avalanche Susceptibility*. Geotechnical Engineering Office, Hong Kong. GEO Technical Note TN 1/98.

Evans NC, Huang SW and King JP (1997) *The Natural Terrain Landslide Study: Phase III*. Geotechnical Engineering Office, Hong Kong. GEO Special Project Report SPR 5/97.

Fell R, Ho KKS, Lacasse S and Leroi E (2005) A framework for landslide risk assessment and management. In *Landslide Risk Management* (Hungr O, Fell R, Couture R and Eberhardt E (eds)). Balkema, Rotterdam, pp. 3–26.

Fulton ARG, Jones DKC and Lazzari S (1987) The role of geomorphology in post-disaster reconstruction: the case of Basilicata, Southern Italy. In *International Geomorphology 1986* (Gardiner V (ed.)). Wiley, Chichester, pp. 241–262.

Gardner T (1754) *An Historical Account of Dunwich, Blithburgh and Southwold*. London.

Gazzetta Ufficiale della Repubblica Italiana (1998) Misure urgenti per la prevenzione del rischio idrogeologico ed a favore delle zone col-pite da disastri franosi nella regione Campania. *Gazzetta Ufficiale della Repubblica Italiana* **139(208)**: 53–74 (in Italian).

Gill J (2011) Case Study: Bududa landslide, Uganda – March 2010. Geology for global development. http://geo-development.blogspot.co.uk/2011/03/case-study-bududa-landslide-uganda.html (accessed 16/9/2013).

Godt JW and Savage WZ (1999) El Nino 1997–98: direct costs of damaging landslides in the San Francisco Bay region. In *Landslides* (Griffiths JS, Stokes MR and Thomas RG (eds)). Balkema, Rotterdam, pp. 47–55.

Gokceoglu C, Sonmez H, Nefeslioglu H, Duman TY and Can T (2005) The 17 March 2005 Kuzulu landslide (Sivas, Turkey) and landslide susceptibility map of its near vicinity. *Engineering Geology* **81(1)**: 65–83.

Government of Hong Kong (1972a) *Interim Report of the Commission of Inquiry into the Rainstorm Disasters*. Hong Kong Government Printer, Hong Kong.

Government of Hong Kong (1972b) *Final Report of the Commission of Inquiry into the Rainstorm Disasters*. Hong Kong Government Printer, Hong Kong.

Griffiths JS, Hutchinson JN, Brunsden D, Petely DJ and Fookes PG (2004) The reactivation of a landslide during the construction of the Ok Ma tailings dam, Papua New Guinea. *Quarterly Journal of Engineering Geology & Hydrogeology* **37(3)**: 173–186.

Guzzetti F (2000) Landslide fatalities and evaluation of landslide risk in Italy. *Engineering Geology* **58**: 89–107.

Halcrow Asia Partnership (1999) *Stage 2 Detailed Site-specific QRA Report*. Geotechnical Engineering Office, Hong Kong.

Hand D (2000) *Report on the inquest into the deaths arising from the Thredbo landslide*. Attorney Generals Department, Office of the NSW Coroner, Glebe.

Harp EL and Jibson RL (1995) *Inventory of Landslides Triggered by the 1994 Northridge, California Earthquake*. US Geological Survey, Renton, VA.

Harp EL and Jibson RL (1996) Landslides triggered by the 1994 Northridge, California earthquake. *Seismological Society of America Bulletin* **86**: S319–S332.

Harp EL, Jibson RL and Wieczorek GF (1981) *Landslides from the February 4 1976 Guatemala Earthquake*. US Geological Survey, Renton, VA.

Health and Safety Executive (2001) *Reducing risks, protecting people*. HMSO, London.

Hendron AJ Jr and Patton FD (1985) *The Vaiont Slide: A Geotechnical Analysis Based on New Geologic Observations of the Failure Surface*. US Army Corps of Engineers, Vicksburg, MI.

Herd DG and the Comite de Estudios Vulcanologies (1986) The 1985 Ruiz volcano disaster. *Eos* **67**: 457–460.

Hewitt K (1997) *Regions of Risk: A Geographical Introduction to Disasters*. Addison Wesley Longman, London.

Highways Agency (1997) *Design Manual for Roads and Bridges. Volume 13. Economic Assessment of Road Schemes, Section 2. Highways Economic Note No. 2*. Highways Agency.

Ho K, Leroi E and Roberds B (2000) Quantitative risk assessment: application, myths and future direction. *Proceedings of the Geo-Eng Conference*, Melbourne, Publication 1, pp. 269–312.

Ikeya H (1976) *Introduction to Sabo Works: The Preservation of Land against Sediment Disaster*. Japan Sabo Association, Tokyo.

International Federation of the Red Cross and Red Crescent Societies (1999) *World Disasters Report*. Edigroup, Chene-Bourg.

International Pacific Salmon Fisheries Commission (1980) *Hell's Gate Fishways*. New Westminster, British Colombia.

Jibson RW (1992) The Mameyes, Puerto Rico, landslide disaster of October 7, 1985. *Reviews in Engineering Geology* **9**: 37–54.

Jibson RW (2006) The 2005 La Conchita, California, landslide. *Landslides* **3**: 73–78.

Jones DKC (1992) Landslide hazard assessment in the context of development. In *Geohazards: Natural and Man-made* (McCall GJH, Laming DJC and Scott SC). Chapman and Hall, London, pp. 117–141.

Jones DKC (1995) The relevance of landslide hazard to the International Decade for Natural Disaster Reduction. In *Landslides Hazard Mitigation*. Royal Academy of Engineering, London, pp. 19–33.

Jones DKC and Lee EM (1994) *Landsliding in Great Britain*. HMSO, London.

Jones DKC, Lee EM, Hearn G and Genc S (1989) The Catak landslide disaster, Trabzon Province, Turkey. *Terra Nova* **1**: 84–90.

Jones-Lee MW (1989) *The Economics of Safety and Physical Risk*. Blackwell, Oxford.

Keefer DK (1984) Landslides caused by earthquakes. *Geological Society of America Bulletin* **95**: 406–421.

Keefer DK (2002) Investigating landslides caused by earthquakes – a historical review. *Surveys in Geophysics* **23**: 473–510.

Kiersch GA (1964) Vaiont Reservoir disaster. *Civil Engineering* **34(3)**: 32–39.

King JP (1997) *Natural Terrain Landslide Study: The Natural Terrain Landslide Inventory*. Geotechnical Engineering Office, Hong Kong. GEO Report No. 74.

King JP (1999) *Natural Terrain Landslide Study: The Natural Terrain Landslide Inventory*. Geotechnical Engineering Office, Hong Kong. GEO Technical Note TN 1/97.

Kingdon-Ward J (1952) *My Hill So Strong*. Jonathan Cape, London.

Kingdon-Ward J (1955) Aftermath of the Great Assam Earthquake of 1950. *Geographical Journal* **121**: 290–303.

Lacey GN (1972) Observations on Aberfan. *Journal of Psychometric Research* **16**: 257–260.

Lin CW, Shieh C-L, Yuan B-D, Shieh Y-C, Liu S-H and Lee S-Y (2003) Impact of Chi-Chi earthquake on the occurrence of landslides and debris flows: example from the Chenyulan River watershed, Nantou, Taiwan. *Engineering Geology* **71**: 49–61.

Lipman PW and Mullineaux D (eds) (1981) *The 1980 Eruptions of Mount St Helens*. US Geological Survey, Renton, VA.

Lotter M, Charman JH, Lee EM, Hengesh JV, Shilston DT and Poscher G (2005) Geohazard assessment of a major crude oil pipeline system – the Turkish section of the BTC Pipeline route. In *Terrain and Geohazard Challenges Facing Onshore Oil and Gas Pipelines* (Sweeney M (ed.)). Thomas Telford, London, pp. 301–310.

McIntyre T (2005) Standing legacy: ghost towns preserve the Ottawa Valley's rich history. *Canadian Geographic*, September/October.

Malone AW (1998) Risk management and slope safety in Hong Kong. *Proceedings of the Seminar on Slope Engineering in Hong Kong*. Balkema, Rotterdam, pp. 3–17.

Marin A (1992) *Costs and Benefits Of risk Reduction. Appendix in Risk: Analysis, Perception and Management*. Royal Society, London.

Mathers C, Vos T and Stevenson C (1999) *The Burden of Disease and Injury in Australia*. AIHW, Canberra.

Miller J (1974) *Aberfan – A Disaster and its Aftermath*. Constable, London.

Ministry of Agriculture, Fisheries and Food (1993) *Project Appraisal Guidance Notes*. MAFF, London.

Mooney GM (1977) *The Valuation of Human Life*. Macmillan, London.

Murray JG (2000) The NZ landslide safety net. In *Landslides: In Research, Theory and Practice* (Bromhead EN, Dixon N and Ibsen ML). Thomas Telford, London, pp. 1075–1080.

Nieto AS and Schuster RL (1999) Mass wasting and flooding induced by the 5 March 1987 Ecuador earthquakes. In *Landslides of the World* (Sassa K (ed.)). Kyoto University Press, Kyoto, pp. 220–223.

OECD (2011) *Valuing Mortality Risk Reductions in Regulatory Analysis of Environmental, Health and Transport Policies: Policy Implications*. OECD, Paris.

Olser T (1993) Injury severity scoring: perspectives in development and future directions. *American Journal of Surgery* **165(2A)**: 435–515.

O'Riordan NJ and Milloy CJ (1995) *Risk Assessment for Methane and Other Gases from the Ground*. Construction Industry Research and Information Association (CIRIA) Report 152. CIRIA, London.

Parker DJ, Green CH and Thompson PM (1987) *Urban Flood Protection Benefits: A Project Appraisal Guide*. Gower Press, London.

Pearce DW, Cline WR, Achanta AN, Fankhauser S, Pachauri RK, Tol RSJ and Vellinga P (1995) The social costs of climate changes: Greenhouse damage and the benefits of control. In *Climate Change 1995: Economic and Social Dimensions of Climate Change*. Contribution of Working Group III to the Second Assessment Report of the IPCC (Bruce JP, Lee H and Haites EF (eds)). Cambridge University Press, Cambridge, pp. 183–224.

Pearce D, Atkinson G and Mourato S (2006) *Cost–benefit Analysis and the Environment: Recent Developments*. OECD, Paris.

Peek-Asa C, Ramirez M, Shoaf K and Seligson H (2002) Population-based case-control study of injury factors in the Northridge earthquake. *Annals of Epidemiology* **12**: 525–526.

Penning-Rowsell EC, Green CH, Thompson PM, Coker AM, Tunstall SM, Richards C and Parker DJ (1992) *The Economics of Coastal Management: a Manual of Benefits Assessment Techniques*. Belhaven Press, London.

Pethick JS (1996) Coastal slope development: temporal and spatial periodicity in the Holderness cliff recession. In *Advances in Hillslope Processes* (Anderson MG and Brooks SM (eds)). Wiley, Chichester, vol. 2, pp. 897–917.

Petley DN (2012) Global patterns of loss of life from landslides. *Geology* **40**: 927–930.

Petley DN, Dunning SA and Rosser NJ (2005) The analysis of global landslide risk through the creation of a database of worldwide landslide fatalities. In *Landslide Risk Management* (Hungr O, Fell R, Couture R and Eberhardt E (eds)). Balkema, Rotterdam, pp. 367–374.

Plafker G and Ericksen GE (1978) Nevados Huascaran avalanches, Peru. In *Rockslides and Avalanches – 1 Natural Phenomena* (Voight B (ed.)). Elsevier, Amsterdam, pp. 277–314.

Pradham EK, West KP Jr, Katz J, LeClerq SC, Khatry SK and Shrestha SR (2007) Risk of flood-related mortality in Nepal. *Disasters* **31**: 57–70.

Regione Basilicata (1987) *Gli Interventi per Senise, per il Consolidamento e il Transferimento di Insedimenti Abitati in Basilicata*. Potenza.

Rozier IT and Reeves MJ (1979) Ground movements at Runswick Bay, North Yorkshire. *Earth Surface Processes and Landforms* **4**: 275–280.

San Francisco Chronicle (1983) Highway 50 reopens and Tahoe rejoices. *San Francisco Chronicle*, 24 June, p. 2.

Sanchez C, Lee TS, Batts D, Benjamin J and Malilay J (2009) Risk factors for mortality during the 2002 landslides in Chuuk, Federated States of Micronesia. *Disasters* **33**: 705–20.

Sapir DG and Misson C (1992) The development of a database on disasters. *Disasters* **16**: 74–80.

Sassa K, Fukuoka, H and Carreno R (2009) Landslide investigation and capacity building in the Machu Picchu – Aguas Calientes Area (IPL C101-1). In *Landslides – Disaster Risk Reduction* (Sassa K and Canuti P (eds)). Springer-Verlag, Berlin, pp. 229–248.

Schuster RL (1983) Engineering aspects of the 1980 Mount St Helens eruptions. *Bulletin of the Association of Engineering Geologists* **20(2)**: 125–143.

Schuster RL and Fleming RW (1986) Economic losses and fatalities due to landslides. *Bulletin of the Association of Engineering Geologists* **23(1)**: 11–28.

Shilston DT, Lee EM, Pollos-Pirallo S, Morgan D, Clarke J, Fookes PG and Brunsden D (2005) Terrain evaluation and site investigation for design of the trans Caucasus oil and gas pipelines in Georgia. In *Terrain and Geohazard Challenges Facing Onshore Oil and Gas Pipelines* (Sweeney M (ed.)). Thomas Telford, London, pp. 283–300.

Smallwood ARH, Morley RS, Hardingham AD, Ditchfield C and Castleman J (1997) Quantitative risk assessment of landslides: case histories from Hong Kong. In *Engineering Geology and the Environment* (Marinos PG, Koukis GC, Tsiambaos GC and Stournaras GC (eds)). Balkema, Rotterdam, pp. 1055–1060.

Smith K (2001) Environmental *Hazards: Assessing Risk and Reducing Disaster*, 3rd edn. Routledge, London.

Stalin Benitez A (1989) Landslides: extent and economic significance in Ecuador. In *Landslides: Extent and Economic Significance* (Brabb EE and Harrod BL (eds)). Balkema, Rotterdam, pp. 123–126.

Suleman MS, N'Jai A, Green CH and Penning-Rowsell EC (1988) *Potential Flood Damage Data: a Major Update*. Flood Hazard Research Centre, Middlesex University, London.

Tavenas F, Chagnon JY and LaRochelle P (1971) The Saint-Jean Vianney landslide: observations and eyewitness accounts. *Canadian Geotechnical Journal* **8**: 463–478.

Taype V (1979) Los desastres naturals como probleme de al defensa civil. *Bol. Soc. Geolog. del Peru* **61**: 101–111.

Ulusay R, Aydan O and Kilic R (2007) Geotechnical assessment of the 2005 Kuzulu landslide (Turkey). *Engineering Geology* **89**: 112–128.

University of Utah (1984) *Flooding and Landslides in Utah – An Economic Impact Analysis*. University of Utah Bureau of Economic and Business Research. Utah Department of Community and Economic Development and Utah Office of Planning and Budget, Salt Lake City, UT.

US Geological Survey (1989) Notes about the Armenian earthquake of 7 December 1988. *Earthquakes and Volcanoes* **21**: 68–78.

USAID (2000) *Venezuela Factsheet, February, 2000*. Office of Foreign Disaster Assistance, US Agency for International Development, Washington, DC.

Valentin H (1954) Der landverlust in Holderness, Ostengland von 1852 bis 1952. *Die Erde* **6**: 296–315.

Van Westen CJ, Van Asch TWJ and Soeters R (2006) Landslide hazard and risk zonation – why is it still so difficult? *Bulletin of Engineering Geology and Environment* **65(2)**: 167–184.

Viscusi W (2003) The value of a statistical life: a critical review of market estimates throughout the world. *Land Economics* **70**: 145–54.

Walkinshaw J (1992) Landslide correction costs on US state highway systems. *Transportation Research Record* **1343**: 36–41.

Wang WN, Wu HL, Nakamura H, Wu S-C, Ouyang S and Yu M-F (2003) Mass movements caused by recent tectonic activity: the 1999 Chi-chi earthquake in central Taiwan. *The Island Arc* **12**: 325–334.

Wieczorek GF, Larsen MC, Eaton LS, Morgan BA and Blair JL (2001) *Debris-flow and Flooding Hazards Associated with the December 1999 Storm in Coastal Venezuela and Strategies for Mitigation*. US Geological Survey, Renton, VA. Open File Report 01-0144.

Wong HN and Ho KKS (1996) Travel distance of landslide debris. In *Landslides* (Senneset K (ed.)). Balkema, Rotterdam, vol. 1, pp. 417–423.

Wong HN, Ho KKS and Chan YC (1997) Assessment of consequence of landslides. In *Landslide Risk Assessment* (Cruden D and Fell R (eds)). Balkema, Rotterdam, pp. 111–149.

Yoshimatsu H (1999) A large rockfall along a coastal highway, in Japan. In *Landslides of the World* (Sassa K (ed.)). Kyoto University Press, Kyoto, pp. 203–204.

Landslide Risk Assessment
ISBN 978-0-7277-5801-9

ICE Publishing: All rights reserved
http://dx.doi.org/10.1680/lra.58019.375

Chapter 10
Quantifying risk

10.1. Introduction

Risk is the likelihood of different magnitudes of adverse consequences. In the case of landslide risk assessment it is typically expressed as the product of the likelihood of a hazard (e.g. a damaging landslide event) and its adverse consequences:

$$\text{risk} = P(\text{landslide event}) \times \text{adverse consequences}$$

This simple equation can be expanded to include spatial and temporal exposure to landslide events and the vulnerability of the elements at risk:

$$\text{risk} = P(\text{landslide event}) \times P(\text{spatial}) \times P(\text{temporal}) \times \text{vulnerability} \times \text{adverse consequences}$$

The main aim of quantitative risk assessment is to express risk as mathematical values, preferably a single mathematical value. This, it is often argued, facilitates better communication and aids appreciation, thereby improving decision-making.

Reducing risk to a single mathematical value also makes it possible to compare landslide 'risk' with other risks, such as flooding, in order to determine their relative significance. However, it is important to bear in mind that the same, or similar, computed values of 'risk' can be the product of very different combinations of probability and consequence. For example, in a given year, a 1 in 2 chance ($P = 0.5$) of landslide movement (slow creep) causing £1000 worth of damage to a footpath generates the same 'risk' as a 1 in 100 chance ($P = 0.01$) of a debris flow causing £50 000 of damage, or a 1 in 10 000 chance ($P = 0.0001$) of a catastrophic large run-out landslide with losses of £5 million; that is, all three have *mathematical expectation values* of £500 within the time period. Thus, although the mathematically computed measures of 'risk' may be the same, the 'perceived threats' are rather different (see Chapter 2). As a result, the attitudes of decision-makers may differ significantly from those of the public regarding these three threats because of the differing scale of the potential losses, resulting in contrasting views as to their prioritisation and the way that they may be managed, thereby necessitating the process of risk evaluation (see Chapter 11).

Bearing in mind the uncertainties and assumptions involved in the process of developing hazard models, estimating the probability of landsliding, exposure and vulnerability, and placing values on the adverse consequences, it is often misleading to present a single point estimate of risk. Sensitivity testing can be used to examine how sensitive the estimated risk value is to changes in the input parameters and underlying assumptions. Typically, it is useful to repeat the analysis using a range of landslide magnitudes and probabilities, from a 'worst case' estimate to a lower-bound estimate. Similarly, the values of the adverse consequences could be varied to reflect uncertainties in the price base that might have been used.

10.2. Current annual risk

In many instances it will be sufficient to provide an estimate of the landslide risk under current conditions, i.e. what is the risk at the moment? This can be expressed as an annual value, as illustrated in the following examples.

Example 10.1
A hilltop community located on a small plateau is vulnerable to major first-time landslide events along one unstable margin, which could result in the sudden loss of a strip of land up to 100 m wide. Such events have occurred in the past, and their historical frequency suggests an annual probability for a further event of 1 in 400 (0.0025). A total of 25 buildings situated within 100 m of the edge would be destroyed by the landslide, resulting in losses of £3.75 million at current values. It is expected that there would be further indirect losses to the local community, resulting from a decline in tourist numbers and disruption to businesses; these indirect losses are expected to total £5 million. The risk, expressed as an annual value, is:

$$\text{annual risk} = P(\text{landslide event}) \times (\text{direct losses} + \text{indirect losses})$$

$$= 0.0025 \times (3.75 + 5)$$

$$= £0.022 \text{ million}$$

Example 10.2
A large town has grown up within an area of deep-seated landsliding. Throughout its history it has been subjected to very slow ground movements, with less frequent episodes of more active movement. The landslide hazard is associated with the reactivation of the landslides, and varies from almost continuous deep-seated movement, of the order of millimetres per year, to infrequent, short periods of significant ground movement that can result in widespread surface cracking and heave. Reactivation events are triggered by high groundwater levels that coincide with wet winters (i.e. as determined from historical analysis). The wetter the winter, the greater the resulting ground movement; minor events are regular occurrences, while large events are infrequent. It is assumed, therefore, that there is a continuous range of discrete reactivation events (scenarios 1–5 in Figure 10.1), with each magnitude of event associated with a different return period (i.e. probability) of winter rainfall.

Figure 10.1 Example 10.2: the relationship between winter rainfall and landslide reactivation

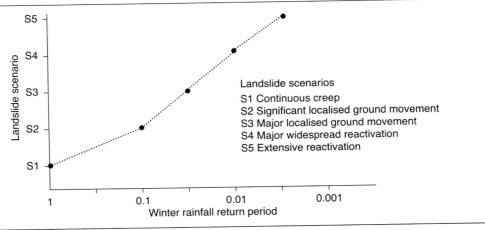

Quantifying risk

Table 10.1 Example 10.2: relationship between winter rainfall events and landslide damage

Reactivation scenario	Winter rainfall event (return period)	Property damage and destruction: £ million	Traffic disruption: £ million
1	1 in 1	0.25	0.1
2	1 in 10	0.5	0.2
3	1 in 50	1	0.5
4	1 in 100	5	1
5	1 in 500	10	5

The losses associated with reactivation can be identified as a combination of repairable damage and property destruction, together with traffic disruption, with the severity of losses related to the size of the reactivation event, as shown in Table 10.1.

From this information it is possible to construct a *landslide reactivation damage curve* that relates losses to the probability of the scenario (Figure 10.2). As any of the scenarios could occur in a given year, depending on the winter rainfall, it is necessary to calculate the average annual damage so as to take account of every possible combination of scenario probability and loss (Table 10.2). In Figure 10.2, the average annual risk (£0.68 million) equates to the area below the damage curve, calculated as a series of slices between event probabilities.

Example 10.3

A seabed gas field is exposed to a series of different landslide scenarios that potentially threaten the various wells, manifolds, flowlines and export pipelines. The landslide scenarios range from shallow

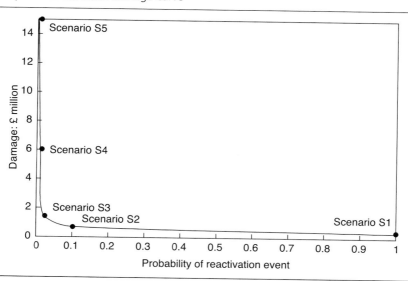

Figure 10.2 Example 10.2: landslide damage curve

Landslide Risk Assessment

Table 10.2 Example 10.2: annual risk calculation (see Figure 10.2 for the landslide damage curve)

Damage category	Return period (years) and annual probability of reactivation event: damage (£ million)				
	1 1.000	10 0.100	50 0.020	100 0.010	500 0.002
Residential property	0.25	0.5	1	5	10
Industrial property	0	0	0	0	0
Indirect losses (e.g. tourism)	0	0	0	0	0
Traffic disruption	0.1	0.2	0.5	1	5
Emergency services	0	0	0	0	0
Other	0	0	0	0	0
Total damage	0.35	0.7	1.5	6	15
Area (damage × frequency)		0.47	0.09	0.04	0.08
Average annual risk (area beneath curve): £ million					0.68

Area (damage × frequency) for between the 1 in 1 (scenario 1) and 1 in 10 (scenario 2) year events is calculated as: area (damage × frequency) = $(P1 - P2) \times$ (total damage 2 + total damage 1)/2; average annual risk = $\sum$ area(damage × frequency)

slab slides to debris flows and retrogressive spreading failures, and are generally associated with the combination of high pore pressures and earthquake events.

A hazard identification review was undertaken to identify all the landslide scenarios that might threaten each of the individual field components. A simple conditional probability model was then used to generate estimates of the likelihood of a *top event*, i.e. damage to a particular field component. For example, the probability of damage to a particular well (w) from an individual landslide scenario (a 'damage event') was calculated as:

$$P(\text{damage well w}) = P(\text{landslide}) \times P(\text{hit}|\text{landslide}) \times P(\text{damage}|\text{hit})$$

As this well could be damaged by a number of different landslide scenarios, the overall probability of that particular 'outcome' (damage to well w) is:

$$P(\text{outcome}) = \sum P(\text{damage to well events 1 to } n)$$

Each outcome was given its own unique reference number, relating to the seabed unit and the particular component affected. A schedule of the outcomes, outcome probabilities and the estimated losses expected to be incurred is presented in Table 10.3. The results were compiled as a *frequency–magnitude of loss curve* (F–M curve, an adaptation of the F–N curves used to plot societal risk; Table 10.4, Figure 10.3).

Example 10.4

A mountain road between two small towns in the Scottish Highlands is regularly blocked by rockfalls, and the traffic is forced to follow a lengthy diversion around the affected area. In Example 9.6, a consequence model was presented to estimate the traffic disruption costs associated with three rockfall scenarios.

Table 10.3 Example 10.3: the schedule of outcomes

Outcome No.	Description	Total loss: $ million	Annual probability of outcome
LTZ17A	Well centre 4: part loss of manifold	7.5	3.415E-05
P1	Loss of flowline (100 m)	15	8.892E-05
LH1	Loss of export pipeline (100 m)	15	6.054E-04
S4	Loss of export pipeline (100 m)	15	1.797E-04
LT3B	Well centre: loss of manifold	17.5	9.762E-05
LTZ4B	Well centres 1 and 2: loss of manifold	17.5	9.942E-05
LTZ12B	Well centre 3: loss of manifold	17.5	9.942E-05
LTZ17B	Well centre 4: loss of manifold	17.5	6.972E-04
P2	Loss of flowline (2500 m)	22.5	2.010E-11
P3A	Well unit: loss of tree	22.5	7.365E-06
LTZ5A	Well centre 1: loss of tree	22.5	3.298E-05
LTZ8A	Well centre 2: loss of tree	22.5	3.298E-05
LTZ13A	Well centre 3: loss of tree	22.5	3.298E-05
LTZ18A	Well centre 4: loss of tree	22.5	3.298E-05
LT5A	Well centre: loss of 2 trees	32.5	3.059E-05
LTZ6A	Well centre 1: loss of 2 trees	32.5	2.928E-05
LTZ9A	Well centre 2: loss of 2 trees	32.5	2.928E-05
LTZ14A	Well centre 3: loss of 2 trees	32.5	2.928E-05
LTZ19A	Well centre 4: loss of 2 trees	32.5	2.928E-05
LH2	Loss of export pipeline (5000 m)	32.5	1.800E-06
LTZ7A	Well centre 1: loss of 3 trees	42.5	2.928E-05
LTZ10A	Well centre 2: loss of 3 trees	42.5	2.928E-05
LH3	Complete loss of export pipeline (17 100 m)	102.5	1.920E-07
LTZ5B	Well centre 1: loss of well and tree	137.5	9.505E-05
P4	Complete loss of well unit and flowline	157.5	4.452E-07
LT5B	Well centre: loss of 2 well and trees	262.5	8.939E-05
LTZ19B	Well centre 4: loss of 2 wells and trees	262.5	8.557E-05
LTZ7B	Well centre 1: loss of 3 wells and trees	387.5	8.557E-05
LTZ20B	Well centre 4: loss of 3 wells and trees	387.5	8.557E-05
LTZ16B	Well centre 3: loss of seabed components and wells	527.5	8.623E-05
LTZ21B	Well centre 4: loss of seabed components and wells	527.5	8.623E-05
LTZ11B	Well centres 1 and 2: loss of seabed components and wells	1027.5	8.623E-05
LTZ22	Well centres 1–4: complete loss	2107.5	2.184E-06
D1	Complete loss of development	2825	4.000E-07

- Scenario 1. A small rockfall (100 m^3) which results in the road being closed for one day, during which the debris is removed by the road maintenance crews (estimated annual probability of 0.3).
- Scenario 2. A relatively small rockfall (1000 m^3) which results in the road being closed for five days because of the need for repairs to the road surface (estimated annual probability of 0.05).
- Scenario 3. A large rockfall (10 000 m^3) which results in the road being closed for 50 days, during which time emergency works are undertaken to stabilise the unstable rockfall backscar and repair the road surface (estimated annual probability of 0.01).

Table 10.4 Example 10.3: the *F–M* pairs, derived from the schedule of outcomes in Table 10.3

Loss *M*: $ million	Cumulative probability *F*
7.5	2.952E-03
15	2.918E-03
17.5	2.044E-03
22.5	1.050E-03
32.5	9.112E-04
42.5	7.616E-04
102.5	7.031E-04
137.5	7.029E-04
157.5	6.078E-04
262.5	6.074E-04
387.5	4.324E-04
527.5	2.613E-04
1027.5	1.750E-04
2107.5	2.584E-06
2825	4.000E-07

Figure 10.3 Example 10.3: *F–M* curve for seabed landslide damage

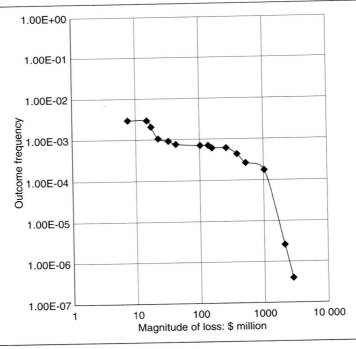

Quantifying risk

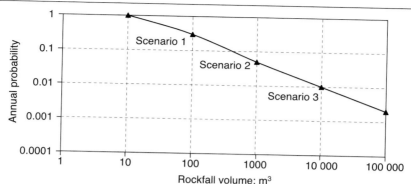

Figure 10.4 Example 10.4: rockfall magnitude/probability distribution

As explained in Example 9.6, a potential loss of £11 900 occurs for each day that the road is blocked.

Ignoring any threat to the road users and clean-up or repair costs, the risk associated with the traffic disruption resulting from each individual scenario is:

Risk(scenario s) = P(event) × total disruption cost

= P(event) × daily disruption cost × length of delay

Risk(scenario 1) = 0.3 × 11 900 = £3570

Risk(scenario 2) = 0.05 × (11 900 × 5) = £2975

Risk(scenario 3) = 0.01 × (11 900 × 50) = £5950

Simply adding together the annual risk associated with each of these three scenarios (£12 495) would underestimate the overall risk. This is because, in reality, the three scenarios are part of the magnitude–frequency distribution for rockfall events in this area (Figure 10.4), in which there are numerous potential combinations of event size and probability, each with a slightly different total disruption cost. However, by using the three scenarios it is possible to construct a landslide damage curve that relates losses to the probability of the event (Figure 10.5). As in the previous example, any of the rockfall events could occur in a given year. By calculating the *average annual damage*, account is taken of every possible combination of rockfall probability and loss (Table 10.5). The resulting value of £26 190 turns out to be double the value obtained by simply adding the risk associated with the three scenarios.

10.3. Cliff recession risk

Where assets are threatened by retreating clifflines, as on the coast or some river meander scars, losses are inevitable at some point in the future because of the ongoing recession. Decision-makers will tend to view the severity of the recession problem in terms of when the assets are likely to be lost and the value of the expected losses. Risk, therefore, needs to be expressed either in terms of which year the losses will occur or as the chance of loss in a particular year in the future.

Figure 10.5 Example 10.4: landslide damage curve

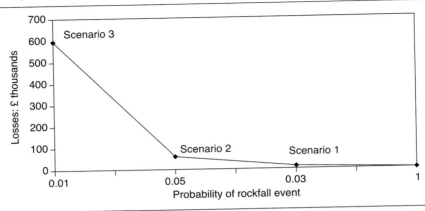

The approach is different from the assessment of current annual risk. An important concept that must be appreciated is that the economic value of any asset does not remain constant for the timescale over which the risk is being considered. If an asset is lost in year 1 it will normally have a higher economic value than if it were to be lost in year 10. One reason for this is that people discount the future, preferring their benefits now rather than some time in the future; a phenomenon known as 'time preference' ('goods now or goods in the future').

The second reason is the *productivity of capital*. If landslide management were not undertaken, the resources could be diverted elsewhere and, over time, would show a return. For example, if the

Table 10.5 Example 10.4: annual risk calculation (see Figure 10.5 for the landslide damage curve)

Damage category	Return period (years) and annual probability of landslide event: damage (£ thousands)			
	1 / 1.0	3.3 / 0.3	20 / 0.05	100 / 0.01
Residential property	0	0	0	0
Industrial property	0	0	0	0
Indirect losses (e.g. tourism)	0	0	0	0
Traffic disruption	0	11.9	59.5	595
Emergency services	0	0	0	0
Other	0	0	0	0
Total damage	0	11.9	59.5	595
Area (damage × frequency)		4.16	8.94	13.09
Average annual risk (area beneath curve): £ thousands				26.19

Area (damage × frequency) for between the 1 in 1 (scenario 1) and 1 in 3.3 (scenario 2) year events is calculated as: area (damage × frequency) = (P1 − P2) × (total damage 2 + total damage 1)/2; average annual risk = $\sum$ area(damage × frequency)

Quantifying risk

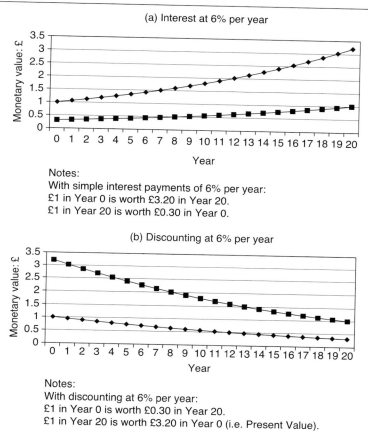

Figure 10.6 The effect of discounting to achieve a present value (PV)

(a) Interest at 6% per year

Notes:
With simple interest payments of 6% per year:
£1 in Year 0 is worth £3.20 in Year 20.
£1 in Year 20 is worth £0.30 in Year 0.

(b) Discounting at 6% per year

Notes:
With discounting at 6% per year:
£1 in Year 0 is worth £0.30 in Year 20.
£1 in Year 20 is worth £3.20 in Year 0 (i.e. Present Value).

return from investing the resources in industry were 5% per annum in real terms, then a £100 sum invested now would yield the equivalent of £105 in one year's time. Therefore, having £100 now and £105 in one year are equivalent. For an asset that gives no return, £100 in one year's time is worth less than £100 now; in other words, it has depreciated (see Figure 10.6).

To achieve this modification of future asset values, it is necessary to express all future losses in terms of their present value (PV) by discounting. The discount factor (D) applicable to a particular year n can be calculated as follows:

$$D_n = (1+r)^{-n} \quad \text{or} \quad D_n = \frac{1}{(1+r)^n}$$

where r is the discount rate, expressed as a decimal, and D_n is the discount factor applicable for year n.

The sum of the discounted flows (i.e. present value) is expressed at the mid-point of year 0.

The effect of using a discount rate is to reduce the value of predicted future losses to their value as seen from the present day. Thus, if the present value of an asset is £1 million in year 0 and a 6% discount rate is used, then the value will have declined to £0.56 million in year 10 and £0.06 million in year 50. In this economic model, it is assumed that inflation of the asset value does not occur because of the existence of a threat, where 'inflation' has its everyday meaning of the price of a resource increasing without its relative value also increasing.

Two different examples are presented below which show the potential application of this approach. The first can be described as a deterministic method, in that the asset is assumed to be lost in a particular year. The second is a probabilistic method, developed by Hall et al. (2000), which recognises that there is a chance that the assets may be lost in any year.

Example 10.5
A small coastal community is threatened by cliff recession (see Example 9.1). Inspection of historical topographic maps of different dates has indicated that the cliffline has been retreating at an average annual rate of 1.25 m/year over the last 150 years. Continued recession over the next 50 years will threaten a number of properties and important services, including a gas main and sewer pump station. It is assumed that the future pattern of recession can be determined by simply projecting the historic average annual rate into the future; that is, in 10 years time 12.5 m will have been lost, while in 50 years the cliff will have retreated an average of 62.5 m.

The risk assessment involves extrapolating the average recession rate in order to estimate the year in which particular properties will be lost. The present value risk of the lost assets is the summation of the discounted asset losses over the period under consideration (in this case 50 years). For each year the calculation is (assuming a 6% discount rate):

$$\text{risk (year } T) = \text{market value} \times \text{discount factor (year } T)$$

and over 50 years:

$$\text{risk (years 0–49)} = \sum (\text{market value} \times \text{discount factor (years 0–49)})$$

The discount factor is calculated as follows:

$$\text{discount factor (year } n) = 1/(1 + \text{discount rate})^n$$

$$\text{discount factor (year 25)} = 1/(1 + 6\%)^{25}$$

$$= 0.223$$

Note that the market value is the price for which an asset is bought or sold in an open market and is used as a measure of the economic value of the asset.

The results are presented in Table 10.6 (note that, although the time period under consideration is 50 years, no losses were predicted to occur after year 25, and so the table has been shortened to save space). Thus, although the current market value of the assets is £6 226 000, the present value risk is lower at £2 952 100. For example, the properties 1–5 Acacia Avenue are currently valued at £350 000, but are expected to be lost in year 8. Applying the discount factor for year 8 (0.627) produces a present value of £219 600.

Table 10.6 Example 10.5: present value risk calculation

Year	Discount factor	Property lost due to recession	Market value: £ thousands	Present value risk: £ thousands
0	1.000		0	
1	0.943		0.0	
2	0.890		0.0	
3	0.840		0.0	
4	0.792		0.0	
5	0.747		0.0	
6	0.705		0.0	
7	0.665	Trunk gas main	450	299.3
8	0.627	1–5 Acacia Avenue	350	219.6
9	0.592		0.0	
10	0.558	Cafe	120	67.0
11	0.527	Sunnyview and Dunswimmin	180	94.8
12	0.497	Sewage pump station	2300	1143.0
13	0.469	Hightrees House	1500	703.3
14	0.442		0.0	
15	0.417		0.0	
16	0.394		0.0	
17	0.371	2–8 Acacia Avenue	400	148.5
18	0.350	The Saltings	76	26.6
19	0.331		0.0	
20	0.312		0.0	
21	0.294	14–20 Rocco Boulevard	850	250.0
22	0.278		0.0	
23	0.262		0.0	
24	0.247		0.0	
25	0.233		0.0	
Total			6226	2952

Discount rate = 6%; present value risk (year T) = discount rate (year T) × market value (property lost, year T)

Example 10.6

The Holderness cliffs, UK, range in height from less than 3 m to around 40 m. The cliffs are formed in a sequence of glacial tills, predominantly silty clays with chalky debris and lenses of sand and gravel. They are subject to severe marine erosion but remain unprotected for most of their length. Long-term recession rates are of the order of 1.2–1.8 m/year (Pethick and Leggett, 1993; Valentin, 1954). At the example site, ongoing recession threatens cliff-top land and property, including a caravan site.

A probabilistic model was developed by Jim Hall (presented in Lee and Clark (2002)) to simulate the episodic cliff-recession process on the cliffline. The statistical model was based on generating random sequences of landslide event sizes and of durations between events, with statistics that conform to the measured values at the sites. From each random sequence, the annual recession distance between years 1 and 49 was extracted. A large number of simulations (in this case 10 000) was used to generate a

Landslide Risk Assessment

histogram or annual recession distance. A kernel density estimation method was then used to obtain a smooth probability density estimate from the histogram.

The analysis involves estimating both the levels of damages/losses that could result from a particular event (i.e. the loss of cliff-top land) and the probability that such an event occurs in a particular year. The potential losses were calculated for each 1-m wide strip of cliff-top land between the present cliff edge and 200 m inland. As soon as the cliff edge encroaches within a strip containing part of an asset, the asset is assumed to be useless/completely lost.

For example, considering the strip of land between the cliff edge and 1 m inland (strip 1), the present value (PV) of the losses associated with recession in a particular year (year T) was calculated as follows:

$$\text{PV losses (strip 1; year } T) = P(\text{loss strip 1; year } T) \times \text{asset value (within strip 1)}$$
$$\times \text{ discount factor (year } T)$$

As strip 1 could be lost (or not) in any year over a 50-year period, the PV of losses associated with the event (i.e. loss of strip 1) is the sum of the annual losses (years 0–49) for strip 1.

For another 1-m wide strip of land, located, say, 9–10 m from the present cliff edge (i.e. Strip 10):

$$\text{PV losses (strip 10; year } T^*) = P(\text{loss strip 10; year } T^*)$$
$$\times \text{ asset value (within strip 10)}$$
$$\times \text{ discount factor (year } T^*)$$

The overall risk is the sum of the PV losses for each 1-m wide strip of cliff-top land over a 50-year period; that is, the risk, per unit of market value (MV), can be calculated directly from:

$$\text{PV(damage risk)} = \text{MV} \sum_{i=0}^{j} \frac{f_{XT}(i|X=x)}{(1+r)^i}$$

where j is the appraisal period, r is the discount rate and $f_{XT}(I|jX=x)$ is the time to recede a given distance.

The results are presented in Table 10.7 (note that for ease of reproduction the timescale has been reduced to 25 years and that the recession distance limited to 15 m). The sum of the risk for the part of the example presented in Table 10.7 is £523 990.

Note the following points for Table 10.7.

- Row A (*asset values £ thousands at given distance*) has been determined from estate agent valuations.
- Row B (*sum of annual probability times discount rate*) is calculated from

    sum (1 m recession) = $\sum P$ receding 1 metre × discount rate (years 1–25)

- Row C (*risk at a given distance*) is calculated for each 1-m wide strip from

    risk (strip 1) = row A × row B (strip 1)

Quantifying risk

Table 10.7 Example 10.6: present value risk calculation

		Recession distance: m															
		0	1	2	3	4	5	6	7	8	9	10	11	12	13	14	15
A	Asset values at given distance: £ thousands	0	0	0	0	0	0	0	0	0	0	140	0	200	0	350	140
B	Sum of annual probabilities times discount rate	0.91	0.86	0.84	0.82	0.80	0.78	0.76	0.74	0.72	0.70	0.68	0.67	0.65	0.63	0.61	0.60
C	Risk at given distances: £ thousands, i.e. A × B	0.00	0.00	0.00	0.00	0.00	0.00	0.00	0.00	0.00	0.00	95.58	0.00	129.7	0.00	214.9	83.79

Year	Discount factor	Simulated probability of receding given distance in given year: m															
		0	1	2	3	4	5	6	7	8	9	10	11	12	13	14	15
1	0.94	0.65	0.24	0.23	0.16	0.12	0.08	0.07	0.05	0.03	0.02	0.02	0.02	0.01	0.01	0.01	0.01
2	0.89	0.20	0.34	0.25	0.22	0.18	0.15	0.12	0.10	0.08	0.06	0.05	0.04	0.03	0.03	0.02	0.02
3	0.84	0.08	0.20	0.20	0.20	0.19	0.18	0.15	0.13	0.12	0.10	0.08	0.07	0.06	0.05	0.04	0.03
4	0.79	0.04	0.11	0.13	0.15	0.16	0.16	0.14	0.14	0.13	0.12	0.11	0.10	0.08	0.07	0.06	0.06
5	0.75	0.02	0.05	0.08	0.10	0.12	0.13	0.14	0.14	0.14	0.13	0.12	0.11	0.10	0.09	0.08	0.07
6	0.70	0.01	0.03	0.05	0.07	0.08	0.10	0.11	0.12	0.13	0.13	0.12	0.12	0.11	0.10	0.10	0.09
7	0.67	0.01	0.02	0.03	0.04	0.06	0.07	0.08	0.09	0.10	0.11	0.12	0.12	0.11	0.11	0.10	0.10
8	0.63	0.00	0.01	0.02	0.02	0.04	0.05	0.06	0.07	0.08	0.09	0.10	0.10	0.10	0.10	0.10	0.10
9	0.59	0.00	0.00	0.01	0.01	0.02	0.03	0.04	0.05	0.06	0.07	0.08	0.08	0.09	0.09	0.10	0.10
10	0.56	0.00	0.00	0.00	0.01	0.01	0.02	0.03	0.04	0.05	0.05	0.06	0.07	0.08	0.08	0.08	0.09
11	0.53	0.00	0.00	0.00	0.00	0.01	0.01	0.02	0.03	0.04	0.04	0.05	0.05	0.06	0.07	0.07	0.08
12	0.50	0.00	0.00	0.00	0.00	0.01	0.01	0.01	0.02	0.03	0.03	0.04	0.04	0.05	0.05	0.06	0.06
13	0.47	0.00	0.00	0.00	0.00	0.00	0.01	0.01	0.01	0.02	0.02	0.03	0.03	0.03	0.04	0.05	0.05
14	0.44	0.00	0.00	0.00	0.00	0.00	0.00	0.01	0.01	0.01	0.02	0.02	0.02	0.03	0.03	0.04	0.04
15	0.42	0.00	0.00	0.00	0.00	0.00	0.00	0.00	0.01	0.01	0.01	0.02	0.02	0.02	0.02	0.03	0.03
16	0.39	0.00	0.00	0.00	0.00	0.00	0.00	0.00	0.01	0.01	0.01	0.01	0.01	0.02	0.02	0.02	0.02
17	0.37	0.00	0.00	0.00	0.00	0.00	0.00	0.00	0.00	0.01	0.01	0.01	0.01	0.01	0.02	0.02	0.02
18	0.35	0.00	0.00	0.00	0.00	0.00	0.00	0.00	0.00	0.00	0.00	0.01	0.01	0.01	0.01	0.01	0.01
19	0.33	0.00	0.00	0.00	0.00	0.00	0.00	0.00	0.00	0.00	0.00	0.00	0.00	0.00	0.01	0.01	0.01
20	0.31	0.00	0.00	0.00	0.00	0.00	0.00	0.00	0.00	0.00	0.00	0.00	0.00	0.00	0.00	0.01	0.01
21	0.29	0.00	0.00	0.00	0.00	0.00	0.00	0.00	0.00	0.00	0.00	0.00	0.00	0.00	0.00	0.00	0.00
22	0.28	0.00	0.00	0.00	0.00	0.00	0.00	0.00	0.00	0.00	0.00	0.00	0.00	0.00	0.00	0.00	0.00
23	0.26	0.00	0.00	0.00	0.00	0.00	0.00	0.00	0.00	0.00	0.00	0.00	0.00	0.00	0.00	0.00	0.00
24	0.25	0.00	0.00	0.00	0.00	0.00	0.00	0.00	0.00	0.00	0.00	0.00	0.00	0.00	0.00	0.00	0.00
25	0.23	0.00	0.00	0.00	0.00	0.00	0.00	0.00	0.00	0.00	0.00	0.00	0.00	0.00	0.00	0.00	0.00
Overall probability		1.00	1.00	1.00	1.00	1.00	1.00	1.00	1.00	1.00	1.00	1.00	1.00	1.00	1.00	1.00	1.00

The sum of the risk is the sum of row C = £523 900

A: This row contains the market value of properties within each 1-m wide strip of cliff-top land (i.e. a £140 000 house lies 10 m from the cliff edge)
B: This row contains values for the product of the discount rate × probability of loss for each year from 1 to 25
C: This row contains risk values calculated as: risk (strip 10) = row A (strip 10) × row B (strip 10) = 140 × 0.6827 = 95.58

- The *simulated probabilities of receding a given distance in a given year (m)* have been derived from the probabilistic model.
- The *discount rate* used is 6%.

10.4. Comparing the risks associated with different management options

Risk-based methods can also be used to compare the level of risk associated with different management strategies, i.e. as *project* or *options appraisal* tools. One approach could be simply to compare the annual risk between continuing with the current situation and that expected to occur with the proposed management strategy in place:

risk (current situation) = P(landslide) × losses

compared with

risk (with management strategy) = P(landslide) × losses

This could be extended if there were more than one possible management strategy, and would identify which is the most effective strategy. However, the main purpose of making such a comparison is usually to address the question 'Is it worth it?' to make the investment in a particular management strategy (i.e. does the risk reduction achieved justify the cost?):

risk reduction benefits = 'without project' risk − 'with project' risk

The decision to invest in landslide-management activity should depend on a thorough appraisal of the benefits of risk reduction over the expected lifetime of the scheme/project. The simple comparison between the annual risk associated with the current situation and that achieved following a stabilisation project can significantly underestimate the value of the risk reduction achieved. This is because landslide stabilisation reduces risk in every year until the end of its design life (and often beyond), although major costs are only incurred at the time of construction (year 0). Maintenance and repair costs will, of course, be spread over the lifetime of the scheme. The benefits of management are the difference between the value of the losses that could be expected to be incurred without a scheme/project and the value of the losses that would be incurred when the scheme fails and landslide activity is renewed. An alternative way of looking at this is that the management scheme reduces the probability of ground movement, and thereby reduces the risk over its expected lifetime, not just in a single year.

For example, a stabilisation scheme with a design life of 50 years will reduce the risk over a period of 50 years. Thus the benefits of this management option are:

landslide management benefits = 'without-project' risk (years 0–49) − 'with-project' risk (years 0–49)

As highlighted in the previous section, the economic value of risk reduction in future years is worth less than that achieved at present. It is, therefore, necessary to express all future risks in terms of their present value (PV), by discounting:

'without-project' risk (years 0–49) = $\sum$ (P event × losses × discount factor (years 0–49))

'with-project' risk (years 0–49) = $\sum$ (P event × losses × discount factor (years 0–49))

A variety of options can be compared by decision-makers in order to select the most 'desirable'. To make this comparison it is necessary to establish a baseline against which the various options, including continuing with the current management practice, can be assessed. This baseline is the so-called *do-nothing* option, which should involve no active landslide management whatsoever, simply walking away and abandoning all maintenance, repair or management activity. In most cases, it should be easy to demonstrate that continuing the current practices are better than the do-nothing case.

Example 10.7

A cliff-top community is vulnerable to a landslide event with an annual probability of 0.01. The event could result in estimated total losses of £25 million. A combined seawall and slope-stabilisation scheme is proposed that would prevent further cliff-foot erosion and reduce the probability of a major event to an estimated 1 in 1000 (0.001).

The comparison of the 'without-project' case (i.e. do nothing) and the 'with-project' case (i.e. the scheme) is presented in Table 10.8. Note that, to reduce the space required for this table, calculations for only 21 years of the 50-year design life are shown, and that:

PV(without-project) risk (year T) = discount factor (year T)

$\times$ P(landslide occurring in year) $\times$ losses (£25 million)

PV(with-project) risk (year T) = discount factor (year T)

$\times$ P(landslide occurring in year) $\times$ losses (£25 million)

PV risk reduction = without-project risk (year 0–20) − with-project risk (years 0–20)

It should be noted that for events which can be assumed to occur only once at a particular site (i.e. one-off events), it is necessary to take account of the fact that the event may have already occurred in year 1 and, hence, could not occur in year 2 (and so on). Thus, the annual probability for year 2 (and subsequent years) is modified, as follows:

P(event; year T) = annual $P \times$ (P event has not already occurred by year $(T-1)$)

From Table 10.8, the risk reduction achieved by the scheme is:

PV without-project risk	PV with-project risk	PV risk reduction
£2.88 million	£0.31 million	£2.57 million

Example 10.8

Periodic reactivation of an extensive area of deep-seated landsliding causes a combination of repairable damage and property destruction, with the severity of damage related to the scale/intensity of the reactivation event. It is proposed to undertake extensive stabilisation works to reduce the frequency of reactivations. It is assumed that the scheme will reduce the event probability by a factor of 10; thus, a 1 in 15 year event, for example, would become a 1 in 150 year event with the scheme in place.

The average annual risk can be calculated from the area beneath reactivation damage curves developed for both the 'without-project' and the 'with-project' cases. The PV risks for both cases can be calculated from:

risk (years 0–49) = $\sum$ (average annual risk $\times$ discount factor (years 0–49))

Table 10.8 Example 10.7: present value risk calculation (without-project and with-project cases)

Year	Discount factor	Without-project case			With-project case			PV risk	
		Probability of a landslide	Probability of a landslide occurring in year	Probability that a landslide has not occurred	Cumulative probability of landslide	Probability of a landslide	Probability of a landslide occurring in year	Probability that a landslide has not occurred	Cumulative probability of landslide	PV without-project losses: £ million	PV with-project losses: £ million
0	1.00	0.010	0.010	0.990	0.010	0.001	0.001	0.999	0.001	0.25	0.03
1	0.943	0.010	0.010	0.980	0.020	0.001	0.001	0.998	0.002	0.23	0.02
2	0.890	0.010	0.010	0.970	0.030	0.001	0.001	0.997	0.003	0.22	0.02
3	0.840	0.010	0.010	0.961	0.039	0.001	0.001	0.996	0.004	0.20	0.02
4	0.792	0.010	0.010	0.951	0.049	0.001	0.001	0.995	0.005	0.19	0.02
5	0.747	0.010	0.010	0.941	0.059	0.001	0.001	0.994	0.006	0.18	0.02
6	0.705	0.010	0.010	0.932	0.068	0.001	0.001	0.993	0.007	0.17	0.02
7	0.665	0.010	0.009	0.923	0.077	0.001	0.001	0.992	0.008	0.15	0.02
8	0.627	0.010	0.009	0.914	0.086	0.001	0.001	0.991	0.009	0.14	0.02
9	0.592	0.010	0.009	0.904	0.096	0.001	0.001	0.990	0.010	0.14	0.01
10	0.558	0.010	0.009	0.895	0.105	0.001	0.001	0.989	0.011	0.13	0.01
11	0.027	0.010	0.009	0.886	0.114	0.001	0.001	0.988	0.012	0.12	0.01
12	0.497	0.010	0.009	0.878	0.122	0.001	0.001	0.987	0.013	0.11	0.01
13	0.469	0.010	0.009	0.869	0.131	0.001	0.001	0.986	0.014	0.10	0.01
14	0.442	0.010	0.009	0.860	0.140	0.001	0.001	0.985	0.015	0.10	0.01
15	0.417	0.010	0.009	0.851	0.149	0.001	0.001	0.984	0.016	0.09	0.01
16	0.394	0.010	0.009	0.843	0.157	0.001	0.001	0.983	0.017	0.08	0.01
17	0.371	0.010	0.008	0.835	0.165	0.001	0.001	0.982	0.018	0.08	0.01
18	0.350	0.010	0.008	0.826	0.174	0.001	0.001	0.981	0.019	0.07	0.01
19	0.331	0.010	0.008	0.818	0.182	0.001	0.001	0.980	0.020	0.07	0.01
20	0.312	0.010	0.008	0.810	0.190	0.001	0.001	0.979	0.021	0.06	0.01
									Total	2.88	0.31

PV risk (year T) = discount factor × P(landslide occurring in year T) × losses (£25 million)

From Table 10.9, the risk reduction achieved by the scheme is:

PV without-project risk	PV with-project risk	PV risk reduction
£2 090 000	£355 000	£1 735 000

Example 10.9

Cliff recession is threatening a small cliff-top community. A proposed seawall has been designed to safeguard the properties and delay any losses for the design life of the scheme, which is 50 years (see Examples 9.1 and 10.5 for further details).

The risk reduction achieved by the scheme can be assessed using the deterministic approach developed by the Ministry of Agriculture, Fisheries and Food (MAFF, 1993). This considers the risk in terms of an annual value of the assets for each year until they are lost by cliff retreat. The logic of this approach becomes clear when the method is used to compare options for reducing the retreat rate, whereby the benefits of intervention are the resulting increase in asset value.

The benefit associated with each year that a given cliff-top asset remains usable is calculated by considering risk-free market value of the cliff-top asset. The risk-free market value (MV) can be thought of as being equivalent to the present value of n equal annual payments of A (*the equivalent annual value*), where n is the life of the asset. If the annual payments occur from year 1 to year n, then:

$$A = \frac{MV \times r}{1 - D_n}$$

where r is the discount rate (e.g. 6%) and D_n is the discount factor (see earlier in this section).

More usually, the value of a cliff-top asset is thought to extend from year 0 to year $n-1$, in which case:

$$A = \frac{MV}{1 - D_n} \times \frac{r}{1+r}$$

It follows that A can be approximated by

$$A = MV \times r$$

for annual payments starting in year 1.

With reference to Table 10.10, the row of houses 1–5 Acacia Avenue has a total market value of £350 000 and an equivalent annual value (A) of

$$A = MV \times r = 350\,000 \times 0.06 = £21\,000$$

For the 'without-project' case, it is assumed that these houses would be lost in year 8. The asset value that would be lost is equivalent to eight years of an annual payment equal to the equivalent annual value (A) of £21 000. As each of these payments has to be brought back to a present value, the 'without-project' risk is:

$$\text{'without-project' risk} = \sum (A \times \text{discount factor}) \text{ for years 1–8 (excluding year 0)}$$

$$= \sum (21\,000 \times \text{discount factor}) \text{ for years 1–8} = £130\,000$$

Landslide Risk Assessment

Table 10.9 Example 10.8: present value risk calculation (without-project and with-project cases)

Damage category	Return period and annual probability of reactivation event									Total PV risk: £ thousands
	5	10	15	25	50	100	150	250	Infinity	
	0.200	0.100	0.067	0.040	0.020	0.010	0.007	0.004	0	
Without project										
Residential property			600	750	1000	2500	6000	7000	8500	2090
Industrial property									0	0
Indirect losses (e.g. tourism)									0	0
Total damage: £ thousands	0	0	600	750	1000	2500	6000	7000	8500	
Area (damage × frequency)		0.00	10.00	18.00	17.50	17.50	14.17	17.33	31.00	
Average annual risk (area beneath curve): £125 500; PV factor, 15.650; present value risk £2 090 000										
With project										
Residential property							600	750	8500	355
Industrial property									0	0
Indirect losses (e.g. tourism)									0	0
Total damage: £ thousands	0	0	0	0	0	0	600	750	8500	
Area (damage × frequency)		0.00	0.00	0.00	0.00	0.00	1.00	1.80	18.50	
Average annual risk (area beneath curve): £21 300; PV factor, 16.650; present value risk £355 000										

Area (damage × frequency) for between the 1 in 5 (scenario 1) and 1 the 1 in 10 (scenario 2) year events is calculated as:
area(damage × frequency) = (Prob 1 − Prob 2) × (total damage 1)/2; average annual risk = $\sum$ area(damage × frequency)
PV factor = $\sum$ discount factor(years 0–49)

Table 10.10 Example 10.9: present value risk calculation (without-project and with-project cases)

Year	Property	MV: £ thousands	Utilities	MV: £ thousands	Equivalent annual value A	Asset value: £ thousands Without project	Asset value: £ thousands With project
0		0		0	0	0	0
1		0		0	0	0	0
2		0		0	0	0	0
3		0		0	0	0	0
4		0		0	0	0	0
5		0		0	0	0	0
6		0		0	0	0	0
7		0	Trunk gas main	450	27	151	434
8	1–5 Acacia Avenue	350		0	21	130	338
9		0		0	0	0	0
10	Cafe	120		0	7	53	116
11	Sunnyview and Dunswimmin	180		0	11	85	175
12		0	Sewage pump station	2300	138	1157	2238
13	Hightrees House	1500		0	90	797	1462
14		0		0	0	0	0
15		0		0	0	0	0
16		0		0	0	0	0
17	2–8 Acacia Avenue	400		0	24	251	392
18	The Saltings	76		0	5	49	75
19		0		0	0	0	0
20		0		0	0	0	0
21	14–20 Rocco Boulevard	850		0	51	600	836
22		0		0	0	0	0
23		0		0	0	0	0
24		0		0	0	0	0
25		0		0	0	0	0
Totals		3476		2750		3274	6066

Equivalent annual value (A) = MV × discount factor (0.06); asset value (without project) = $\sum A$ × discount factor (years 1 to T), where T is the expected year of loss; asset value (with project) = $\sum A$ × discount factor (years 1 to T) 50), where T is the expected year of loss

The overall 'without-project' risk is the sum of the present value of assets lost in different years over the time period under consideration (in this case 50 years, although, in order to save space, only 25 years are presented in Table 10.10). This value may appear anomalously small, but it has to be appreciated that this approach measures reduction of risk in terms of increased asset value.

Assuming that the scheme is implemented in year 0 (i.e. immediately), a seawall with a 50-year design life would increase the life of all assets by 50 years (i.e. the life of 1–5 Acacia Avenue would be extended to 58 years). The 'with-project' risk is calculated as follows:

'with-project' risk $= \sum (A \times$ discount factor) for years 1–58 (excluding year 0)

$= \sum (21\,000 \times$ discount factor) for years 1–58

$= £338\,000$

The overall 'with-project' risk is the sum of the present value of assets over the time period under consideration (see Table 10.10).

The reduction in risk (i.e. the scheme benefits) achieved by the seawall is expressed in terms of the increase in asset value generated by extending the asset life:

Risk reduction (PV) = PV 'with-project' asset value − PV 'without-project' asset value

For 1–5 Acacia Avenue:

risk reduction (PV) = £338 000 − £130 000 = £208 000

For the whole scheme, the risk reduction is:

PV asset value – with project	PV asset value – without project	PV risk reduction
£6 066 000	£3 274 000	£2 792 000

Example 10.10

In studies of eroding coastal cliffs it is generally true that an approach involving the probabilistic appraisal of cliff recession and the potential benefits of landslide management is preferable to the use of deterministic methods, because it takes a more 'realistic' account of the large uncertainties associated with the recession process. A probabilistic methodology for economic evaluation has been developed by Hall et al. (2000). The method takes the write-off value of the threatened assets and evaluates the loss associated with the probability that cliff retreat will result in the assets being written off in a given year (see Examples 9.1 and 10.9).

The probability of property loss varies with time and location. The probability density function (p.d.f.) $f_{XT}(x, i)$ is the probability of damage at distance x from the cliff edge during year i. The function f is a function of a distance random variable X and a time random variable T. For the purposes of benefit assessment, T is considered to be a discrete random variable measured in years.

For example, a single house is at threat from gradual cliff recession (albeit at an irregular rate). The setting is shown diagrammatically in Figure 10.7. Profile a shows the current cliff position; three of the many possible future locations are labelled as u, v and w. If erosion has advanced to profile w, then the asset will have been destroyed, whereas if it has proceeded only as far as profile u or v, it will not.

The p.d.f. of cliff location in a particular year $(T = i)$, $f_{XT}(x|T = i)$ is shown on the same horizontal scale as the cliff cross-section. In a probabilistic analysis, there is no single conventional erosion contour for year i but instead a band of potential erosion limits or risk. The contour of predicted average recession lies at the mean of the distribution. The probability p_i of damage to the house

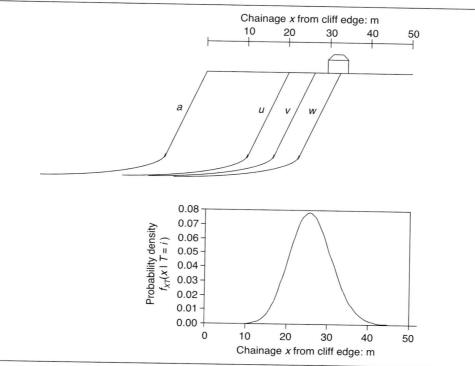

Figure 10.7 Example 10.10: a schematic representation of the probability of cliff recession reaching a single property. (After Hall et al., 2000)

before or during year i is given by:

$$p_i = \int_0^d f_{XT}(x|T=i)\,dx$$

where d is the distance of the house from the cliff edge.

The more convenient representation for the purposes of benefit assessment (see Example 10.9) is the distribution of the predicted year of loss for assets at a given distance x from the cliff edge. The p.d.f. $f_{XT}(i|X=x)$ can be entered directly into the discounting table to obtain the probability weighted sum of the damage risk:

$$\text{PV(damage)} = \text{MV} \sum_{i=0}^{j} \frac{f_{XT}(i|X=x)}{(1+r)^i}$$

where j is the appraisal period.

Thus, the present value (PV) of the loss in any year is calculated as follows:

PV(damage year i) = probability of loss × asset value × discount factor (year i)

The overall present value losses are:

'without-project' PV(damage years 0–49) = $\sum$ probability of loss × asset value

× discount factor (years 0–49)

Table 10.11 shows an example of the probabilistic discounting procedure for a house with a risk-free market value of £100 000 situated 10 m from the edge of an eroding cliff, using an illustrative probability distribution (obtained using the two-distribution probabilistic method described in Example 6.9). According to the probabilistic discounting procedure, the PV damage risk is £29 720.

A coast-protection scheme would delay the recession scenario by a length of time equivalent to the scheme design life (e.g. 50 years). The 'with-project' losses can be calculated as follows:

'with-project' PV(damage years 50–99) = $\sum$ probability of loss × asset value

× discount factor (years 50–99)

Note that between years 0 and 49 the probability of loss would be zero, i.e. the scheme had delayed the recession losses by 50 years.

An alternative approach would be to consider the scheme as reducing the probability of loss in each year up to the end of its design life:

'with-project' PV(damage years 0–49) = $\sum$ revised probability of loss × asset value

× discount factor (years 0–49)

In either case, the reduction in risk achieved by the scheme is:

risk reduction (PV) = 'without-project' PV (damage) − 'with-project' PV (damage)

Where multiple cliff-top assets are at risk, the above methodology should be repeated for each of the assets. Table 10.12 summarises the analysis for a hypothetical case in a similar format to the deterministic approach illustrated in Table 10.10. It shows how predictions of risk for various erosion-control scenarios, ranging from the 'without-project' option to a high standard of protection in option 4, could be presented.

10.5. Individual risk

The risk specific to humans, such as the general public or particular activity groups, is the frequency with which individuals within such groupings are expected to suffer harm (Health and Safety Executive, 1992; Royal Society, 1992). This notion is usually referred to as *individual risk*.

Individual risk is a somewhat abstract concept and must not be confused with the risk faced by a specific individual, which is best termed *personal risk*. A formal definition of individual risk is the frequency with which an individual within a specific group or population may be expected to sustain a given level of harm from the realisation of a specific hazard or particular combination of hazards (IChemE, 1992). In the case of landsliding, it can be considered to be the risk of fatality or injury to individuals who live within a zone liable to be impacted by a landslide or who follow a particular pattern of life that might subject them to the adverse consequences of a landslide (Fell and Hartford, 1997). Individual risk has traditionally been taken to be the risk of death, expressed as the probability

Table 10.11 Example 10.10: probabilistic discounting for a single cliff-top asset

Year i	Discount factor	Probability of damage in year i	PV damage: £ thousands
0	1.00	0.000	0.00
1	0.94	0.002	0.18
2	0.89	0.002	0.19
3	0.84	0.005	0.39
4	0.79	0.006	0.47
5	0.75	0.010	0.73
6	0.70	0.011	0.75
7	0.67	0.013	0.89
8	0.63	0.018	1.11
9	0.59	0.019	1.10
10	0.56	0.024	1.34
11	0.53	0.023	1.20
12	0.50	0.024	1.20
13	0.47	0.028	1.32
14	0.44	0.032	1.42
15	0.42	0.032	1.34
16	0.39	0.033	1.30
17	0.37	0.035	1.29
18	0.35	0.035	1.22
19	0.33	0.035	1.14
20	0.31	0.037	1.15
21	0.29	0.033	0.98
22	0.28	0.035	0.97
23	0.26	0.033	0.87
24	0.25	0.032	0.79
25	0.23	0.030	0.70
26	0.22	0.031	0.68
27	0.21	0.029	0.60
28	0.20	0.030	0.58
29	0.18	0.027	0.50
30	0.17	0.025	0.44
31	0.16	0.022	0.36
32	0.15	0.021	0.33
33	0.15	0.021	0.30
34	0.14	0.019	0.26
35	0.13	0.018	0.23
36	0.12	0.017	0.21
37	0.12	0.016	0.19
38	0.11	0.013	0.14
39	0.10	0.013	0.13
40	0.10	0.012	0.12
41	0.09	0.009	0.08
42	0.09	0.011	0.10
43	0.08	0.008	0.06
44	0.08	0.009	0.07
45	0.07	0.008	0.06
46	0.07	0.007	0.04
47	0.06	0.006	0.04
48	0.06	0.006	0.04
49	0.06	0.004	0.02
		Total PV risk: £ thousands	29.72

From Hall et al. (2000)
Risk-free market value of asset = £100 000. Discount rate = 6%. Distance of asset from cliff edge = 10 m. PV damage (year i) = discount factor (year i) × P(damage (year i)) × asset value

Table 10.12 Example 10.10: summary of discounted asset values for multiple assets and different erosion control options (see Example 10.9 for background)

Property/utility	Distance from cliff edge: m	Market value: £ thousands	Damage risk value with different protection options: £ thousands				
			Without project	Option 1	Option 2	Option 3	Option 4
Trunk gas main	7	450	252	183	152	84	12
1–5 Acacia Avenue	8	350	184	132	111	62	7
Cafe	10	120	62	43	30	26	1
Sunnyview Boulevard	11	180	81	61	49	29	4
Sewage pump station	12	2300	940	675	555	315	47
Hightrees House	13	1500	585	417	343	195	25
2–8 Acacia Avenue	17	400	136	99	75	44	7
The Saltings	18	76	26	17	14	8	0
14–20 Rocco Boulevard	21	850	165	122	100	67	9
Total PV risk: £ thousands			2431	1749	1429	830	112
Erosion-control benefit: £ thousands				682	1002	1601	2319

From Hall *et al.* (2000)

per year, although in the case of the transport sector it is normally expressed as 'per journey' or 'per passenger mile or kilometre'. It is the risk deemed to be experienced by a notional single individual in a given time period, and generally reflects the amount of time for which the individual is exposed to the hazard and the severity of the hazard.

Individual risk is a problematic concept, and discussions in the literature are often contradictory. It is essential, therefore, when computing values for individual risk, to specify at the outset whether the calculated frequencies relate to those most at risk from a given activity (say, as a result of their home location, work, recreation or time periods for which they remain vulnerable) or whether they relate to an 'average' or 'shared value' representative of all potentially affected individuals. The use of an average value is only strictly appropriate where the risk is relatively uniformly distributed over the affected population. Otherwise, this measure can be highly misleading, because the existence of a few individuals exposed to high risk levels is concealed when risk is averaged over a large number of people, most of whom are at relatively low risk.

There are typically three different types of individual risk (e.g. IChemE, 1992; Kauer *et al.*, 2002):

1. *Location-specific individual risk* (*LSIR*). The risk for an individual who is present at a particular location for the entire period under consideration, which may be 24 hours per day, 365 days per year or during the entire time that a risk-generating plant, process or activity is in operation. The LSIR can be a misleading risk measure, as few individuals remain at the same location all the time or are constantly exposed to the same risk. However, it is useful in spatial planning, as it is a property of the location in question, rather than the behaviour patterns of the population (e.g. Bottelberghs, 2000):

    $$LSIR = P(\text{landslide}) \times P(\text{spatial}) \times P(\text{temporal}) \times \text{vulnerability}$$

Figure 10.8 Individual risk associated with a series of debris-flow events

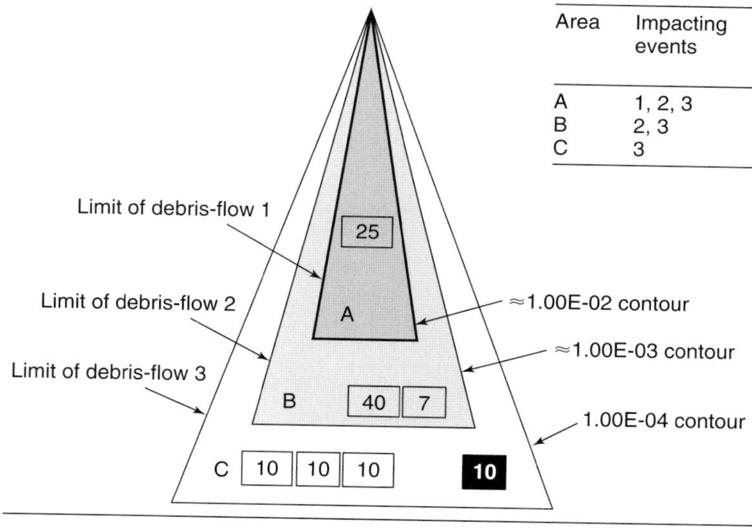

Area	Impacting events	Combined event probability	Vulnerability	Individual risk (LSIR)
A	1, 2, 3	0.0111	1	1.11E-02
B	2, 3	0.0011	1	1.10E-03
C	3	0.0001	1	1.00E-04

Event	*P* (event)
Debris-flow 1	0.01
Debris-flow 2	0.001
Debris-flow 3	0.0001

25 Building and number of occupants

10 Pump station and number of occupants

P(spatial) is the probability that the path of the landslide intersects the location where the individual could be (i.e. in the 'Wrong Place'; see Chapter 7), and *P*(temporal) is the probability that the individual is in the path of the landslide (landslide danger zone) at the time of landslide occurrence (i.e. at the 'wrong time'; see Chapter 7). For LSIR, both *P*(spatial) and *P*(temporal) = 1.

Figure 10.8 shows the LSIR associated with three possible debris-flow events with different run-out distances. The LSIR in each of the three zones on the map represents the total risk from three different debris-flow events. The generic representation of the risk equation for such multiple events (1 to *n*) is:

$$\text{LSIR (zone } z) = \sum \text{LSIR, events 1 to } n$$

Area *A* would be impacted by all three events, so the LSIR is calculated (assuming an individual would be killed) as:

$$\begin{aligned}\text{LSIR} &= [P(\text{debris flow 1}) \times \text{vulnerability}] + [P(\text{debris flow 2}) \times \text{vulnerability}] \\ &\quad + [P(\text{debris flow 3}) \times \text{vulnerability}] \\ &= (0.01 \times 1) + (0.001 \times 1) + (0.0001 \times 1) \\ &= 0.0111 \text{ per year}\end{aligned}$$

2 *Individual-specific individual risk* (*ISIR*). The risk for an individual who is present at different locations during different periods. The ISIR can be a more realistic measure than the LSIR.

$$\text{ISIR} = P(\text{landslide}) \times P(\text{spatial}) \times P(\text{temporal}) \times \text{vulnerability}$$

For ISIR, *P*(temporal) is variable depending on the time spent in the 'danger zone'. If an individual spends an hour a day, 250 days of the year, walking through area C (*P*(spatial) = 1) in Figure 10.8, then the risk associated with debris-flow 3 is (assuming a vulnerability of 1, i.e. death) is:

$$\text{ISIR} = 0.0001 \times 1 \times [(1/24) \times (250/365)] \times 1$$

$$= 0.0001 \times 1 \times (250/8760) \times 1$$

$$= 2.85 \times 10^{-6} \text{ per year (2.85E-06)}$$

3 *Average individual risk (AIR)*. The AIR can be calculated from historical data of the number of fatalities per year divided by the number of people at risk.

$$\text{AIR} = \frac{\text{number of fatalities/year}}{\text{exposed population}}$$

AIR is related to the potential loss of life (PLL per year):

$$\text{AIR} = \frac{\text{PLL}}{\text{exposed population}}$$

Note that the *exposed population* could be either the total population or the average population expected to be within the danger zone during a given time period, such as an hour or a day (see Chapter 7).

In the example shown in Figure 10.8, assuming that the occupants are present all the time and would be killed if a debris flow occurs, then the PLL for the entire danger zone (areas A–C) is:

$$\text{PLL} = \text{occupants (area A)} \times \text{LSIR (area A)} + \text{occupants (area B)} \times \text{LSIR (area B)}$$

$$+ \text{occupants (area C)} \times \text{LSIR (area C)}$$

$$= (25 \times 0.0111) + (47 \times 0.0011) + (40 \times 0.0001)$$

$$= 0.333 \text{ per year}$$

$$\text{AIR} = \text{PLL/number exposed}$$

$$= 0.333/112$$

$$= 0.003 \text{ per year}$$

However, the AIR is highest in area A, where fewer people (25) are exposed to all three events:

$$\text{AIR (area A)} = \frac{\text{occupants (area A)} \times \text{LSIR (area A)}}{\text{number exposed}}$$

$$= (25 \times 0.0111)/25$$

$$= 0.0111 \text{ per year}$$

This is the same as the LSIR for area A.

It is important to recognise that the AIR per year for an area may conceal significant variability, from 0.0111 in area A to 0.0001 in area C.

A measure of the average individual risk can also be derived from the societal risk (see Section 10.6) divided by the number of people at risk:

$$\text{AIR} = \frac{\text{societal risk}}{\text{exposed population}}$$

Locations with equal individual risk can be shown on a map by means of so-called *risk contours* (see Figure 10.8). These contours can be constructed using a commercial package, although terrain unit boundaries can provide a more realistic framework for distinguishing between different zones than simply relying on mathematical contouring.

Individual risk can also be expressed by means of the *fatal accident rate* (*FAR*), which is the number of fatalities per unit number of hours of exposure (e.g. Bedford and Cooke, 2001). FARs can be more convenient and are often more readily understandable than the individual risk per year. In the oil and gas industry, for example, the FAR is the number of fatalities per 10^8 exposed hours (e.g. Spouge, 1999; Vinnem, 2007). The number of 10^8 exposed hours is roughly equivalent to the number of hours at work in 1000 working lifetimes. The FAR measure was developed to describe onshore occupational risks, which apply only during working hours. Hence, in onshore studies, 'exposed hours' is taken to mean 'hours at work', and the FAR is defined as:

Onshore FAR = fatalities at work × 10^8/person-hours at work

When applied offshore, some risks apply only during working hours, but others apply during the whole time the worker is offshore, including the time spent off-duty. Therefore, in offshore studies, 'exposed hours' are usually taken to be 'hours spent offshore':

Offshore FAR = fatalities offshore × 10^8/person-hours offshore

FARs are convenient for describing the risk in individual activities (e.g. walking along a promenade, playing rugby, rock climbing, flying in a helicopter), and are used to describe the risk faced by certain groups of people who undertake certain activities or lifestyles (*group risk*). However, FARs may be misleading if used more generally, because they represent a rate of risk *per unit time in the activity*. Hence, in contrast to individual risks per year, FARs cannot necessarily be added together. For example, the FAR in helicopter travel to an offshore site may be in the range 200–400 (as it involves high risks during a short time period), while the total FAR in offshore activities may be only 10–20 (with the helicopter risk averaged over the whole time period offshore).

The FAR per 10^8 exposed hours and the potential loss of life per year (PLL) are closely related:

FAR = PLL × $10^8/H$

where H is the number of hours per year spent at work or offshore per individual.

The conversions from the various measures of individual risk to the FAR are:

FAR = ISIR × 10^8/exposed hours per year

For a typical offshore worker the exposed hours is 24 hours/day for 20 weeks/year, i.e. 3360 hours per year, of which 1680 would actually be on shift:

FAR = ISIR × 10^8/3360 hours per year

The conversion of the location-specific individual risk (LSIR) to the FAR is based on the exposure time implicit in LSIRs of 24 hours/day for 365 days/year:

$$FAR = LSIR \times 10^8/8760 \text{ hours per year}$$

The pump station in Figure 10.8 (the shaded building in area C) is permanently staffed by the equivalent of 10 people (i.e. a manning level of 10). Assuming that all 10 staff would be killed if the pump station was hit by the debris flow, then

$$PLL = LSIR \text{ (area C)} \times \text{exposed individuals}$$

$$= 0.0001 \times 10$$

$$= 0.001 \text{ per year}$$

However, each individual only works an 8-hour shift, 250 days of the year, and hence the annual number of hours worked by any individual is 2000 out of the 8760 hours in a year. The average individual risk associated with debris-flow 3 is:

$$AIR = \frac{PLL \text{ (pump station)}}{\text{manning level} \times (8760/\text{exposed hours per year})}$$

$$= 0.001/[10 \times (8760/2000)]$$

$$= 2.28 \times 10^{-5} \text{ per year}$$

The fatal accident rate is:

$$FAR = PLL \times 10^8/\text{total exposed hours}$$

$$= PLL \times 10^8/\text{manning level} \times 8760$$

$$= 0.001 \times 10^8/8760$$

$$= 1.14 \text{ per } 10^8 \text{ exposed hours}$$

The FAR can also be calculated from the AIR:

$$FAR = AIR/(\text{hours worked per individual} \times 10^{-8})$$

$$= 2.28 \times 10^{-5}/(2000 \times 10^{-8})$$

$$= 1.14 \text{ per } 10^8 \text{ exposed hours}$$

A variant is the *death per unit activity*, where the time unit is replaced by a unit measuring the amount of activity. The risks of travel by car, train or aeroplane are often expressed in the form of the number of deaths per kilometre travelled, the number of journeys or the hours of travel (Table 10.13). Interested parties tend to choose the form of presentation that suits their own purposes. The air-transport industry, for example, tends to choose a per kilometre basis, as most fatalities occur on landing and take-off, while the overall travel distances are very large. Bus companies might select fatalities per number of journeys or hours of travel, as the risks are uniformly spread. In this way, both are able to demonstrate that theirs is the safest form of transport.

Quantifying risk

Table 10.13 Fatalities associated with different modes of transport in the UK

Transport mode	Fatalities/billion kilometres	Fatalities/billion journeys	Fatalities/billion hours
Air	0.05	117	30.8
Bus	0.4	4.3	11.1
Rail	0.6	20	30
Car	3.1	40	130
Foot	54.2	40	220
Motorcycle	108.9	1640	4840

Ford (2000)

Example 10.11

Large hurricanes in the Gulf of Mexico can generate waves large enough to cause significant sea-floor pressures and initiate submarine mudslides. For example, in 1969 Hurricane Camille generated 70 foot high waves that triggered a mudslide in the South Pass Block 70 area of the Mississippi delta. This mudslide destroyed the recently installed Shell Oil South Pass 70B platform and the Gulf Oil Co. South Pass 61 platform (Sterling and Strobeck, 1973). As all personnel had been evacuated from the platforms ahead of the storm, no lives were lost. As a result of this incident, mudslide risk to the workforce is assessed as part of the major accident risk process for developments in the Gulf of Mexico.

A new manned platform is to be located on the Mississippi delta. Based on a programme of geotechnical studies and probabilistic stability analysis, the annual probability of a landslide hitting the platform and causing it to collapse was estimated to be 0.0001 (1 in 10 000), using the following input probabilities:

$$P(\text{collapse}) = P(\text{mudslide}) \times P(\text{spatial}) \times \text{platform vulnerability}$$
$$= 0.01 \times 0.04 \times 0.25$$
$$= 0.0001$$

The platform is designed to have an average manning level of 220 workers, each of whom has an annual number of 3000 exposure hours offshore (on duty and resting).

If it is assumed that 30% of the workforce would be killed if the platform collapsed (or, a probability of 0.3 that any individual would be killed), then the location-specific individual risk (LSIR) is:

$$\text{LSIR} = P(\text{collapse}) \times \text{workforce vulnerability}$$
$$= 0.0001 \times 0.3$$
$$= 0.00003 \text{ per year}$$

The potential loss of life (PLL) per year would be:

$$\text{PLL} = \text{LSIR} \times \text{exposed population}$$
$$= 0.00003 \times 220$$
$$= 0.0066$$

Landslide Risk Assessment

The fatal accident rate (FAR) per 10^8 exposed hours would be:

$\text{FAR} = \text{PLL} \times 10^8/\text{exposed hours}$

$= 0.0066 \times 10^8/(\text{manning level} \times 8760 \text{ hours})$

$= 0.342 \text{ per } 10^8 \text{ exposed hours}$

The average individual risk (AIR) per year would be:

$\text{AIR} = \text{PLL}/\text{exposed individuals}$

$= \text{PLL}/[\text{manning level} \times (8760/\text{offshore hours per individual})]$

$= \text{PLL}/[220 \times (8760/3000 \text{ hours})]$

$= 0.0000102 \text{ per year } (1.02\text{E-}05)$

$\text{AIR} = \text{FAR} \times \text{offshore hours per individual} \times 10^{-8}$

$= 0.000\,010\,2 \text{ per year } (1.02\text{E-}05)$

This assessment considers the threat from a single mudslide scenario. If other mudslide scenarios are included in the analysis, then the PLL is:

$\text{PLL} = \sum (\text{LSIR} \times \text{exposed population})$ for scenarios 1 to n

Most assessments will also include the PLL associated with other types of mudslide incident on the platform and during the time spent travelling to and from the platform (other accidents) in order to generate a total FAR and AIR for the workforce (Table 10.14). The FAR and AIR values can also be calculated for different groups associated with specific areas of the platform, in which case the exposed population is the number of staff in the particular group.

Table 10.14 Example 10.11: measures of mudslide risk to the workforce of an offshore platform

Accident	P(mudslide)	P(spatial)	Platform vulnerability	P(collapse)	LSIR[a]	Workforce	PLL
Scenario 1	0.01	0.04	0.25	0.0001	0.00003	220	0.0066
Scenario 2	0.005	0.05	0.5	0.000125	0.0000375	220	0.00825
Scenario 3	0.002	0.1	0.75	0.00015	0.000045	220	0.0099
Scenario 4	0.001	0.2	1	0.0002	0.00006	220	0.0132
Scenario 5	0.0005	1	1	0.0005	0.00015	220	0.033
Other accidents							0.386
						Total PLL	0.45695
						FAR	23.71
						AIR	0.00071

[a]The workforce vulnerability during a platform collapse incident is assumed to be 0.3 (i.e. 30% fatality rate)
P(collapse) = P(mudslide) × P(spatial) × platform vulnerability
LSIR = P(collapse) × workforce vulnerability
PLL = LSIR × workforce

Example 10.12

The seaside promenade at Brighton, UK, includes a 500 m long section located directly beneath a high chalk cliff (see Examples 7.1. and 7.7). A range of landslide events present a threat to promenade users, ranging from pebble-sized falls to very-large-mass failures (see Table 8.10). The risks to the public were estimated in order to provide a baseline against which possible risk-reduction measures could be compared (Brighton and Hove City Council, 2006).

Considering a very-large-mass failure (1000 m^3, with a notional width of 100 m and an estimated annual probability of 0.01), the LSIR for a hypothetical person permanently present on the promenade (i.e. 24 hours/day, 365 days/year) is:

$$\text{LSIR} = P(\text{landslide}) \times P(\text{spatial}) \times P(\text{temporal}) \times \text{vulnerability}$$

In which:

$$P(\text{spatial}) = \text{landslide width/promenade length}$$
$$= 100/500$$
$$= 0.2$$

$$P(\text{temporal}) = 1$$

From Table 8.10, the vulnerability of an individual to a very-large-mass failure is estimated to be 1.

$$\text{LSIR} = 0.01 \times 0.2 \times 1 \times 1$$
$$= 0.002 \text{ per year}$$

The risk to a single pedestrian walking along this 500-m section at 2.5 km/h (i.e. the ISIR) was calculated as:

$$\text{ISIR} = P(\text{landslide}) \times P(\text{spatial}) \times P(\text{temporal}) \times \text{vulnerability}$$

In which:

$$P(\text{temporal}) = \text{time in danger zone (hours)/time in year (hours)}$$
$$= (\text{length of cliff section/distance travelled in 1 hour})/8760$$
$$= (500/2500)/8760$$
$$= 2.283 \times 10^{-5}$$

$$\text{ISIR} = 0.01 \times 0.2 \times 2.283 \times 10^{-5} \times 1$$
$$= 4.566 \times 10^{-8} \text{ per year}$$

This is a measure of the individual risk for a person making a single journey along the promenade. If the same person does the same journey twice a day (i.e. there and back again), 200 days/year, at exactly the same speed, the ISIR becomes:

$$\text{ISIR} = 4.566 \times 10^{-8} \times (2 \times 200)$$
$$= 1.826 \times 10^{-5} \text{ per year}$$

If a different individual made the journey once a month (12 journeys/year), cycling at 10 km/h, then the ISIR for that person would be:

$$\text{ISIR} = 0.01 \times 0.2 \times ((500/10000)/8760) \times 1 \times 12$$

$$= 1.369 \times 10^{-7} \text{ per year}$$

This clearly illustrates that the risk to individuals can vary dramatically, depending on their activity patterns and exposure.

To determine the potential loss of life (PLL) from this type of incident, it is necessary to consider the potential for one or more people being present when the landslide event occurs and of being killed. This involves estimating the number of people who might be in the 100 m long section of the promenade when the landslide occurs, and requires the development of a population model from promenade-user statistics (see Chapter 7).

For this example, a number of different scenarios were developed, each with a different number of people present when the landslide occurs (Table 10.15). The scenario probabilities were estimated from analysis of user count data, and it was assumed that each of the individuals had the same level of exposure (i.e. was walking along the promenade at 2.5 km/h). It follows, therefore, that the ISIR for each of these people would be the same (as above):

$$\text{ISIR} = P(\text{landslide}) \times P(\text{spatial}) \times P(\text{temporal}) \times \text{vulnerability}$$

$$= 4.566 \times 10^{-8} \text{ per year}$$

Table 10.15 Example 10.12: calculation of potential loss of life (PLL) and average individual risk (AIR)

	Fatalities/incident				
	1	2	3–5	5–10	10–20
P(scenario)	0.7	0.2	0.05	0.04	0.01
Average fatalities	1	2	4	7.5	15
Scenario × fatalities	0.7	0.4	0.2	0.3	0.15
ISIR	4.57E-08	4.57E-08	4.57E-08	4.57E-08	4.57E-08
PLL	3.20E-08	1.83E-08	9.13E-09	1.37E-08	6.85E-09
Total PLL	7.99E-08				
AIR (15 000)	5.33E-12				
AIR (3850)	2.08E-11				

P(scenario) is the estimated probability that a particular number of fatalities would happen if the landslide occurred
Average fatalities is the average number of fatalities for each scenario
Scenario × fatalities = P(scenario) × average fatalities
The ISIR is as calculated in the text for an individual walking along the promenade at 2.5 km/h
PLL = ISIR × scenario × fatalities
AIR (15 000) is the average individual risk for each of 15 000 journeys (AIR = total PLL/15 000)
AIR (3850) is the average individual risk for each of 3850 users (AIR = total PLL/3850)

The PLL for each scenario was calculated as (see Table 10.15):

PLL (scenario 1) = ISIR × [number of fatalities × P(scenario 1)]

$$= 4.566 \times 10^{-8} \times (1 \times 0.7)$$

$$= 3.196 \times 10^{-8}$$

The total PLL is the sum of the PLL for each of the scenarios (see Table 10.15):

Total PLL = $\sum$ PLL(scenarios 1 to n)

$$= 7.99 \times 10^{-8} \text{ per year}$$

The average individual risk (the average risk shared between all promenade users) was calculated from the Total PLL:

AIR = total PLL/exposed population

Determining the exposed population can be problematic. The available user count data for the promenade only records individual journeys and not the total number of separate users (many users may make multiple journeys). If the AIR is calculated for each individual journey (15 000 per year):

AIR = $7.99 \times 10^{-8}/15\,000$

$$= 5.33 \times 10^{-12} \text{ per year}$$

However, if it is assumed that 75% of the journeys are made by 100 multiple users and 25% by single users, the exposed population is 3850:

AIR = $7.99 \times 10^{-8}/3850$

$$= 2.08 \times 10^{-11} \text{ per year}$$

This example has focused on the very-large-mass failures that could occur along the section of cliffline. The promenade users are also exposed to other landslide events (see Table 8.10). To determine the total risk to the users it would be necessary to repeat the above calculations for all the different sizes of landslide events, each with different values of P(landslide), P(spatial), vulnerability and PLL scenarios.

Example 10.13

A major highway through mountainous terrain in British Columbia is susceptible to rockfalls that can cause delay, damage, injury and death to road users. Along a particular section, where the road passes through a deep rock cutting, maintenance records and rockfall impact marks on the carriageway suggest a minimum rockfall frequency of 2.2 incidents per year (Bunce *et al.*, 1997). The road carries an average of 4800 vehicles per day (on average, 200 per hour), at speeds of around 80 km/h.

The binomial distribution can be used to model the probability of a vehicle being hit by a falling rock (see Example 6.4 for details). Each rockfall is represented by a separate trial with two possible outcomes: collision or no collision. The probability of one or more collisions is related to the probability of the rockfall in a specific trial hitting a vehicle (i.e. a vehicle being in the 'wrong place') and the number of falls per year:

$P(\text{collision}) = 1 - [1 - P(\text{vehicle in the 'wrong place'})]^{\text{number of falls/year}}$

To estimate the probability of a *specific vehicle* being hit while stationary in traffic, say for half an hour, it is necessary to calculate the probability of it being in the 'wrong place' at the 'wrong time', as well as the probability of the trial (i.e. the rockfall) resulting in a collision outcome:

$$P(\text{spatial}) = \frac{\text{length of vehicle}}{\text{length of road cutting}}$$

$$= \frac{5.4}{476}$$

$$= 0.011$$

$$P(\text{temporal}) = \frac{\text{length of stay (hours)}}{\text{length of year (hours)}}$$

$$= \frac{0.5}{8760}$$

$$= 5.7 \times 10^{-5}$$

As there is more than one rockfall in a single year, the probability of a collision between a rockfall and the specific vehicle ($P(\text{collision})$) can be calculated from the binomial model:

$$P(\text{collision}) = 1 - [1 - P(\text{spatial})]^{\text{number of falls/year}}$$

$$= 1 - (1 - 0.011)^{2.2} = 0.025$$

$$\text{Annual } P(\text{collision}) = P(\text{temporal}) \times P(\text{collision})$$

$$= 5.7 \times 10^{-5} \times 0.025 = 1.4 \times 10^{-6}$$

Thus, the annual probability of a specific vehicle being hit by a falling rock is 1.4×10^{-6}, that is 0.000 001 4.

To calculate the probability of *any vehicle* in the line of stationary traffic being hit, it is assumed that the jam extends for the full length of the cutting and lasts for 30 minutes, with vehicles 'bumper to bumper', in which case:

$$P(\text{spatial}) = \frac{476}{476}$$

$$= 1$$

$$P(\text{temporal}) = \frac{\text{length of stay (hours)}}{\text{length of year (hours)}}$$

$$= \frac{0.5}{8760}$$

$$= 5.7 \times 10^{-5}$$

$P(\text{collision}) = 1$ (i.e. if a rockfall occurs then it will hit a vehicle)

$$\text{Annual } P(\text{collision}) = P(\text{temporal}) \times \text{frequency of rockfalls}$$

$$= 5.7 \times 10^{-5} \times 2.2 = 0.000\ 125\ 4 \text{ or } 1.25 \times 10^{-4}$$

Note that the 'individual risk' to all vehicles is the same, as can be shown by calculating the number of vehicles in the cutting and dividing by the above number, as follows:

$$\text{number of vehicles} = \frac{\text{length of cutting}}{\text{length of vehicle}}$$

$$= \frac{476}{5.4}$$

$$= 88.15$$

$$\text{'individual risk'} = \frac{1.25 \times 10^{-4}}{88.15}$$

$$= 0.0000014 \text{ or } 1.4 \times 10^{-6}$$

The same principles apply when estimating the probability of a moving vehicle being hit. When vehicles are in motion, the proportion of time for which a part of the highway is occupied by a vehicle is:

$$P(\text{spatial}) = \frac{\text{number of vehicles/hour}}{\text{speed}(\text{m/hour})} \times \text{length of vehicle}$$

$$= \frac{200 \times 5.4}{80\,000}$$

$$= 1.35 \times 10^{-2}$$

$$P(\text{collision}) = 1 - [1 - P(\text{spatial})]^{\text{number of falls/year}}$$

$$= 1 - (1 - 0.0135)^{2.2} = 1 - (0.9865)^{2.2} = 2.95 \times 10^{-2}$$

The probability of an accident on a single trip through the road cutting can be approximated by the annual probability of a collision, divided by the total number of trips per year (4800×365):

$$\text{Annual } P(\text{single trip collision}) = 2.95 \times 10^{-2}/(4800 \times 365)$$

$$= 1.7 \times 10^{-8}$$

This is equivalent to the 'individual risk' shared between all vehicles. Bunce *et al.* (1997) assumed that the probability of a fatality following a collision (i.e. boulder impact) is 0.125 and 0.2 for stationary vehicles and moving vehicles, respectively. The probability of one or more deaths is:

$$P(\text{death}) = \text{annual } P(\text{collision}) \times P(\text{fatality})$$

	Stationary traffic for 30 minutes (specific vehicle)	Stationary traffic for 30 minutes (any vehicle)	Moving traffic (any vehicle)	Moving traffic (single trip)
$P(\text{death})$	1.75×10^{-7}	1.56×10^{-5}	5.9×10^{-3}	3.4×10^{-9}

10.6. Societal risk

Societal risk is the frequency and the number of people suffering a given level of harm from the realisation of specified hazards (IChemE, 1992). It usually refers to the risk of death, and is expressed as risk per year.

The term 'societal risk' is generally taken to refer to members of the general public. In many industries, where the workers are isolated and members of the public are unlikely to be affected, the term 'group risk' is used.

Potential loss of life (PLL), equivalent to the *expected value of the number of deaths per year* (see Vrijing and van Gelder, 1997), is used as a measure of the risk to all individuals exposed to the full range of landslide events that might occur in an area (i.e. the *societal risk from landsliding*). As described earlier, to calculate PLL it is necessary to estimate, for each event and its possible outcome, the frequency per year (f) and the associated number of fatalities (N). The PLL is the sum of the outcome of multiplying f and N for each event:

$$\text{PLL} = \sum f_1 N_1 + f_2 N_2 + \ldots + f_n N_n$$

Societal risk can also be calculated from individual risk (see Section 10.5):

societal risk = individual risk × exposed population

Frequency and number of fatalities data are usually presented as F–N curves, which show the cumulative frequency (F) of all event outcomes with N or more fatalities. The advantage of F–N curves is that they provide a framework for comparing the societal risk associated with landsliding, or other sources of risk, against *risk criteria* (see Chapter 11).

Example 10.14
Natural slopes in Hong Kong are often strewn with large boulders, especially when mantled with colluvium or beneath rock cliffs. Boulder falls are a common occurrence, especially during or following intense rainstorms, and can result in property damage and fatalities. Between 1984 and 1995, there were 169 reported rockfall and boulder-fall incidents, causing three injuries but no fatalities (ERM-Hong Kong, 1998b). However, over the 69 years between 1926 and 1995 there were three fatal incidents.

- Elliot Pumping Station, Pok Fu Lam, in 1926 (five deaths)
- Shau Kei Wan squatter area in 1976 (three deaths)
- Kings Road in 1981 (one death).

By way of contrast, there have been 117 fatal landslide incidents associated with man-made slopes, with the 'worst-case' event being the collapse of a low-rise building in Po Hing Fong during 1917, which caused 73 deaths.

These incidents, along with fatal landslide events on man-made slopes, are presented as F–N data in Table 10.16. The average historical PLL associated with boulder falls is 0.13 fatalities per year (i.e. nine deaths in 69 years). Figure 10.9 presents the F–N curves derived from these statistics. To express this societal risk as an economic value, it is necessary to assign a *value of statistical life* (VOSL), typically assumed to be around £2 million (see Chapter 9):

annual societal risk = value of life × PLL

= £2 million × 0.13

= £260 000

Quantifying risk

Table 10.16 Example 10.14: F–N data for historical landslide incidents and boulder falls in Hong Kong

Number of fatalities, N	Boulder falls 1926–1995 (69 years)		Landslides on man-made slopes 1917–1995 (78 years)	
	Number of events with N or more fatalities	Frequency F of N or more fatalities	Number of events with N or more fatalities	Frequency F of N or more fatalities
1	3	0.0435	76	0.974
2			31	0.397
3	2	0.0290	19	0.244
4			11	0.141
5	1	0.0145	9	0.115
6			8	0.103
8			7	0.0897
16			5	0.0641
18			4	0.0513
67			3	0.0385
71			2	0.0256
73			1	0.0128

Modified from ERM-Hong Kong (1998b)

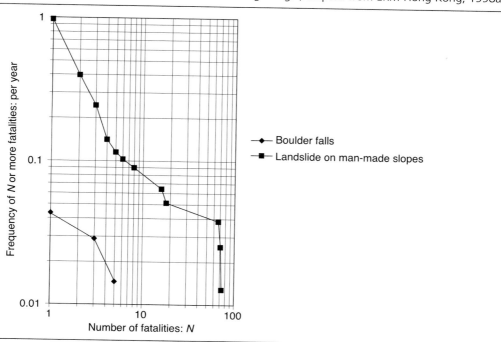

Figure 10.9 Example 10.14: F–N curve for boulder falls in Hong Kong. (Adapted from ERM-Hong Kong, 1998a)

411

Example 10.15
During the early hours of 13 August 1995, a cut slope failed along the Fei Tsui Road, Hong Kong, triggered by heavy rainfall. The road along the base of the slope was totally engulfed by landslide debris up to about 6 m thick; a man was slightly injured, but his son was trapped in the debris and killed. An assessment of the societal risk was undertaken as part of the post-event investigations, in order to establish a reliable indication of the risk posed to the affected community (Ho et al., 2000; Wong et al., 1997). The objective of the assessment was to consider what might have happened rather than simply to focus on what had happened. Indeed, the loss of life could have been higher if the landslide had occurred during daytime, rather than at 1.15 a.m., when traffic flows were very low.

The assessment involved estimating the potential loss of life (PLL) associated with a number of consequence scenarios (Table 10.17), together with the development of an F–N curve for the event (Figure 10.10). The following example considers only the threat to road users. Note, however, that the societal risk assessment presented in Wong et al. (1997) also considers the threat to an area of open space, a playground, a Baptist Church and a kindergarten.

The risk assessment involved the following steps.

1 *Estimating the probability of the landslide event.* Analysis of rainfall records suggested that the storm that preceded the event had a return period of around 100 years. The probability of the landslide was assumed, therefore, to be 0.01 (1 in 100).
2 *Development of consequence scenarios.* A range of scenarios was considered, each with different numbers of people exposed along the road through the landslide area. The extreme scenario was for 200 people to be at the base of the cut slope; this might occur if a traffic jam were to be caused by a road accident. The probability of the consequence scenarios was estimated from available traffic data and by analogy with traffic-flow conditions on similar roads.
3 *Estimating the vulnerability of the people exposed to the landslide.* A vulnerability factor of 0.85 was used, based on an assessment of the proximity of the road to the cut face and the travel distance of the landslide debris (see Chapter 8 and Example 9.7 for further details of the general approach to defining vulnerability factors).
4 *Estimating the probable number of fatalities for each consequence scenario.* This was calculated as follows:

   probable fatalities (scenario s) = exposed population (scenario s) × vulnerability factor

   So, for scenario 7 (see Table 10.17):

   probable fatalities (scenario 7) = exposed population (scenario 7) × vulnerability factor

   $$= 200 \times 0.85 = 170$$

5 *Calculating the potential loss of life associated with each consequence scenario.* This was calculated as follows:

   PLL (scenario s) = P(scenario s) × probable fatalities (scenario s)

   So, for scenario 7:

   PLL (scenario 7) = P(scenario 7) × probable fatalities (scenario 7)

   $$= 0.000\,02 \times 170 = 0.0034$$

Quantifying risk

Table 10.17 Example 10.15: societal risk associated with the 1995 Fei Tsui landslide

Probability of landslide event	Consequence scenario	No. of people exposed	Probability of consequence scenario	Vulnerability factor	Probable fatalities N	PLL	Risk	Frequency of event F	Frequency of N or more fatalities $\geq N$
0.01	1	0	0.4989	0.85	0	0.000000	0.000000	4.99E-03	1.00E-02
0.01	2	1	0.1875	0.85	0.85	0.159375	0.001594	1.88E-03	5.01E-03
0.01	3	5	0.2225	0.85	4.25	0.945625	0.009456	2.23E-03	3.14E-03
0.01	4	10	0.0875	0.85	8.5	0.743750	0.007438	8.75E-04	9.11E-04
0.01	5	30	0.0033	0.85	25.5	0.084150	0.000842	3.30E-05	3.60E-05
0.01	6	100	0.00028	0.85	85	0.023800	0.000238	2.80E-06	3.00E-06
0.01	7	200	0.00002	0.85	170	0.003400	0.000034	2.00E-07	2.00E-07
					Total	1.960100	0.019601		

Adapted from Wong et al. (1997)
Probable fatalities N = exposed population × vulnerability factor
PLL = P(consequence scenario) × probable fatalities
Frequency F = P(landslide event) × P(consequence scenario)
Frequency $\geq N$ = (frequency N) + (frequency $> N$)

Landslide Risk Assessment

Figure 10.10 Example 10.15: *F–N* curve for road users affected by the Fei Tsui landslide, Hong Kong (Wong *et al.*, 1997)

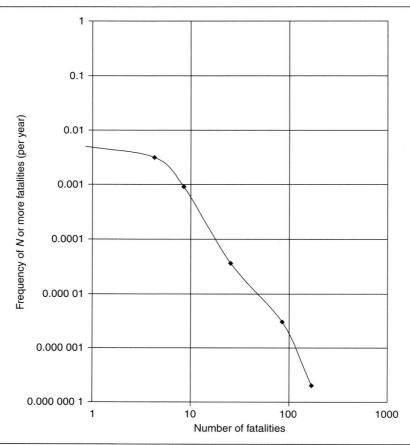

The overall potential loss of life is:

PLL (scenarios 1–7) = $\sum$ PLL (scenarios 1–7) = 1.96

This provided a measure of the consequences, given that the landslide has occurred. To establish the risk associated with each scenario, it was necessary to take account of the probability of the landslide event:

risk (scenario *s*) = *P*(event) × *P*(scenario *s*) × probable fatalities (scenario *s*)

Risk (scenario 7) = *P*(event) × *P*(scenario 7) × probable fatalities (scenario 7)

$$= 0.01 \times 0.000\,02 \times 170$$

$$= 0.000\,034 \text{ (i.e. } 3.4 \times 10^{-5})$$

The overall risk (per annum) was:

$$\text{Risk (scenarios 1–7)} = \sum \text{Risk (scenarios 1–7)}$$
$$= 0.0196 = 1.96 \times 10^{-2}$$

6   *Compiling an F–N curve for the landslide event.* This involved calculating the event frequency (F) and the frequency of N or more fatalities ($>N$):

$$\text{frequency } (F, \text{ scenario } s) = P(\text{event}) \times P(\text{scenario } s)$$

$$\text{frequency } (F, \text{ scenario } 7) = P(\text{event}) \times P(\text{scenario } 7)$$
$$= 0.01 \times 0.000\,02$$
$$= 0.000\,000\,2 = 2 \times 10^{-7}$$

The frequency of N or more fatalities was calculated as follows, using the results presented in Table 10.17:

$$\text{frequency } (\geq 85) = \text{frequency (85 fatalities)} + \text{frequency (170 fatalities)}$$
$$= 0.000\,002\,8 + 0.000\,000\,2$$
$$= 0.000\,003 = 3.0 \times 10^{-6}$$

The F–N curve presented in Figure 10.10 was compiled by plotting the calculated values of F and $>N$ for each consequence scenario.

Example 10.16

Landslides present a risk to highway traffic in Hong Kong. For example, historical data for the 14-km stretch of Castle Peak Road indicates that between 1984 and 1996 there were 32 recorded landslide incidents (2.38 incidents/year). The risk to life was estimated by ERM-Hong Kong (1999), using the landslide consequence model (see Chapter 9 and Example 9.7), which relates potential fatalities to the landslide volume. Considering those situations where the road is at the base of a cut slope and threatened by debris flows and landslide run-out (87% of slope sections along the road are below cut slopes, the remainder are fill slopes), the PLL was estimated as:

$$\text{PLL} = \sum \text{event frequency} \times \text{consequence (for all events)}$$

Table 10.18 Example 10.16: slope height–volume distribution for landslides on cut slopes along Castle Peak Road, Hong Kong

Volume: m³/Height: m	<20	20–50	50–500	500–2000	>2000
<10	0.23	0.1	0.05	0.05	0
10–20	0.15	0.1	0.05	0.01	0
>20	0.05	0.08	0.08	0.05	0

Based on ERM-Hong Kong (1999)

1 *Event frequency*. The historical frequency of past incidents along Castle Peak Road was used to generate a probability distribution for the height/volume of recorded slides (the *slope height–volume distribution*; Table 10.18).
   The annual frequency of events of a particular size was calculated as follows:

   frequency = total incident frequency × proportion of road below cut slopes

   × height/volume probability

   Thus, for an event of between 500 and 2000 m^3 on a slope higher than 20 m:

   frequency = 2.38 × 0.87 × 0.05 = 0.104

2 *Consequences*. As explained in Example 9.7, the consequence is the product of the *expected fatalities* for a reference landslide, given the volume of traffic and the number of lanes, the *vulnerability* for each lane and a scale factor (with respect to the reference landslide).

The results are presented in Table 10.19, and indicate a PLL from upslope failures along this section of Castle Peak Road of 0.98 fatalities per year.

10.7. Statistics are signs from God?

The reduction of a wide range of landslide hazard and multiple-consequence scenarios to a mathematical expectation value has considerable advantages. On one level it provides a rational framework for a decision-making process in which risk levels are compared against a predetermined set of criteria (e.g. benefit/cost ratios, risk acceptance criteria; see Chapter 11). At another level it supports the view that the future is manageable, thereby reducing any feelings of helplessness in the face of capricious nature: 'constituting something as a statistically describable risk makes possible the ordering of the future through the use of mathematical probability calculus' (Knights and Vurdubakis, 1993).

However, it is important not to lose sight of the fact that quantitative risk assessment is neither a neutral nor an entirely objective process. It is, in part, subjective and, as a consequence, the results can be value-laden and biased. Individuals and groups who do not share the judgements and assumptions of the assessors may see the results of the risk-assessment process as invalid, flawed or irrelevant (Stern and Fineberg, 1996). Judgements that can be the source of conflict include the following.

- The way in which hazard models are framed can influence which adverse consequences are analysed or ignored. For example, a landslide hazard model that focuses on rainfall or basal erosion as the prime cause of instability can direct attention away from the significance of leaking swimming pools and water pipes or the excavation of building plots.
- The focus on readily measurable or easily valued adverse consequences can lead to other consequences being excluded, especially those that are *close to home* to many of the affected community; for example, the effect of the risk assessment on property values and the availability of insurance cover, disruption of the local social framework and adverse impact on the character of a neighbourhood. When confronted with a statistical risk assessment, people often reframe the question in terms of 'What does it mean for me or my family?' (Plough and Krimsky, 1987; Siegal and Gibson, 1988). By ignoring such direct and personal questions, risk assessment can end up being misunderstood and mistrusted by the local community.
- The use of discounting techniques to analyse future risks is highly contentious, and confusing to many. This practice can have the effect of reducing almost to zero the significance of risks that

Table 10.19 Example 10.16: assessment of potential loss of life for cut slopes along Castle Peak Road, Hong Kong

Slope height: m	Slide volume: m	Landslide frequency/year	Slope proportion	Height/volume factor	Vulnerability factor	Expected fatality	Scale factor	Consequence factor	PLL
<10	<20	2.3800	0.8700	0.2300	0.0455	1	0.4000	0.0182	0.0087
10–20	<20	2.3800	0.8700	0.1500	0.0685	1	0.4000	0.0274	0.0085
>20	<20	2.3800	0.8700	0.0500	0.835	1	0.4000	0.0334	0.0035
<10	20–50	2.3800	0.8700	0.1000	0.3500	1	0.7000	0.2450	0.0507
10–20	20–50	2.3800	0.8700	0.1000	0.4350	1	0.7000	0.3045	0.0630
>20	20–50	2.3800	0.8700	0.0800	0.4850	1	0.7000	0.3395	0.0562
<10	20–50	2.3800	0.8700	0.0500	0.5850	1	1.5000	0.8775	0.0908
10–20	20–50	2.3800	0.8700	0.0500	0.6650	1	1.5000	0.9975	0.1033
>20	20–50	2.3800	0.8700	0.0800	0.6950	1	1.5000	1.0425	0.1727
<10	500–2000	2.3800	0.8700	0.0500	0.8900	1	2.0000	1.7800	0.1843
10–20	500–2000	2.3800	0.8700	0.0100	0.9500	1	2.0000	1.9000	0.0393
>20	500–2000	2.3800	0.8700	0.0500	0.9500	1	2.0000	1.9000	0.1967
<10	>2000	2.3800	0.8700	0.0000	0.9500	1	2.5000	2.3750	0.0000
10–20	>2000	2.3800	0.8700	0.0000	0.9500	1	2.5000	2.3750	0.0000
>20	>2000	2.3800	0.8700	0.0000	0.9500	1	2.5000	2.3750	0.0000
Total			1.0						0.9778

Based on ERM-Hong Kong (1999)
Landslide frequency: recorded events/year
Slope proportion: proportion of failure from cut slopes (0.87) and fill slopes (0.13)
Height/volume factor: see Table 10.18
Vulnerability factor: see Table 9.4
Expected fatality: see Table 9.3 for road with heavy vehicular or pedestrian traffic density
Scalar factor: see Table 9.6
Consequence factor = vulnerability factor × expected fatality × scale factor
PLL = landslide frequency × slope proportion × height/volume factor × consequence factor

lie more than a generation or two into the future. Many people consider that notions of sustainability deem it appropriate to use a low or zero discount rate so as to ensure that future risks or environmental damage are given sufficient weight in any analysis.
- The use of loss-of-life statistics as a measure of risk can be a source of controversy. Treating all fatalities as equal involves a judgement. It is assumed that the deaths of the old and the young are the same; deaths that occur during an event are treated the same as deaths that follow a protracted and painful period of hospitalisation. No value is placed on people who were exposed to a particularly traumatic event and spent many years in constant fear of another similar incident. Few realise that individual risk is an abstract statistic and does not provide a realistic measure of the risk to *me or my family*. The situation is even worse when 'value-of-life' statistics are used, for there is widespread misunderstanding of this abstract measure (see Chapter 9). Once again the controversy arises because of the widespread misinterpretation of a risk measurement as representing the actual worth or value of an individual person.

Many people see risk in a completely different manner from the risk analyst. As discussed in Chapter 2, research has shown that the way in which people react to risk is dominated by two dimensions (e.g. Slovic *et al.*, 1980):

- *dread*, that is the horror of the hazard and its outcomes, the feeling of lack of control, fatal consequences, catastrophe potential
- *the unknown* nature of the hazard and the resulting adverse consequences.

The quantitative risk assessment process generates results that are an expression of probability and loss of life or monetary value. It is important to appreciate, therefore, that such risk assessments do not deliver results that are directly relevant to many peoples' perception of risk. As Stern and Fineberg (1996) state:

> Conflicts over 'risk' may reflect differences between specialists in risk analysis and others on their definitions of the concept. In this light, it is not surprising that citations about 'actual risks' often do little to change most people's attitudes and perceptions. Non-specialists factor complex, qualitative considerations into their estimates of risk, including judgements about uncertainty, dread, catastrophic potential, controllability, equity, and risk to future generations.

The solution is not to weight risks to conform to the majority values of the affected community, but rather to recognise that the quantitative risk assessment process produces one type of risk measure, and not to be so presumptuous as to suggest that it delivers the only valid measure:

> When lay and expert values differ, reducing different kinds of hazard to a common metric (such as number of fatalities per year) and presenting comparisons only on that metric, have great potential to produce misunderstanding and conflict and to engender mistrust of expertise. (National Research Council, 1989).

REFERENCES

Bedford T and Cooke RM (2001) *Probabilistic Risk Analysis: Foundations and Methods*. Cambridge University Press, Cambridge.

Bottelberghs PH (2000) Risk analysis and safety policy developments in the Netherlands. *Journal of Hazardous Materials* **71**: 59–84.

Brighton and Hove City Council (2006) Undercliff walk. Note to Council Meeting, 26 July 2006.

Bunce C, Cruden DM and Morgenstern NR (1997) Assessment of the hazard from rock fall on a highway. *Canadian Geotechnical Journal* **34**: 344–356.

ERM-Hong Kong (1998a) *Landslides and Boulder Falls from Natural Terrain: Interim Risk Guidelines*. Geotechnical Engineering Office, Hong Kong. GEO Report No. 75.

ERM-Hong Kong (1998b) *Quantitative Risk Assessment of Boulder Fall Hazards in Hong Kong: Phase 2 Study*. Geotechnical Engineering Office, Hong Kong. GEO Report No. 80.

ERM-Hong Kong (1999) *Slope Failures along BRIL Roads: Quantitative Risk Assessment and Ranking*. Geotechnical Engineering Office, Hong Kong. GEO Report No. 81.

Fell R and Hartford D (1997) Landslide risk management. In *Landslide Risk Assessment* (Cruden D and Fell R (eds)). Balkema, Rotterdam, pp. 51–108.

Ford R (2000) Risk, perception and the cold numbers. *Modern Railways*, October.

Hall JW, Lee EM and Meadowcroft IC (2000) Risk-based assessment of coastal cliff recession. *Proceedings of the ICE: Water and Maritime Engineering* **142**: 127–139.

Health and Safety Executive (HSE) (1992) *The Tolerability of Risk from Nuclear Power Stations (revised)*. HMSO, London.

Ho K, Leroi E and Roberds B (2000) Quantitative risk assessment: application, myths and future direction. *Proceedings of the Geo-Eng Conference, Melbourne*, Publication 1, pp. 269–312.

IChemE (Institution of Chemical Engineers) (1992) *Nomenclature for hazard and risk assessment in the process industries*. IchemE, Rugby.

Kauer R, Fabbri L, Giribone R and Heerings J (2002) Risk acceptance criteria and regulatory aspects. *Operation, Maintenance, Materials Issues* **1(3)**: 1–11.

Knights D and Vurdubakis T (1993) Calculations of risk: towards an understanding of insurance as a moral and political technology. *Accounting, Organisations and Society* **18(7–8)**: 729.

Lee EM and Clark AR (2002) *Investigation and Management of Soft Rock Cliffs*. Thomas Telford, London.

MAFF (Ministry of Agriculture, Fisheries and Food) (1993) *Project Appraisal Guidance Notes*. MAFF Publications, London.

National Research Council (1989) *Improving Risk Communication*. Committee on Risk Perception and Communication. National Academy Press, Washington, DC.

Pethick JS and Leggett D (1993) The geomorphology of the Anglian coast. In *Coastlines of the Southern North Sea* (Hillen R and Vergagen HJ (eds)). American Society of Civil Engineers, Reston, VA, pp. 52–56.

Plough A and Krimsky S (1987) The emergence of risk communication studies: social and political context. *Science, Technology and Human Values* **12**: 4–10.

Royal Society (1992) *Risk: Analysis, Perception and Management*. Report of a Royal Society Study Group. Royal Society, London.

Siegal K and Gibson WC (1988) Barriers to the modification of sexual behaviour among heterosexuals at risk from acquired immune deficiency syndrome. *New York State Journal of Medicine* **14**: 66–70.

Slovic P, Fischhoff B and Lichtenstein S (1980) *Facts and fears: understanding perceived risk*. In *Societal Risk Assessment: How Safe is Safe Enough?* (Shwing R and Albers W (eds)). Plenum, New York, pp. 181–214.

Spouge J (1999) *A Guide for Quantitative Risk Assessment for Offshore Installations*. CMPT, Aberdeen.

Sterling GH and Strobeck EE (1973) The failure of the South Pass 70B platform in Hurricane Camille. *Proceedings of the Offshore Technology Conference*, Houston, TX, Paper 1898.

Stern PC and Fineberg HV (eds) (1996) *Understanding Risk: Informing Decisions in a Democratic Society*. National Academy Press, Washington, DC.

Valentin H (1954) Der landverlust in Holderness, Ostengland von 1852 bis 1952. *Die Erde* **6**: 296–315.

Vinnem JE (2007) *Offshore Risk Assessment: Principles, Modelling and Application of QRA Studies*, 2nd edn. Springer, Berlin.

Vrijing JK and van Gelder PHA JM (1997) Societal risk and the concept of risk aversion. In *Advances in Safety and* Reliability **1**: 45–52.

Wong HN, Ho KKS and Chan YC (1997) Assessment of consequence of landslides. In *Landslide Risk Assessment* (Cruden D and Fell R (eds)). Balkema, Rotterdam, pp. 111–149.

Landslide Risk Assessment
ISBN 978-0-7277-5801-9

ICE Publishing: All rights reserved
http://dx.doi.org/10.1680/lra.58019.421

Chapter 11
From risk estimation to landslide management strategy

11.1. Introduction to landslide risk management

The decisions whether or not to reduce the risk posed by landsliding, and how best to reduce the risk, involve consideration of a range of views, interests and factors. The results obtained from the risk assessment process (see Chapter 2), irrespective of their form and how they may have been generated, provide an indication of the level of risk and the likelihood of adverse outcomes. However, it is then necessary to ask the question *how much does it matter?* before going on to address a second question, *what should be done about it?*

Risk evaluation addresses the first of these questions. It is here that estimations of *risk*, as depicted by measures of the likelihood and severity of future adverse outcomes are considered by individuals, groups and relevant bodies, including the potentially affected population (stakeholders). The risk estimations have to be critically reviewed in terms of the *assumptions* that may have been made and the *levels of uncertainty* that are involved, in order to establish *confidence* in the results. The risks have also to be compared with other prevailing risks of concern, for the purposes of prioritisation.

Stakeholders (individuals, groups or the public more generally) have to be consulted in order to find out their views on the risks that they are, or could be, exposed to, and it is here that significant problems may be encountered. Many people are not well versed in the use of probabilities, and are, therefore, unable to appreciate the meaning of such terms as 'annual probability', 'return period' or 'recurrence interval'. Each and every person has different formulations of risk based on their individual perceptions and experience, with the result that their toleration of different risks often shows little or no relationship with statistically based risk estimations, especially where deaths or catastrophic outcomes are possible. However, good communications with the general public, including consultation from an early stage of a project or exercise, can significantly reduce potential tensions by limiting the extent to which risk management strategies are perceived to have been 'imposed' on stakeholders (see Chapters 1 and 2).

The results of *risk evaluation* feed into the final *risk assessment* stage where the second question *what should be done about it?* is addressed and decisions taken regarding the most appropriate risk management strategy. A wide range of factors must be considered at this stage, and options evaluated in terms of their technical feasibility, economic viability, environmental acceptability and political desirability. However, it is important to emphasise that just because there is an identified physical problem (landsliding) does not necessarily mean that there has to be a physical solution involving engineering and the application of technology.

Humans have three main options when faced by a geohazard, such as landsliding.

- Accept the consequences and *bear* the costs (*loss bearing* and *do nothing*).
- Respond by abandoning a site, relocating elsewhere to safer ground or changing the use of a site so as to reduce risk (*choose change* and *risk avoidance*).
- *Take* active steps to reduce risk by limiting hazard potential and/or the potential to suffer loss (*adjustment*).

Only in the case of adjustment is landslide management involving engineering works an option, and even here it is but one of the three main approaches outlined by Smith (2001), which are as follows.

- *Modification of loss burden*, which involves spreading the potential losses as widely as possible, through such measures as insurance. This is essentially a loss-sharing approach with limited emphasis on *loss reduction*, so the total risk remains roughly the same but the financial exposure of individuals, groups, etc., is reduced because it is shared between a large number of participants.
- *Modification of hazard events*, which involves reducing the potential for loss by the use of hazard-resistant designs and engineered structures so as to safeguard lives and property and, if possible, to physically suppress the hazard potential of the geohazard concerned (mitigation).
- *Modification of human vulnerability*, which focuses on reducing losses through land use planning programmes that seek to relate *land use zonation and building codes/ordinances* to *hazard zonation*, together with the development of preparedness programmes that aim to limit losses, especially human casualties, through the installation of monitoring networks linked to forecasting and warning systems that translate into effective emergency actions.

A more detailed division of these landslide risk management approaches is shown in Figure 11.1, which is based on the work of Burton *et al.* (1978) but subsequently modified for use in Jones (1991, 1996) and the Royal Society (1992). The five categories of action recognised are as follows:

1. Actions designed to *affect the cause* of risk from landsliding by limiting the potential for slope failure through the use of land use management (e.g. soil conservation and afforestation; Fannin *et al.*, 2005; Phillips and Marden, 2005; Sidle *et al.*, 1985) and development control, slope stabilisation measures, slope drainage (e.g. Bromhead, 2005; Holz and Schuster, 1996; Hutchinson, 1977; Wyllie and Norrish, 1996), and erosion control structures in rivers and along the coast (e.g. Lee and Clark, 2002). The objective is to *reduce the likelihood of landsliding*.
2. Actions designed to reduce risk by *modifying or constraining* slope failure so as to limit adverse impacts. Such measures assume that landsliding will occur but seek to limit its ability to cause detriment by predetermining pathways and run-out areas and, if possible, reducing landslide frequency, volume and velocity. Technical measures include covering unstable rock slopes in mesh, using barriers to stop rolling rocks and nets to catch falling rocks, erection of rock shelters and chutes over roads and railways, specially constructed debris flow channels/chutes around villages and under transport routes, and check dams in gullies to inhibit debris flow development, storage basins (e.g. Bromhead, 2005; Costa and Wieczorek, 1987; Fookes and Sweeney, 1976; Wyllie and Norrish, 1996).
3. Actions designed to *modify loss potential* through improved forecasting and prediction (*prognostication*) and better education about the nature of hazards and possible adverse consequences, together with the development of warning systems (*risk communication*), the establishment of emergency action plans and procedures, the development of building

From risk estimation to landslide management strategy

Figure 11.1 Classification of adjustment choices or management options, illustrating the differences between 'hazard management', 'vulnerability management' and 'risk management. (Developed from Burton et al., 1978; Jones, 1996)

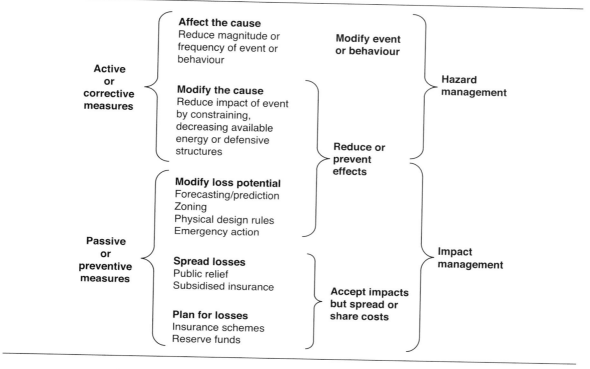

codes/building ordinances designed to improve the resistance of structures to slope movements and thereby limit damage, and the creation of a planning framework that seeks to relate geographical patterns of hazardousness (*hazard zonation*) with patterns of land use (*land use zonation*) so as to limit vulnerability (e.g. Clark et al., 1996; Malone, 2005; Schuster and Kockelman, 1996). Landslide forecasting is extremely difficult except where conditions lead to repeated activity (e.g. debris flows, lahars). The classic example of a landslide warning system is that at Tessina, in the Italian Dolomites, where a system of survey prisms and tiltmeters is automatically monitored to protect the village of Funes (Angeli et al., 1994).

4 Actions designed to *spread the losses*, usually taken in the wake of a serious impact when humanitarian concern makes it inevitable that aid be given to those that either could not or did not protect themselves against loss. This type of *passive loss sharing* takes the form of emergency aid or disaster relief provided both within a nation as well as between nations. National support is obtained from private donations, the work of charitable non-governmental organisations (NGOs) and expenditure from central funds via governmental organisations and agencies (e.g. the National Disaster Relief Arrangements (NDRAs) in Australia and the work of the Federal Emergency Management Agency (FEMA) in the USA, which may be overridden by a Presidential Disaster Declaration). International relief and aid is also sometimes obtained from private donations, but the vast majority is in the form of *bilateral aid* (from government to government or indirectly through NGOs) and *multilateral aid* (through international bodies such as the EU, World Bank, Asian Bank and UN agencies, most especially the Disaster Relief

Organisation (UNDRO)). With the passage of time it is becoming increasingly difficult to distinguish between such disaster aid and longer-term development aid.

5   Actions designed to *plan for losses* through insurance (e.g. Paus, 2005) and the establishment of reserve funds (e.g. the New Zealand Earthquake Commission, which provides natural disaster insurance cover to residential property owners; Murray, 2000). These measures can be termed *active loss sharing*, for the potential victims take deliberate actions in advance of an impact in order to protect themselves financially and ensure the potential for recovery should an impact occur; a process even carried-out by the insurance industry through the process of re-insurance.

From a different perspective, the groupings of management options shown in Figure 11.1 into *active/corrective* measures and *passive/preventative* measures emphasise whether the focus of activity is on the threat posed by the hazard (i.e. landsliding) or on the potential for impact on human society, and provide the basis of the fundamental division into *hazard management* and *impact* or *vulnerability management* (Jones, 1996), both of which form major components within the broad field of risk management. In this context, it is important to recognise the significance of the measures listed under *modify loss potential* in reducing risk, most especially risk communication and the crucial role of prognostication, without which the human population would continually be 'surprised' by hazardous events.

The fivefold division shown in Figure 11.1 has to be recognised as imperfect because not all actions and activities fit neatly into only one of the categories and because of evolving management practices. For example, insurance is no longer a purely loss-sharing activity but can be used to actively encourage the establishment of land use zonation policies and the adoption of building codes. In other instances, properties adversely affected by hazards may be purchased by the state rather than reconstructed or restored, so that loss sharing is used to achieve a reduction in future vulnerability. This is the case in the USA, where the Hazard Mitigation Grant Program (HMGP) was created in November 1988, by Section 404 of the Robert T. Stafford Disaster Relief and Emergency Assistance Act. The HMGP assists US states and local communities in implementing long-term hazard mitigation measures following a major disaster declaration. In 1998, the US Federal Emergency Management Agency (FEMA) and the California Governor's Office of Emergency Services provided a $1.3 million grant to the property owners in Humboldt County, California (Lee-Youngren and Hodgson, 1998). The money was for the purchase of 17 residential properties in the Big Lagoon landslide area that had been threatened by erosion as a consequence of the El Niño storms. The grant represented 75% of the appraised value. Any structures on the properties were demolished, and the land, to be maintained by the county, is to be kept as open space.

Similarly in France, the Law Barnier (2 February 1995; e.g. Vallet, 2004) authorises appropriation by the government (with compensation) of all property threatened by natural risks when the remedial works are too expensive to undertake. Compensation is funded from a state surcharge of 12% that is added to all property insurance premiums. A risk prevention plan (PPR) determines the areas where a natural risk is foreseeable. The PPR is intended to allow action to be taken in advance by the proprietor and the local authority.

Landslide risk management is the broad field of activities that covers all the strategies outlined above, including the passive acceptance of loss (e.g. the repeated repair of roads distorted by minor movements) and the 'giving-up' or abandoning of sites (e.g. the abandonment of Lemieux, Ontario, between 1989 and 1991, after it was found that the town was built on unstable quick clay, and was in danger of experiencing a landslide similar to the one that had destroyed the town of Saint-Jean-Vianney, Quebec, in 1971; McIntyre, 2005). Indeed, abandonment is now seen to be an increasingly preferred policy option for some eroding coastlines in the context of progressively rising sea levels due to global warming (e.g. Lee and Clark, 2002). While landslide risk estimations inform all of these strategies, it has to be emphasised that the role of science and engineering figures prominently only in reducing the likelihood of landsliding, modifying or constraining slope failures when they do occur and, to a lesser extent, modifying the loss potential.

It is beyond the scope of this book to examine the landslide risk management process in detail or to debate the contentious nature of risk management (e.g. Dai *et al.*, 2002; Leroi *et al.*, 2005; McInnes, 2005; Royal Society, 1992). Instead, attention will be focused on the important issues that determine whether or not landslide management is likely to be the preferred choice and how choices are made between different management strategies.

11.2. Assessment criteria

In the event that the risk evaluation process reveals that the risk from landsliding is significant and needs to be reduced, then a choice has to be made between various management options. The nature and scale of the problem, together with the value of the elements at risk, will greatly influence the decision, although the level at which the decision is made is also a crucial factor. Individuals tend to view the adverse consequences associated with geohazards as *imposed risks*, and are, therefore, generally less tolerant of them than they are of *chosen risks* (see Chapter 1). In the developed world there is also a well-established and growing view that the application of science, technology and engineering should protect people from the harmful aspects of the physical environment and that people should be fully informed of the risks they face. As a consequence, a property owner whose house is threatened by a developing landslide will usually insist on slope stabilisation (landslide management) funded from elsewhere (it must be someone else's fault/responsibility, so they should pay); the authority, on the other hand, may well conclude that abandonment is the best option on the grounds of cost. However, where a large number of properties are involved, the pressure for tangible evidence of protection is greatly increased.

How governments respond to landslide risks is dictated by the legislative framework, as it provides the context for what can and cannot be achieved (e.g. Lee, 2002; Palm, 1990). In the UK, for example, the powers to provide state-funded coastal cliff protection works are permissive, not mandatory – that is, authorities have the power to undertake works but are not obliged to do so. The selection process requires that consideration should only be given to those strategies that can deliver an *acceptable level of risk reduction*, while being both *economically viable* and *environmentally acceptable*, in the broadest sense. Pressures exerted from interest groups and the broader socio-economic and political context will also influence the management response. For example, there can be powerful lobbying for new coastal defence works (Lee *et al.*, 2001). As Penning-Rowsell *et al.* (1986) have suggested, 'the invisible political power of influence through social and political connection is far more significant than the publicly documented expressions of pressures on decision-making'.

Landslide management often involves the planning of *public* expenditure to increase social welfare by reducing land instability losses. As only a minority of the tax-paying community (i.e. the nation) is affected, the use of public funds can be seen as a subsidy (e.g. extending the property life and safeguarding investments). Investment in landslide management can, in some instances, be viewed as a means of safeguarding the vulnerable within society and helping towards the redistribution of wealth. There are, of course, other mechanisms for delivering improved social welfare (e.g. education, health and efficient infrastructure), all of which compete for resources.

Allocation of *public resources* for landslide management is, therefore, influenced by the need to find an acceptable balance between investments in a wide range of competing public services. Three tests are usually applied to decision-making about the allocation of public expenditure.

- The *scarcity* of resources requires that investments give the highest returns from the relevant perspective (i.e. national, regional or local).
- Decisions to invest public funds must be *accountable and justifiable*.
- Decisions must be based on a *rational comparison* between the available options.

Economic evaluation provides a mechanism for comparing the benefits of landslide management with the costs incurred, so as to determine

- whether, and by how much, the benefits exceed the costs
- the strategy that is expected to deliver the greatest economic return – that is, the most efficient use of resources
- the anticipated 'loss' to be incurred if it is decided to proceed with an 'uneconomic' strategy.

Not all landslide management activity is funded by national or local government, as individuals or organisations may wish to undertake works to protect their own property or assets. For private enterprise, the goal of risk management is often to improve business performance by avoiding surprises and reducing the frequency of adverse outcomes. It is claimed that successfully managing risks can enhance business success and increased shareholder value (but see Hubbard, 2009). In contrast, poor risk management can lead to business incidents and shareholder impacts. For example, poor identification and assessment of risk can create many issues for an organisation, including

- serious injury or fatalities/significant financial losses or reputational damage
- lack of common understanding of risks at a site, resulting in conflicting management strategies or ineffective allocation of resources.

There are important differences between economic evaluation, which seeks to examine the returns to the community at large, and financial appraisal, which examines whether the investment is worthwhile to an individual or organisation. For example, for an individual developer, the decision whether to protect a proposed hotel site from debris flow activity will be influenced by the additional profits to be generated after the implementation of mitigation measures. However, from a national perspective, the new hotel may simply divert visitors from other hotels in the country or even in the neighbourhood. From this perspective, the national benefits of landslide mitigation might be minimal.

Managing landslide risk cannot be viewed as solely an economic or financial issue, as the environment has become an increasingly important factor in determining the preferred option and the level of risk that is acceptable. This is because landslide mitigation works may result in environmental losses. To the individuals directly affected by landsliding or the threat of landsliding, the benefits of mitigation may far outweigh these losses. To others, the losses can represent an unacceptable price to pay for subsidising the lifestyle of a few.

11.3. Risk acceptance criteria: legal frameworks

A cornerstone of risk management is the concept that there is a degree of risk that is tolerable and that this can be defined through the use of risk acceptance criteria. However, the form and nature of these criteria need to be viewed in the context of the prevailing legal system. There are distinct differences between two of the most widely established legal systems in the developed world (Ale, 2005; Hartford, 2009):

- *The common law system* (e.g. Britain and its former colonies), in which statutes and codes are interpreted in the context of the common law tradition, which includes consideration of precedents. What is not explicitly allowed is forbidden, unless it can be justified to the regulator or, where necessary, in court after the fact. Injured parties can sue for damages caused by the wrongful acts of others, both intentional and unintentional, associated with negligence. The decision on compensation for damage is made in court in terms of an adversarial intellectual argument between legal counsel for the parties to the action. This has led to the courts considering situations where the potential for damage cannot be eliminated and where it is necessary for one party to take a certain level of risk with respect to the safety of others.
- *The Roman/Napoleonic legal code systems* (e.g. France and many countries occupied by the French during the Napoleonic Wars such as Italy, the Netherlands and Spain, and elsewhere) in which unlawful or unjust acts are defined in the law along with a specified penalty, be it imprisonment, a fine, or the payment of compensation. Everything that is not explicitly forbidden is allowed.

Under the common law system, the ALARP principle (as low as reasonably practicable – see below) applies, meaning that the (marginal) costs of safety improvements should be incurred until they outweigh the marginal benefits with respect to lives and property saved (HSE, 2001, but see below). Risk analysis is only the starting point of a discussion between the operator, the authorities and the UK Health and Safety Executive (HSE) about the tolerability of the risk; the accuracy of the analysis is not as critical as the effort that goes into the demonstration of ALARP (Ale, 2005). Ultimately, it is up to the courts to decide whether the operators have complied with their obligations.

A principle of the Roman/Napoleonic code is that costs and benefits of risk reduction must be balanced (Ale, 2005). In the Netherlands, for example, the ALARP principle would simply be a 'token statement' and the acceptability criteria are 'the end of the discussion' (Ale, 2005). The notion of tolerability of risk does not really apply because the legally enshrined criteria define the political acceptability of the risk. As the risk acceptance criteria are set in law, the role of the risk analyst is to demonstrate compliance (Hartford, 2009). Dutch courts have frequently stated that if the government wants more risk reduction then it should put stricter levels in the law. This role of the authorities is to ensure that these minimum requirements are met, rather than striving for the maximum achievable risk reduction. If risk is reduced to the prescribed limit, the operator can have confidence that their legal obligations have been met.

11.4. Acceptable or tolerable risks: the ALARP principle in the UK

During the 1980s the term 'acceptable' came to be progressively replaced by 'tolerable' as research revealed that in many instances people do not accept risks but merely tolerate them (e.g. Royal Society, 1992). Central to this shift was the work of the UK HSE (1988, 1992) on the *tolerability of risk*, when a framework was developed for making decisions on the tolerability of risk arising from any practice, activity, action or location (Figure 11.2), based on the recognition of three levels of risk. Above a certain threshold the risks might be considered intolerable or unacceptable. Below another much lower threshold, the risk might be considered to be so small that it is broadly acceptable. The zone between the unacceptable and broadly acceptable regions is the 'tolerable region', where the level of risk is typical of the risks posed by activities that people are prepared to tolerate in order to secure benefits (HSE, 2001). These benefits typically include employment, lower cost of production, personal

Figure 11.2 Risk tolerability and the ALARP concept

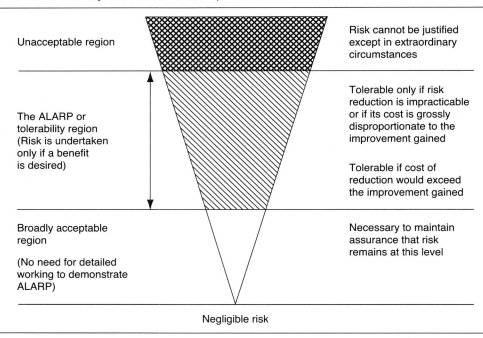

convenience or the maintenance of general social infrastructure such as the production of electricity or the maintenance of food or water supplies.

In the tolerable region, it is generally expected that the level of risk should be reduced to a level which is *as low as reasonably practicable* (the so-called **ALARP** principle). Within this region, risks may be tolerated; however, tolerability does not mean 'acceptability' (HSE, 1988):

> To tolerate a risk means that we do not regard it as negligible or something we might ignore, but rather as something we need to keep under review and reduce still further if we can. For a risk to be 'acceptable' on the other hand means that for purposes of life or work, we are prepared to take it pretty well as it is.

In England, the definition of *reasonably practicable* has been established by case law (Judge Asquith, Edwards *v.* National Coal Board, *All England Law Reports*, vol. 1, p. 747 (1949)):

> Reasonably practicable is a narrower term than 'physically possible' and seems to me to imply that a computation must be made by the owner in which the quantum of risk is placed on one scale and the sacrifice involved in the measures necessary for averting the risk (whether in money, time or trouble) is placed in the other, and that, if it be shown that there is a gross disproportion between them – the risk being insignificant in relation to the sacrifice – the defendants discharge the onus on them.

This case established that a balance must be made between the level of risk and the sacrifice (e.g. time or money) involved in averting the risk (HSE, 2001):

> Risk is tolerable only if risk reduction is impracticable or if its cost is grossly disproportionate to the improvement gained.

Residual risk is tolerable only if further risk reduction is impracticable or requires action that is gross disproportionate in time, trouble and effort to the reduction in risk achieved.

The ALARP principle forms the basis of the approach used by the UK HSE in its regulation of the major hazardous industries, such as the nuclear, chemical and offshore oil and gas industries. The concept, as illustrated in Figure 11.2, implies that

- if the risk is unacceptable it must be avoided or reduced, irrespective of the cost, except in extraordinary circumstances
- if the risk falls within the ALARP or tolerability region, then cost is taken into account when determining how far to pursue the goal of minimising risk or achieving safety. As a consequence, the ALARP or tolerable region (see Figure 11.2) includes a continuous spectrum of conditions ranging from where the cost of risk reduction would exceed the improvement gained adjacent to the lower boundary to where the risk is only tolerable if risk reduction is impracticable or if the cost of risk reduction is grossly disproportionate to the improvement gained (McQuaid and Le Guen, 1998). Thus, risk does not have to be reduced to *as low as possible* employing *best available techniques* (BAT), as this will almost certainly involve excessive cost. The benefits to be gained from a reduction in risk are normally expected to exceed the costs of achieving such a reduction. This comparison leads to the important concepts of *as low as reasonably achievable* (ALARA) and *best available technique not entailing excessive cost* (BATNEEC) that underpin the ALARP principle.

11.5. Individual risk criteria

Individual risk addresses the safety of individuals who are most at risk in an existing or proposed development. Individual risk criteria are intended to demonstrate that workers or members of the public are not exposed to excessive risk. In the UK, the HSE (2001) has suggested that, in terms of *average individual risk* (see Chapter 10), the boundaries of the ALARP region could be as follows.

- The boundary between the broadly acceptable and tolerable regions: 10^{-6} fatalities per year (1 in 1 million) for both workers and the public. This figure is regarded as 'extremely small' when compared with the background level of risks that people are exposed to over their lifetimes (typically a risk of death of 10^{-2} per year averaged over a lifetime).
- The boundary between the unacceptable and tolerable regions: the HSE suggests separate criteria for this boundary, depending on who is exposed to the risks. A value of 10^{-4} fatalities per year (1 in 10 000) is suggested for members of the public who have a risk imposed on them 'in the wider interests of society'. For workers in industry, the HSE suggested that the boundary should be around 10^{-3} fatalities per year (1 in 1000).

The maximum tolerable criterion of 10^{-3} fatalities per year was used by the HSE because it approximates to the risk experienced by high-risk groups in mining, quarrying, demolition and deep-sea fishing (HSE 1992; Table 11.1; this is often termed the 'group risk' because these groups of workers have more dangerous occupations). Table 11.2 shows how many risky activities an individual would need to undertake in 1 year to reach an IRPA of 10^{-3} fatalities per year. This illustrates that the 10^{-3} per year upper limit to the tolerable region is actually quite high. However, in practice, few modern facilities with proactive risk reduction strategies have risk levels approaching 10^{-3} fatalities per year. This tends to be recognised in company risk tolerability standards, where design standards are often set for new facilities in the region of 3×10^{-4} to 1×10^{-4} fatalities per year (Lewis, 2007).

Table 11.1 Individual risk levels for workers in different industries

Industry	Individual risk per annum (IRPA) level for employees
Deep sea fishermen on UK-registered vessels	1 in 750
Extraction of mineral oil and gas	1 in 999
Coal extraction	1 in 7100
Construction	1 in 10 200
Agriculture	1 in 13 500
Metal manufacturing	1 in 17 000

From Vinnem (2007)

The Australian National Committee on Large Dams (ANCOLD, 2003) has adopted the ALARP principle and has established individual risk criteria to define unacceptable levels of risk to any individual, from all causes of dam failure.

- For existing dams, an individual risk to the person or group most at risk of higher than 10^{-4} per year is unacceptable, except in exceptional circumstances.
- For new dams or major augmentations of existing dams, an individual risk to the person or group most at risk of higher than 10^{-5} per annum is unacceptable, except in exceptional circumstances.

In the Netherlands, legally binding *location-specific individual risk* criteria (see Chapter 10) are used in planning, where the Dutch Public Safety Decree (BEVI) regulates land use around hazardous installations. Risk contours around the site are developed and compared against acceptability criteria. From 2010, any 'vulnerable objects' (e.g. hospitals, houses and schools) will not be allowed within the 10^{-6} contour around the installation. Less vulnerable objects such as industrial zones, office buildings or recreational facilities are permitted between the 10^{-5} and 10^{-6} contours (Basta *et al.*, 2007).

Over the last decade or so there has been increasing interest in the development and application of individual risk criteria for landslide management. In 1997, the Hong Kong Geotechnical Engineering Office (GEO) proposed interim risk guidelines for natural terrain landslides (ERM-Hong Kong, 1998;

Table 11.2 Activities corresponding to a 1×10^{-3} risk of fatalities/year

Activity	Number of activities in 1 year that equals the criteria of 10^{-3} fatalities per year
Hang-gliding	116 flights
Surgical anaesthesia	185 operations
Scuba diving	200 dives
Rock climbing	320 climbs

From HSE (2001) and Lewis (2007)

Table 11.3 Individual risk criteria for landslides and snow avalanches in Iceland (Iceland Ministry for the Environment, 2000)

Risk zone	Lower criteria	Upper criteria	Restrictions
C	3×10^{-4}/year		No new buildings, except for summer houses (if the risk is less than 5×10^{-4}/year), and buildings where people are seldom present
B	1×10^{-4}/year	3×10^{-4}/year	Industrial buildings may be built without reinforcements. Homes have to be reinforced and hospitals, schools, etc., can only be enlarged, and have to be reinforced. The planning of new housing areas is prohibited
A	3×10^{-5}/year	1×10^{-4}/year	Houses where large gatherings are expected, such as schools, hospitals, etc., have to be reinforced

Ho et al., 2000; Reeves et al., 1999; note that the guidelines remain interim in 2013). The following average individual risk levels were proposed to mark the upper limit of the tolerable region

- for new developments: maximum $<10^{-5}$/year for individuals most at risk
- for existing developments: maximum $<10^{-4}$/year for individuals most at risk.

The Australian Geomechanics Society guidelines for landslide risk management include identical individual risk criteria (AGS, 2007; note that these criteria do not represent a regulatory position). Acceptable risks are considered to be one order of magnitude lower than the tolerable risks.

In 2009, the District of North Vancouver (DNV), Canada, formerly adopted landslide risk criteria (DNV, 2009)

- maximum 1×10^{-4} risk of fatality per year for redevelopments involving an increase to the gross floor area on the property of less than or equal to 25%
- maximum 1×10^{-5} risk of fatality per year for new developments and for redevelopments involving an increase to gross floor area on the property of greater than 25%.

This policy gives the district's chief building official the discretion to apply the criteria to building permits, subdivision and development applications for sites exposed to landslide and debris flow hazards.

In Iceland, the Ministry of the Environment (2000) defined and implemented acceptable individual risk levels for landslides and snow avalanches in a national regulation (Table 11.3). The risk criteria are used to support land use zoning and the requirement for increased safety provided by mitigation measures (e.g. Iceland Ministry for the Environment 2000).

11.6. Societal risk criteria

It is widely believed that society generally tends to be more concerned about multiple fatalities in a single event than in a series of smaller events that collectively kill the same number of people (e.g. Ball and Floyd, 1998; Horowitz and Carson, 1993). While low-frequency high-consequence events might represent a very small risk to an individual, they may be seen as unacceptable when a large number of people are exposed. For business, such incidents can significantly impact shareholder value, and, in some cases, the company may not recover (Knight and Pretty, 2002).

Societal risk criteria originated with the work of the UK Atomic Energy Authority in the 1960s in response to the need to ensure that the chances of a major accident were minimised and that nuclear power plants were located away from centres of population. This led to the development of the 'Farmer curve' which identifies an authorised area and a forbidden area on either side of a curve plotted on a graph of probability versus consequences, with the consequences expressed as levels of radioactive iodine release. One of the first societal risk criterion was suggested by the UK Advisory Committee on Major Hazards (ACMH) following the chemical plant explosion at Flixborough in 1974 during which 28 people were killed and 36 seriously injured: 'that in a particular plant a serious accident was unlikely to occur more often than once in 10 000 years (i.e. 10^{-4}/year) ... this might perhaps be regarded as just the borderline of acceptability' (ACMH, 1976).

Research into the 'tolerability of risk' from nuclear power stations was commissioned by the UK HSE after the Sizewell B Inquiry in the mid-1980s. The report (HSE 1988, 1992) is generally regarded as having set out the foundations for risk control in the UK, and has influenced safety policy around the world.

Societal risk criteria are often presented on F–N curves (see Chapter 10), where two criteria lines divide the space into three regions – where risk is unacceptable/intolerable, where it is broadly acceptable and where it requires further assessment and risk reduction as far as is reasonably practicable.

As society tends to have an aversion to multiple fatalities, the criteria lines generally slope steeply away from an individual risk value ($N = 1$), reflecting this. A slope of -1 ($y = bx^{-1}$) is commonly regarded as 'risk neutral', in that the weighting in preference of preventing large accidents is proportional to N, that is, the permitted probability of an accident that results in 100 people (or more) should be ten times lower than one that kills ten people (or more). Criteria lines are described as 'risk adverse' when they slope at -2 or higher powers of N. For a slope of -2, the permitted probability of a 100 (or more) fatality accident is 100 times lower than for the ten (or more) fatality accident.

Unlike individual risk criteria, there are no single 'one-size-fits-all' criteria for societal risks in use by operators and regulators in the major hazardous industries world-wide. Indeed, there is wide variation in regulatory criteria, as shown by the upper tolerability criterion lines in Figure 11.3. The UK (for the transportation of dangerous goods) and Hong Kong (for potentially hazardous installations) have adopted a gradient of -1 for the slope of the criteria line, whereas the Netherlands (for fixed installations) has used a more risk adverse gradient of -2 ($y = bx^{-2}$). The difference between the curves is related to the different legal context in the UK and the Netherlands, as described earlier. Although the Dutch criteria appear to promote greater safety, the UK approach has been to seek further risk reduction below the specified tolerability limits, in accordance with the ALARP principle (Ball and Floyd, 1998). The Hong Kong criteria incorporate a 'consequence cut-off' at 1000 fatalities, indicating that no incident involving this number of fatalities (or more) would be acceptable irrespective of the frequency/probability.

In the absence of agreed risk criteria for landslides, Fell and Hartford (1997) considered dam safety to be a good analogy to landsliding. Figures 11.4 and 11.5 present F–N curves for dams developed by

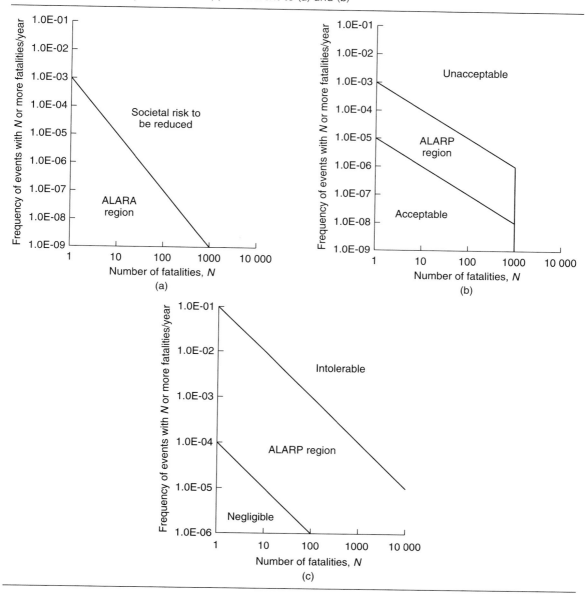

Figure 11.3 Societal risk criteria – upper tolerability F–N criteria: (a) The Netherlands (fixed installations); (b) Hong Kong (potentially hazardous installations); (c) UK (Transportation of Dangerous Goods). (Based on ERM-Hong Kong, 1998). Note: the y-axis scale on (c) is different to (a) and (b)

1. British Columbia Hydro (BC Hydro; a major dam owner and operator), which recognises 'tolerable' and 'intolerable regions' (Figure 11.4; BC Hydro, 1993)
2. The Australian National Committee on Large Dams (ANCOLD, 2003), which identifies separate tolerability lines for new and existing dams. Risks ten times higher are tolerated for existing dams than new dams (Figure 11.5). Of interest, the ALARP region is truncated horizontally at a 10^{-6} per year failure probability, because ANCOLD felt that it was unrealistic

Landslide Risk Assessment

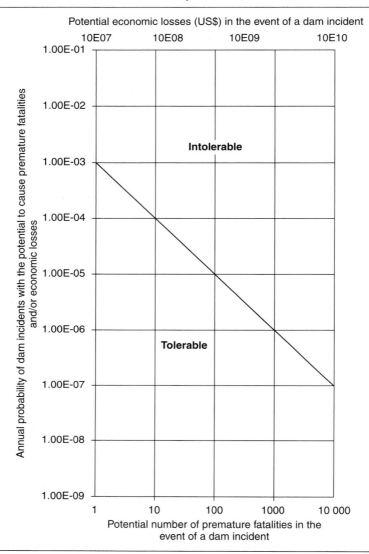

Figure 11.4 Societal risk criteria for dam failures: BC Hydro (1993)

to design a dam with a failure probability lower than this figure. However, it tends to imply that it is no more unacceptable for 10 000 or more to die in a failure incident than 100 people.

The Hong Kong Government interim risk guidelines for natural terrain landslide hazards for trial use include societal risks (ERM-Hong Kong, 1998; Ho et al., 2000; Reeves et al., 1999). Two options are being tested (Figure 11.6). The first option involves a conventional three-tier system incorporating an unacceptable region, a broadly acceptable region and an intervening ALARP region. The second option involves a two-tier system comprising an unacceptable region and an ALARP region. When the risk level is assessed to be within the ALARP region, cost–benefit calculations need to be carried out to demonstrate that all cost-effective and practicable risk mitigation measures are being

Figure 11.5 Interim societal risk criteria: ANCOLD (2003)

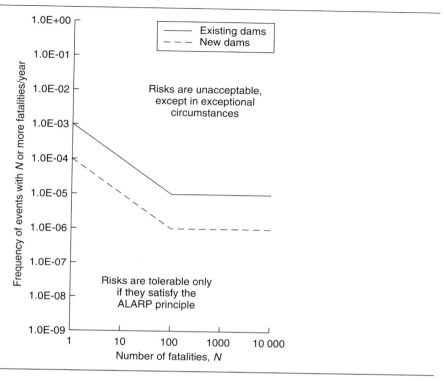

undertaken. An *intense scrutiny* zone has been included in both options, and is intended to reflect society's aversion to events with 1000 or more fatalities.

11.7. Corporate risk management: major accident risk criteria

From a business perspective, a company must make a judgement as to how frequently high-fatality incidents/accidents ('major accidents') would need to occur before the company's survival is put severely at risk due to the adverse reaction of shareholders, the regulator, media and the public (e.g. Lewis, 2007). For example, a company might believe that its future survival would be severely threatened if an accident causing ten or more fatalities occurred more regularly than once every 10 years across all of its facilities, and if an accident causing 100 or more fatalities occurred more regularly than a rate equivalent to once every 300 years. A straight line can then be drawn between these two points and extrapolated to higher values of N. Furthermore, if the company operated 30 facilities, it might decide to allocate its risk evenly between each facility. The resulting company upper criterion is shown in Figure 11.7, together with the single-facility criterion line if the company operated 30 facilities. In practice, the criterion line may be a 'group reporting line', above which a higher level of corporate scrutiny would be applied. Only then can the decision be made by senior corporate management to proceed with the project or continue existing operations (e.g. BP, 2007).

The energy company BP, for example, has developed a major accident risk (MAR) process that applies to all the company's operations with the potential to give rise to an incident that causes multiple

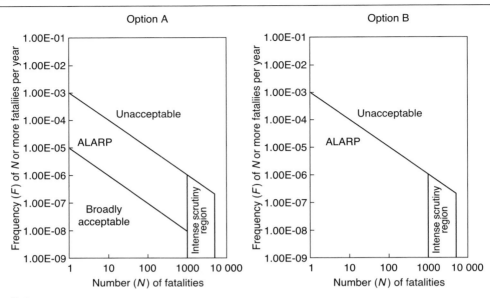

Figure 11.6 Proposed societal risk criteria for landslides and boulder falls from natural terrain in Hong Kong. (From Ho et al., 2000)

Notes.
1. The above societal risk criteria are to be used in conjunction with a reference toe length of the natural hillside of 500 m (Reeves et al., 1999).
2. If a development is affected by more than 500 m toe length of natural terrain, an appropriate linear scaling factor should be used to scale up the risk criteria. For example, in the case of a large development affected by natural terrain with a toe length of 5 km, then the above societal risk criteria should be increased by one order of magnitude.
3. If the development is affected by less than 500 m toe length of natural terrain, then the same criteria as proposed above are taken to apply (i.e. the criteria will not be scaled down).
4. The societal risk criteria are intended to aid decision-making and not intended to be mandatory.

fatalities (e.g. Considine and Hall, 2008). The process follows the ALARP principle and recognises the need for continuous risk reduction. All assets and operations within BP are required to quantify the societal and environmental risks from potential major accidents. These risks are then compared against a 'group reporting line' (GRL). Each asset has been allocated its own GRL, determined from a consideration of

- company sustainability
- regulatory precedents for establishing risk criteria
- industry experience of major accidents
- scale of the operation.

The GRL does not seek to set a level of 'acceptable' or 'tolerable' risk across all activities, as the MAR approach requires that efforts should be sought to reduce all risks as part of a process of continuous improvement. Rather, activities above the GRL are considered as representing a disproportionately high level of risk to the sustainability of the company, and hence need to be brought to the attention of, and monitored at, the group level. For any activity found to be above the GRL, both the risks and a

Figure 11.7 Hypothetical corporate *F–N* criteria and the group reporting line

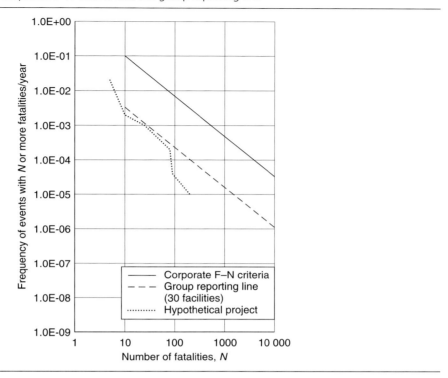

mitigation plan are reported to the group engineering director, whose approval is required if the operations are to continue. Risks below the GRL are reported to the relevant business segment. This is part of a management process in which decisions on risk mitigation are taken at a level in the BP organisation appropriate to the degree of risk. *F–N* curves are used to regularly monitor risk reduction at the segment and group levels to ensure that MAR is on a steady decline.

11.8. Applying the ALARP principle: loss of life

If the risk to people at a particular site or area falls within the ALARP region (see Figure 11.2), it will be necessary to consider introducing further risk reduction measures to drive the remaining, or 'residual', risk downwards. To demonstrate that ALARP has been met, the cost-effectiveness of any proposed mitigation measures will have to be determined. A comparison therefore needs to be made between the *costs* of an option and the *benefits* it is expected to deliver.

The risk reduction associated with each option is usually expressed in terms of potential loss of life (PLL; see Chapter 10). For discrete events, such as landslides, PLL can be calculated as follows:

   PLL = event probability × expected number of deaths

Therefore, reducing the likelihood of landsliding reduces the PLL, as does reducing the magnitude or suddenness of events. Thus, if the implementation of mitigation measures upslope of a resort hotel would reduce the annual probability of debris avalanches likely to kill ten people from 0.01 to 0.005, then the PLL risk reduction would be

$$\text{PLL risk reduction} = \text{PLL ('do nothing')} - \text{PLL (post-stabilisation)}$$
$$= (10 \times 0.01) - (10 \times 0.005)$$
$$= 0.1 - 0.05$$
$$= 0.05$$

However, in order to make direct comparisons with the costs, it is necessary to establish a monetary value to the reduction in PLL. This involves determining the *value of life* or the *value of a statistical life* (VOSL), from the amount that people would be willing to pay for a very small change in risk (i.e. their *willingness to pay*; see Chapter 9):

$$\text{VOSL} = \frac{\text{willingness to pay}}{\text{risk reduction}}$$

For example, a VOSL of £2 million is equivalent to saying that people would pay £200 for a reduction in the risk of death of 1 in 10 000 (0.0001):

$$\text{VOSL} = \frac{200}{0.0001}$$
$$= 2 \text{ million}$$

Society tends to be more averse to a single large multiple death event than to a series of events that yield a comparable cumulative death toll. This has led some organisations to use 'adversion factors' that result in higher value of life figures in large multiple death events.

Useful indicators of the economic efficiency of measures to reduce loss of life are:

- *The maximum justifiable expenditure*. This provides a guide to the upper limit of annual investment in risk reduction measures:

    $$\text{maximum expenditure} = \text{total PLL} \times \text{VOSL} \times \text{aversion factor}$$

- *The implied cost of averting a fatality* (ICAF). This provides a means of comparing options according to the benefits delivered over the lifetime of the measures:

    $$\text{ICAF} = \frac{\text{cost of option}}{\text{PLL risk reduction} \times \text{lifetime of measures}}$$

    It is similar to the BCR (benefit : cost ratio), but provides a monetary value that can be compared to the VOSL. If the ICAF is less than the VOSL, then the option may be considered to be cost-effective.

Example 11.1
A small town lies within an ancient landslide complex. Periodic reactivation causes significant economic damage, but poses little threat to public health and safety. However, there remains a potential for more dramatic first-time failure of the landslide backscar area. Such an event could be sudden and be accompanied by rapid movements. A number of fatalities could be expected, mainly due to falling masonry.

An expert panel has estimated that the first-time failure has an annual probability of the order of 0.001 (1 in 1000). Analogues suggest that this type of first-time failure in an urban area could cause up to 10 deaths. The PLL is as follows:

PLL = event probability × number of deaths

$= 0.001 \times 10 = 0.01$

Using a VOSL of £2 million (no aversion factor has been applied because of the limited number of anticipated fatalities), the maximum justifiable expenditure per year for this event is:

maximum annual expenditure = PLL × VOSL

$= 0.01 \times 2$ million

$= £0.02$ million

Any proposed mitigation measures would involve a 'one-off' cost, but would remain effective over a 50-year design lifetime. Therefore, the maximum 'one-off' expenditure would be

maximum 'one-off' expenditure $= 0.02 \times 50$

$= £1$ million

Example 11.2
A new road is under construction through landslide-prone, mountainous terrain. The risk to potential users has been established, and, in places, lies within the ALARP region but close to the unacceptable threshold. A variety of landslide mitigation measures have been proposed, including the use of boulder fences, check dams and retaining walls.

Table 11.4 sets out the costs and benefits of each of the various options, with a 'do nothing' option providing a baseline for the comparison. For each option, the scheme benefits are the risk reduction, expressed in monetary terms:

benefits = PLL reduction × VOSL

Using a VOSL of £2 million, the only option with a BCR above 1.0 and with an ICAF less than £2 million is option A – that is, the use of boulder fences. The other two options are not cost-effective. Had the site fallen within the area of 'intense scrutiny' on Figure 11.6 and an *aversion factor* of 20 been applicable, then all three options would have been cost-effective.

11.9. Applying the ALARP principle: economic risk
The *benefit* of a landslide management strategy is the reduction in risk, expressed in monetary terms, compared with a 'do nothing' case (see Chapter 10). The *costs* should include all the expenditure incurred during the investigation, planning and design, construction and operation of the strategy. Both benefits and costs should be considered over the strategy lifetime and, hence, need to be brought back to their present value by discounting.

A range of strategy options should be evaluated, including

- the 'do nothing' case involving no active landslide management – simply walking away and abandoning all maintenance, repair or management activity

Landslide Risk Assessment

Table 11.4 Example 11.2: the BCR and implied cost of averting a fatality for a landslide mitigation scheme

Option	Scheme type	Scheme life	PLL	PLL reduction	Value of life: £ thousands	Scheme lifetime benefits: £ thousands	Scheme costs: £ thousands	BCR	ICAF: £ thousands	ICAF <value of life
Do nothing			0.008		2000					
A	Boulder fence	40	0.0073	0.0007	2000	56	25	2.2	893	Yes
B	Check dam	40	0.007	0.001	2000	80	100	0.8	2500	No
C	Retaining wall	40	0.006	0.002	2000	160	250	0.6	3125	No

Adapted from Kong (2002)

Table 11.5 Indicative standards of coast protection used in England

Land use band		Annual probability of failure	Return period: years
A	Intensively developed urban areas	0.003–0.01	100–300
B	Less intensive urban areas with some high-grade agricultural land or environmental assets	0.005–0.02	50–200
C	Large areas of high-grade agricultural land and/or environmental assets; some property at risk	0.01–0.10	10–100
D	Mixed agricultural land with occasional properties at risk	0.05–0.40	2.5–20
E	Low-grade agricultural land with isolated properties	>0.20	<5

From Ministry of Agriculture, Fisheries and Food (MAFF) (1999)

- a 'do minimum' case which might involve limited intervention aimed at attempting to reduce, rather than control, the problems and provide a minimum level of protection – for example, promoting the build-up of a beach in front of an unprotected cliff, preventing water leakage on unstable slopes, or the provision of early warning systems for cliff instability
- a variety of combinations of mitigation works that provide different levels of risk reduction.

The benefit : cost ratio (BCR) is a widely used measure of economic cost-effectiveness. Often, decision-makers seek to identify strategies that maximise the BCR while ensuring that the level of residual risk would be acceptable, given the current or proposed land use.

Other useful measures include the net present value (NPV) and the incremental BCR, which represents the change in present value (PV; see Chapter 10) costs and benefits between options:

$$NPV = \text{risk reduction} - \text{costs}$$

$$\text{incremental BCR} = \frac{\text{risk reduction(option B)} - \text{risk reduction(option A)}}{\text{costs(option B)} - \text{costs(option A)}}$$

Maximisation of the BCR is often the aim of project appraisal for landslide management works. However, it is common for the option with the greatest BCR to fall short of providing an acceptable standard of protection (e.g. Table 11.5) or risk reduction. In Britain, a decision rule is used to help identify the most economic option, and involves the following steps (MAFF, 1993).

1. Examine the BCR of all options. If none is above 1.0, then the project is uneconomic.
2. Identify the option with the greatest BCR that is at least 1.0. If this option delivers an acceptable risk reduction, it should be the final choice. If not, then it is necessary to examine other options (step 3).
3. Determine whether an increase in standard of protection would be economically efficient. If the incremental BCR of the next option exceeds 1.0, then this option will be economic and should be chosen.
4. If the choice under steps 2 and 3 falls short of delivering an acceptable standard of protection, then the option that approaches the standard should be chosen, provided the BCR is at least 1.0 and its incremental BCR exceeds 1.0.

Example 11.3

A major public building has started to show signs of cracking and settlement. Investigations have shown that it had been built within an ancient landslide complex, prone to periodic reactivation. A range of landslide management options have been proposed, including

- 'do nothing'
- option A – 'do the minimum', ensuring that water supply and sewerage pipes are monitored and where necessary repaired so as to prevent leakage into the landslide
- option B – the installation of a network of surface and deep drains
- option C – the construction of a combined toe-weighting and drainage scheme.

Table 11.6 sets out the PV (see Chapter 10) costs and benefits (risk reduction) associated with each option, for a 50-year design life.

The risk reduction achieved is the 'do nothing' risk minus the residual risk associated with a particular option.

Table 11.6 Example 11.3: costs and benefits

	Do nothing	Option A	Option B	Option C
PV costs: £ thousands		100	250	2500
PV risk: £ thousands	5250	5000	4000	500
PV risk reduction		250	1250	4750
NPV		150	1000	2250
Average BCR		2.5	5	1.9
Incremental BCR			6.67	1.56

PV cost is the present-day value of the option costs, i.e. discounted at a rate of 6% per year (see Section 10.3 on cliff recession risk). For example, for a £1 million scheme which will be implemented in year 5:

PV cost = scheme cost × discount factor (year 5) = 1 × 0.747 = £747 000

PV risk is the PV of the risk associated with each option, over a 50-year period:

'do nothing risk' (years 0–49) = $\sum[P(\text{event}) \times \text{losses} \times \text{discount factor (years 0–49)}]$

'with project' risk (years 0–49) = $\sum[P(\text{event}) \times \text{losses} \times \text{discount factor (years 0–49)}]$

PV risk reduction is the difference in risk between the 'with project' risk (i.e. options A–C) and the do nothing risk:

PV risk reduction = 'do nothing' risk (years 0–49) – 'with project' risk (years 0–49)

The NPV is the difference between the option benefits (i.e. risk reduction) and the option costs:

NVP = PV risk reduction – PV costs

The average BCR is the ratio of the option benefits (i.e. risk reduction) and the costs:

BCR = PV risk reduction/PV costs

Incremental BCR represents the change in PV costs and PV benefits between the options:

$$\text{Incremental BCR} = \frac{\text{risk reduction(option B)} - \text{risk reduction(option A)}}{\text{costs(option B)} - \text{costs(option A)}}$$

Table 11.7 Example 11.4: costs and benefits

Option	Standard of protection (maximum return period)	Benefits: £ million	Costs: £ million	BCR	Incremental BCR
A	1	4.5	1.5	3.0	
B	2.5	7.25	1.95	3.7	6.1
C	10	11.75	2.5	4.7	8.2
D	25	14.5	2.875	5.0	7.3
E	50	16	3.250	4.9	4.0
F	100	16.75	3.625	4.6	2.0
G	200	17.25	4.750	3.6	0.4
H	300	17.5	5.5	3.2	0.2

Adapted from MAFF (1999)

The option with the greatest NPV is C – the construction of a combined toe-weighting and drainage scheme – and might be considered to be the most cost-effective. However, it has a lower BCR and incremental BCR than option B. If the *decision rule* is applied to this example, the first preference would be option B – surface and deep drains – as it has the highest BCR. This option should be selected, provided it is expected to deliver an acceptable reduction in the risk. Were this not to be the case, then option C could be selected as it is also economically efficient, with both a BCR and incremental BCR above 1.0.

Example 11.4

The lower slopes of a mountain range are prone to debris flow activity. A variety of diversion and control works have been proposed to reduce the risk to downslope properties. Each of the eight options would provide a particular level of protection and risk reduction, corresponding to the return period event it is designed to provide protection against. For example, option A will only provide protection against a 1 in 1-year debris flow (Table 11.7).

The choice of option will reflect both the level of risk that can be accepted and the economic efficiency of the option. The cost of improving the defence standard from 1 in 100 years to 1 in 200 years would be an extra £1.125 million, but would only reduce the risk by a further £0.5 million. In economic terms, this would not be an efficient use of resources.

Figure 11.8 illustrates how the interplay between benefits and costs can lead to difficult choices. In this example, the 'acceptable' standard of protection to the semi-urban developments at the mountain foot corresponds to the 1 in 100-year event. However, the maximum BCR coincides with the option that delivers protection against the 1 in 25-year event (option D). An increase in the standard of protection beyond this level can be justified by the incremental BCR of 2.0 for the option that delivers protection against the 1 in 100-year event (option F). Further increases in the standards of protection would not be justifiable because of the low incremental BCRs.

11.10. Environmental protection

Landslides can be important in creating and sustaining internationally important environmental resources (e.g. Clark *et al.*, 1996). In Britain, for example, some landslides have been designated

Figure 11.8 Example 11.4: costs and benefits and BCRs for different schemes (each for protection against different return period debris flow events)

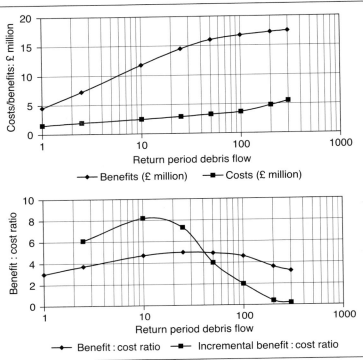

Sites of Special Scientific Interest (SSSI) for their geomorphological importance for earth science research and training (e.g. Alport Castles in the Peak District). Landslides can also create unique landscapes and habitats, as in the Landslip Nature Reserve on the east Devon coast. Most cliffs are shaped by and dependent on landslide processes, a feature brought dramatically into focus when waste from the drilling of the Channel Tunnel was dumped as a platform of spoil at the base of the chalk cliffs just south of Dover, leading to a public outcry when a section of the famous 'white' cliffs locally turned greenish-grey. Coastal erosion also maintains exposures of geological features, some of which may be internationally important stratigraphic or fossil reference sites (e.g. the Jurassic Coast of Dorset, UK, which has been designated a World Heritage Site). Cliff recession can have an important role in supplying sediment to beaches, sand dunes and mudflats on neighbouring stretches of coastline. These landforms absorb wave and tidal energy arriving at the coast, and can form important components of flood defence or coast protection solutions elsewhere (Lee, 1995, 2000). Disruption of the supply of sediment from eroding cliffs will invariably lead to the starvation of some coastal depositional landforms, and, hence, may lead to increased erosional risks elsewhere.

Some approaches to landslide management can present significant threats to these environmental resources. For example, many coast protection schemes have had significant impacts on the environment (Lee and Clark, 2002; Lee et al., 2001). Seawalls or rock revetments have been built that stop the recession process. Cliff faces have been stabilised by drainage works, regraded and landscaped. As a result, geological exposures have become obscured, hardy grasses of little or no conservation value have replaced bare soil and early pioneer stages, and wet areas have dried out. In Britain, a significant proportion of the soft rock cliff habitat resource has been adversely affected, with consequent loss or degradation of biological sites of national and international conservation value.

The potential threats to specific environments need to be identified and taken into account in the appraisal of landslide management options, as the environmental losses or mitigation measures are additional project costs over and above the costs associated with a particular option. Environmental impact assessment (EIA) is a systematic procedure for collecting information about the potential environmental impacts of a project or policy, and for measuring these impacts. As such, it is an essential input into cost–benefit analysis, even though it ignores costs (Pearce *et al.*, 2006), and is widely used to identify the likely significance of the various options being considered (for a good introduction, see Wood, 1995). The key factors that will need to be investigated in an EIA include the potential impacts on flora, fauna, population, amenity, cultural heritage, property and the built environment, landscape and geological/geomorphological features. Other effects might include

- the impact of construction traffic
- impact on access
- impacts due to construction noise and emissions
- health and safety
- water quality implications.

It is important to carry out a scoping study at an early stage of the appraisal of possible management options, in order to pinpoint the key environmental issues and concerns that will need to be addressed in a more formal EIA. In many countries, an EIA will need to be undertaken where landslide management is likely to have a significant effect. The EIA should help identify which options are best from an environmental perspective. Its key findings should be incorporated in the design and construction process.

In order to fully quantify the risks associated with a management option, it is important that the environmental costs, such as a reduction in habitat area or quality, are included in the assessment of risks and economic evaluation. The need to value environmental resources in monetary terms, in order that they can be included in an economic evaluation, presents major difficulties. A number of measures have been proposed, including

- direct-use values – that is, the direct use of the environmental resource by humans (recreation, fishing, etc.)
- indirect-use values, covering the value to humans of 'background' environmental resources (flood regulation, soil fertility, etc.)
- option or future-use values, including the desire for preservation of environmental resources for possible future use as enshrined in notions of sustainability (e.g. plant communities for possible use in medicine)
- existence and non-use values, representing human preferences for the preservation of the environment, over and above use and option values.

Perhaps the most pragmatic and lowest-cost approach is to estimate a proxy value for the resource, based on

- the cost of creating a similar site elsewhere of equivalent environmental value (e.g. a maritime woodland habitat)

- the cost of relocating a resource to another site (e.g. relocation of a specially protected species)
- the cost of local protection in situ (e.g. the construction of a viewing chamber for access to a geological exposure).

This would provide an indication of the minimum environmental value. Other methods, such as contingent valuation (e.g. Bateman and Willis, 1999; Bateman et al., 2002; Department for Transport, Local Government and the Regions, 2002; Mitchell and Carson, 1989; Pearce et al., 2006; Penning-Rowsell et al., 1992), may be necessary if an estimate of the full environmental value is needed. Recent developments in terms of 'ecosystem services' will be of relevance here (see Chapter 12).

Replacement costs should include land acquisition, planning, design and implementation, and ongoing monitoring and management of the site. As for other aspects of economic valuation, the costs should be adjusted to their PV by discounting.

A limitation of economic evaluation is that it seeks to compare alternative options in terms of a single objective – their economic efficiency. As it is difficult to give environmental losses their full significance in monetary terms, it is useful to also set out environmental objectives for a project. Possible options can then be compared in terms of their ability to deliver both economic efficiency and environmental objectives.

Example 11.5

An urban development sited on mountain footslopes is vulnerable to the impact of channelised debris flows. Possible mitigation measures include the construction of concrete-lined debris chutes to carry the flows through the developed area, or the use of a combination of upstream diversion barriers and storage basins. Environmental studies identified two important objectives.

- Ensuring public access and enjoyment of the natural views along a well-used right-of-way next to the urban channel.
- Ensuring that there was no net loss of nationally important orchid and insect species that occur within the proposed storage basin area.

Consultation with local residents and nature conservation bodies revealed that the loss of 'naturalness' along the urban channel banks could not be satisfactorily overcome in scheme design, but it might be acceptable to relocate the orchids and insects by recreating a similar habitat in what was currently an area of grazing land.

From Table 11.8, it is clear that the option with the highest BCR and NPV is option A – the construction of debris chutes through the urban area. However, this option fails to satisfy the environmental objectives of ensuring access and unspoilt naturalness along the channel banks. The favoured option then becomes option B2, involving the recreation of the orchid and insect habitats at an additional cost of £150 000. This achieves the environmental objectives and would still be economically efficient, with a BCR greater than 1.0.

Example 11.6

Over the next 50 years a small group of houses located on a cliff top will be destroyed as a result of cliff recession. Environmental scoping studies have identified that proposals to prevent recession by constructing a seawall at the cliff foot, together with slope stabilisation measures, would lead to degradation of an internationally important geological exposure and the loss of maritime cliff habitats.

Table 11.8 Example 11.5: costs and benefits

Option	PV risk: £ million	PV risk reduction: £ million	PV costs: £ million	BCR	NPV: £ million	Satisfies environmental objectives
Do nothing	2.5					
Option A: debris chutes	0.5	2	1	2	1	No
Option B1: diversion barriers and storage basin	0.4	2.1	1.25	1.7	0.85	No
Option B2: diversion barriers and storage basin – with habitat replacement	0.4	2.1	1.4	1.5	0.7	Yes

As a result, the proposals were considered to be unacceptable from an environmental perspective. It was recognised that the geological and habitat value of the cliffs could be maintained if the rate of recession was reduced from its present value, rather than prevented altogether.

It was agreed that the objectives of the management scheme would be to reduce the level of risk to the cliff-top community to an acceptable level, while maintaining the scientific quality of the geological site with unrestricted access and continuing to provide high biodiversity habitats on the coastal slopes.

Table 11.9 summarises the economic evaluation for three alternative options, along with their environmental benefits and losses. On economic grounds, the most efficient option would be option A, as it has the highest BCR and NPV. However, option B is preferred, as it delivers a reduction in the environmental losses while remaining economically efficient.

11.11. Environmental acceptability

In the past, it has often proved possible to find a compromise solution that delivers what is perceived to be an acceptable balance between risk reduction and environmental impacts. Hence, mitigation measures have been put forward and then modified to address environmental objections. This balancing act has become increasingly difficult as environmental conservation has attained greater and greater significance (see Chapter 12).

11.12. Corporate risk management: the risk matrix approach

Many major corporations have developed risk assessment and management processes to provide a structured approach to managing the risks to which they are exposed. The Shell Group, for example, developed the Hazards and Effects Management Process (HEMP), prompted by incidents such as the *Exxon Valdez* oil spill (1989) and the disposal of Brent Spar (1995). Part of this process involves the use of a risk assessment matrix to demonstrate that risks are managed to the ALARP principles (Energy Institute, 2008a,b; Johnson, 2012). Similar procedures have been developed by other major companies, including BP (Morrison, 2012), ExxonMobil (e.g. Liew *et al.*, 2002) and Rio Tinto (2009). For example, the BP Group Risk Standard requires the same risk matrix be utilised by every operating entity in the company (Morrison, 2012).

Table 11.9 Example 11.6: costs and benefits

Option	Environmental benefits	Environmental losses	PV risk: £ million	PV risk reduction: £ million	PV costs: £ million	BCR	NPV: £ million	Satisfies environmental objectives
Do nothing	Ongoing geological exposure and habitats		3.5					
Option A: concrete seawall and slope stabilisation		Geological exposure obscured Degradation of habitats	0.15	3.35	2.5	1.3	0.85	No
Option B: beach management and rock revetment	Natural processes largely preserved	Partial reduction in exposure Dynamic nature of habitats reduced, but not lost	1.25	2.25	2	1.1	0.25	Yes

Typically, the matrix structure is defined by the frequency/probability of a damaging event or scenario (i.e. a 'top event'; see Chapter 2) and the severity level of the consequences (Figure 11.9). However, in addition to providing a mechanism for assessing risks, these matrices are also constructed to reflect group reporting line criteria for health and safety, business and financial losses, impact on reputation and environmental risks (Table 11.10). Often the matrix is intended for use throughout the project cycle, from pre-feasibility to operations and subsequent decommissioning. As a result, the input data on frequency and severity can be qualitative (i.e. based on expert judgement) or quantitative (i.e. where the results of a detailed risk analysis are simplified to match the matrix classes). The risk classes within the matrix are usually linked to clearly defined actions, ranging from the requirement to demonstrate continuous risk reduction (i.e. ALARP) to the requirement for a board-level decision by the group on whether the operation should continue after the development of an appropriate risk mitigation plan (e.g. Table 11.11).

11.13. Future uncertainty: implications for landslide management

Landslide risk management decisions will have to be made despite uncertainty as to future risk levels due to changes to environment, climate and sea levels. Decisions taken now will have implications in the future in terms of the vulnerability of society to landslide hazards. This is not the place to consider the likely nature and scale of such changes (see Chapter 12), but rather to indicate the existence of a number of available strategies for managing landslide problems in the face of uncertain change, especially possible climate change.

- Using so-called *no-regret* or *low-regret* options that are worthwhile pursuing irrespective of how the climate changes: for example, improving drainage systems in urban landslide areas (e.g. McInnes, 2000).
- Making sound decisions, based on a thorough analysis of the information available, while ensuring that future generations are not committed to inflexible and/or excessively expensive landslide management practices (*adaptive management*). The approach focuses on flexibility as a key basis for effective risk management, and leaves scope for amending decisions at a later date as improved information becomes available on changing conditions.
- *Delaying action* if the assessment of current and future risks suggests that problems are not likely to become significant unless and until certain climate change thresholds are reached: for example, in mountain regions prone to debris flows, the decision to replace sacrificial bridges across debris flow chutes with more permanent structures.
- *Delaying and buying time*, through the implementation of a short-term solution to the current slope problems and delaying the point at which significant investment or relocation decisions have to be made. For example, the A3055 road on the Isle of Wight, UK, lies immediately inland of the crest of 70 m-high chalk cliffs, and is threatened by coastal erosion. Construction of erosion control measures at the cliff foot is not considered appropriate, because of the environmental value of the site. In order to extend the life of this important tourist route, a monitoring system was installed to provide early warning of ground movement that could affect the safety of road users (Fort and Clark, 2002).
- *Changing the land use* to one where the future risks were likely to be more acceptable: for example, a change from allocation of land for housing in an area where climate change might lead to an increase in landslide activity to a lower-risk use, such as playing fields or public open space. In some places, *site abandonment* might be the best option.
- *Contingency planning*, involving making plans and provisions for a possible increased frequency of extreme events and climatic 'surprises'. For example, this might involve establishing a

Landslide Risk Assessment

Figure 11.9 Example risk matrix for corporate risk management

Severity level (see Table 11.11)	Damaging event frequency/probability								Approach to estimating damaging event frequency/probability
	1	2	3	4	5	6	7	8	
	$<1 \times 10^{-6}$	$>1 \times 10^{-6}$ to 1×10^{-5}	$>1 \times 10^{-5}$ to 1×10^{-4}	$>1 \times 10^{-4}$ to 1×10^{-3}	$>1 \times 10^{-3}$ to 1×10^{-2}	$>1 \times 10^{-2}$ to 1×10^{-1}	$>1 \times 10^{-1}$ to 0.25	>0.25	Estimated annual probability
	$<1 \times 10^{-6}$/year	$>1 \times 10^{-6}$ to 1×10^{-5}/year	$>1 \times 10^{-5}$ to 1×10^{-4}/year	$>1 \times 10^{-4}$ to 1×10^{-3}/year	$>1 \times 10^{-3}$ to 1×10^{-2}/year	$>1 \times 10^{-2}$ to 1×10^{-1}/year	$>1 \times 10^{-1}$ to 1/year	>1/year	Historical frequency
A									
B									
C									
D									
E									
F									
G	An event that would be unlikely in the industry. Only a remote possibility of occurrence	A similar event has never occurred in the industry	Similar event has occurred somewhere in the industry	Similar event has occurred somewhere in the business	Similar events have occurred in other locations, on average <1/year	Event expected to occur 1–2 times over the facility lifetime	Event expected to occur multiple times within facility lifetime	A common occurrence, at least 1/year at the location	Qualitative event frequency

Legend:
- I — Tolerability to be endorsed by Group management.
- II — Tolerability to be endorsed by operations management.
- III — Risk reduction to ALARP.
- IV — Continuous risk management.

Table 11.10 Example risk matrix severity levels (from various sources)

Level	Harm to people	Financial/business loss	Environmental effects	Reputational damage
A	Slight health/safety incident; first aid required or single/multiple over-exposures causing noticeable irritation but no actual health effects	Slight damage (<$50 000)	Slight effect; slight damage contained within facility, e.g. leaks that evaporate from a small spill	Slight impact e.g. local public awareness but no discernible concern; no media coverage
B	Minor injury or health effect	Minor damage ($50 000–100 000), brief disruption to operations	Minor effect; minor damage, but lasting effect	Minor effect; local public concern and local media coverage
C	Major incident, with potential for 1–2 fatalities, >10 injuries or health effects	Moderate damage ($100 000–1 million), partial shutdown	Moderate effect; limited environmental damage that will persist and require clean-up	Moderate impact; significant impact in region or country; regional public concern, local stakeholders aware (e.g. NGOs, industry); extensive attention in local media
D	Very major incident, with potential for 3–10 fatalities, >30 serious injuries and/or health effects	Major damage ($1 million–1 billion), up to 2 weeks' shutdown	Major effect; severe environmental damage that will require extensive measures to restore beneficial use of the environment	Major impact; likely to escalate and affect group reputation; national public concern and impact on local and national stakeholder relations; extensive attention in national media, some international coverage
E	Catastrophic health/safety incident causing 10–50 fatalities within and outside the facility	Massive asset damage ($1–5 billion), partial loss of operation	Massive environmental damage lasting 1–5 years	Massive impact; severe impact on group reputation; international public concern and international media attention, high level of concern by government and action by international NGOs
F	Catastrophic health/safety incident causing 50–200 fatalities within and outside the facility	Severe asset damage ($5–10 billion), loss of operation	Extensive environmental damage lasting >5 years	Prolonged international media coverage; public outrage; regional brand damage
G	Comparable to the most catastrophic health/safety incidents in the industry; potential for >200 fatalities	Catastrophic damage (>$10 billion), loss of operation	Catastrophic damage to the environment lasting >5 years	Global outrage; major impact on share price

Table 11.11 Example risk matrix required actions (from various sources)

Scenario risk level	Required action
I	Tolerability to be endorsed by the group management For continued operation, the appropriate group vice-president (VP) shall be promptly notified A short-term risk management plan shall be implemented promptly, while the VP's approval is required for a longer-term risk mitigation plan
II	Tolerability to be endorsed by operations management For continued operation, the operations manager shall be promptly notified A short-term risk management plan shall be implemented promptly, while the operations manager's approval is required for a longer-term risk mitigation plan
III	Risk reduction to ALARP Short-term risk mitigation shall be implemented while the appropriate operations manager's approval is required to manage the risks as part of an overall programme of continuous risk reduction
IV	Continuous risk management Continuous risk management must be embedded in the business group's health, safety and environment plan

coordinated approach to disaster management, such as the US Federal Emergency Management Agency (FEMA) that supervises the Hazard Mitigation Grant Program.
- *Making allowance for climate change* in the design of slope stabilisation and erosion control measures. This can be achieved through the use of design specifications that allow for uncertainty or variability in the design parameters or loadings, and by ensuring that structures can be modified at a later date if the allowances are found to be inadequate.

All of these strategies should be supported by monitoring of climate/environment/sea-level changes and their impact on landslide risk. In addition to improving the understanding of changing conditions, monitoring also allows adaptive decisions themselves to be modified in the light of improved knowledge. However, the need for monitoring should not be used as an excuse to delay more direct intervention where the existing, or future, level of risk is considered to be unacceptable.

11.14. Risk assessment, decision-making and consultation

Risk assessment is part of a process that begins with the recognition and formulation of a problem – the likelihood of damage or harm – and ends with a decision about how best to manage the problem (see Figure 2.1). The decision should take account of the results of the risk assessment process, but not be determined solely by them. There are a wide range of other factors that will usually need to be considered in reaching a decision, including environmental issues, financial constraints and the broader socio-economic and political context.

In many instances the participation in the decision-making process may extend wider than a regulatory authority or a corporate management team. There has been a long-running debate as to whether the traditional approach of restricting decision-making on risk issues to a few scientifically well-informed individuals (narrow participation) should be replaced by broader involvement (see Chapter 2 and

Royal Society, 1992). It is now generally appreciated that broad participation is preferable (see Pidgeon et al., in Hood and Jones, 1996), because the information base is widened, the accountability of the technical decision-makers is increased and the acceptability of the final decision rises, if those that are affected have actually been involved in the process. Thus, pressures exerted from homeowners, landowners, interest groups, business groups and environmentalists will also influence the choice of landslide risk management strategy.

Consultation must, therefore, be seen to be an essential part of the decision-making process, despite the fact that wider involvement can increase costs and cause delay. Stern and Fineberg (1996) suggest that effective consultation should involve

- ensuring that all interested parties are identified and invited to participate in the consultation process, preferably at an early stage of the assessment process
- ensuring that the participants are aware of the legal status of the consultation process and how their representations will be used by the decision-makers
- making information readily available to the consultees and, where appropriate, providing access to technical expertise and other resources for groups that lack these resources.

Getting the broad range of groups and interests that may be affected by landslide problems to accept risk-based decisions is often critical to the successful implementation of landslide risk management strategies. Such acceptance is dependent on the establishment of trust, and this, in turn, is dependent on openness, involvement and good communications (see Section 2.7 on risk communication). It follows, therefore, that the risk assessment process should seek to inform all stakeholders, addressing their questions in a form that they can understand. Herein lies the true importance of risk communication, a process which is not simply confined to the development of forecasts and the issuing of warnings but embraces all the diverse ways by which information on risks is disseminated among all those involved in order to improve safety and security. All those involved in risk assessment must learn from the consequences of the 2009 L'Aquila earthquake (central Italy), which resulted in six scientists and one governmental official being found guilty of manslaughter, not for poor science but for failing to adequately evaluate and communicate risk and presenting, or failing to correct, falsely reassuring findings to the public (*Nature Geoscience*, 2013).

REFERENCES

ACMH (Advisory Committee on Major Hazards) (1976) *First Report*. London, HMSO.
AGS (Australian Geomechanics Society) (2007) Practice note guidelines for landslide risk management. *Australian Geomechanics* **42(1)**: 63–114.
Ale BJM (2005) Tolerable or acceptable: a comparison of risk regulation in the United Kingdom and in the Netherlands. *Risk Analysis* **25(2)**: 231–241.
ANCOLD (Australian National Committee on Large Dams) (2003) *Guidelines on risk assessment*. Australian National Committee on Large Dams Incorporated, Melbourne.
Angeli MG, Gasparetto P, Menotti RM, Pasuto A and Silvano S (1994) A system of monitoring and warning in a complex landslide in North-Eastern Italy. *Landslide News* **8**: 12–15.
Ball DJ and Floyd PJ (1998) *Societal Risks: Final Report*. Health and Safety Executive, Bootle.
Basta C, Neuvel JMM, Zlatanova S and Ale B (2007) Risk-maps informing land-use planning processes. A survey on the Netherlands and the United Kingdom recent developments. *Journal of Hazardous Materials* **145**: 241–249.
Bateman I and Willis KG (1999) *Valuing Environmental Preferences*. Oxford University Press, Oxford.

Bateman IJ, Carson RT, Day B, Hanemann M, Hanley N, Hett T, Jones-Lee, M, Loomes G, Mourato S, Ozdemiroglu E, Pearce DW, Sugden R and Swanson J (2002) *Economic Valuation with Stated Preference Techniques: A Manual*. Edward Elgar, Cheltenham.

BC Hydro (1993) *Interim Guidelines for Consequence-based Dam Safety Evaluation and Improvements*. BC Hydro Hydroelectric Engineering Division, Report No. H2528, August.

BP (2007) *Annual Report 2006. Health, Safety and the Environment*. http://www.bp.com/liveassets/bp_internet/norway/norway_norwegian/STAGING/local_assets/downloads_pdfs/h/hse_annual_report_2006.pdf (accessed 22/9/2013).

Bromhead EN (2005) Geotechnical structures for landslide risk reduction. In *Landslide Hazard and Risk* (Glade T, Anderson MG and Crozier MJ (eds)). Wiley, Chichester, pp. 549–594.

Burton I, Kates RW and White GF (1978) *The Environment as Hazard*. Oxford University Press, Oxford.

Clark AR, Lee EM and Moore R (1996) *Landslide Investigation and Management in Great Britain: A Guide for Planners and Developers*. HMSO, London.

Considine M and Hall SM (2008) The major accident risk (MAR) process – developing the profile of major accident risk for a large multinational oil company. *Process Safety and Environmental Protection* **87(1)**: 59–63.

Costa JE and Wieczorek GF (eds) (1987) Debris flows/avalanches: processes, recognition and mitigation. In *Reviews in Engineering Geology*, vol. VII. Geological Society of America, Boulder, CO.

Dai FC, Lee CF and Ngai YY (2002) Landslide risk assessment and management: an overview. *Engineering Geology* **64(1)**: 65–87.

Department for Transport, Local Government and the Regions (2002) *Economic Valuation with Stated Preference Techniques: A Summary Guide. DTLR Appraisal Guidance*. HMSO, London.

DNV (District of North Vancouver) (2009) *Natural Hazards Risk Tolerance Criteria*. Report to Council, 10 November 2009. http://www.dnv.org/upload/documents/Engineering/CDNV_DISTRICT_HALL-%231311308-v1-risk_tolerance_criteria_-_RTC.PDF (accessed 22/9/2013).

Energy Institute (2008a) *The Risk Assessment Matrix: Bringing it to Life*. Energy Institute, London. http://www.eimicrosites.org/heartsandminds/ram.php (accessed 22/9/2013).

Energy Institute (2008b) The risk assessment matrix: bringing it to life. A sample presentation. Energy Institute, London. http://www.eimicrosites.org/heartsandminds/userfiles/file/RAM/RAM%20-%20PPT%20presentation.pps (accessed 22/9/2013).

ERM-Hong Kong (1998) *Landslides and Boulder Falls from Natural Terrain: Interim Risk Guidelines*. Geotechnical Engineering Office, Hong Kong. GEO Report No. 75.

Fannin RJ, Moore GD, Schwab JW and VanDine DF (2005) Landslide risk management in forest practices. In *Landslide Risk Management* (Hungr O, Fell R, Couture R and Eberhardt E (eds)). Balkema, Rotterdam, pp. 299–320.

Fell R and Hartford D (1997) Landslide risk management. In *Landslide Risk Assessment* (Cruden D and Fell R (eds)). Balkema, Rotterdam, pp. 51–108.

Fookes PG and Sweeney M (1976) Stabilisation and control of local rockfalls and degrading rock slopes. *Quarterly Journal of Engineering Geology* **9**: 37–56.

Fort DS and Clark AR (2002) The monitoring of coastal landslides: a management tool. In *Instability – Planning and Management* (McInnes RG and Jakeways J (eds)). Thomas Telford, London, pp. 479–486.

Hartford DND (2009) Legal framework considerations in the development of risk acceptance criteria. *Structural Safety* **31**: 118–123.

Ho K, Leroi E and Roberds B (2000) Quantitative risk assessment: application, myths and future direction. *Proceedings of the Geo-Eng Conference*, Melbourne, Publication 1, pp. 269–312.

Holz RD and Schuster RL (1996) Stabilization of soil slopes. In *Landslides: Investigation and Mitigation* (Turner AK and Schuster RL (eds)). National Academy Press, Washington, DC, pp. 439–473. TRB Transportation Research Board Special Report 247.

Hood C and Jones DKC (eds) (1996) *Accident and Design: Contemporary Debates in Risk Management*. UCL Press, London.

Horowitz JK and Carson RT (1993) Baseline risk and preference for reductions in risk-to-life. *Risk Analysis* **13**: 457–462.

HSE (Health and Safety Executive) (1988) *The Tolerability of Risk from Nuclear Power Stations*. HMSO, London.

HSE (1992) *The Tolerability of Risk from Nuclear Power Stations*, revised. HMSO, London.

HSE (2001) *Reducing Risks, Protecting People*. HMSO, Norwich.

Hubbard DW (2009) *The Failure of Risk Management: Why It's Broken and How to Fix It*. Wiley, Somerset, NJ.

Hutchinson JN (1977) Assessment of the effectiveness of corrective measures in relation to geological conditions and types of movement. *Bulletin of the International Association of Engineering Geology* **16**: 131–155.

Iceland Ministry for the Environment (2000) Regulation on hazard zoning due to snow and landslides, classification and utilisation of hazard zones, and preparation of provisional hazard zoning. 6 July 2000. http://andvari.vedur.is/snjoflod/haettumat/reglugerd_505_2000_e.pdf (accessed 22/9/2013).

Johnson AD (2012) Qatar Shell GTL implementation of process safety management. *SPE International Production and Operations Conference and Exhibition*, Doha.

Jones DKC (1991) Environmental Hazards. In *Global Challenge and Change* (Bennett RJ and Estall RC (eds)). Routledge, London, pp. 27–56.

Jones DKC (1996) Anticipating the risks posed by natural perils. In *Accident and Design* (Hood C and Jones DKC (eds)). UCL Press, London, pp. 14–30.

Knight RF and Pretty DJ (2002) *The Impact of Catastrophes on Shareholder Value*. Oxford University Business School, Oxford.

Kong WK (2002) Risk assessment of slopes. *Quarterly Journal of Engineering Geology and Hydrogeology* **35**: 213–222.

Lee EM (1995) Coastal cliff recession in Great Britain: the significance for sustainable coastal management. In *Directions in European Coastal Management* (Healy MG and Doody JP (eds)). Samara Publishing, Swansea, pp. 185–194.

Lee EM (2000) The management of coastal landslide risks in England: the implications of conservation legislation and commitments. In *Landslides: In Research, Theory and Practice* (Bromhead EN, Dixon N and Ibsen ML (eds)). Thomas Telford, London, pp. 893–898.

Lee EM (2002) A dynamic framework for the management of coastal erosion and flooding risks in England. In *Instability: Planning and Management* (McInnes RG and Jakeways J (eds)). Thomas Telford, London, pp. 713–720.

Lee EM and Clark AR (2002) *Investigation and Management of Soft Rock Cliffs*. Thomas Telford, London.

Lee EM, Brunsden D, Roberts H, Jewell S and McInnes R (2001) *Restoring Biodiversity to Soft Cliffs*. English Nature, Peterborough. Report 398.

Lee-Youngren T and Hodgson J (1998) FEMA's check is in the mail! *North Coast Journal Weekly*, 3 December.

Leroi E, Bonnard C, Fell R and McInnes (2005) Risk assessment and management. In *Landslide Risk Management* (Hungr O, Fell R, Couture R and Eberhardt E (eds)). Balkema, Rotterdam, pp. 159–198.

Lewis S (2007) *Risk Criteria – When is Low Enough Good Enough?* Risktec Solutions, Warrington.

Liew KC, Zainuddin ZM and Samad AB (2002) Experience in integrating occupational health into safety risk matrix. *SPE International Conference on Health, Safety and Environment in Oil and Gas Exploration and Production*, Kuala Lumpur.

McInnes RG (2000) *Managing Ground Instability in Urban Areas: A Guide to Best Practice*. Cross Publishing, Ventnor.

McInnes R (2005) Instability management from policy to practice. In *Landslide Hazard and Risk* (Glade T, Anderson MG, Crozier MJ (eds)). Wiley, Chichester, pp. 401–428.

McIntyre T (2005) Standing legacy: ghost towns preserve the Ottawa Valley's rich history. *Canadian Geographic*, September/October.

McQuaid J and Le Guen JM (1998) The use of risk assessment in government. In *Risk Assessment and Risk Management. Issues in Environmental Science and Technology*, vol. 9 (Hester RE and Harrison RM (eds)). The Royal Society of Chemistry, London.

MAFF (Ministry of Agriculture, Fisheries and Food) (1993) *Project Appraisal Guidance Notes*. MAFF, London.

MAFF (1999) *FCDPAG3 Flood and Coastal Defence Project Appraisal Guidance: Economic Appraisal*. MAFF Publications, London.

Malone AW (2005) The story of quantified risk and its place in slope safety policy. In *Landslide Hazard and Risk* (Glade T, Anderson MG and Crozier MJ (eds)). Wiley, Chichester, pp. 643–674.

Mitchell RC and Carson RT (1989) *Using Surveys to Value Public Goods: The Contingent Valuation Method*. Resources for the Future, Washington, DC.

Morrison R (2012) *Safety and Operational Risk: Progress and Perspectives*. Rio Oil and Gas, Rio de Janeiro.

Murray JG (2000) The NZ landslide safety net. In *Landslides: In Research, Theory and Practice* (Bromhead EN, Dixon N and Ibsen ML (eds)). Thomas Telford, London, pp. 1075–1080.

Nature Geoscience (2013) Communication at Risk: The L'Aquila earthquake trial tragically highlights that risk communication is integral to Earth science training. *Nature Geoscience* **6(2)**: 77.

Palm RI (1990) *Natural Hazards: An Integrative Framework for Research and Planning*. John Hopkins University Press, Baltimore.

Paus HL (2005) Reply of insurance industry to landslide risk. In *Landslide Hazard and Risk* (Glade T, Anderson MG and Crozier MJ (eds)). Wiley, Chichester, pp. 251–284.

Pearce D, Atkinson G and Mourato S (2006) *Cost–Benefit Analysis and the Environment: Recent Developments*. Organisation for Economic Co-operation and Development, Paris.

Penning-Rowsell EC, Parker DJ and Harding DM (1986) *Floods and Drainage*. Allen and Unwin, London.

Penning-Rowsell EC, Green CH, Thompson PM *et al.* (1992) *The Economics of Coastal Management: A Manual of Benefits Assessment Techniques*. Belhaven Press, London.

Phillips C and Marden M (2005) Reforestation schemes to manage regional landslide risk. In *Landslide Hazard and Risk* (Glade T, Anderson MG and Crozier MJ (eds)). Wiley, Chichester, pp. 517–548.

Reeves A, Ho KKS and Lo DOK (1999) Interim risk criteria for landslides and boulder falls from natural terrain. *Proceedings of the Seminar on Geotechnical Risk Management*, Geotechnical Division, Hong Kong Institution of Engineers, Hong Kong, pp. 127–136.

Rio Tinto (2009) *Risk Policy and Standard*. Rio Tinto, London.

Royal Society (1992) *Risk: Analysis, Perception and Management*. Royal Society, London.

Schuster RL and Kockelman WJ (1996) Principles of landslide hazard reduction. In *Landslides: Investigation and Mitigation* (Turner AK and Schustereds RL (eds)). National Academy Press, Washington, DC, pp. 91–105. Transportation Research Board Special Report 247.

Sidle RC, Pearce AJ and O'Loughlin CL (1985) *Hillslope Stability and Land Use.* American Geophysical Union, Washington, DC.

Smith K (2001) Environmental *Hazards: Assessing Risk and Reducing Disaster*, 3rd edn. Routledge, London.

Stern PC and Fineberg HV (eds) (1996) *Understanding Risk: Informing Decisions in a Democratic Society.* National Academy Press, Washington, DC.

Vallet S (2004) Insuring the uninsurable: the French natural catastrophe insurance system. In *Catastrophe Risk and Reinsurance* (Gurenko E (ed.)). Risk Books, London, pp. 199–215.

Vinnem JE (2007) *Offshore Risk Assessment: Principles, Modelling and Application of QRA Studies*, 2nd edn. Springer, London.

Wood C (1995) *Environmental Impact Assessment: A Comparative Review.* Longman, London.

Wyllie DC and Norrish NI (1996) Stabilization of rock slopes. In *Landslides: Investigation and Mitigation* (Turner AK and Schuster RL (eds)). National Academy Press, Washington, DC, pp. 474–506. Transportation Research Board, Special Report 247.

Landslide Risk Assessment
ISBN 978-0-7277-5801-9

ICE Publishing: All rights reserved
http://dx.doi.org/10.1680/lra.58019.459

Chapter 12
Future challenges

12.1. Introduction

Ten years have passed since the first edition of this book was prepared. During that time the application of landslide risk assessment has increased dramatically, from a relatively few pioneering studies to a burgeoning array of case histories and research projects. Our own experience of risk assessment has changed through project work, discussions with colleagues and exposure to the emerging 'risk cultures' in many organisations. In this chapter, we try to identify some of the lessons learnt and emerging issues that could shape the landslide risk agenda over the next decade. The issues concern

- the practicability and value of landslide-risk zonation maps
- balancing the needs of risk assessment and the demands of project schedules
- risk assessment in the context of extreme events
- striking a balance between landslide risk management and conservation
- the valuing of intangibles, especially within the environment
- future uncertainty regarding landslide risk due to changes in environmental conditions, climate change and sea-level change.

12.2. Quantitative risk zonation: lessons from Ventnor, Isle of Wight

In Chapter 10, the focus was on methods for quantifying risk to a given facility, site or linear asset (e.g. a road or pipeline), or within a particular area affected by landslides. However, it is widely recognised that risk assessment also may be of value in supporting land-use planning, especially in areas where the development of housing in landslide-prone areas is poorly controlled by local planning. To this end, efforts have been made to develop landslide risk zonation maps, where areas are subdivided into units having similar levels of risk (e.g. Cascini *et al.*, 2005; Fell *et al.*, 2008a,b). Dai *et al.* (2002) imply that this is essentially a straightforward GIS-based task, combining a variety of multi-scale and temporal databases, including geology and topography, socio-economic parameters, landslide inventories and monitoring. The risk within a particular unit can then be determined as the product of the landslide probability, exposure, vulnerability and the value of the elements at risk.

To date, however, zonation maps have tended to be based on qualitative assessment, typically showing areas of high, medium and low risk (see Chapter 4), such as the risk maps of the Liri-Garigliano and Volturno river basins in central southern Italy (e.g. Cascini *et al.*, 2005). This is because experience has shown that highly localized landslide processes, combined with complex spatial variations in the exposure and vulnerability of multiple, stationary and non-stationary assets, together with temporal changes in the nature, value and extent of the elements at risk, cannot be adequately represented within wide-area mapping approaches. This mismatch of scales is a major reason for the minimal uptake and application of risk maps in developing countries, as observed by Opadeyi *et al.* (2005) and Zaitchik *et al.* (2003). There are also major challenges involved in integrating the different risks associated with all the various landslide types, volumes and run-out potential across an area in order to calculate total risk. Indeed, van Westen et al. (2006) have suggested that quantitative risk zonation maps remain a 'step too far, especially at medium scales' i.e. 1 : 10 000 to 1 : 15 000. They have followed Finlay and Fell (1997) in questioning whether effective urban planning requires quantitative risk maps: 'it may not be required for local authorities to know the exact amount of losses expected in monetary values, even more because the level of risk due to landsliding is often several orders of magnitude lower than for other hazards, such as traffic accidents, fires and diseases' (van Westen et al., 2006).

Time will tell, but landslide risk maps may well prove less valuable to land-use planners than the establishment of zones based simply on landslide hazard, such as the maps developed for the Ventnor Undercliff, UK. Ventnor lies within an extensive, ancient landslide complex. There has been a history of problems associated with the degradation or reactivation of the landslide complex over the last 100 years or so (e.g. Hutchinson, 1991; Hutchinson and Bromhead, 2002; Hutchinson and Chandler, 1991; Hutchinson et al., 1991).

In 1988, the UK Government (Department of the Environment) commissioned an assessment of landslide hazard in the Ventnor area as part of its land-use planning research programme (Lee and Moore, 1991; Lee et al., 1991a,b; Moore et al., 1991). The aim of the study was to define the nature and extent of the landslide problems and identify ways in which landslide hazard information could be used to assist local land-use planners. The original terms of reference included maps of landslide hazard and landslide risk as key deliverables. However, as the project progressed, the steering committee, which included members from both the UK Department of the Environment and the local authority (South Wight Borough Council), came to the view that risk maps were unlikely to be the best way of summarising the landslide issues for the local planners, developers and other stakeholders. Instead, 1 : 2500 scale 'planning guidance maps' were to be developed to summarise the landslide hazard information in terms that are directly relevant to forward planning and development control.

Developments in Great Britain require planning permission. Local planning authorities are required and empowered under the Town and Country Planning Act 1990 to control most forms of development, and are responsible, under the Building Regulations and Housing Acts, for ensuring standards of construction of development. When considering an application for planning permission, local authorities in England and Wales, have a duty to take into account a range of material considerations, which include potential land instability problems (e.g. ground movement and landsliding). However, the responsibility for determining whether land is physically suitable for a proposed development and the appropriate technical measures to protect that development, lies with the developer and/or the landowner.

In addressing all instability matters throughout the Isle of Wight, the council's approach is set out in their development plan which states: 'Development of areas known to suffer from instability will not normally be permitted, unless the local planning authority can be satisfied that the site can be developed and used safely and not add to the instability of the site or adjoining land'.

Thus the determination of applications for development in the Undercliff takes possible ground-movement problems into account. The council may reject outright any built development proposals in areas considered to be unsuitable, or insist on particular conditions being met before granting planning permission. In all instances, it is the developer's responsibility to investigate the potential problems on and around a site and to satisfy the planning authority that adequate attention has been paid to ground movement in the proposed building design.

These policies are supported by the planning guidance maps. The maps indicate that different areas of the landslide complex need to be treated in different ways for both land-use policy formulation and the review of planning applications (Table 12.1). Areas are identified that are likely to be physically capable of development, along with areas that are either subject to significant constraints or are likely to be unsuitable. The nature of investigation required in support of a planning application will depend on the type of development that is proposed and the potential instability problems anticipated at the site, taking into account the hazard zonation shown on the maps.

The planning guidance maps have proved to be an effective tool for the control of development and land use in this area of urban landsliding (e.g. McInnes, 2005). They are relatively simple to use by land-use planners. Advice is also provided on the level of stability information that should be presented with applications for development in different areas of the Undercliff. This zonation has been sufficient to ensure that the on-going development in the Undercliff takes account of the types of ground movement that could be expected in different areas, either through improved building designs and/or the provision of site specific mitigation works. Quantitative risk assessment methods have been applied in the Undercliff, but these have been developed with the specific purpose of evaluating the potential effectiveness of particular landslide management options (e.g. Lee and Moore, 2007).

12.3. Landslide risks and major project schedules: lessons from the Camisea pipeline, Peru

Projects are often exposed to a variety of risks from landslides, including

- the potential impact on capital expenditure, through the requirement for design changes or the provision of mitigation measures
- the potential impact on project schedule, through the need for additional investigation, laboratory testing, stability analyses and design of mitigation measures
- the operational risks over the project lifetime ('enduring risks') posed by future landsliding.

These different risks can place conflicting demands on project managers. For example, the need to assess and respond to the operational risks can have significant impacts on project cost and schedule. In many instances, these schedules are driven by commercial pressures rather than the terrain conditions. As Sweeney (2005) wrote, 'one of the disadvantages of working on major (oil and gas pipeline) projects in a large organisation is that project programmes appear as if by magic from the planning and commercial back office'. The result is that managers are faced with the challenge of finding room within a tight schedule for landslide risk studies that almost inevitably lead to unplanned route changes and additional costs.

The Camisea pipeline system, Peru, comprises two buried pipelines within the same right of way: a natural gas line (714 km) and a natural gas liquids (NGL) line (540 km). The pipelines originate at the Camisea gas field near Malvinas in the Amazon jungle (Selva) and continue across the Andes (Sierra) to the Pacific coast (Costa). During the first 30 months of operation, the NGL line ruptured six times, four of these incidents being caused by landslide movements (Lee et al., 2009). This prompted the Peruvian Ministry of Energy and Mines to commission an investigation into the causes of the failures, together with an independent audit whether these failures were unrelated events or an

Table 12.1 Ventnor Undercliff: planning guidance

Zonation	Development plan	Development control
A	Areas likely to be suitable for future development in accordance with the development plan Rationale: In the past, properties situated in these areas have been largely unaffected by ground movement	Ground movement does not impose significant constraints, although some mitigation/stabilisation measures may be required to ensure the stability of the site and surrounding land An *outline stability report* would normally be required, prepared by a competent person. The report should be based primarily on a desk study involving a review of available information relating to instability problems in and around the proposed development site
B	Areas likely to be suitable for future development in accordance with the development plan provided the developer undertakes appropriate mitigation and stabilisation measures Rationale: In the past, most properties situated in these areas have been largely unaffected by ground movement. However, in places, the cumulative effect of this movement has resulted in moderate and slight damage to property	Ground movement imposes significant constraints that would generally require mitigation/stabilisation measures to ensure the stability of the site and surrounding land A *standard stability report* would normally be required, prepared by a competent person. The report should be based on an inspection of the site and surrounding area to assess the geomorphological context of the proposed development and to identify any recent ground cracking or structural damage to property. A ground investigation and subsurface investigations involving trial pitting, boreholes and groundwater monitoring *may be required* in certain areas
C	Area unsuitable for built development Rationale: In the past, properties situated in these areas have been affected by differential shear, distortion and tilting. The cumulative effect of this movement has resulted in serious and severe damage to property	Ground movement imposes severe constraints that probably could not be overcome by cost-effective and environmentally acceptable mitigation or stabilisation measures to ensure the stability of the site and surrounding land A *detailed stability report* would normally be required, prepared by a competent person. The report should be based on a thorough investigation of surface and subsurface conditions at the site and surrounding area, together with appropriate ground movement and groundwater monitoring
D	Areas that *might* be suitable for future development in accordance with the development plan provided the developer presents adequate evidence of stability Rationale: Little is known of the past ground movement in these areas because of the absence of existing property. It is necessary to be cautious about the potential for landslide problems	Ground movement *may* impose major constraints that would generally require large-scale mitigation/stabilisation measures to ensure the stability of the site and surrounding land A *detailed stability report* would normally be required, prepared by a competent person

From Lee (2003)

indication of a general problem with the planning, design and construction of the pipeline system. The audit findings included (Lee *et al.*, 2009)

- there had been a general failure to recognise landsliding as a dominant risk along the pipeline route and this should have required special attention in final route selection and the design of mitigation measures
- that a 36-month (September 2001 to August 2004) project timetable and associated financial penalties had forced the contractor to minimise field geological investigations and geotechnical analyses – a more robust investigation programme could have improved the definition and characterisation of geohazards and thereby reduced the risk of failure

The Transportadora de Gas del Peru (TgP) consortium was awarded the BOOT (build, own, operate, transfer) contract for the pipeline system in December 2000. The government of Peru set a pipeline start date of 1 August 2004, and established a system of fines for failure to meet completion deadlines (US $90 million for a delay of commissioning of at least 90 days). A key milestone was the submission of the downstream environmental impact assessment (EIA) in September 2001. Geological field investigations along the pipeline route could proceed only after the EIA was submitted and had been approved by the Peruvian Government. Construction started in April 2002, and the system was commissioned in August 2004, following a two-year construction period.

A 3-km wide pipeline corridor had been preselected by the Ministry of Mines and Energy for the project. The corridor provided a general constraint within which the pipeline contractor was required to select and characterise a final route for construction. A process of micro-routing was carried out to define the route centreline within the corridor, taking into account a variety of constraints, such as archaeological, social, environmental, topographic and geohazards issues. However, in the Selva and parts of the Sierra, the options for routing within a 3-km wide corridor were severely restricted by the steep, dissected nature of the terrain. In many places, the preferred route (e.g. along very narrow ridge crests or plunging spurs) was the only feasible option within the corridor.

Given the very rugged nature of the terrain, there were numerous geohazard issues that should have been characterised and resolved during the micro-routing stage of the project. In an effort to compete the project within the project schedule mandated by the Ministry, this requirement was not fully achieved. However, the evaluation of geohazard issues along mountain pipelines requires considerable investment in geological and geotechnical studies prior to construction (e.g. Sweeney, 2005). This can have major schedule implications. International best practice indicates that this type of work for a remote region pipeline crossing challenging terrain would typically require 2–3 years of geological investigation, from reconnaissance-level overviews to support corridor evaluation, to detailed assessments of individual geohazards (e.g. landslides, soil erosion, seismic hazard, river-bed instability), as reported by Hengesh *et al.* (2004) and Lotter *et al.* (2005). On the Camisea project, the requirement that fieldwork could be undertaken only after completion of the downstream EIA (September 2001; 6 months prior to the start of construction, including a rainy season) severely restricted the time available for geological and geotechnical investigation. The time allocated for the pre-construction geological investigations was of the order of 3–4 months. It was inevitable that the project entered the construction phase without a clear definition of the requirements to mitigate geohazard risks along the selected right of way.

Following the rupture incidents, the project implemented a comprehensive field monitoring, maintenance and remediation effort to stabilise the pipeline right of way and prevent further damage to the pipeline. This has cost around $50 million (to January 2007) and involved establishing an ongoing programme of right-of-way maintenance and the development of a detailed landslide and soil-erosion monitoring programme. A risk matrix approach has been developed to identify and prioritise critical sites (e.g. Pettinger and Sykora, 2011).

The priority for the Camisea project was the supply of gas to Lima within 36 months. The use of gas instead of other fossil fuels by domestic and industrial users is reported to have contributed to improved air quality in the metropolitan area of Lima (IDB, 2007). However, this achievement came at the expense of sufficient landslide hazard studies and mitigation works to ensure pipeline system integrity. Clearly, the benefits of early gas supply need to be set against the consequences of adopting this approach, including the loss of revenue and environmental damage to an area of the Amazon rainforest, together with reputational damage to all of the companies involved and the costs of ongoing maintenance, monitoring and corrective measures.

The Camisea project also illustrates how landslide risk can be caused by both the nature of the terrain and the project process. It can be argued that the decision of the client to set what, in hindsight, could be regarded as an unrealistic schedule and associated penalty clauses, reflected a lack of awareness of the implications of their strategy for project delivery. As Baynes (2010) has written:

> Potent geotechnical risks often develop at a very early stage (of a project) if appropriate project-wide geotechnical risk mitigation measures are not implemented. When these risks develop it is usually because high-level decisions have been made by people who are overworked and under-resourced and/or who do not appreciate the importance of geotechnical risks, through a lack of experience, education or training; those people, unwittingly, become the source of the geotechnical risk.
> When the contract and accompanying documentation is inadequate, the source of the risk must be the project staff responsible for managing the procurement and production of the documentation. The reason that this occurs is usually an inadequate understanding of the importance of the geo-engineering aspects of the contract on the part of the project staff, or a limitation placed on those staff by a higher-level project management decision.

12.4. Risk assessment and management of extreme events: lessons from Fukushima, Japan

On 11 March 2011, Japan was struck by the moment magnitude (Mw) 9.0 Tohoku earthquake, the largest ever recorded in Japan and the fourth largest recorded worldwide since 1900. The earthquake generated a tsunami up to 15 m high that caused over 19 000 deaths and triggered the fuel meltdown and significant offsite release of radiation at the Fukushima Daiichi nuclear power plant.

The operating reactor units 1–3 were automatically shut down in response to the earthquake, as designed. As all six external power supply sources were lost due to earthquake damage, the emergency diesel generators in the turbine building basements began operating to cool the reactors. However, all power was lost when the site was flooded by the tsunami. This disabled 12 of 13 back-up generators on site, resulting in a complete loss of AC power for the units (i.e. station blackout). The plant was equipped with DC batteries to compensate for the station blackout; however, the batteries in units 1 and 2 were flooded and inoperable (the batteries in unit 3 continued to function for about 30 hours). The seawater pumps and their motors, which were responsible for transferring heat extracted from the reactor cores to the ocean (the 'ultimate heat sink') and also for cooling most of the emergency diesel generators, were built at a lower elevation than the reactor buildings. They were flooded and completely destroyed.

Over the next three days, the three reactor units lost core-cooling capability. Without cooling, the water in the reactor pressure vessels boiled, uncovering the fuel, which subsequently melted. As the reactors overheated and the fuel melted, highly flammable hydrogen was generated, leading to explosions in units 1 and 3. A large quantity of radioactivity from the damaged fuel escaped into the environment.

It is now widely recognised that the scale of the disaster was due, at least in part, to the underestimation of earthquake and tsunami risk to the site (e.g. Acton and Hobbs, 2012; Stein *et al.*, 2012). The Fukushima Daiichi reactors were built in the late 1960s and early 1970s. The design-basis tsunami was estimated to have a maximum height of 3.1 m above mean sea level. As a result, the plant's owners, TEPCO (Tokyo Electric Power Company), located the seawater-intake buildings at 4 m above sea level and the main plant buildings were 10 m above sea level.

A number of failings in the risk assessment process have been identified (Acton and Hibbs, 2012; Kingston, 2012):

- *An inadequate tsunami magnitude-frequency model*: the original design basis of 3.1 m for tsunami defences had been selected because this was the height of the tsunami generated by the 1960 Chile earthquake (Mw 9.5) (Onishi and Glanz, 2011). However, there was evidence for larger tsunamis on the Japanese coast. Since 1498, Japan and the adjacent Kuril Islands (Russia) have experienced 16 tsunamis higher than 10 m. The historical frequency of 16/513 years (0.0312/year) indicates that there is a 60% chance of a >10 m high tsunami occurring within a 30-year period (using a binomial model, see Chapter 5) (Mohrbach *et al.*, 2011).
- *Inadequate tsunami run-up modelling*: TEPCO's modelling indicated that the 3.1 m design-event tsunami would not run up significantly. Following improvements in run-up modelling in 2002, the design event was revised to 5.7 m, although no improvements were made to the defences. In 2008, the company performed preliminary modelling of an historical tsunami event (869 AD) which suggested that the run-up threat from the 5.7 m high event had been underestimated, potentially reaching over 15 m. However, the results were not considered reliable, and no mitigation action to improve the tsunami defences was undertaken.

A week before the disaster, TEPCO and two other utilities had lobbied Japan's Earthquake Research Committee to tone down the wording in a report warning that a massive tsunami could hit the Tohoku coast. Apparently the committee agreed to modify the report in accord with concerns expressed by the utilities that a stark warning about the possibility of a major tsunami might cause 'misunderstanding' among the public (*Japan Times*, 27 February 2012).

Concerns have also been raised about the role of the regulator (the Nuclear Industry Safety Agency (NISA)). The regulatory guidelines did not set out risk-acceptability criteria or what measures TEPCO should undertake to protect the plant. Acton and Hibbs (2012) wrote that:

> the clearest warning signs of potential risk before the accident were procedural: Japan's methodology for assessing tsunami risks lagged markedly behind international standards, TEPCO did not even implement that methodology in full, and NISA showed little concern about the risks from tsunamis.
> NISA lacked independence from both the government agencies responsible for promoting nuclear power and also from industry. In the Japanese nuclear industry, there has been a focus on seismic safety to the exclusion of other possible risks. Bureaucratic and professional stovepiping made nuclear officials unwilling to take advice from experts outside of the field. Those nuclear professionals also may have failed to effectively utilise local knowledge. And, perhaps most importantly, many believed that a severe accident was simply impossible.

Kingston (2012) has suggested that a 'myth of safety' contributed to the way the disaster developed:

> The myth that nuclear reactors could be operated with absolute, 100% safety embraced and promoted by what the Japanese call their 'nuclear village' of pro-nuclear power advocates made it taboo to question safety standards and militated against sober risk assessment and robust disaster emergency preparedness. Those responsible for operating or regulating nuclear reactors bought into a myth of 100% safety and this collective failure left them unprepared to deal with an accident or worst-case scenario. Paradoxically, this safety myth explains why TEPCO lacked a culture of safety and why it's crisis response was so deficient.

Lessons had not been learnt from near misses experienced by the industry elsewhere. In December 1999, a storm surge exceeded the design-basis flood scenario for the Blayais Nuclear Power Station in France, causing flooding at two units and a partial loss of power (Mattéi et al., 2001). Investigations revealed that the defences had been too low and that rooms containing emergency equipment were insufficiently protected from flooding. It was recognised that there had been a systemic failure in hazard assessment across the nuclear industry in France. Power plants were ordered by regulators to identify all phenomena that could cause a flood, and to reassess site-specific flood management protection with regard to loss of off-site power, communications and heat sinks. Plant safety was also reassessed for the postulated case that a combination of extreme natural phenomena could simultaneously threaten any given plant.

Although it was triggered by the 11 March earthquake and tsunami, the Fukushima disaster occurred because of a failure in risk assessment and management. Stein *et al.* (2012) have suggested that it may have been a 'Black Swan' because of the unexpected nature of the event. With hindsight, it is clear that the potential for such an event had been ignored, despite the available historical evidence. This reluctance to entertain the possibility of extreme events is unlikely to have been unique to the Japanese nuclear industry. It is almost certain that landslide-risk specialists will be challenged by their colleagues or clients to limit the assessment to those events that are 'credible' under current circumstances, and ignore those that are believed to be 'simply impossible'. The probability of these Black Swans, Perfect Storms and Dragon Kings (see Chapter 1) may be impossible to predict but their consequences are not; they should have a place on the risk register.

12.5. Landslide management and the environment: lessons from Easton Bavents

The Easton Bavents cliffs on the Suffolk coast, UK, are developed in Norwich Crag sediments, consisting of the mixed sand–shingle of the early Pleistocene Westleton Beds and the clay with sand laminae of the Easton Bavents clay. The cliffs form part of the Suffolk coastal Site of Special Scientific Interest (SSSI) due to their geological conservation interest. The cliffline has experienced very severe recession over the last century, with average recession rates of the order of 2–3 m/year (e.g. Lee, 2005).

By the end of the last century it had become clear that a number of cliff-top properties would be lost within the next decade or so. Government-funded coast-protection works were not considered viable on economic grounds. However, a local homeowner, Peter Boggis, decided to protect his own property and those of his neighbours at his own expense. Between 2002 and 2005, Mr Boggis built an 8 m high soil-fill revetment that extended about 15 m seawards from the cliff foot, using a combination of building rubble, excess soil from nearby road construction works and other sources of material. The seaward edge of the revetment was intended to erode and maintain the supply of coarse sediment inputs to the nearby beaches of Southwold. The fill material lost each year would be replaced as part of an ongoing maintenance programme.

However, he did not have planning consent for the revetment which, in the UK, is viewed as a 'development' and therefore requires permission from the local planning authority, and, because it potentially affected an SSSI, also requires consent from Natural England, the government's conservation advisors on the natural environment. Natural England objected to the works because of the potential impact on both the geological interest of the cliffline and the nature conservation interest of the adjacent low-lying marsh and lagoon areas. Following a series of complex court cases and appeals, Mr Boggis was ordered to allow the cliffs to erode naturally.

In October 2009, the Appeal Court judge Lord Justice Sullivan stated (Judd, 2009; Savill, 2009):

> I am not unsympathetic to the plight of Mr Boggis and the other residents who can see the cliff face remorselessly approaching the boundaries of their properties. The lawful course would be to apply for planning permission and coastal protection consent so that all material considerations, including their human rights and the SSSI, could be taken into account.

Ever the realist, Mr Boggis responded after the hearing by saying (Judd, 2009):

> I am not walking into the trap of wasting tens of thousands of pounds on a beleaguered planning application that Natural England will convince the council not to give.

The dispute at Easton Bavents illustrates how environmental conservation has become an increasingly important factor in determining the way in which coastal landslide risks are managed in England. Lee (2000) considered that the UK's commitment to the EC Habitats and Species Directive (Council Directive 92/43/EEC) and the Convention on Biological Diversity (the Rio Convention) has effectively introduced a 'no net loss' policy for maritime cliff and slope habitats, with the aspiration of actually achieving, over time, a 'net gain'. These commitments and aspirations could lead to the situation where some existing coast-protection and slope-stabilisation works will have to be removed in the future in order to ensure that biodiversity targets are met. Indeed, a coast-protection scheme that might affect the integrity of the habitats would only be approved if there were imperative reasons of overriding public interest. In such circumstances, compensation measures would be required as part of the scheme; that is, the creation of replacement vegetated sea-cliff habitat. As the only viable option for recreating vegetated sea cliffs of international quality is to 'restore' natural habitats at previously protected cliffs, for every length of new defences there would have to be an equivalent abandonment of existing defences (e.g. Lee *et al.*, 2001).

A major obstacle to the UK achieving its environmental targets is the contrasting attitudes to coastal erosion within British society. Many feel that loss of land is unacceptable and needs to be resisted by public investment in coast protection. This *fortress Britain* attitude is in marked contrast to the view that the erosion process is necessary for maintaining the natural beauty of the coastline (the *living coast* view). To this latter group, coastal defence leads to environmental degradation, and should only be contemplated where there is an overriding national need. The 'living coast' view reflects current legislation and government policies. However, the 'fortress Britain' view has considerable popular support, ensuring that there will be a continued demand for coast-protection schemes and a resistance to the abandonment of current defences (e.g. Lee, 2000, 2002; Lee *et al.*, 2001).

As described in Chapter 11, conventional approaches to risk reduction generally involve seeking an acceptable balance between the costs of the management measures and the benefits (i.e. prevention of loss of life, injury or economic losses). Risk criteria, such as the ALARP principle, are often used to indicate whether or not a risk should be tolerated. The level of expenditure on risk-reduction measures should be proportional to the level of risk. However, the increasing importance of environmental legislation and international commitments appears to be placing further constraints on landslide management. The environmental evaluations that form part of the ecosystem-services approach described in the following section may provide a solution to this problem.

12.6. Future uncertainty: valuation of environmental resources

Over the past decade or so, there has been a major shift in focus towards the comprehensive examination of the ways in which the 'natural environment' is of benefit to humankind and, as a result, the way in which landslide risks might be managed. In this context, the term 'natural' indicates non-human or non-anthropogenic, rather than that it is in its natural state or pristine, which in the majority of areas it most clearly is not (a better term is 'biophysical').

The fundamental unit of the 'natural'/biophysical environment is the *ecosystem*; that is 'a natural unit within which there exists a complex inter-relationship among organisms and between organisms and their physical surroundings'. Ecosystems provide benefits or services to human populations in a great variety of ways, hence the growth in the use of the term *ecosystem services*, meaning 'ecosystem processes and functions that have value to individuals and society'. These 'services' provide outputs or outcomes that directly and indirectly affect human well-being. They take many forms, but the four main groups of 'services' identified are (Defra, 2007):

- *provisioning* (i.e. products obtained from ecosystems), e.g. production of food, fibre, biochemicals, genetic materials and ornamental resources (e.g. flowers)
- *regulating*, e.g. the role of ecosystems in controlling climate, pollution and disease, beneficial influences on water quality, run-off rates (flooding), soil-erosion control and 'natural hazard' protection (landslides)
- *supporting*, e.g. soil formation and retention, water and nutrient cycling, crop pollination, oxygen formation, creation of habitat
- *cultural*, e.g. recreational, aesthetic (including landscape), artistic, spiritual and cultural (heritage) benefits.

To which a fifth can be added:

- *preserving*, e.g. maintenance of biodiversity.

While there is, as yet, no single, agreed method of categorising all ecosystem services, the United Nations (UN) Millennium Ecosystem Assessment (MA) (World Resources Institute, 2005) framework presents a useful starting point. The MA was established in 2001 following the publication of an earlier UN report *People and Ecosystems: The Fraying Web of Life* (World Resources Institute, 2000), which concluded that the world's major ecosystems were in decline, and that even beginning to tackle this serious problem was hindered by limited information on ecosystem functioning. In response, the MA addressed five key questions.

- How have ecosystems and their services changed over the recent past?
- What has caused these changes?
- How have these changes affected human well-being?
- How might ecosystems change in the future, and the implications for human well-being?
- What options exist to enhance the conservation of ecosystems and their services to maintain human well-being?

The resulting *MA Synthesis Report* (World Resources Institute, 2005) reflects the work of over 1300 scientists from 95 countries, and reported significant and escalating impact. More specifically, it reported that

- over the past 50 years, humans have changed ecosystems more rapidly and extensively than in any comparable time in human history, largely to meet rapidly growing demands for food, fresh water, timber and fuel
- the changes that have been made to ecosystems have contributed to substantial net gains to human well-being and economic development, but these gains have been achieved at growing costs in the form of the degradation of many ecosystem services, increased risks of non-linear change, and the exacerbation of poverty for some groups of people; the costs and benefits are increasingly skewed
- the degradation of ecosystem services could grow significantly by 2050 and is a barrier to achieving each of the Millennium Development Goals (MDGs)
- reversing the degradation of ecosystems while meeting the increased demands for their services can be partially achieved by some of the future scenarios considered, but only if there are significant changes in policies, institutions and practices that are not currently under way.

Set against the background thrust of *sustainable development*, the MA made for gloomy reading. In many nations there were moves to incorporate measures for ecosystem evaluation and conservation in planning, policies and practice, but to do this effectively required some progress on valuation. For some years now research has been underway to develop methods of valuing the different types of service (Pearce et al., 2006) with a view to estimating the *total economic value* (TEV) of an environmental asset and, therefore, the cost (or benefit) of environmental change. TEV decomposes into 'use' and 'non-use' (passive) values, but does not encompass intrinsic values (defined as residing 'in' the asset and unrelated to human observation or preference). The stated aspiration is that 'any project or policy that destroys or depreciates an environmental asset needs to include in its costs the TEV of the lost asset. Similarly, in any project or policy that enhances an environmental asset, the change in the TEV of the asset needs to be counted as a benefit' (Pearce et al., 2006). Within the UK the result was that an introductory guide to valuing was issued by the Department of Environment, Food and Rural Affairs (Defra) in 2007, with additional detail contained within HM Treasury's *Green Book: Appraisal and Evaluation in Central Government*.

Despite these advances, there still remains considerable uncertainty as to how ecosystems function internally, and what they do in the way of life-support systems, so the ultimate goal of environmental evaluation may still be some way off. For example, the UK has only recently published its ecosystem assessment showing what major ecosystems exist, how they benefit society and the extent to which they have changed over the recent past (UK Natural Ecosystem Assessment, 2011), but work on valuation is only just beginning. Calculation of TEV is also problematic, involving as it does consideration of two main components: *use* and *non-use values* (Defra, 2007).

1. The *use value*, consists of three components:
   - the *direct-use value*, which consists of the value of consumptive use (e.g. resource extraction) and non-consumptive use (e.g. recreation, landscape amenity)
   - the *indirect use value*, which is the benefit obtained from ecosystem services not generally recognised (i.e. taken for granted) until they are reduced or lost (e.g. climate regulation, pollution filtering, soil retention, water regulation, pollination)

- the *option value*, which is the value that people place on having the option to use a resource in the future, or preserve a resource for the future, even though they may never have actually used the resource (it can be thought of as a form of insurance and figures prominently in sustainable development).
2. The *non-use value* is derived from the mere knowledge that the existence of the biophysical environment is maintained, and consists of three parts:
    - the *bequest value*, where individuals place value on the fact that the environmental resource will be passed on to future generations
    - the *altruistic value*, where individuals place value on the contemporary availability of the resource to others
    - the *existence value*, the willingness to pay displayed by people to ensure the mere continued existence of an ecosystem resource, despite the fact that they have no planned use of it (whales are the classic example).

Defra acknowledges that non-use values are 'relatively challenging to capture since individuals find it difficult to "put a price" on such values as they are rarely asked to do so' (Defra, 2007). It also indicates that 'non-use values' often exceed 'use values', including one reported instance where they formed 99% of the overall value!

Finally, consideration has to be given to *quasi-option value*, which is the 'value of information secured by delaying a decision where outcomes are uncertain and where there is opportunity to learn by delay' (Defra, 2007). Although outside the TEV framework, Pearce et al. (2006) have concluded that the quasi-option value is especially relevant in the context of ecosystems where there is uncertainty due to lack of knowledge, irreversibility and a major chance to learn through scientific investigation exactly what specific ecosystems do and how they behave.

Currently, the UK is at the stage of researching values for individual ecosystem components and services. However, when valuations are achieved, incorporation of environmental valuations in risk assessment is likely to follow. Although including the valuation of ecosystem services may well prove to be a problematic and time-consuming extension of landslide risk assessment from the perspective of the practitioner, it is undoubtedly important that the needs of environmental conservation become included on a similar basis as other elements at risk.

12.7. Future uncertainty: global change

One of the major challenges facing landslide specialists is the increasing need to take account of medium- to longer-term change in environmental conditions in risk assessments. This challenge has become of increasing importance over the past two decades, as it is now generally recognised that the global environment has been changing, the pace of change is predicted to accelerate, and the future is very uncertain. As a consequence, since the 1990s, the term 'global environmental change' has been progressively replaced by 'global change', meaning 'The full range of global issues and interactions concerning natural and human-induced changes in the Earth's environment.'

This emphasis on the complexity of interaction between societal and biophysical processes has resulted in the recognition of two broad sets of influences, both of which owe much to human (anthropogenic) activity (Turner et al., 1990).

- *Systemic* changes, which directly influence the operation of the global climatic system by affecting the global energy balance (e.g. changing levels of greenhouse gases or changing

reflectivity (albedo) due to land-cover changes such as deforestation and urbanisation, or variations in aerosol concentrations in the atmosphere)
- *Cumulative* changes due soil loss, local pollution, resource depletion (i.e. forest clearance), urbanisation, etc.

The systemic influences operate at a global scale (i.e. global climate change) and result in 'top-down' changes that are imposed on areas irrespective of the extent to which they may have contributed to the change. The cumulative influences, on the other hand, are essentially local but may exacerbate the effects of climate change on particular areas or even, if extensive, contribute to additional climate change ('bottom up').

It is important to appreciate this interrelationship between the 'global' and the' local' in order to understand both the causes and consequences of environmental change. Global change is the complex product of a wide range of anthropogenic and 'natural' biophysical influences and its affects are similarly varied over time and space. What this means is that 'headline' predictions of global-level changes in temperature, rainfall, etc., by specific future dates have little value when considering changes at specific locations. Indeed, it must also be recognised that future local changes to environmental conditions, such as deforestation, agricultural development and urbanisation, may have greater significance in terms of future slope stability conditions within an area than those changes imposed by the climate-change regime; sometimes overwhelming it, sometimes nullifying it and sometimes exacerbating it.

12.8. Evidence for climate change

The global climate has displayed a general warming trend since about 1910 (IPCC, 1990, 1995a,b, 2001, 2007, 2012), often largely attributed to the accumulation of so-called 'greenhouse gases' (GHGs), such as CO_2, in the atmosphere. Evidence from Antarctic ice cores, supplemented by direct measurements since the mid-1950s, reveals a gradual rise in GHG concentrations from the late 1700s, changing to a more rapid and accelerating rise post-1850. Current additions to GHG concentrations continue to be dominated by CO_2, which has increased from a pre-industrial revolution figure of about 280 ppm by volume (bv) (275–284) to approaching 400 ppm bv (392 in 2011) and currently rising at >2 ppm/year. However, to this has to be added the combined effect of the other human-produced/accentuated GHGs (methane (CH_4), nitrous oxide (N_2O) and chlorofluorocarbons (CFCs)) which, when expressed in terms of CO_2 equivalence (CO_2e), raise the current (2010) level to 435 ppm. But just how much of the global temperature increase of about >0.8°C since 1910 is due to anthropogenic GHGs (i.e. anthropogenic global warming) remains disputed. Some have argued that anthropogenic influences on climate began 8000 years ago, while others point to the industrial revolution as the starting point, but, either way, the question remains as to why does it only become apparent post-1910?

Use of the term 'anthropogenic global warming' has proved problematic. The current phase of warming (post-1910) displays a non-linear relationship between GHG accumulation and temperature rise. Indeed, the global temperature curve displays marked irregularity, with pronounced warming over the years 1910–1940, then a slight fall back of temperature until the late 1970s (the 'mid-20th century pause'), followed by renewed warming from the late 1970s to about 2000, at which point there began a second 'pause' or 'plateau' to the present. As GHG concentrations have displayed a fairly regular, if progressive, increase since the beginning of the twentieth century, then why have there been these two surges of warming and two pauses?

The first phase of warming (1900–1940) has been shown to owe much (~60%) to 'natural factors', including increased energy receipts from the Sun (solar insolation). The pause in global temperature rise from the 1940s to the 1970s is claimed to be a period when the effects of GHG-induced global warming were wholly masked by the cooling influence of high concentrations of sulphur aerosols in the lower atmosphere, partly assisted by a slight fall in solar insolation. As a consequence, the effects of anthropogenic global warming only appear to become marked since the late 1970s, although some argue that a small part of this increase was again due to increased solar radiation receipts. The relatively abrupt termination of this second warming phase and the commencement of the so-called 'post-1998 plateau', appears to indicate the continuing influence of biophysical processes, and has recently (Kaufmann et al., 2011) been explained as being due to a combination of

- the formation of extensive smoke-generated smogs in China, which have increased albedo (reflectivity) due to sulphate particles (aerosols)
- the onset of a reduced solar radiation phase ('quiet sun')
- the switch from predominantly El Niño (warming) to La Niña (cooling) conditions

to which has recently been added the possibility of increased absorption of energy by the oceans.

It is clear that separating natural influences from those produced by anthropogenic GHGs remains problematic, pointing to

- anthropogenic GHG accumulation not being the sole influence on recent climate change
- the complex relationship that exists between GHG forcing and resultant climate response, especially the effect of feedback mechanisms, some of which counteract the effects of GHGs (negative feedback), while others can greatly reinforce the effects (positive feedback).

Thus, predicting future climate change is problematic, even if future GHG concentrations could be determined, because of the existence of other factors and the complexity of the climate system. How near-future global increases in temperature are distributed spatially is likely to be very variable. Some areas may become much warmer, others a little warmer, while yet others experience little change and some (e.g. parts of Europe) could even, under certain circumstances, become cooler. The timing of change may also not be synchronised across the globe. This variability in response is likely to be even more pronounced in the case of future precipitation amounts, which are greatly influenced by local- and regional-scale factors. Hence future climate change is likely to continue to be both non-linear and spatially variable.

Evidence of this spatial variation has recently be shown with respect to the early twentieth century warming phase (i.e. 1910–1940), which appears to owe much to the disproportionately rapid warming of North America and the Arctic (Bronnimann, 2009), while the recently completed Berkeley Earth Surface Temperature Study (Muller et al., 2011), which independently examined all available temperature records, concluded that the air over the land areas has warmed by an average of $0.911°C$ since the mid-1950s, effectively settling the arguments over the validity of temperature records and the existence of global warming. The latter result clearly shows why the temperature increase in the northern hemisphere ($\sim 0.95°C$, 1910–2005) has been greater than that of the ocean-dominated southern hemisphere ($\sim 0.65°C$).

12.9. Future uncertainty: climate model predictions

Climatic modelling has improved rapidly over recent decades (see Houghton, 2009), and there exist numerous published multi-coloured maps of future predictions of temperature, precipitation, etc., which give an impression of exactitude and accuracy (see World Bank, 2010). However, it must be recognised that, in reality, this is still a far from a perfect science because of uncertainty and ignorance. According to Schiermeier (2010), there are four main areas of uncertainty with respect to scientific predictions.

- An inability to agree on the pattern and causes of climate change over the last 1000 years. This is crucial, because all numerical models have to be tested against the recent past (usually to 1860 at present) to prove their credibility, so if we cannot explain the past how can we possibly predict the future?
- Imperfect scientific knowledge regarding the operation of the geosystem, i.e. exactly how the atmospheric and oceanic systems operate, interact and respond to change.
- Model deficiencies.
- Uncertainty regarding human behaviour and the future production of GHGs. This is crucially important. The Inter-Governmental Panel on Climate Change (IPCC) employs a range of 'emission scenarios' (35), divided into four main groups (families) which seek to cover the range of possible spatial and temporal patterns of population growth, economic growth and response to the global warming threat in terms of take-up of mitigation.

Until recently there currently about 30 global climate numerical models, which could be fed with any one of 35 emissions scenarios, differing values of geosystem behaviour and one of three climatic sensitivities (i.e. the global average temperature rise produced by a doubling of CO_2: 'low' = 1.5°C, 'best estimate' = 2.5°C and 'high' = 4.5°C, reflecting the possible significance of reinforcing (feedback) mechanisms within the geosystem). When these were combined they result is huge uncertainties as to the likelihood of outcomes over time, so that the mapped results must be treated as an *indication* rather than *fact*. Indeed, there is a growing tendency to produce maps of the averaged results of a number of models, or simply to show averaged values for areas where a significant proportion of model predictions agree. The adoption of a new set of scenarios (Representative Concentration Pathways – RCPs) for IPCC (2013) will reduce but not eliminate these problems.

Bearing all of the above in mind, the current predictions of relevance to landsliding are as follows.

- According to the IPCC (2013), the mean global temperature by 2100 AD will have risen by a further 0.3–4.8°C (likely range 1–3.7°C, with most agreement in the range 2.5–3°C, representing a downward shift since the last report). Note that land areas will warm more rapidly than the oceans, and that warming at the higher end of the likely range would shift some climatic zones pole-ward by about 550 km.
- Warming is likely to be spatially variable, but there is increasingly good evidence that high latitudes in the northern hemisphere will experience the greatest warming, so that the decrease in ice, snow and permafrost will be most marked in these areas (IPCC, 2012). The record reduction in Arctic sea ice in August 2012 is clear evidence of such changes; changes that will have significant implications for slope stability due to the progressive melting of permafrost to produce increasingly thick active layers within the ground and increasing frequency of freeze–thaw cycles. Within these areas, falls, debris flows (Rist and Phillips, 2005; Zimmermann *et al.*, 1997) and solifluction movements will all increase in frequency and magnitude.
- High mountain belts are likely to experience similar conditions, and display increased frequencies of falls and slides, as well as increased glacial lake outbursts. According to a major review (IPCC, 2012), the main drivers are glacier ice-mass loss, permafrost degradation and increasing intensity of precipitation, with evidence for an increase in the frequency of large rock slides, but, as yet, there is limited evidence for the anticipated increase in debris-flow activity (but see below).

- At latitudes of 45° or greater, the trend will be to warmer and wetter regimes, with annual precipitation increasing by 100–300 mm, mainly due to increased winter precipitation (i.e. northern Europe, Russia, China, northern and central USA, Canada and the southern extremes of South America). Heavy precipitation events will also increase in frequency in high latitudes and in northern mid-latitudes in winter (IPCC, 2012).
- At lower latitudes (5–45°), annual precipitation could well decrease by 100–700 mm in many areas (i.e. parts of southern and western Australia, southern Africa, southern USA, western South America, Central America and Mexico, north Africa, the Mediterranean region, the Middle East and India), and many of these areas may become increasingly drought-prone (IPCC, 2012) (i.e. southern Africa, central USA, north eastern Brazil, central America and Mexico, the Mediterranean region, central Europe), although there is much uncertainty. Many areas will experience an increase in the proportion of rainfall coming from heavy precipitation events, especially in tropical regions, over much of North America and in those areas impacted by tropical cyclones (IPCC, 2012).
- Within 5° of the equator there is currently some uncertainty. Both the Andes and east Africa are likely to become wetter, while west Africa and the eastern Amazonian basin could become drier (IPCC, 2012).

However, it is not just the climate that could change – the weather may also change too. For many years it has been assumed that a warmer atmosphere capable of holding more water vapour would generate more latent heat release on condensation, and thereby result in increases in the frequency, power and spatial distribution of tropical revolving storms (i.e. hurricanes, typhoons, cyclones). However, the IPCC (2012) found limited evidence for this, and supported Knutson *et al.* (2010) to conclude that 'it is *likely* that the mean maximum wind speed and near-storm rainfall rates of tropical cyclones will increase with projected twenty-first century warming, and it is *more likely than not* that the frequency of the most intense storms will increase substantially in some basins, but it is likely that overall global tropical cyclone frequency will decrease or remain essentially unchanged'. However, as shown above, the IPCC (2012) did find evidence for the increasing occurrence of 'extreme weather' (i.e. storms and intense or prolonged rainfall events), and this clearly has implications for landslide risk. Therefore, there are likely to be spatially variable changes to rainfall intensity (a rather general phenomenon), proneness to drought (southern Africa, central South America, central America and western USA, Australia, the Mediterranean region, although confidence in these predictions is low) and amounts of run-off (increase in high latitudes and mountainous areas, possible major decrease in the Mediterranean region and much of southern Africa).

The predicted changes in climate and weather detailed above will undoubtedly lead to changes in the probability of landsliding in many areas, and hence landslide risk. Two main aspects are worthy of further consideration: increasing rainfall and sea-level change.

12.10. Increased rainfall

Precipitation has increased by 0.5–1% per decade in the twentieth century over the northern hemisphere continents, together with varying changes in seasonality and a 2–4% increase in the frequency of heavy precipitation events, and the IPCC (2012) confirms that extreme weather conditions will increase in frequency into the future. Such changes in rainfall totals, intensity and seasonality will have implications for landslide hazard and risk, but to evaluate the changes in hazard will require detailed predictions at regional and local scales, because rainfall intensities and totals display marked local variations as determined by predominant wind direction and topography. Such predictions are increasingly available, and, although they suffer from problems, they will improve over time (see

Osborn and Hulme, 2002; Senior *et al.*, 2002). However, it is also important to recognise that the triggering of shallow slides is the consequence of varying combinations of patterns of rainfall (both spatial and temporal), character of rainfall (intensity and duration) and antecedent rainfall conditions (Iverson, 2000; Sidle and Ochiai, 2006; Wieczorek *et al.*, 2005), so that in certain areas intensity and duration are most important (Jakob and Weatherly, 2003), while in others it is antecedent rainfall (Kim *et al.*, 1991; Glade, 1998). Thus, simply applying increased precipitation may not be sufficient.

There are, however, major uncertainties associated with the prediction of future rainfall characteristics due to climate change, especially at the local scale, arising from limitations of knowledge, limitations of available global circulation models and the coarseness of the input grid, choice of emission scenarios, the 'climate sensitivity' used, the vagaries of future human behaviour and the variable influence of local factors. For example, the predictions for southern England are for a trend towards wetter winters and drier summers by 2060. Analysis of UK rainfall records for 1961–2000 confirm this trend to be very strong, with increased winter rainfall being due to increased wet-day amounts rather than increased numbers of wet days, while the summer results show both fewer wet days and reduced wet-day amounts (Osborn and Hulme, 2002). However, it is unclear whether these changes are due to climate forcing or changing regional-scale atmospheric circulation patterns, possibly influenced by climate forcing, i.e. the extent to which they represent a long-term trend (Osborn and Hulme, 2002).

Such uncertainty is compounded by speculations about the possible impact on slope stability and future levels of landslide hazard and, as a result, landslide risk. However, it can be argued that the changes in landslide hazard due to global climate change will probably be less (at least in the short term) than those arising from population growth, development changes and alterations in land use. Nevertheless, three situations can be envisaged where landslide risk could increase.

- Increased wetter periods (i.e. cumulative rainfall) in especially sensitive areas such as urban zones developed on old and extensive landslides (i.e. slid areas) prone to reactivation.
- Higher rainfall totals and increased intensity of rainfall events raising the frequency and magnitude of flood discharges of rivers (Cox *et al.*, 2002), thereby causing undercutting of unprotected bluffs or the reactivation of landslides. It is interesting to note, however, that a recent review of flood risk in Europe (Kundzewics, 2012) found that, while damages are undoubtedly rising, 'no unambiguous, general and significant changes in observed flood flows can be detected, even at a national scale'. While not refuting the general link between increasing rainfall and increasing flooding, this does indicate the prominent role of flood-control management in limiting high flows in regulated rivers.
- Landsliding could be increased by accelerated run-off generated by intense rainfall on deforested or otherwise human-modified slopes. Two examples from Campania, Italy, illustrate this phenomenon. First, some 300 shallow soil slides and debris flows were generated during the winter of 1996–1997 (Calcaterra and Guarino, 1999) which, when added to the historical record of landsliding in the area, revealed a pattern of 'migration' from district to district, indicative of human involvement (Anon., 2010). The second, and more serious, were the widespread and large debris flows that were generated by intense rain falling on pyroclastic mantled limestone on 4 and 5 May 1998, some of which extended 3–4 km onto the adjacent lowlands, severely damaging the towns of Sarno, Quindici, Bracigliano and Siano, and killing at least 161 people, apparently partly due to the creation of trackways (Del Prete *et al.*, 1998).

Taken together, these show that landslide hazard will change in the future in response to changing rainfall conditions (intensity, frequency and duration), sometimes quite dramatically. But it is unwise simply to assume that changes in landslide hazard will only result from increases in rainfall amounts. Areas that become increasingly drought-prone could also display increased hazard due to changes in landslide type, e.g. a trend towards the increased frequency of debris flows and mud floods if the reduced precipitation were to be associated with increased intensity of precipitation falling on ground with reduced vegetation cover. The devil is in the detail!

12.11. Sea-level change

The prospect of rising sea levels is generally perceived to be the primary threat posed by global warming, due to the possible inundation of low-lying coastal areas and accelerated rates of cliff retreat. In these cases, the concern is with *relative sea level* or the relationship between the surface level of the sea/ocean and the adjacent land. Clearly, if the relative sea level rises, then coastal cliffs and defences will come under increasing attack, and slope failures will increase in magnitude and frequency, thereby increasing landslide risk.

The current prediction (IPCC, 2013) is that global sea level will rise by 0.26–0.98 m by 2100 AD (most likely range 0.4–0.62 m), figures based on assessing the various possible contributions to sea-level change by scenario and then evenly distributing the effect across the Earth's oceans, But relative sea-level change does not work like this. It is location specific, so a global figure of say 0.6 m is misleading. This is because change in relative sea-level (RSL) is the result of the interaction between two variables (Figure 12.1): the changing mean level of the sea surface (loosely termed *eustatic* movements), and the vertical movements of the adjacent land (loosely termed *tectonic* movements).

The *eustatic* movements are actually influenced by seven components, but only six are relevant for a 100-year time-frame.

- *The changing volume of water in the ocean basins* due mainly to the growth and shrinkage of ice sheets (*glacio-eustasy*). This can be very rapid – during the last deglaciation phase, the global sea level is considered to have risen 110 m in 14 000 years (~7.9 mm/year but with peak rates of up to 42 mm/year), largely due to the release of glacial melt water – but currently is <1 mm/year.
- *The displacement of water due to sediment deposition in oceans (sedimentary eustasy)*.
- *The changing density of ocean water due to heating or cooling*, which produces changes in surface elevation because of volumetric changes (small but important *steric* effect (i.e. change due to the rearrangement of molecules) estimated to cause 30–55% of rise by 2100 (IPCC, 2013)).
- *Minor spatial changes in the strength of gravity* due to tectonic forces (poorly understood *steric* effect).

Figure 12.1 Components of relative sea-level change

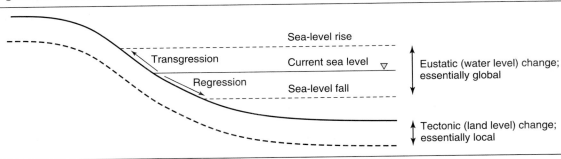

- *Change in terrestrial water storage* due to climate change, volumetric reduction of frozen stores, land-cover change, more efficient surface drainage, reduction in water storage and exploitation of underground water reserves. It is currently positive (i.e. is contributing to eustatic rise).
- *Atmospheric influences*, such as changes in atmospheric pressure (a *steric* effect where 1 mbar change in pressure = 1 cm change in water surface elevation), or the accumulation/removal of water along coastlines due to winds or changes in the pattern and strength of ocean currents (all of which are particularly noticeable in the changing short-term sea-level response of the Pacific to the El Niño Southern Oscillation).

The *tectonic* movements that affect coastlines, on the other hand, result from

- *tensional and compressional movements of the crust* (widespread and spatially variable in terms of rate and direction, sometimes abrupt)
- *isostatic movements* of the crust in response to loading (e.g. sedimentation (delta accumulation), growth of volcanic edifices, ice accumulation) and unloading (e.g. ice-sheet melt (a regional phenomenon currently up to 10 mmpa), erosion)
- *diapiric* movements
- *subsidence* due to the withdrawal of underground fluids or gas, or compaction due to loading.

The variable interaction of these factors results in complex patterns of contemporary sea-level change (RSL) in terms of both the *rate* and *direction* of change. For example, if the *sea-surface level* factors in an area combine to produce a positive trend (say $+X$ mm/year), and the coast is stable, then sea-level change (RSL) will be $+X$ mm/year and display a *transgressive* tendency. However, should there be an adjacent short section of coastline beyond an active fault which is physically sinking at $-Y$ mm/year, then the apparent rate of RSL rise in this area will be $X + Y$ mm/year (i.e. *more rapidly transgressive*). But if, in the other direction, the stable section of coast passes laterally into a stretch of coast experiencing tectonic deformation characterised by increasing rates of uplift with distance, then the RSL will appear to decline with distance up to the point where rates of eustatic rise and tectonic rise are equal (i.e. there is no net change (*still-stand*)). Beyond this point, rates of tectonic uplift are increasingly greater than the rate of eustatic rise, so the net effect is apparently falling RSL (*regression*) at progressively increasing rates with distance. Change in RSL has traditionally been measured by means of tide-gauge installations located at specific points along coastlines, which record the changing elevation of still water surfaces within structures. The continuous records produced take the form of graphs showing the effects of tides, surges, atmospheric-pressure changes and wind effects on the sea-surface level, which are analysed to yield annual maximum, minimum and mean values for each station. Time series of annual values for each station yield further irregular graphs due to climate variability, from which the rates and directions of sea-level change (mean values) and high-water (maximums) can be determined. Comparing the results for stations reveals marked spatial variability, which is sometimes due to human activities near particular sites that alter the coastal hydrography, resulting in changes in the patterns of tidal flows and high-tide levels. Despite such problems, tide-gauge records have revealed three things.

- That removal of deduced tectonic influences yields a picture of a globally rising sea level (transgression). Indeed, the rise has accelerated from ~1.12 mm/year in the mid-20th century, to an average of 1.8 mmpa (1.3–2.3 mm/year) for the period 1961–2003 and 3.2 mm/year (2.8–3.6 mm/year) for 1993–2010 (IPCC, 2013), indicating a rate of approximately 3 mm/year at the start of the current century, although Jevrejeva *et al.* (2006) estimate the 1993–2000 rate as ~2.4 mm/year.

- General agreement that the rise in global mean sea-surface elevation during the 20th century was 0.17 m, despite a 2007 study concluding that this should be reduced to 0.13 m following a reappraisal of tide-gauge sites using GPS data.
- That three main zones of RSL change can be identified (Clark et al., 1978; Peltier, 1987):
  - formerly glaciated areas suffering isostatic rebound are characterised by RSLs that are falling, stationary or rising slower than the prevailing global eustatic rate (e.g. Scotland, Scandinavia, northern North America)
  - broad belts bordering the above areas are characterised by physical sinking due to the collapse of the peripheral bulges bordering the former ice sheets, so that relative sea-level rise is amplified (e.g. south-east England and Holland)
  - the remaining areas where 'eustatic' rise is variably modified by local tectonic conditions.

However, recent work has reassessed tide-gauge data and revealed some interesting conclusions (Holgate, 2007; Jevrejeva et al., 2006; Woodworth et al., 2009).

- Holgate (2007) has shown that sea-level rise (transgression) over the 20th century (Figure 12.2) was actually irregular, with periods of decelerated rise culminating in five brief periods of minor fall (Figure 12.3); irregularities possibly arising from the impact of climatic oscillations such as El Niño.
- Jevrejeva et al.'s (2006) estimate of the 1993–2000 rise as ~2.4 mm/year actually turns out to be less than the peak rate of 2.5 mm/year that they reported for the period 1920–1945. Holgate's (2007) sea-level curve (see Figure 12.2) also shows that global sea-level rise was actually faster prior to 1960 than post-1960, although the detailed analysis showed that, despite the highest decadal rate of sea-level rise being centred on 1980, a longer-term consideration shows that the rate of rise was actually greater between 1904 and 1953 (~2.03 mm/year).

Thus tide-gauge records present a conflicting record of sea-level change.

Figure 12.2 The mean global sea-level record over the period 1904–2003. (From Holgate, 2007)

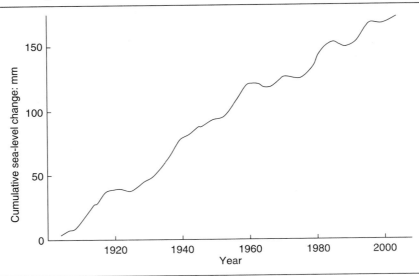

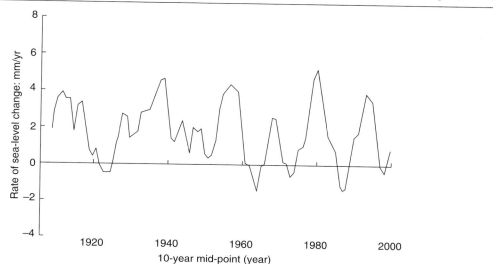

Figure 12.3 The global mean decadal rates of sea-level change over the period 1904–2003. (From Holgate, 2007)

Since 1993, satellites have taken over the provision of global information on sea level. This new coverage is not tied to point data on coastlines but extends over entire oceans, thereby allowing a much more complete picture of how whole oceans are responding to change. However, satellites record changes in *absolute sea level* (i.e. the level of the sea surface with respect to the centre of the Earth), so adjustments have to be made regarding relative sea-level change. The main results that have emerged are

- the dynamic character of sea level over the oceans due to atmospheric and thermo-haline influences, hence the emergence of the phrase 'sea level is not level'(Gehrels and Long, 2008)
- that gravitational effects mean that melt water from Antarctica or Greenland is not distributed evenly over the oceans but has a limited, or even negative, effect on local relative sea levels because it mainly accumulates in the opposite hemisphere or in the north-western Pacific (see Milne et al., 2009).

Nevertheless, measurements of ocean surface levels over the period 1993–2003 indicate a rate of rise of at least 3.1 mm/year (2.4–3.8 mm/year), which is rather greater than the tide-gauge figures. The satellite measurement (~2.5 mm/year) was used to indicate the minimum rate of sea-surface rise at the turn of the century was clearly too low, as subsequent annual estimates have risen rapidly to ~3.8 mm/year for 2010. Indeed, Church and White (2011) have shown that satellite-measured sea levels continue to rise at the upper end of the IPCC (2007) projections. The extent to which these changing 'best estimations' of contemporary global sea-level change actually show the true acceleration of sea-level rise over the past 100 years, rather than being the product of different recording techniques, remains to be resolved by the reconciling of satellite and tide-gauge records.

Explaining the contributory factors that have led to observed rises in sea level has proved difficult, not least because of disagreements regarding the actual pattern of past sea-level rise. The IPCC (2007) produced estimates of differing contributions to the recorded rise observed over the periods 1961–2003 and 1993–2003 (satellite only) (see Table 12.2). In the case of the period 1961–2003 there was a

Table 12.2 Observed rate of sea-level rise (mm/year) and estimated contributions from different sources (IPCC, 2007)

Source of sea-level rise	1961–2003	1993–2003
Thermal expansion	0.42 ± 0.12	1.6 ± 0.5
Glaciers and ice caps	0.50 ± 0.18	0.77 ± 0.22
Greenland ice sheet	0.05 ± 0.12	0.21 ± 0.07
Antarctic ice sheet	0.14 ± 0.41	0.21 ± 0.35
Sum of individual climate contributions to sea-level rise	1.1 ± 0.5	2.8 ± 0.7
Observed total sea level rise	1.8 ± 0.5	3.1 ± 0.7
Difference (observed minus sum of estimated climate contributions)	0.7 ± 0.7	0.3 ± 1.0

Data prior to 1993 are from tide gauges, and after 1993 are from satellite altimetry

significant gap between the rise produced by estimated contributions and the observed rise (known as the 'attribution problem'), which suggested that too much emphasis had been placed on the significance of thermal expansion during this early phase, and that increasing ocean mass had become more important than previously thought. Initially, this was wholly explained by underestimations of the contribution of ice and snow melt, especially from Greenland and Antarctica, where early information is limited, particularly as the deficiency is much less over the period 1993–2003. However, a recent study (Pokhrel et al., 2012) estimated that increased run-off due to urbanisation and deforestation, combined with the unsustainable exploitation of groundwater resources, could have increased ocean surface levels by 0.77 mm/year over the period 1961–2003. These uncertainties regarding exactly what caused sea-level change in the recent past clearly reveal the problems of predicting sea-level change into the future.

There are three sets of questions that need to be addressed as the basis for coastal landslide risk assessment:

- *What will the eustatic component of change be?* The IPCC can only realistically use the climate-change predictions arising from the various scenarios to determine the likely changes in ocean volume arising from steric and eustatic factors, and then use this to produce an overall figure for the global change in ocean-surface level per annum, or by specific future dates. The IPCC (2007) concluded that the sea level in 2090–2099 will be 0.16–0.59 m higher than it was in 1980–1999, since when various studies have suggested greater increases, for example, 0.47–1.00 m (Horton et al., 2008), 0.5–1.40 m (Rahmstorf, 2007), 0.75–1.90 m (Vermeer and Rahmstorf, 2009) and 0.9–1.30 m (Grinsted et al., 2010). All these predictions must be considered as unreliable because of the huge uncertainties: How much atmospheric warming will occur? How much ice in Greenland and Antarctica will melt? Will ice-sheet dynamics take over from climate as the main influence on sea-level change? Pfeffer et al. (2008) considered that sea-level rise of more than 2 m by 2100 is physically implausible, and that 0.8 m is more plausible. Thus a value of 0.6–0.8 m by 2100 would appear to be reasonable for risk-assessment purposes and is in line with (IPCC, 2013).

  However, it has long been known that the redistribution of melt water from Antarctica is far from uniform (Clark and Primus, 1987), with local reductions in sea-surface level adjacent to Antarctica due to loss of ice mass causing gravitational attraction to weaken, but increasing values of rise away from these areas, with the greatest net gains (up to 125% of the overall averaged contribution) occurring in the southern North Atlantic and the north-west Pacific. The same is now known to be true of Greenland, so melting here results in local falls or minimal rises in Arctic sea-surface levels, but ~120% net rises in the north-west Pacific, south-east Pacific and south Atlantic. So, for example, if these two ice masses were each to lose ice equivalent to 10 cm of global eustatic rise, absolute sea levels in the Arctic and Antarctic would rise by a few centimetres at most, and locally might actually fall, while sea-surface levels in the south-east Pacific and south Atlantic could rise by >20–23 cm and those of the north-west Pacific by >25 cm.

In recent years these effects have been modelled with increasing sophistication, most recently by Perrette et al. (2013), who have shown that sea-level rises due to ice-melt and steric contributions in the 'western' Pacific, southern Atlantic and Indian Ocean could all be 10–20% above the global mean value. No doubt increasingly sophisticated new predictions will soon emerge, but this is a helpful starting point for coastal landslide risk assessment.

- *Adjustment for local tectonic conditions.* To be meaningful, these figures have to be converted into *relative sea-level change* by reference to *prevailing local values* of crustal movements. Figures for isostatic movements are available, as are values for tectonic movements produced by satellite measurements. In areas experiencing isostatic rebound (i.e. uplift due to deglaciation), the rates of RSL rise will be less than the global average, and, indeed, some areas will start by showing a fall in RSL. However, by the end of the 21st century, the accelerating rate of absolute sea-level rise will have equalled/exceeded the rate of crustal rebound in most areas, so that few locations will still display regression and most will show relatively weak transgression. Other tectonic influences will be variable, depending on local conditions, and need to be established from local sources.

- *Other adjustments*; By the end of the twenty-first century virtually all the world's coastlines will be experiencing varying rates of RSL rise or *transgression*, some with *local rates* possibly in region of 10 mm/year. This has a number of implications:
  - Near-shore topographic conditions will be changed, which could result in alterations to tidal configuration, meaning that high-tide levels are frequently increased at a faster rate than that displayed by cumulative RSL change This is especially true in estuaries, where both the height and frequency of extreme high sea-water levels is expected to increase dramatically.
  - Changes in near-shore water depth could mean that wave heights are increased (potential for overtopping defences) and the breaker zone moved shorewards (increased cliff instability due to attack).
  - The possible effects of extreme sea levels have to be considered. Brief examples are produced by severe weather events (storm surges) and rare tectonic/volcanic events (tsunamis). The former could increase in frequency and magnitude in the future if storminess were to increase more rapidly than predicted, although there is current uncertainty about this (IPCC, 2012).
  - Extreme sea levels are also produced by the short-term climatic oscillations that affect most ocean basins, most especially ENSO, which affects the Pacific (Menedez and Woodworth, 2010; Merrifield et al., 2007). There is also a growing literature on examples of the longer-term influences of climate variability on regional sea levels. Ullmann et al. (2007, 2008) have shown that maximum annual sea levels in the Camargue rose twice as fast as the mean sea level during the 20th century, due to the increased occurrence of southerly (onshore) winds, while two tide gauges in British Columbia both recorded similar results of the sea level rising at double the expected rate for the period 1979–2003, due to the influence of the Pacific decadal oscillation (Abeysirigunawardena and Walker, 2008).

Table 12.3 Current regional net sea-level rise allowances for Great Britain (Defra, 2006)

Region	Previous Allowance (mm/year)	Assumed vertical land movement (mm/year)	Net sea-level rise (mm/year)			
			1990–2025	2025–2055	2055–2085	2085–2115
East England South-east England	6	−0.8	4.0	8.5	12.0	15.0
South-west England Wales	5	−0.5	3.5	8.0	11.5	14.5
North-west England, north-east England Scotland	4	+0.8	2.5	7.0	10.0	13.0

These varied local conditions have the potential to further greatly raise both extreme and mean sea levels along coastlines, as well as to profoundly influence the apparent rate of RSL (thereby revealing one reason why tide-gauge records are so variable).

Clearly, with such a range of factors influencing sea-level change, advice needs to be sought at national/regional scales. In the case of Great Britain, current advice is provided by the Department for Environment, Food and Rural Affairs (Defra, 2006) (Table 12.3) as to anticipated rates of sea-level rise, by major region, for 30-year periods until 2115. Obviously these are generalisations and further detail has to be gained from other sources,

It is clear that accelerating regional sea-level rise (RSLR) is anticipated for the majority of coastlines and has to be planned for, including the prospect of accelerating rates of cliff recession (Bray and Hooke, 1997; Clayton, 1989), which could threaten existing properties and infrastructure. Investigations have tended to resort to multiplying the historical recession rate by the ratio of the future RSLR to the historical RSLR, or applying the Bruun rule (Bruun, 1962; Lee, 2005; Lee and Clark, 2002). Bray and Hooke (1997) used both approaches to model the effects of RSLR on the soft-rock cliffs of southern England, and found that historic recession rates would be raised by 22–133% by 2050, depending on location, assuming a RSLR value of 6 mmpa.

To test this assertion, Lee (2011) undertook a detailed study of the rapidly eroding (>1 mpa), 3–40 m high glacial till cliffs of the 60 km long Holderness coast (east Yorkshire, UK). Using the annual data produced for the period 1951–2004 by measurements of cliff recession from 120 marker posts (repositioned when threatened), he calculated that there has been a decadal scale increase in recession of 5 mmpa (from ~1.2 m pa in the early 1950s to ~1.5 m pa by 2000), but with no evidence of a marked increase in recession rate over the past two decades to coincide with the reported accelerating pace of global sea-level rise (Figure 12.4). Indeed, the only relationship found was between the 10-year moving mean of cliff recession and the decadal rates of sea-level change identified by Holgate (2007) (see Figure 12.3). However, the actual rate of local RSLR is surprisingly low (1.11 mmpa between 1960 and 1995 at the nearby Immingham tide gauge), especially as this coastline is thought to be gently

Figure 12.4 Overall annual recession rates for the Holderness coast, 1952–2004. (From Lee, 2011)

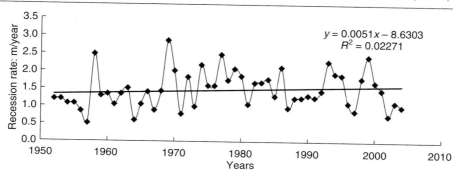

subsiding at ≤0.5 mmpa (Shennan et al., 2009). Clearly some marine and/or atmospheric influences are contriving to minimise the RSLR, so it will come as little surprise that, in this instance, the dominant influence on cliff recession appears to be the 18.61-year tidal cycle.

Applying the recorded recession rate for 1951–1990, the predicted recession for 1990–2004 was calculated using both the historical projection and Bruun rule approaches (see Table 12.4 for an explanation), and by simply extrapolating the historical rates. The results were then compared with the actual record (see Table 12.4). The method that gives the best prediction for the period 1990–2004 would have been simple extrapolation of past recession rates. The historical projection approach

Table 12.4 Holderness Erosion Post 59 recession predictions for the period 1990–2004 (from Lee, 2011)

Prediction method	Historical recession rate m/year (1951–1990)	Predicted recession rate m/year (1990–2004)	Predicted recession distance (1990–2004): m	Actual recession distance (1990–2004): m
Extrapolation[a]	2.16	2.16	32.4	33.85
Historical projection[b]	2.16	9.73	146	33.85
Bruun rule[c]	2.16	2.71	40.65	33.85

[a] Predicted recession rate = historical recession rate
[b] Predicted recession rate = historical recession rate × predicted RSLR/historical RSLR = 2.16 × (5/1.11)
The predicted RSLR is 5 mm/year (MAFF, 1991)
The nearest tide gauge to Holderness is at Immingham, on the Humber estuary. The historical RSLR rate at this gauge is 1.11 mm/year (Immingham, 1960–1995; standard error ±0.52 mm; Woodworth et al., 2009)
[c] Predicted recession rate = $R_1 + Sc((L)/P(B+H))$
R_1 = historical recession rate = (2.16 m/year)
Sc = change in rate of sea-level rise (m) i.e. $0.005 - 0.0011 = 0.0039$ m
P = sediment overfill (the proportion of sediment eroded that is sufficiently coarse to remain within the equilibrium profile) ($P = 0.25$)
B = cliff height = 16.7 m
H = closure depth = 12 m
L = length of cliff profile (to the closure depth) = 1000 m

and the Bruun rule overestimate the actual recession for this period by over 400% and 20%, respectively. Lee's paper (2011) is important because it

- reveals the importance of obtaining local records/estimates of sea-level change
- shows that tectonic influences are not the only important 'local' influences on RSL change
- shows that coastal erosion rates have accelerated since the 1950s, despite the unusually low local RSLR
- reveals that the simple application of the historical projection and Bruun rule approaches can result in exaggerated rates of predicted cliff recession and elevated risk assessments.

12.12. Uncertainty and risk assessment

Risk assessment should present a view of the world that recognises uncertainty in the future rather than presenting an overconfident 'this will happen' view of what is known. Uncertainty will inevitably be present because of incomplete knowledge of either, or both, the probability of events and their consequences (see Chapter 5). As illustrated in Figure 12.5, where there is good knowledge of both it is possible to characterise the risk, both quantitatively and reliably. In most instances, however, there will be gaps in the available knowledge and the characterisation of the risk will be less reliable.

Uncertainty tends to increase as time frames of analysis extend further and further into the future, or the past. It may be possible to assess the main features of landslide hazard over the next few decades and use this knowledge to develop quantitative assessments of risk. However, as uncertainties accumulate in the future, especially so-called *non-probabilistic* uncertainties such as socio-economic change, so it becomes less and less appropriate to attempt to construct the probability distributions or consequence models that are needed to support quantitative risk assessment.

Stern and Fineberg (1996) suggest that risk assessments often present misleading information about uncertainty. For example, they might give the impression of more scientific certainty or agreement

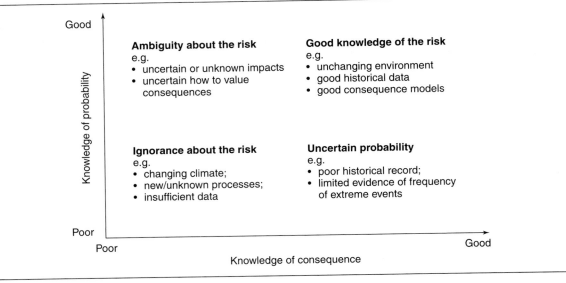

Figure 12.5 Impact of uncertainty on risk assessment. (Adapted from Willows et al., 2000)

than is the case, or suggest that the uncertainty is a reflection of data availability, when in fact there are differences in interpretation of the ground conditions and in judgement about the significance of the features that have been identified. It is also possible to give the impression that particular risks do not exist, when in fact they simply have not been analysed.

Risk assessment needs to be supported by a clear statement of the uncertainties in order to 'inform all the parties of what is known, what is not known, and the weight of evidence for what is only partially understood' (Stern and Fineberg, 1996). However, this is not a straightforward process, as it is difficult to characterise uncertainty without making the risk appear larger or smaller than the experts believe it to be (Johnson and Slovic, 1995). Careful and elaborate characterisation of the uncertainties might be incomprehensible to non-specialists and unusable by decision-makers. Clearly a balance needs to be found between providing sufficient information on uncertainty to enable decision-makers and other participants in the risk-management process to be aware of the issues, and diverting attention away from the reality of the situation by dwelling on the unknown.

REFERENCES

Abeysirigunawardena DS and Walker IJ (2008) Sea level response to climate variability and change in northern British Columbia. *Atmosphere-Ocean* **47(1)**: 41–62.

Acton JM and Hibbs M (2012) Why Fukushima was Preventable. The Carnegie Papers, Nuclear Policy. http://www.carnegieendowment.org/pubs (accessed 24/9/2013).

Anon. (2010) Urban geology in the Neapolitan area. http://www.virtualexplorer.com.eu (accessed 24/9/2013).

Baynes F (2010) Sources of geotechnical risk. *Quarterly Journal of Engineering Geology and Hydrogeology* **43**: 321–331.

Bray MJ and Hooke JM (1997) 'Prediction of soft-cliff retreat with accelerating sea-level rise. *Journal of Coastal Research* **13**: 453–467.

Bronnimann S (2009) Early Twentieth Century Warming. *Nature Geoscience* **2(11)**: 735–6.

Bruun P (1962) Sea-level rise as a cause of shore erosion. *Journal of the Waterways and Harbour Division, ACSE* **88**: 117–130.

Calcaterra D and Guarino PM (1999) Morphodynamics and recent landslides in the Neapolitan slopes (western sector). *Geologia Technica ed Ambientale* **2(99)**: 11–17.

Cascini L, Bonnard C, Corominas, J, Jibson R and Montero-Olarte J (2005) Landslide hazard and risk zoning for urban planning and development. In *Landslide Risk Management* (Hungr O, Fell R, Couture R and Eberhardt E (eds)). Balkema, London, pp. 199–236.

Church JA and White NJ (2011) Sea-level rise from the late 19th to the early 21st century. *Surveys of Geophysics* **32(4–5)**: 585–602.

Clark JA and Primus JA (1987) Sea level changes resulting from future retreat of ice sheets: an effect of CO_2 warming of the climate. In *Sea Level Changes* (Tooley MJ and Shennon I (eds)). Blackwell, Oxford, pp. 356–370.

Clark JA, Farell WE and Peltier WR (1978) Global changes in postglacial sea-level: a numerical calculation. *Quaternary Research* **9**: 265–287.

Clayton KM (1989) Implications of climate change. In *Coastal Management*. Thomas Telford, London, pp. 165–176.

Cox D, Hunt J, Mason P, Wheater H and Wolf P (eds) (2002) Flood risk in a changing climate. *Philosophical Transactions of the Royal Society, Mathematical, Physical and Engineering Sciences* **360**: 1796.

Dai FC, Lee CF and Ngai YY (2002) Landslide risk assessment and management: an overview. *Engineering Geology* **64(1)**: 65–87.

Defra (Department for Environment, Food and Rural Affairs) (2006) *Flood and Coastal Defence Appraisal Guidance*. FCDPAG3 Economic appraisal, supplementary note to operating authorities – Climate change impacts. Defra, London.

Defra (2007) *An Introductory Guide to Valuing Ecosystem Services*. Defra, London.

Del Prete M, Guadagno FM and Hawkins AB (1998) Preliminary report on the landslides of 5 May 1998, Compania, southern Italy. *Bulletin of Engineering Geology and the Environment* **57(2)**: 113–129.

Fell R, Corominas J, Bonnard C et al. on behalf of the JTC-1 Joint Technical Committee on Landslides and Engineered Slopes (2008a) Guidelines for landslide susceptibility, hazard and risk zoning. *Engineering Geology* **102**: 85–98.

Fell R, Corominas J, Bonnard C et al. on behalf of the JTC-1 Joint Technical Committee on Landslides and Engineered Slopes (2008b) Commentary: guidelines for landslide susceptibility, hazard and risk zoning. *Engineering Geology* **102**: 99–111.

Finlay PJ and Fell R (1997) Landslides: risk perception and acceptance. *Canadian Geotechnical Journal* **34(6)**: 169–188.

Gehrels R and Long A (2008) Sea-Level is not Level. *Geography* **93(1)**: 11–16.

Glade T (1998) Establishing the frequency and magnitude of landslide-triggering rainstorm events in New Zealand. *Environmental Geology* **55(2)**: 160–174.

Grinsted A, Moore AJ and Jevrejeva S (2010) Reconstructing sea level from paleo and project temperatures 200 to 2100 AD. *Climatic Dynamics* **34(4)**: 461–472.

Hengesh JV, Angell M, Lettis WR and Bachhuber JL (2004) A systematic approach for mitigating geohazards in pipeline design and construction. *Proceedings of IPC 2004 International Pipeline Conference*, Calgary. IPC04- 0147.

Holgate SJ (2007) On the decadal rates of sea level change during the twentieth century. *Geophysical Research Letters* **34**: L01602. doi:10.1029/2006GL028492.

Horton R, Herweijer C, Rosenweig C et al. (2008) Sea level rise projections for current generation CGCMs based on the semi-empirical method. *Geophysical Research Letters* **35**: L02715.

Houghton J (2009) *Global Warming: The Complete Briefing*, 4th edn. Cambridge University Press, Cambridge.

Hutchinson JN (1991) The landslides forming the South Wight undercliff. In *Slope Stability Engineering: Development and Applications* (Chandler RJ (ed.)). Thomas Telford, London, pp. 157–168.

Hutchinson JN and Chandler MP (1991) A preliminary landslide hazard zonation of the undercliff of the Isle of Wight. In *Slope Stability Engineering: Development and Applications* (Chandler RJ (ed.)). Thomas Telford, London, pp. 197–205.

Hutchinson JN and Bromhead EN (2002) Isle of Wight landslides. In *Instability – Planning and Management* (McInnes RG and Jakeways J (eds)). Thomas Telford, London, pp. 3–70.

Hutchinson JN, Brunsden D and Lee EM (1991) The geomorphology of the landslide complex at Ventnor, Isle of Wight. In *Slope Stability Engineering, Developments and Applications* (Chandler RJ (ed.)). Thomas Telford, London, pp. 213–218.

IDB (Inter-American Development Bank) (2007) *Camisea Project Benefits*. http://idbdocs.iadb.org/wsdocs/getdocument.aspx?docnum = 980596 (accessed 24/9/2013).

IPCC (Inter-Governmental Panel on Climate Change) (1990) *Climate Change, the IPPC Scientific Assessment* (Houghton JT, Jenkins GJ and Ephraums JJ (eds)). Cambridge University Press, Cambridge.

IPCC (1995a) *Climate Change 1995: The Science of Climate Change* (Houghton JT, Meira Filho LG, Callender BA, Harris N, Kattenberg A and Maskell K (eds)). Cambridge University Press, Cambridge.

IPCC (1995b) *Climate Change 1995: Impacts, Adaptations and Mitigation of Climate Change: Scientific-Technical Analysis* (Watson RT, Zinyowera MC and Moss RH (eds)). Cambridge University Press, Cambridge.

IPCC (2000) Quantifying uncertainties in practice. In *IPCC Good Practice Guidance and Uncertainty Management in National Greenhouse Gas Inventories.* http://www.ipcc-nggip.iges.or.jp/public/gp/english (accessed 24/9/2013).

IPCC (2001) *The IPCC Third Assessment Report. Summary for Policy Makers.* http://www.ipcc.ch/index.html (accessed 24/9/2013).

IPCC (2007) *4th Assessment Report and Summary for Policymakers.* http://ipcc-wg1.ucar.edu/wg1/wg1-report.html (accessed 24/9/2013).

IPCC (2012) *Managing the Risks of Extreme Events and Disasters to Advance Climate Change Adaptation.* http://ipcc-wg2.gov/SREX/report (accessed 24/9/2013).

IPCC (2013) *Climate Change 2013: The Physical Science Basis Summary for Policy Makers.*

Iverson RM (2000) Landslide triggering by rain infiltration. *Water Resources Research* **36(7)**: 1897–1910.

Jakob M and Weatherly H (2003) A hydroclimatic threshold for landslide initiation on the North Shore Mountains of Vancouver, British Columbia. *Geomorphology* **54(3–4)**: 137–156.

Jevrejeva S, Grinsted A, Moore J and Holegate S (2006) Nonlinear trends and multiyear cycles in sea-level records. *Journal of Geophysical Research* **111**: C09012. doi:10.1029/2005JC003229.

Johnson BB and Slovic P (1995) Presenting uncertainty in health risk assessment: initial studies of its effects on risk perception and trust. *Risk Analysis* **15(4)**: 485–494.

Judd T (2009) Cliff erosion row victory for conservation watchdog. *The Independent*, 20 October 2009.

Kaufmann RK, Kauppi, H, Mann ML and Stock JH (2011) Reconciling anthropogenic climate change with observed temperature 1998–2008. *Proceedings of the National Academy of Science* **108(29)**: 11790–11793.

Kim SK, Hong WP and Kim YM (1991) Prediction of rainfall-triggered landslides in Korea. *Proceedings of the 6th International Symposium on Landslides, Christchurch, New Zealand* (Bell DH (ed.)). Balkema, Rotterdam, pp. 989–994.

Kingston J (2012) Mismanaging risk and the Fukushima nuclear crisis. *The Asia-Pacific Journal* **10(12)**: 19.

Knutson TR, McBride JL, Chan J *et al.* (2010) Tropical cyclones and climate change. *Nature Geoscience* **3(3)**: 157–163.

Kundzewicz ZW (ed.) (2012) *Changes in Flood Risk in Europe.* IAHS, Wallingford.

Lee EM (2000) The management of coastal landslide risks in England: the implications of conservation legislation and commitments. In *Landslides: In Research, Theory and Practice* (Bromhead EN, Dixon N and Ibsen ML (eds)). Thomas Telford, London, pp. 893–898.

Lee EM (2002) A dynamic framework for the management of coastal erosion and flooding risks in England. In *Instability: Planning and Management* (McInnes RG and Jakeways J (eds)). Thomas Telford, London, pp. 713–720.

Lee EM (2003) *A Quick Guide to the Undercliff Planning Guidance Maps.* Technical Note to the Undercliff Landslide Management Committee. Department of the Environment, London.

Lee EM (2005) Benacre to Easton Bavents SSSI: Prediction of coastal change. In *Coastal evolution in Suffolk: an evaluation of geomorphological and habitat change. English Nature Research Reports* (Rees SM (ed.)), No. 647, pp. 41–56.

Lee EM (2011) Reflections on the decadal-scale response of coastal cliffs to sea-level rise. *Quarterly Journal of Engineering Geology and Hydrogeology* **44**: 481–489.

Lee EM and Clark AR (2002) *Investigation and Management of Soft Rock Cliffs*. Thomas Telford, London.

Lee EM and Moore R (1991) *Coastal landslip potential assessment: Isle of Wight Undercliff, Ventnor*. Department of the Environment, London.

Lee EM and Moore R (2007) Ventnor Undercliff: development of landslide scenarios and quantitative risk assessment. In *Landslides and Climate Change: Challenges and Solutions* (McInnes R, Jakeways J, Fairbank H and Mathie E (eds)). Balkema, Rotterdam, pp. 323–334.

Lee EM, Moore R, Brunsden D and Siddle HJ (1991a) The assessment of ground behaviour at Ventnor, Isle of Wight. In *Slope Stability Engineering: Developments and Applications* (Chandler RJ (ed.)). Thomas Telford, London, pp. 207–212.

Lee EM, Moore R, Burt N and Brunsden D (1991b) Strategies for managing the landslide complex at Ventnor, Isle of Wight. In *Slope Stability Engineering: Developments and Applications* (Chandler RJ (ed.)). Thomas Telford, London, pp. 219–225.

Lee EM, Brunsden D, Roberts H, Jewell S and McInnes R (2001) *Restoring Biodiversity to Soft Cliffs*. Report 398. English Nature, Peterborough.

Lee EM, Audibert JME, Hengesh JV and Nyman DJ (2009) Landslide-related ruptures of the Camisea pipeline. system, Peru. *Quarterly Journal of Engineering Geology and Hydrogeology* **42**: 251–259.

Lotter M, Charman JH, Lee EM et al. (2005) Geohazard assessment of a major crude oil pipeline system – the Turkish section of the BTC Pipeline route. In *Terrain and Geohazard Challenges Facing Onshore Oil and Gas Pipelines* (Sweeney M (ed.)). Thomas Telford, London, pp. 301–310.

McInnes R (2005) Instability management from policy to practice. In *Landslide Hazard and Risk* (Glade T, Anderson MG, Crozier MJ (eds)). Wiley, Chichester, pp. 401–428.

MAFF (Ministry of Agriculture, Fisheries and Food) (1991) Advice on allowances for sea level rise. Issued November 1991.

Mattéi JM, Vial E, Rebour V, Liemersdorf H and Türschmann M (2001) Generic Results and Conclusions of Re-evaluating the Flooding in French and German Nuclear Power Plants. Eurosafe Forum 2001. http://www.eurosafe-forum.org/files/semb1_7.pdf (accessed 24/9/2013).

Menedez M and Woodworth PL (2010) Changes in extreme high water levels based on a quasi-global tide-gauge dataset. *Journal of Geophysical Research* **115**: C10011.

Merrifield MA, Firing YL and Marra JJ (2007) Annual climatologies of extreme water levels. In *Aha Hulikoa: Extreme Events, Proc. Hawaiian Winter Workshop, Univ. of Hawaii at Manoa, 23–26 Jan 2007*. SOEST, Manoa, pp. 27–32.

Milne GA, Gehrels WR, Hughes CW and Tamisiea ME (2009) Identifying the causes of sea-level change. *Nature Geoscience* **2(7)**: 471–478.

Mohrbach L, Linnemann T, Schäfer G and Vallana G (2011) *Earthquake and Tsunami in Japan on March 11, 2011 and Consequences for Fukushima and Other Nuclear Power Plants*. VGB Power Tech, http://www.vgb.org/vgbmultimedia/News/Fukushimav15VGB.pdf (accessed 24/9/2013).

Moore R, Lee EM and Noton N (1991) The distribution, frequency and magnitude of landslide movements at Ventnor, Isle of Wight. In *Slope Stability Engineering: Developments and Applications* (Chandler RJ (ed.)). Thomas Telford, London, pp. 213–218.

Muller J et al. (2011) *Berkeley Earth Surface Temperature Study*. http://www.berkleyearth.org (accessed 24/9/2013).

Onishi N and Glanz J (2011) Japanese rules for nuclear plants relied on old science. *New York Times*, 26 March 2011.

Opadeyi S, Ali S and Chin F (2005) *Status of Hazard Maps, Vulnerability Assessments and Digital Maps in the Caribbean*. Caribbean Disaster Emergency Response Agency (CEDRA), Georgetown.

Osborn TJ and Hulme M (2002) Evidence for trends in heavy rainfall events over the UK. *Philosophical Transactions: Mathematical, Physical and Engineering Sciences* **360(1796)**: 1313–1325.

Pearce D, Atkinson G and Mourato S (2006) *Cost–benefit Analysis and the Environment: Recent Developments*. Organisation for Economic Co-operation and Development (OECD), Paris.

Peltier WR (1987) Mechanisms of relative sea-level change and the geophysical responses to ice-water loading. In *Sea Surface Studies: A Global View* (Devoy RJN (ed.)). Croom Helm, New York, pp. 264–293.

Perrette M, Landerer F, Riva R, Frieler K and Meinshausen M (2013) A scaling approach to project regional sea level rise and its uncertainties. *Earth System Dynamics* **4**: 11–29.

Pettinger AM and Sykora DW (2011) Landslide risk assessment for pipeline systems in mountainous regions. *Journal of Pipeline Engineering*, 10 September.

Pfeffer WT, Harper JT and O'Neel S (2008) Kinematic constraints on glacier contributions to 21st-century sea-level rise. *Science* **321(5894)**: 1340–1343.

Pokhrel YN, Hanasaki N, Yeh Pat JF *et al.* (2012) Model estimates of sea-level change due to anthropogenic impacts on terrestrial water storage. *Nature Geoscience* **5(6)**: 389–392.

Rahmstorf S (2007) A semi-empirical approach to projecting future sea-level rise. *Science* **315(5810)**: 329–343.

Rist A and Phillips M (2005) First results of investigations on hydrothermal processes within the active layer above alpine permafrost in steep terrain. *Norsk Geografisk Tidsskrift* **59(2)**: 177–183.

Savill R (2009) King Canute told by courts he cannot hold back tide. *The Daily Telegraph*, 20 October.

Schiermeier Q (2010) The real holes in climate science. *Nature* **463**: 284–287.

Senior CA, Jones RG, Lowe JA, Durman CF and Hudson D (2002) Predictions of extreme precipitation and sea-level rise under climate change. In *Philosophical Transactions of the Royal Society London, Series A* **360**: 1301–1311.

Shennan I, Milne G and Bradley SL (2009) Late Holocene relative land – and sea-level changes: providing information for stakeholders. *GSA Today* **19**: 52–53.

Sidle RC and Ochial H (2006) *Landslides; Processes, Prediction and Land Use*. Water Resources Monograph 18. American Geophysical Union, Washington, DC, p. 312.

Stein S, Geller RJ and Liu M (2012) Why earthquake hazard maps often fail and what to do about it. *Tectonophysics* **562–563**: 1–25.

Stern PC and Fineberg HV (eds) (1996) *Understanding Risk: Informing Decisions in a Democratic Society*. National Academy Press, Washington, DC.

Sweeney M (2005) Terrain and geohazard challenges facing onshore oil and gas pipelines: historic risks and modern responses. In *Terrain and Geohazard Challenges Facing Onshore Oil and Gas Pipelines* (Sweeney M (ed.)). Thomas Telford, London, pp. 37–51.

Turner BL, Kasperson RE, Meyer WB, Dow KM, Golding D, Kasperson JX, Mitchell RC, and Ratick RS (1990) Two types of global environmental change: definitional and spatial-issues in their human dimensions. *Global Environmental Change* **1**: 14–22.

UK Natural Ecosystem Assessment (2011) *Synthesis of Key Findings*. UK Natural Ecosystem Assessment, Cambridge.

Ullmann A, Pirazzoli PA and Tomasin A (2007) Sea surges in Camargue: trends over the 20th century. *Continental Shelf Research* **27(7)**: 922–934.

Ullmann A, Pirazzoli PA and Moron V (2008) Sea surges around the Gulf of Lions and atmospheric conditions. *Global and Planetary Research* **63(2–3)**: 203–214.

Van Westen CJ, Van Asch TWJ and Soeters R (2006) Landslide hazard and risk zonation – why is it still so difficult? *Bulletin of Engineering Geology and Environment* **65(2)**: 167–184.

Vermeer M and Rahmstorf S (2009) Global sea level linked to global temperature. *Proceedings of the National Academy of Sciences* **106(51)**: 21527–21532.

Wieczorek GF, Glade T, Jakob M and Hungr O (2005) Climatic factors influencing occurrence of debris flows. *Debris-Flow Hazards and Related Phenomena* (Jakob M and Hungr O (eds)). Springer-Verlag, Berlin, pp. 325–362.

Willows RI, Meadowcraft IC and Fisher J (2000) *Climate Adaptation Risk and Uncertainty: Draft Discussion Framework*. Environment Agency, London. Report No. 21.

Woodworth PL, White NJ, Jevrejeva S et al. (2009) Review: Evidence for the acceleration of sea level on multi-decade and century timescales. *International Journal of Climatology* **29**: 777–789.

World Bank (2010) Reducing human vulnerability: helping people help themselves. *World Development Report 2010*, Ch. 2, pp. 87–123. World Bank, Washington, DC.

World Resources Institute (2000) *A Guide to World Resources 2000–2001. People and Ecosystems: The Fraying Web of Life*. Island Press, Washington, DC.

World Resources Institute (2005) *Ecosystems and Human Well-being: Synthesis*. Island Press, Washington, DC.

Zaitchik BF, van Es HM and Sullivan PA (2003) Variability and scale in the application of a physical slope stability model for landslide evaluation in Honduras. *Journal of the Soil Science Society of America* **67**: 268–278.

Zimmermann M, Mani P and Romang H (1997) Magnitude–frequency aspects of alpine debris flows. *Eclogae Geologicae Helvetiae* **90(3)**: 415–420.

Glossary of terms

Acceptable risk   The level of risk that individuals and groups are prepared to accept at a particular point in time, as further expenditure in risk reduction is not considered justifiable (e.g. the costs of risk reduction exceed the benefits gained). The term has largely come to be replaced by 'tolerable risk'.

Adverse consequences   The adverse effects, losses or harm from a human perspective, resulting from the realisation of a hazard. They can be expressed quantitatively (e.g. deaths, economic losses) or qualitatively (loss of amenity) and occur in the short or longer term.

Adverse outcomes   See *Adverse consequences*.

Benefit : cost ratio   The ratio of the present value of benefits to the present value of costs.

Consequences   The adverse effects and benefits arising from the realisation of a hazard.

Consequence assessment   The identification and quantification of the full range of adverse consequences arising from the identified patterns and sequences of hazard.

Deterministic method   A method in which precise, single values are used for all variables and input values, giving a single value as the output.

Detriment   A numerical measure of the expected harm or loss associated with an adverse event and, therefore, an important ingredient in benefit–cost and risk–benefit analyses. Increasingly used to describe adverse consequences in general.

Direct and indirect economic losses   Those losses capable of being given monetary values because of the existence of a market, with all other losses classified as intangibles. Direct economic losses arise principally from the physical impact of a landslide on property, buildings, structures, services and infrastructure. Indirect economic losses are those that subsequently arise as a consequence of the destruction and damage caused by primary hazard (i.e. the landslide itself), secondary hazards or follow-on hazards.

Disaster   An imprecise term which should only be applied to situations where the level of adverse consequences is sufficiently severe so that either outside assistance is required to facilitate the recovery process or the detrimental effects are long-lasting and debilitating.

Discounting   The procedure used to arrive at the sum of either costs or benefits over the lifetime of a project, using a discount rate to scale down future benefits and costs. The effect of using a discount rate is to reduce the value of projected future costs or benefits to their values as seen from the present day.

Economic risk   The risk of financial loss due to potential hazards causing loss of production, damage or other adverse financial consequences.

Elements at risk   The population, buildings and engineering works, economic activities, public services utilities, artefacts, valued possessions, infrastructure and environmental features in any area that are valued by humans and potentially adversely affected by landslides.

Environmental hazard   The threat potential posed to humans or nature by events originating in, or transmitted by, the natural or built environment.

Environmental risk   An amalgam of the probability and scale of exposure to loss arising from hazards originating in, or transmitted by, the physical and built environments.

Event tree   A form of logic diagram designed to specify the steps by which the range of possible adverse outcomes can arise after the occurrence of a selected initiating event.

Exposure   The proportion of time that an asset or person is exposed to the hazard. Exposure involves the notion of being in 'the wrong place' (i.e. the 'danger zone' where a landslide impacts – the spatial probability) at the 'wrong time' (i.e. when the landslide occurs – the temporal probability).

Fault tree   A form of logic diagram designed to work backwards from a particular event or outcome (known as the top event) through all the chains of possible events that could be precursors of the top event. Its purpose is to analyse why a particular outcome occurred or could occur.

F–N **curve**   A plot showing the frequency–magnitude relationship of adverse consequences arising from different types of hazards. Usually, the consequence referred to is the number of deaths (n), and the cumulative frequency of incidents with n or more deaths is plotted against number of fatalities (F_n).

Frequency   A measure of likelihood expressed as the number of occurrences of an event in a given time.

Global risk assessment   A misleading term for the procedure to determine the overall risk faced by a community.

Hazard   A property or situation that in particular circumstances could lead to harm from a human perspective. For a hazard to exist situations have to arise or circumstances occur where human value systems might be adversely impacted. Hazards are threats to humans and what they value: life, well-being, material goods and environment. A primary hazard event (i.e. a major landslide) may generate three other types of hazard.
   Post-event hazards, which occur after the initial sequence and are a product of the specific system returning towards stability, as is the case with the aftershocks following major earthquakes.
   Secondary hazards, which are different geohazards generated by the main hazard event sequence; for example, destructive tsunami generated by earthquakes and major landslides and the floods caused by the failure of landslide-generated dams.
   Follow-on hazards   generated by the primary events but arising due to failures of infrastructure and management systems; for example, fire caused by overturned stoves, electrical short-circuits and broken gas pipes, localised flooding caused by broken supply mains or sewers and disease.

Individual risk   The risk specific to humans, such as the general public, road users or particular activity groups, is the frequency with which individuals within such groupings are expected to suffer harm. Measures of individual risk include the following.
  Location-specific individual risk (LSIR)   The risk for an individual who is present at a particular location for the entire period under consideration, which may be 24 hours per day, 365 days per year or during the entire time that a risk-generating plant, process or activity is in operation.
  Individual-specific individual risk (ISIR)   The risk for an individual who is present at different locations during different periods.
  Average individual risk (AIR)   This can be calculated from historical data of the number of fatalities per year divided by the number of people at risk. Alternatively, a measure of the average individual risk can be derived from the societal risk divided by the number of people at risk.

Intangible losses   The vague and diffuse adverse consequences that arise from an event and which cannot easily be valued in economic terms because there is no market. They include effects on the environment, nature conservation, amenity, local culture, heritage, aspects of the local economy, recreation and peoples' health, as well as their attitudes, behaviour and sense of well-being.

Likelihood   Used as a qualitative description of probability or frequency.

Market value   The price for which individual goods are bought or sold in the market.

Natural hazard   Those elements of the physical environment harmful to humans and caused by forces extraneous to human society. Because of the scale of human impacts on the environment, the term is increasingly being replaced by 'environmental hazard' and 'geohazard'.

Outcomes   The range of hazards, adverse consequences and benefits that can result from an initial hazard event.

Perceived risk   The combined evaluation that is made by an individual of the likelihood of an adverse event occurring in the future and the magnitude of its likely adverse consequences.

Potential loss of life (PLL)   A measure of the expected number of fatalities for a population exposed to the full range of landslide events that might occur in an area (i.e. the societal risk). To calculate PLL it is necessary to estimate, for each event and its possible adverse outcomes, the frequency per year (f) and the associated number of fatalities (N).

Present value (PV)   The value of a stream of benefits or costs when discounted back to the present time.

Probabilistic method   A method in which the variability of input values and the sensitivity of the result are taken into account to give results in the form of a range of probabilities for different adverse outcomes.

Probability   The likelihood of a specific outcome, measured as the ratio of specific outcomes to the total number of possible outcomes. Probability is expressed as a number between 0 and 1, with 0 indicating an impossible outcome, and 1 indicating that a particular outcome is certain.

Residual risk   The remaining risk after all proposed improvements in management of the system have been made.

Return period   The average length of time separating the occurrence of extreme events of a similar, or greater, magnitude. Also known as the 'recurrence interval'.

Risk   The potential for adverse consequences, loss, harm or detriment from a human perspective. Risk is often expressed as a mathematical expectation value, the product of the probability of occurrence of a defined future hazard and the monetary value of the adverse consequences.

Risk analysis   The use of available information to estimate the risk to individuals or populations, property, or the environment, posed by identified hazards. Risk analyses generally contain the following steps: problem definition, hazard identification and risk estimation.

Risk Assessment   The process by which risk is analysed, estimated and evaluated.

Risk communication   Any purposeful exchange of information about health or environmental risks between interested parties.

Risk control or **risk treatment**   The process of decision-making for managing risk, the implementation, or enforcement, of risk-mitigation measures and the re-evaluation of its effectiveness from time to time, using the results of risk assessment as one input.

Risk estimation   The process used to produce a measure of the level of health, property, or environmental risks being analysed. Risk estimation contains the following steps: hazard assessment, consequence analysis and their integration.

Risk evaluation   The stage at which perceived values and judgements enter the decision together with the importance of the estimated risks and the associated social, environmental and economic consequences, in order to identify a range of alternatives for managing the risks. Public participation and obtaining the views of stakeholders are the keys to success.

Risk management   The process whereby decisions are made to accept a known or assessed level of risk and/or the implementation of actions to reduce the consequences or probability of occurrence.

Risk perception   How people's knowledge, experience, cultural background and attitudes (i.e. their socio-cultural make-up) leads them to interpret the stimuli and information that they receive concerning risk. Risk perception involves people's beliefs, attitudes, judgements and feelings, as well as the wider social or cultural values and dispositions that people adopt towards hazards and their benefits.

Risk register   An auditable record of the project risks, their consequences and significance, together with proposed mitigation and management measures.

Risk scoping   The process by which the spatial and temporal limits of a risk assessment are defined at the outset of a project. These may be determined on purely practical grounds, such as budget constraints, time constraints, staff availability or data availability.

Risk screening   The process by which it is decided whether or not a risk assessment is required or whether it is required for a particular element, or elements, within a project.

Sensitivity testing A method in which the impact on the output of an analysis is assessed by systematically changing the input values.

Site-specific risk assessment A procedure to determine the nature and significance of the hazards and risk levels at a particular site.

Societal risk The number of people within a group, at a location or undertaking an activity, and the frequency at which they suffer a given level of harm from the realisation of specified hazards. It is the shared level of risk and usually refers to the risk of death, expressed as risk per year.

Tolerable risk A risk that society is willing to live with so as to secure certain net benefits in the confidence that it is being properly controlled, kept under review and will be further reduced as and when possible. In some situations, risk may be tolerated because the individuals at risk cannot afford to reduce risk, even though they recognise that it is not properly controlled.

Value of a statistical life (VOSL) The value of a change in the risk of death, not human life itself; that is, how a person's welfare is affected by an increased mortality risk, not what his or her life is worth. If 100 000 people are exposed to an annual mortality risk of $1:100\,000$ there will, statistically, be one death incidence per year. If the risk is reduced to $1:120\,000$, then there will, on average, be 0.833 deaths per year, a saving of 0.167 of a statistical life. The cost of the risk-reduction measures can be used to place a value on a statistical life. Values of £1–2 million are usually applied.

Vulnerability The potential to suffer harm, loss or detriment, from a human perspective. It can be represented by the level of potential damage, or degree of loss, of a particular asset (expressed on a scale of 0 to 1) subjected to a damaging event of a given intensity. For property, the loss will be the value of the damage relative to the value of the property; for persons, it will be the probability that a particular life (the element at risk) will be lost, given that the person(s) is affected by the landslide.

Willingness to pay The amount an individual is prepared to pay in order to obtain a given improvement in utility, or specified reduction in risk.

Index

Page references in italics refer to figures and tables.

abandonment, 424–425, 449
Aberfan, 47, 218, *219*, 220, 313, 326, 338
Abergorchi colliery, 218
accident sequence, 27, *28*, 30, 31, 281, 315, 324, 328, 365
activity states of landslides, 56, *57*
adjustment bias, 173
Africa, 71, 474
ALARP principle, 449, 467
  applying the ALARP principle, economic risk, 439, 441–443, *444*
  applying the ALARP principle, loss of life, 437–439
  Australia, 430, 433
  BP major accident risk process, 435–437, 447
  Hong Kong, 433–434, *436*
  landslide mitigation measures example, 439, *440*
  risk tolerability and the ALARP concept, *428*
  Shell Group, 447
  UK, 427–429
Alaska, 70
Alberta Transportation Landslide Management System, 106
Alborz Mountains, Iran, 245–246
aleatory uncertainty, 7, 129, 143
  see also randomness
Alexander, DE, 297, 300, *301*, *302*, 311, 313, 315, 327, 330, 331, 334, 335, 349, 368
Alika slide, 70
anchoring, 173
Ancona landslide, 335
Andes, 70, 138, *139*, 212, 336–337, 474
apartment block, Hong Kong example, 292–293
Argillite Cut, British Columbia, 215–216
Armenia, 328
Armero mudflows, Colombia, 47, 309, 325
assets, 5, 9, 281–282, 324, *362*, 368
  asset value, 301, 305, 306, 307, 383, 384, 386, *387*, 391, *393*, 394, 395, 396, *397*, *398*

fixed/static assets, 13, 282–286
mobile assets, 13, 282, 286–291
vulnerability of a range of assets to landslide events, *303*
assurance, 84–88
  illustration of how degree of assurance might change with intensity of investigation, *86*
Aswan High Dam, 140, 222, 236
Atlantic, 481
Atlantis, 49, 268–270
Attachie slide, British Columbia, 58–59
Australia, 70, 100, 282. 288–289, 309–310, 327, 430, 431, 433–434, *435*, 474
  disaster relief, 423
Australian Geomechanics Society (AGS), 14, 15, 41, 51, 109, *112*, *113*, *252*, 300, 307, 310, 311, 431
availability bias, 173
average individual risk (AIR), 400–401, 402, *404*, *406*, 407, 493
AVI landslide database, 226

back-water flooding, 68
backward logic, 360
Baguio City, Island of Luzon, 107–109, *110*, *111*
Baku-Supsa oil pipeline, 207
Baku-Tbilisi-Ceyhan pipeline, 368
ball in a landscape, 37–39, *38*
barangays, Baguio City, 109, *111*
Basilicata region, Italy, 335
Bayesian probability *see* degree of belief
Beacher, G, 197
behaviour, landslide, 52–56, *53*, 273
  classification, *259*
  four stages of landslide movement, 53–54
  landslide displacement, in different states of activity, 56
  see also intensity; landslides, nature of; magnitude; velocity

Index

Bell Telephone Laboratory, 190
bench mark, use in modelling, 342
bent trees, 173
Berkely Escarpment, Vancouver, 237–238
best case scenario, 323, 368
Bildudalur village, Iceland, 304, *305*
binomial distribution, 143–148, *149*, 153–154, 168, 181, 208, 225, *228*, 268, 289, 408
biodiversity, 447, 467, 468
Bishop, AW, 267
Black Swans, 6–7, 209, 466
Blayais Nuclear Power Station, 466
Boston soft clay embankment, 174
boulders, impact of falling, 68, *364*, 365, *411*, *436*
    see also cliff sites; roads; rock catch fence; rock fall
bow-tie analysis, 30–31, *31*
BP, 138, 178, 208, 212, 435–437
    Group Risk Standard, 447
Brazil, 368, 474
Brighton, UK, 283, 287–288, 405–407
    Undercliff Walk, 310–311, *312*
Brooks Peninsula, Vancouver Island, 137–138, 161, *162*, 163–164
Building Research Establishment, 349
burial, 24, 64, 68, 308, 311, 338

Cambrian railway, Wales, 360
Camisea natural gas liquids pipeline, 138, 461, 463–464
Canada, 15–16, 47, 50, 55, 58–59, 60, 71, 80, 106, *107*, *108*, 109–110, 112–114, 137–138, 161–162, 163–164, 209, 210–211, 215–216, 237, 327, 335–336, *337*, 352–353, 407–409, 424, 431, 433, *434*, 482
Canary Islands, 71
Casagrande, 172
Caspian Sea, 16–17, 245, 368
Castle Peak Road, 342, 415–416, *417*
Çatak disaster, 284, 323–324
catastrophe potential, 29, *322*, 323, 328, 368
Caucasus pipeline, Georgia, 257–262
Central America, 474
chance, role of, 359–360, 375
    see also randomness
chemicals, 190, 359
China, 47, 54, 68, 70, 138, 173, 309, 472
    see also Hong Kong
Christian, JT, 174, *175*, 182
Cilfynydd colliery, 218, *219*, *220*
civil disorder, 336
climate change, 32, 207, 213, 263, 272, 293, 299, 449, 452, 459, 471–472, 473, 475, 476, 471–472
    see also global warming; IPCC

climate, global, 470–471
climate modelling, 472–474
coalfields, 47, 216, 217, 218–221, *219*, *220*, *221*, 313, 326, 338
    see also mining spoil tip examples
coastal defence study, strategic, 101
coastal sites, 101, *102*, *103*, 104–106, 116–119, 121–124, 137–138, 142, 144, 158, 161, *162*, 163–164, 168, 189–190, 207, 208, 213–215, 237–238, 262–267, 267–268, 268–270, 281, 283, 310–311, *312*, 336, 349–352, 444
    cliff foot, 116, 148, 185, 230, 233, 237, 268, 281, 285–286, *341*
    cliff line, binomial modelling, 145, 146–148, *149*
    cliff line, Poisson distribution, 149–151, 152–153
    cliff recession, 65, 229–236, 344–345, *347*, 381–388, 391–396, *397*, *398*, 446–447, *448*, 466–468, 482–484
    cliff-top, 26, 65, 101, *103*, *105*, 121–122, 124, 213, 215, 231, 233–234, 236, 237, 238, 268, 281, *340*, 344–345, *344*, 345, *347*, 385, 386, *387*, 389, 391, 396, *397*, 446, 447, 466, 446–447, *448*, 466–468
    indicative standards of coast protection used in England, *441*
    see also fjords; rock catch fence; rock fall; seaside promenade; seawalls; submarine slides ; Undercliff; West Nile Delta
Coca River, 212
cognitive distortion, 173
Coledale, Wollongong, Australia, 327
Colombia, 47, 138, 309, 325
Colorado Rockfall Simulation Program (CRSP), 63
communication, 35–40, 494
    key to good communication, 39–40
complex events, 322
complex outcomes, 367–369
compound events, 321
computer generated results, 83
CONCAWE, 138
conceptual models, 184–192, 209, 236–243, 243–252, 252–267, 267–271
concrete, 9, 49, 71, 116, 121, 239, 300, 305, 307, 326, 327, 367, 446, *448*
conditional probabilities, 244, 252, 265, *266*, 299
confidence, 27, *61*, 88, 118, *119*, 124, *125*, *131*, 136, 174, 189, 194, 195, 230, 321, 421, 427, 474, 495
    overconfidence, 174
consensus, 142, 175, 176, 177, 178, *180*, *181*, 257, 264, 265
consequence assessment, 27–31, *28*, 106, *108*, *112*, *113*, *114*, 115, *116*, *117*, 118, 124, 126, 143, 281, 321–369, 375, 412, *413*, 414, 416, 422, 491

categories of adverse consequences, 329–330
distinguishing range of landslide events based on scale and complexity of adverse consequences, 321–322
key factors that determine damage, 323
modelling, 323, 339–359, *346*, *347*, *348*, *351*, 359–367
quantifying possible adverse outcomes, 323
timing, 323–324, 368
uncertainty, 339
consultation, 34, 35, 336, 421, 446, 452, 453
contingency planning, 449, 452
contingent valuation, 446
continuous probability distributions, 154–157, *158*, 194, 267
Convention on Biological Diversity, 467
convergence, 175, 264
corporate risk management, 435–437, 447, 449, *450*, *451*, *452*
costs, 34–35, 192, 281, 329, 330, 422
benefit cost ratio (BCR), 438, 439, *440*, *441*, 442, *443*, *444*, 446, *447*, *448*, 491
environmental costs, 445–446, *448*, *451*
implied cost of averting a fatality (ICAF), 438, 439
maintenance and repairs, 388, 464
opportunity costs, *341*, 354, 355
see also economic factors
cracks, 53, 64, 122, 197, 206, 237, 246, 442–443
Crozier, MJ, 253
Cumbre Vieja volcano, 71
cumulative change, 471
cumulative probability, 144, *145*, *146*, 157, *158*, 161, 220, 223, 224, 270, 271, 390
current annual risk, 376–381
cut slope failure 97, *98–99*, 100, 151, 238–243, *244*, 290, *341*, 343, *344*, *345*, 357–359, 412–416, *417*

damaging events, 4, *24*, *26*, 75, 449, *450*, 495
dams, 68–70, 116, 192, 330
ANCOLD risk criteria, 430, 433–434, *435*
British Columbia Hydro risk criteria, 433, *434*
dam overtopping, *188*
historic landslide dams, *69*
tailings dam, 338
data collection, historical *see* historical data collection
see also dating landslides; incident databases
data uncertainty, 339
dating landslides, 55–56, 139–140, 208, 218, 222, *223*, *224*, 236
decision making, 41–42, *42*, 85, 88, 375, 381–382, 425, 452–453
Deepwater Horizon explosion, 6
defensibility, 84–88
need to ensure judgements are defensible, 88

summary of expected defensibility of different methods for undertaking subjective judgements, *87*
degree of belief interpretation, 136, 142, 142–143, 171, 176, 207, 208
Deixi landslide dam, 68, *69*, 70
delaying action, 449
Delphi panel, 175, 176, 264
design event approach, 15, *115*, *116*, 251
deterministic method, *14*, 129, 192, 194, 229, 230, 236, 267, 268, 271, 273, 359, 384, 391, 394, 396, 491
detriment, 8, 9, 11, 12, 32, 34, 321, 322, 323, 324, 328–329, 368, 422, 491, 494, 495
differential ground movement, 65, 67–68
direct losses, *124*, 330, 334, 335, 336, 337, 376, 491
disaggregation, *59*, 75
disagreement, 175
disaster, 11–12, 491
see also catastrophe potential; humanitarian relief
disaster potential, 29
disaster relief, 423–424, 452
discount factor, 383, 384, *385*, 386, 389, *390*, 391, *393*, 394, 395, 396, *397*, *442*, 491–492
discrete events, *137*, 437
displacement, pipeline, 259
distortion, 65, 67–68
distribution, 143–170, 181, 207, 208, 216–225, *226*, *227*, *228*, 229, 231, *232*, 233, *234*, *235*, 236, 268, 271, 289, 357
do nothing scenarios, 42, 179, 180, *181*, 263, 389, 422, 438, *440*, 439, 442, *447*, *448*
Dorset, 9–10, 72, *73*
downstream flooding, 68, 330
see also flooding
Dragon Kings, 158, 209, 466
drainage, 116, 118, 122, *125*, 156, *158*, 194, 229, *242*, 442
dread, 36, 418
Duncan, JM, 172, 267
Dunwich Suffolk, 336
Durham Fatal Landslide Database, 48, 326, 327
dynamic intensity, 302

early warning, 30, 80, 101, 192, 254, 257, 294, 295, 423, 441, 449
earnings foregone approach, 334
earthquake, 27, 47, 50, 54, 68, 70, 131, 156, 157, 159–160, 169, 176–177, 206, 207, 244, 257, 298, 300, 308, 311, 328, 335, 464–466
probability of earthquake-triggered landslide, 249, 251–252
Easton Bavens, Suffolk coast, 466–468
EC Habitats and Species Directive, 467

Echizen-cho, Japan, 327
ecological factors, 3, 30, 36, 156, 338, 359
  ecological vulnerability, 299
economic factors, 311, *313*, 314, 315, 324, 329–330, 368, *434*, 493
  applying the ALARP principle, 439, 441–443, *444*
  comparing benefits of landslide management with cost incurred, 426
  consequences of cliff recession example, 345
  consequences of damage to coastal property example, 350–352
  consequences of pipe rupture example, 349
  consequences of road blocked by rock fall, Scottish Highlands, additional transportation costs, 353–357
  direct losses, *124*, 330, 334, 335, 336, 337, 376, 491
  financial crisis, 2007, 6, 7
  financial risk, definition, 2
  indirect losses, 330, 336–337, 376, *378*, *382*, *392*, 491
  intangible losses, 329, 330, 337–338, 368, 493
  loss of life or injury as priority of risk assessment, 330–334
  predicting economic losses, buildings, structures and infrastructure, 334–336
    direct losses from geohazards, 335–336
  present value (PV), *383*, 384, *385*, 386, *387*, 388, *390*, 391, *392*, *393*, 394, 395, 396, 439, 441, 446, 447, 491, 493
  productivity of capital, 383
  public expenditure, 425–426
  recession, 381–382
  statistical life, value of a, 34, 332–334, 410, 438, 439, 495
  total economic value (TEV), 469–470
  total loss event, 339, *340–341*, 342
  willingness-to-pay/willingness-to-accept, 333, 438, 470, 495
  see also asset value; costs; coastal sites, cliff recession risk; current annual risk; individual risk; management options, risks associated with; market value; societal risk; statistical analysis
ecosystem services, 446, 468–470
Ecuador, 212
EGIG, 138
electricity pylon example, 286
elements at risk, 9, 10, 11, 12, 33, 43, 78, 102, *103*, 104, *114*, 115, 230, 281, 297, 299, 305, 324, 323, 324, 339, 341, 375, 425, 459, 470, 492
Elm rockslide, 47
EM-DAT database, 326
ENSURE, 299, 300
Environment Agency, UK, 142

environmental acceptability, 425, 447
environmental change, global, 470–471
  see also climate change; climate modelling; global warming
environmental costs, 445–446, *448*, *451*
environmental hazard, term, 11, 492, 493
  see also ball in a landscape
environmental impact assessment (EIA), 445
environmental protection, 443–447, 466–468
  examples, 446–447, *447*
Environmental Protection Agency (USA), 3, 178
environmental resources, valuation of, 468–470
environmental risk, term, 492
environmental uncertainty, 339
environmental vulnerability, 298–299, 324, 330, 368
epistemic uncertainty, 129
EPOCH project, 51, *52*
ERM-Hong Kong, 342, 410, *411*, 415, 430, 434
estimation, 32, 33, 85, 133, 184–185, 186, 206, 207, 209, 212–213, 225–236, 243–252, 254, 256, 258, 259–262, 263, *264*, *265*, *266*, 269, 270, 271, 272, 273, 321, 328, 412, 421–453, 475, 494
Euler-Mascheroni constant, 168
Europe, 5, 51, 71, 138, 139, 328, 467, 474
eustatic movements, 476–477
evaluation, 22–23, 33–35, 41, 42, 49, 81, 82, 84, 97, 101, 102, 104, 108, 142, *175*, 190, 196, 268, 271, 293, 357, 360, 368, 375, 394, 421, 425, 426, 439, 445, 446, 447, 453, 461, 463, 468, 469, 474, 493, 494
event cascade, *24*, 27
event sequence, 27, *28*, 185–186, *206*, 281
  see also event tree; fault tree; rock fall pathways; top event
event size distribution, 233
event timing distribution, 231, 233, 234
event tree, 25, 27, 30, 31, 133, 134, 182, 184, 186, *187*, *188*, 189–190, 208, 255, *259*, 262, 263, *264*, 265, *295*, 360–365, *363*, *364*, 492
exceedence probability, 166–170, 208, 225–229
expert judgement, 171–175, 208, 252–267, 299–300, 362
  seven experts' of additional height of fill to cause failure of the I-95 embankment, *174*
expert panels and their operation, 175–181, 349, 365
  informal procedures, self-assessment question and management/resource cases, 179–181, *180*, *181*
  Stanford/SRI protocol, 178–179
exponential distribution, 156
exposure, 3, 64, 119, 281–296, *322*, 350, 354, 359, 375, 492
  reducing exposure, 294–296
ExxonMobil, 447

F–N curve, 378, 410, *411*, 412, *414*, 415, 432, 437, 492
failure, categories of causal factors promoting, 206
failure, landslide movement, 53–54, 58
  large failures, 101
  small-scale shallow failures, 101
failure, schematic diagram showing redistribution of potential energy after, 76
failure modes, effects and critical analysis (FMECA), 100, 115–124, *125*
failures on pre-existing shear surfaces of non-landslide origin, 58
Falli-Holli, Switzerland, *74*
fat-tailed distributions, 157–160
fatalities, 6, 11, 13, 14, 30, 36, 47, 48, 70, *75*, 81, *113*, 114, 130, 131–134, 143, 173, 185–186, *187*, *188*, 190, 238, 281, 290, 296, 297, 299, 304, 307, *308*, 309, 310, 311, *312*, *313*, 315, 321, 323–324, 325–328, 329, 330–334, 335, 338, 352, 353, 359, 362, *364*, 365–367, 396, 409, 410–411, 464
  applying the ALARP principle, loss of life, 437–439
  best-known high-fatality landslide events, 326–327
  calculation of potential loss of life and average individual risk, *406*, 407
  F–N curve, 378, 410, *411*, 412, *414*, 415, 432, 437, 492
  fatal accident rate (FAR), 401–402, 404
    death per unit activity, 403, *404*
    fatalities associated with different modes of transport, UK, *403*
  fatalities associated with a reference landslide, 342, *343*, *345*
  HSE, UK boundaries of individual risk, 429, *430*
  implied cost of averting a fatality (ICAF), 438, 439
  incidents that caused fewer deaths, 327
  landslide mitigation measures example, 439, *440*
  multiple fatalities, corporate management and major accident risk criteria, 435–437
  multiple fatalities, societal risk criteria, 431–432, *433*, *434*, *435*, *436*
  potential loss of life (PLL), 400, 401, 402, 403, *404*, *406*, 407, 410, 412, *413*, 414, 415, 416, *417*, 418, 437, 438, 439, *440*, 493
  value of preventing a fatality (VPF), 334
fault tree, 31, 182, 184, *190*, 190–192, *191*, *193*, 208, 360, 365, *366*, 367, 492
Fei Tsu Road, Hong Kong, 412–415
financial crisis, 2007, 6, 7
financial risk, definition, 2
  see also costs; economic factors
first-time failure, 54, 56, 58, 65, 67, 72, 143, 206, 249, 253, 281–282, 435–439
fjords, 71, 254–255, 257, *258*
Flims landslide, 58

flooding, 4, 5, 10, 16, 25, 27, *57*, 59, 60, 64, 65, 68, 70–71, 113, 135, 143, 157, 158, 166, *167*, 168, *193*, *210*, 212, 226, 273, 304, 307, 322, 325, 327, 330, 334, 369, 375, 444, 445, 464, 466, 468, 475, 476, 492
flow slide example (old mining tip), hypothetical, 288
flow slides, 50, 52, 55, 58, 217, 218–221, 237, *238*, 288, 352
flow slides, South Wales coalfields *see* coalfields
follow-on hazards, 25, 311, 323, 329–330, 336, 491, 492
forecasting, 80, 181, 246, 422, 423
FORM approach, 195, 268
France, 16, 161
  disaster relief, 424
  legal code, 427
Frank slide, 60
Fraser River salmon fisheries, 337
frequency, term, 492
frequency assessment, historical, 101, 138, 182, 208, 212–216, 226, 227, 238, 239, 259, 272, 283, 284, 376, 415, *450*, 465
  see also historical data collection
frequency-magnitude distributions, 158, 159, 160, 160–166, 207, 208, 216–225, 236, 465
frequentist interpretation, 136–142, 142–143, 176, 207, 209, 236
Fukushima, Japan, 464–466
future threat, 27

gambler's fallacy, 144
gamma distribution, 156, *165*
Gansu mudslide, 47
gas, 4, 16–17, 25, 30, 49, 138, 184, 190, 209, 221, 283, 321, 336, 345, *347*, 384, *385*, *393*, 401, 429, *430*, 461, 463, 464, 492
gas explosion example, 365–367
gas field on seabed example, 377–378, *379*, *380*
Gaussian distribution, 154–156
GDP, 311
Geohazard Assessment Team (GAT), 178, 208
geohazards, 11, 13, 23, 25, 32, 35–36, *38*, 43, 48, 110, 114, 134, 178, 208, 212, 298, 299, 322, 328, 331, 335, 425, 463, 492, 493
geometric intensity, 302
geometric mean risk ratings, *37*
geomorphological analysis, 82, 84, 101, *102*, 104, 239, *241*, *242*, 263, 272, 273, 325, 444, 445, *462*
Georgia, 207, 257–262
geotechnical approach, 8, 14, 15, 16, *82*, *85*, 97, 114, 104, 115, 142, 172, 173, 174, 178, 179, *180*, 182, 192, 208, 209, 212, 215, 238, 239, 252, 263, 267, 272, 337, 463, 464
Germany, 16, 301
Giddens, A, 5, 13

501

GIS technologies, 83, 229, 459
giving-up, 424–425, 449
global change, 470–471
global risk assessment, 41, 492
global warming, 425, 471, 472, 473, 476, 479, 480
Grand Banks slide, 50
Green Book: Appraisal and Evaluation in Central Government, 469
greenhouse gases (GHGs), 471–472, 473
groundwater, 15, 57, 58, 72, 76, 107, 122, 129, 184, 205, 206, 230, 233, 237, 238, 239, *242*, 244, 246, *264*, 270, 271, 376, *462*, 480
group reporting line (GRL), 436–437
group risk, 14, 401, 410, 429, 447
Guatemala, 165
Guinsaugon, Leyte Island, 217
Gulf of Mexico, 6, 16, 49, 268–270, 285–286, 403–404
Gulf Oil, 403
Gumbel distribution, 156, 166, *167*
Gutenberg-Richter law, 159–160, *160*

Halcrow, 72, *247*, *308*
Hawaii Island, 70
hazard, 8–13, 47–88, 297, 492
    assessment, 23–27, *24*, *26*, *82*
    hazard groupings for events of different sizes and assets at different distances from slope base, *362*
    hazard model, *22*, 23, 83–84, 85, 115, 122, 124, 206–207, 216
    hazard spectrum, *11*
    likelihood of hazard, 29–30
    risk as product of hazard and vulnerability, *12*
    secondary hazards, 330, 336
    see also modelling; probability; risk; vulnerability
hazard zoning *see* zoning
health, 3, 5, 39, *120*, 138, 297, 309, *313*, 314, 324, 325, 327, 331, 333, 334, 337–338, 425, 427, 438, 445, 449, *451*, 493, 494
Herodotus Basin, *139*, 221, 222, *224*
historical data collection, 100, 101, 138, 182, 208, 212–216, 225, 226, *227*, *228*, *234*, *235*, 236, 237, 238, 239, 246, 247, 259, 272, 283, 323, 324, 336, 365, 376, *411*, 415, *450*, 465, 483–484
    using the historical record, 324–329
    see also dating landslides; incident databases
Holbeck Hall landslide, 101, 215, 236
Holderness coastline, 65, 336, 385–388, 482–484, *483*
Hong Kong, 14, 15, 56, 61, *62*, 84, 97, *98–99*, 100, 104, 114–115, *116*, 182, *183*, 184, 238–243, *244*, 292–293, 294, 295–296, 300, 308, *309*, 326, 342, *343*, *345*, 345–349, 357–359, 360–362, *363*, 368, 410–411, 412–415, 415–416, *417*, 430–431, *433*, 434–435, *436*

Hong Kong Cut Slope Ranking System, 97, *98–99*, 100
Hong Kong Geotechnical Control Office (GCO), 97
Hong Kong Natural Terrain Landslide Inventory (NTLI), 182, 345, 346, *348*
Hope slide, British Columbia, 47
Housing, 72, 282, 294, 297, 306, 307, 313, 315, 324, 350, 361, *431*, 449, 459, 460
Huascarán disaster, 47, 55, 60, 325, 335
human activities, 324, 368–369, 401–402, *403*, *404*, 410, 429, *430*
human capital, 333
human factor, 33, 79
human systems, *24*
human vulnerability, 307–311, 352, 422
    loss of life or injury as priority of risk assessment, 330–334
human well-being, 468–470, 493
humanitarian relief, 423–424
Hurricane Camille, 403
Hurricane Mitch, 165, 321
Hurricane Tomas, 327

IAEG, *66*, 80
ICE, 85
Iceland, 304–305
    individual risk criteria, *431*
ICI, 30
impact collapse flow-slides, 55
impact score, 119, *120*, *121*, *124*
incident databases, 209–212
Index of Social Vulnerability, 316
Indian Ocean, 481
indirect losses, 330, 336–337, 376, *378*, *382*, *392*
individual risk, 14, 282, 287–288, 332, 362, 418, 429–431, *430*, *431*, 432, 429–431, 493
    economic factors, 396, 398–409, 410
individual-specific individual risk (ISIR), 399–400, 401, 405, *406*, 407, 493
Indus River, 25, 68
information value method, 83
injury, definition of, 331–332
intangible losses, 329, 330, 337–338, 368, 493
    examples of problematic nature of, 338
intensity, 74–79, *75*, *76*, *77*, 302
    intensity scale for different types of landslide, *78*
    theoretical relationship between landslide intensity and vulnerability, *304*
    values for kinetic energy and kinematic intensity parameters, *78*
    see also magnitude; velocity
intention, description of, 23
InterRisk Assess project, Germany, 301

inventories, 82, 97, 182, 208, 225, 345, 346, *348*
IPCC, 26, 40, 171, 178–179, 471, 473, 474, 479, 480
Iran, 58, 245–246, 328
ISRM, 80
ISSMGE, 7, 8, 80, 299
Italy, 71, 226–227, 229, 290–291, 300, 305–307, 322, 327, 335, 423, 459

Jack problem, 173
Japan, 70, 327, 368, 464–466

Kansu, 328
Kates, RW, 11, 13
Knight, F, 2
knowledge, imperfect, 129, 184
Kuzulu, Turkey, 327

Lago di Garda, 290–291
Lake Geneva, 71
Lake Tahoe, USA, 337
land use multipliers, 109, *111*
land use planning, 80, 101, *113*, 449, 459, 560
landslide mechanisms, 49–52
landslide types, 51–52, 299
 categories of, 322–323, *322*
 classification of types, 50, 51, *52*
 diagram of, *50, 51*
landslides, nature of, 63–74
 see also intensity; landslides, behaviour; magnitude; velocity
Landslip Nature Reserve, Devon, 444
lateral displacement, *26*, 64, 261
Lawrence Hargrave Drive, Sydney, 288–289, 309–310
legal aspects, 426–429
 reasonably practicable, 428
 risk tolerability, 427, 428–429
 see also ALARP principle
Lemieux, Ontario, 424
Lie Yue Mun squatter villages, Hong Kong, 295–296, 360–362, *363*
likelihood, 4, 5, 6, 12, 13, 14, 15, 16, 25–26, 29–30, 32, 35, 41, 43, 79, 100, 101, *103*, 106, 109, *112*, 116, 118, *119*, 124, *125*, 130, 134, 143, 144, 169, 171, 173, 176, 192, 196, 229, 233, 234, 244, 245, 247, 253, 257–258, 259, 260, 261, *262*, 263, 265–266, 267, 315, 323, 360, 365, 367, 368, 375, 437, 452, 473, 492, 493
Linda problem, 173
liquefaction, 49, 54–55, 58
Lituya Bay, 70
loading, 64, *189*, 477
location, cause, indicator (LCI) diagram, 116, *117*, 118, *119*, 122, *123*

location-specific individual risk (LSIR), 398–399, 400, 402, 403, *404*, 405, 430, 493
log-normal distribution, 156, 164, 166, 233
logarithmic function, 156–157
logging, 109–110, 112–114, 335
logic gates, 190–191
Loughborough Inlet, British Columbia, 161–162, 164
Lugnez landslide, 76
Lushan, Taiwan, 270–271, 272
Lyme Regis, 72, *73*, 262–267
Lympne, Kent, 74

Machu Picchu, 338
Mad Dog, 49, 268–270
MAFF, 26, 391, *441*
magnitude 29–30, 41, 158, 159, 160, 160–166, 207, 208, 216–225, 249, 251, 252, 299, 322, 359, 365, 368, *381*, 465
 hypothetical diagram of magnitude-frequency distribution of landslides, *48*
 see also intensity; velocity
main-phase hazards, 65–68
 definitions of features, *67*
 morphology, *66*
maintenance, 59, 75, *108*, *113*, *119*, 122, 215, 142, 282, 300, 302, 326, 334, 354, 379, *388*, 389, 407, 428, 439, 463, 464, 466
major accident risk criteria, 435–437
Mam Tor, UK, 56
management, 23, 33–35, 388–396, *397*, 421–453, 466–468, 494
 adjustment choices, 422–424
 classification of adjustment choices, 423
 assessment criteria, 425–426
mapping, 16, 60, 72, 79, *80*, 81, 84, 97, 101, *102*, 104, 115, 182, 186, 211, 212, 215, 217, 226, 227, 246, 263, 286, 306, 384, 459–461
 direct mapping approaches, 82–83
 indirect mapping approaches, 83
margin of stability, 205
market value, 124, 345, *347*, 384, *385*, 386, *387*, 391, 396, *397*, 398, 493
mass liquefaction, 54
matrices, *33*, *38*, 100, 109–110, 112–115, 447, 449, *450*, *451*, *452*, 464
Mayunmarca sturzstrom, 68
media, role of, 328
Mediterranean, 139, 221, 474
Merthyr Vale Colliery spoil heap *see* Aberfan; coalfields
Micronesia, 330
Middle East, 328
Millennium Development Goals, 469

503

mining spoil tip example, 288
    see also coalfields
missile impacts, 64, 68
modelling, 7, 21, *22*, 23–24, 63, 83–84, 85, 100, 115, 122, 124, 129, 134, 140, 143–170, 171, 180–181, 182, 184–197, 205–273, 298, 299, 323, 339–359, 359–367, 385, 375–418, 472–474, 438
    Fukushima, inadequate tsunami magnitude-frequency model, 465
    Fukushima, inadequate run-up modelling, 465
monitoring, 23, 27, 40, 53, 65, 80, 100, 101, 106, *103, 108*, 116, 118, *119*, 121, 179, 180, 181, 210, 253, 257, 263, 294, 326, 337, 422, 423, 436, 437, 442, 446, 449, 452, 459, *462*, 463, 464
Monte Carlo simulation model, 230, 234
motivational bias, 173
Mount Cook, 47
Mount Elgon, 327
Mount Pinatubo, 60
Mount St Helens, 70, 209, 322, 335
Mount Unzen, 70
Mountain footslope channelised debris flow, cost and benefit example, 446, *447*
mountain range prone to debris flow, cost and benefit example, 443, *444*
mudflows, 16, 54, 60, 285, 286, 309, 325, 327, 335
multiple consequence models, 359–367
multiple events, 321–322, 325
multiplication rule for joint probability, 133
myths of nature *see* ball in a landscape

Nanga Parbat Massif, 68
National Landslide Review, 143
Natural England, 467
natural hazard, 3, *11*, 35, 49, 151, 298, 333, 468, 493
Natural Terrain Landslide Study, 182, *183*, 184
Near East, 179
Nepal, 315
Netherlands, *433*
    legal system, 427
    public safety regulations, 430
Nevado del Ruiz volcano, 47, 309, 325
New South Wales, 70, 100
New Zealand, 47, 248–249, *250*, *251*
Newfoundland, 50, 71
Nicaragua, 368
NGOs, 423
non-statistical portrayal of risk, 33
normal distribution, 154–156, *155*, 157, 158, *159*, 271
normalisation of base-rate frequency, 236–243
NORMDIST, 286
North Island, New Zealand, 248–249, *250*, *251*

Norway, 17, 49, 71, 163, 208, 254–255, 257, *258*
nuclear industry, 176, 186, 190, 209, 432, 464–466
numerical grading, 30
numerical modelling, 100

objective risk, 33
occupancy, 9, 15, 79, 130, 134, 186, *187*, 281, 282, 288, 292–294, 295–296, 300, *302*, 315, 327, *341*, *343*, 345, 352, 353, *354*, *355*, 356, *357*, 360, 365–367, *399*, 400
    see also population; population models; population multipliers
OECD, 332, 333, 334
oil, 4, 6, 16–17, 30, 49, 138–139, 157, 179, 190, 207, 209, 210–212, 268–270, 285–286, 336–337, 345–349, 354, 368, 401, 403–404, 429, *430*, 435–437, 447, 461
    measures of mudslide risk to workforce of offshore platform, *404*
open forum, 175, 264
organised systems, 273
outcomes, *266*, 323, 367–369, 493

Pacific, 481
Pakistan, 68
Papua New Guinea, 210, 211–212
    Ok Tedi copper mine waste, 338
    Southern Highlands, 321
Pareto distribution, 158–159
    double Pareto distribution, 164, *165*
peak strength failures, 53–54
perception, 34, 35–40, *37*, 38, 39, 40, 48, 64, 79, 97, 100, *107*, 126, 170, 173, 179, *313*, 328, 350, 375, 418, 421, 493, 494
Perfect Storms, 6–7, 466
Perry Ridge, British Columbia, 109–110, 112–114
Peru, 47, 52, 55, 60, 138, 321, 325, 327, 338, 461, 463–464
petrochemical industries, 209
Philippines, 60, 107–109, *110*, *111*, 217
PHMSA, 209, *210*
PHSA, 176, 177
physical vulnerability, 297, 298, 299–300, 324
    buildings and infrastructure, 300–307
pipelines, 4, 16, 27, 64, 30, 32, 76, 80, 138, 179–181, 184, 207, 209, 210–212, 257–262, 282, *303*, 336–337, *340*, 345–349, 368, 377, *379*, 459, 461, 463–464
    example event sequence involved in generation of landslide risk to pipeline, *206*
    framework for estimating probabilities, pipeline crosses recorded landslide, *260*, 261
    framework for estimating probabilities, pipeline passes upslope of recorded landslide, *261*, 261–262

indicative probability bands used in review of existing chemical products, 262
see also gas; oil
planning, 430
  contingency planning, 449, 452
  land use, 80, 101, *113*, 449, 459
  town planning, 460–461, *462*, 467
Poisson distribution, 148–154, 168, *169*, 208, 216, 225, 226, *227*, 229
population, 9, 14, 30, 39, 47, *59*, 97, 107, 115, 119, *120*, *121*, 156, 237, 239, 282, 297, 304, *308*, 314, 324, 329, 331, 332, 338, *340*, *341*, 361, 365, 368, 369, 396, 398, 400, 402, 403, 404, 406, 407, 410, 412, *413*, 421, 424, 432, 445, 468, 474, 475, 492, 493, 494
  subpopulation, *139*, 182, 184, 213
  see also fatality; human vulnerability; occupancy
population models, 292–294, 295–296, 365
  residential population, 293
  tourism, 293
  traffic, 293–294
population multipliers, 109, *111*
Portuguese Bend landslide, California, 74
post-event hazards, 25, 492
  civil disorder, 336
  return to stability, 71
post-failure movements, 54, 206
potential for landslides *see* susceptibility
power laws, 158–159, 161, 162, 163–164
pragmatism, 271–273
pre-failure movement, 53, 64–65, 65, *119*, 197, 206
precision, 271–273
prediction, 79–80
  for greater detail see susceptibility
preparedness, 14, *31*, 257, 314, 422, 466
present threat, 27
present value, *383*, 384, *385*, 386, *387*, 388, *390*, 391, *392*, *393*, 394, 395, 396, 439, 441, 447, 491, 493
primary site factors, 239–243
probability, 26–27, 104, 106, *107*, *112*, 118, 124, 129–179, 205–273, 412, *413*, 449, *450*, 496
  classification, 253, 254
  features associated with active and inactive landslides, 255
  flow chart for estimating probability, soil slides, 256
  meaning of, 135–136
  rules, 130–135, *131*, *132*
  worksheet used to define conditional probability of an outcome, 266
  see also distribution; susceptibility; risk
probability method, 181–182, 230, 383, 384, 394, 396, 493
progressive failure, 54, 122

promenade, seaside *see* seaside promenade
Puerto Rico, 162–163, 326
pump station examples, *399*, 402, 410

qualitative analysis, *82*, 97–126, 418
  Perry Ridge qualitative risk assessment matrix, *112*
  Saskatchewan Highways and Transport qualitative risk rating approach, *107*, *108*
  South Shore Cliff, Whitehaven, qualitative risk assessment, *105*
  value of, 124, 126
qualitative expressions of likelihood, 26
qualitative expressions of magnitude, 29–30
qualitative map combination, 82–83
qualitative risk estimation, 32, 33
quantitative analysis, *82*, 97, 124, 138, 142–143, 184, 205, 311, 323, 375–418, 459–461, *462*
  value of, 126, 416, 418
quantitative expressions of frequency and probability, 26–27
quantitative risk estimation, 32, 33
quarries, 185, *343*, 362, *364*, 365
quick-clay behaviour, 55, 58, 75, 335–336, 424

railways, 194, *303*, 305, 327, 335, 337, 359–360
rainfall, increased, 474–476
rainfall and reactivation, 246–248, *247*, 257, 376–377, *376*, *377*
rainfall probability, 248–249, *250*, *251*
Rand Corporation, 176
randomness, 129, 273, 385, 416, 418
  see also chance
re-vegetation, 56
reach angle, 60, *61*, 62, 343, 344, 358, 361
reactivation, 55, 58, 71–74, *73*, *74*, 101, 208, 262, 263, 266, *340*, *392*, 438
  rainfall and reactivation, 246–248, *247*, 257, 376–377, *376*, *377*
recurrence interval (RI), 166–170, 216–217, 218, *220*, *221*, 226
Red Cross/Red Crescent, 48, 321, 325, 328
reference landslide, 342, *343*, 345
reliability index, 195, *196*, 268
reliability methods, 182, 192, 194–197, *196*, *197*, 209, 267–271
representation bias, 173
residual risk, 429, 437, 441, 442, 493
resilience, 8, 298, 309, 314
resistance, 53, 54, 79, 194, 205, 302, 303, *304*, 423, 467
return periods, 166–170, 494
Reventador Volcano, Ecuador, 212
Review of Research on Landsliding in Great Britain, 328

Index

Rimac River, Peru, 52
Rio Tinto, 447
risk, 7–8, 281, 282, 494
   categories of, 13–14
   concept of, 4–5
   definition of, 1–4, 104
   hazard to risk, 12–13, *12*
   landslide risk, 7–8
   see also individual risk
risk acceptability, 427–429, 432, 491
   see also tolerable risk
Risk Analysis, Perception and Management (Royal Society), 2
risk assessment, 33–35, 452–453, 484–485, 494
   procedures, 40–41
   process, 21–23, *22*
   purpose of, 14–17
   qualitative and quantitative assessment *see* qualitative analysis; quantitative analysis
   see also decision making; estimation; evaluation; likelihood; monitoring prediction; probability; statistics; susceptibility; uncertainty; vulnerability
risk control, 31, 432, 494
risk registers, 100–102, *102*, *103*, 494
risk reduction, 23, 34, 41, 115, 310, 388, 389, 391, 394, 405, 416, 422, 425, 427, 428–429, 432, *433*, 436, 437–438, 439, 441, *442*, 443, *447*, *448*, 449, *450*, *452*, 467–468, *491*
River Severn, 166, *167*
road blocked by rock fall, Scottish Highlands example, 353–357, 378–381
road construction, new, example, 439, *440*
road cuttings, 151, *341*, 343, *344*, *345*, 408, 409, 412–416, *417*
road fatalities in Hong Kong, reference slide approach, 342, *343*, *345*
road near mining tip flow slide examples, 288, 352
road-side coffee house Çatak, Turkey, example, 284, 323–324
road-side house example, loose uncompacted fill, 287
road-side housing, Nepal, example, 315
road-side restaurant example, intersection with landslide-prone hill slope, 284–285
roads, vulnerability to landslides, *303*, *305*, *306*
   see also boulders
robustness, 8, 300, 309, 313, 314, 331
rock catch fence, 27, 148, 185, 365, 439, *440*
   see also boulders; coastal sites
Rock Fall Hazard Rating System, 100
rock fall hitting an individual at the seaside, probability of, 287–288
rock fall landing on railway example, 359–360
rock fall pathways, 185–186, *187*, *188*, 190, 283, 353
rock fall related traffic accidents, 151, 190, 215–216, 281, 288–291, *310*, 327, *341*, 352–353, 407–409
Rockfall Simulation Program, Colorado (CRSP), 63
   see also boulders
rollover, problem of, 164, *165*
rule sets, 362, *363*
run-out, *61*, 68, 121, 218, *219*, *220*, *221*, 285, 307, 344, 357, 460
Runswick Bay, North Yorkshire coast, 349–352

safety, 3, 14, 15–16, 68, 172, 173, 194–196, *195*, *196*, 197, 205, 209, 245, 261, 267, 268, 269, 270, 271, 272, 301, 309, *314*, 332, 333, 334, 396, 427, 429–431, 438, 445, 449, *451*, 465–466
Saidmarreh landslide, 58
Saint-Jean-Vianney, Quebec, 335–336, 424
San Andreas Fault, California, 249, 251–252
San Francisco Bay, 335
Sarno, Italy, 327
Saskatchewan Highways and Transportation, 106, *107*, *108*
Sau Mau Ping Resettlement Estate, Hong Kong, 326
Scandinavia, 55
scoping, 23, 27, 29, 33, 41, 43, 83, 213, 368, 445, 446, 494
scoring scales, 30, 97, *98*
   confidence scores, *119*, 124, *125*
   criticality scores, 124
   impact score, 119, *120*, *121*, 124
   relative risk scoring or rating, 100, 102, 104–109, *110*, *111*, 119
   risk scores and management responses, Saskatchewan, *108*
Scotland, 17, 71, 353–357, 378–381
screening, 23, 43, 97, 100, 494
sea-level changes, 140, 207, *224*, 236, 292, 452, 459, 474, 476–484
   components of relative sea-level change, *476*
   current regional net sea-level rise allowances for UK, *482*
   global mean decadal rates of sea-level change over period 1904–2003, *479*
   Holderness Erosion Post 59 recession predictions for period 1990–2004, *483*
   mean global sea-level record over period 1904–2003, *478*
   observed rate of sea-level rise, *480*
   overall annual recession for Holderness coast, *843*
seabed, 49, 50, 140, *141*, 152, 166, 178, 208, 377, 377–378, *379* 380
   see also submarine slides
seaside promenade, 283, 287–288, 405–407

Seattle, USA, 225–226, *227*, *228*
seawalls, 116, 122, *125*, 142, 144, 168, 189–190, *189*, 263, *264*, 265, *266*, 267, 389, 391, 393–394, 444, 446, *448*
secondary hazards, 25, 43, 64, 68–71, *295*, 311, 315, 323, 329–330, 336, 361, 491, 492
semi-quantitative risk estimation, 32
sensitivity testing, 194, 375, 495
Sevenoaks Bypass (A21), 72
sewers, 25, *237*, 336, 345, 384, 442, 492
shallow slides, 56, 59, 71, 80, *101*, *103*, *108*, 122, *208*, 212, 227, 245, 261, 283, 307, 335, 342, 345, 377, 475
Shanklin, Isle of Wight, 311
Shansi earthquake, 54, 328
Shasta, California, 58
Shau Kei Wan squatter area, 410
Shell, 30, 403
   Hazards and Effects Management Process (HEMP), 447
simple events, 321
simulation models, 229–236
site-investigation, need for adequate, 85
site-specific risk assessment, 41, 80, 84, 109, 138, 207, 221, 233, 237, 238, 253, 267, 269, 310, 339, *363*, 466, 495
Sites of Special Scientific Interest (SSSI), 444, 466–467
slab slides, 283–284
sled-model, 63
sliding surface liquefaction, 54–55
slope movement, ability to withstand changes, 205–206
slope movement, different stages of, *53*
   see also cut slope failure
Slope Risk Analysis System, 100
snow, 63, 165, 304, 305, 322, *431*, 473, 480
social amplification of risk, 39
social attenuation of risk, 39
social loss, 329
social vulnerability, 297, 311, 313, 314–315, 324
   landslide risk assessment, 315–316
social welfare, 425
societal loss, 329
societal risk, 13–14, 293, 332, 362, 378, 401, 431–435, *433*, *435*, *436*, 493, 495, 401, 409–416, 431–435, 495
   economic factors, 409–416, *411*, *413*, *414*, *415*
societal vulnerability, 297, 311, 313, 315
   key components of societal vulnerability, *314*
   landslide risk assessment, 315–316
soil, 49, 50, *52*, *53*, 54, 58, 59, 60, 76, *98*–*99*, 129–130, 154, 172, 209, 211, 236, *238*, 245, 248, *255*, *256*, 267, 279, 298, *302*, 306, 345, 422, 444, 445, 463, 464, 466, 468, 471, 475
   see also ISSMGE
South America, 71, 474

South Bay, Scarborough, 101–102, *102*, *103*, 121–124, *122*, *123*, *125*, 213–215, 214, 236
South Shore Cliffs, Whitehaven, UK, 104–106
   examples of cliff section risk assessment, *105*
   qualitative risk assessment, *105*
   risk zonation plan, *106*
   summary of cliff instability, *214*
special probability, 286–291
spanning, 64, 259
speed *see* velocity
SSHAC, 175, 176, *177*
St Lucia, 307, 327
Stanford/SRI protocol, 178–179
Starr, C, 35
statistical analysis, 33, 34, 83, 100, 182–184, 208
   estimating landslide exceedence probability, 225–229
   estimating probability of cliff recession through stimulation models, 229–236
   historical frequency assessment, 212–216
   use of incident databases, 209–212
   use of landslide magnitude-frequency curves, 216–225
   value of statistics, 416–418
statistical life, value of a, 34, 332–334, 410, 438, 495
   table of VOSL values, *333*
Storegga slide, 17, 71, 163
stopping distance, vehicle, 291
Storfjord, Norway, 254–255, *257*, *258*
structural damage, *26*, *59*, *76*, *78*–*79*, *112*, *113*, *114*, *116*, *120*, 122, 185–186, *187*, 237, 281, 282, 284–285, 287, 288, 292, 297, 300–307, *305*–*307*, *305*, *306*, 324, 327, 331, 334–336, *340*, *341*, *343*, 349–352, 360–362, *382*, 389, 390–394, *394*–*396*, *397*, *398*, *446*–*447*, 466–468
   buildings, factors affecting vulnerability of, 300
   buildings, factors that influence impact of damage on, 300–301
   classification of damage to structures and infrastructures, 300, *301*
   damage intensity scale, *302*
   public building showing signs of cracking, example, 442–443
   vulnerability of a range of assets to landslide events, *303*
   vulnerability of buildings according to the type of damage, *303*
sturzstroms, 55, 59, 68
subjective approach, 25, 29, 32, 33, 35, 39, 40, *87*, 88, 142, 143, 171, 172, 207, 208, 230, *253*, 299, 330, 338, 339, 416
submarine slides, 17, 43, 49, 50, 55, 58, 63, 70, 75, 139–140, 162, 163, 166, 208, 221–225, 285–286, 403
   submarine landslide influence diagram, *141*
   see also seabed

susceptibility to landslide, 56–58, 79–83, 115
    classification of hazard assessment techniques, *82*
    hazard terms, *81*
    mapping terms, *80*
Sussex cliffs, UK, 231–236
sustainable development, 469
Switzerland, 16, 47, 53, 58, 71, *74*, 76, 300
systemic change, 470–471

Tadzhikistan, 47, *69*
Taiwan, 270–271, *272*, 328
Tajikistan, 54, 328
Tauredunum Event, 71
technical facilitator/integrator (TFI), 177–178
tectonic movement, 477
temporal probability, 282, 286–291
TEPCO, 465, 466
tension cracks, 53, 64, 122, 197, 206, 237, 246
Terzaghi, K, 51, 172
Thistle landslide, Utah, 337
Thredbo ski village, Australia, 327
Tohoku earthquake, 6, 464–466
tolerable risk, 427, 428–429, 495
    societal risk-upper tolerability, *433*
    see also risk acceptability
top event, 4, 30–31, *31*, 190–191, *191*, 192, 210, 281, 360, 378, 449, 492
tourism, 293, 338, *372*, 376, *382*, 392
town planning, 460–461, *462*, 467
TransCanada Keystone Pipeline, 209, 210–211
travel angle, 60, *358*
travel distance, 60–63, 206, 282, 283, 359
    mobility data, *62*
    regression equation, *62*
    relationship between landslide volume and normalised travel distance, *77*
    run-out, *61*
    terminology, *61*
trees, bent, 173
triggering events, estimating consequences, 321, 475
triggering events, estimating probability from, 133, 184–185, 186, 206, 207, 209, 212–213, 243–252, 254, 258, 259–260, 261, 263, *264*, *265*, *266*, 269, 270, 271, 272, 273, 328
tropical conditions, 7, 15, 48, 68, 129, 130, 137, 138, 151, 173, 212, 311, 315, 326, 328, 330, 474
trust, 39, 179, 416, 418, 453
tsunami, 70–71, 254, 255, *257*, *258*, 322, 334, 464–466
turbidity currents, 50, 55, 139–140, 164
    West Nile turbidites recorded, *139*, 222, *223*, *224*, 236
Turkey, 284–285, 323–324, 327, 368
Tversky and Kahneman, 172, 173

Twain, M, 171
two-distribution model, 231, *232*, *233*, 235
two-wedge model, *268*
Typhoon Winnie, 270

Uganda, 327
UK, 3, 9–10, 16, 17, 47, 56, 65, 71, 72, *73*, 74, 101–102, *102*, *103*, 104–106, 121–124, *122*, *123*, *125*, 142, 143, 144, *158*, 166, *167*, 213–215, *214*, 216, 217, 218–221, 231–236, 246–248, 262–267, 267–268, 283, 287–288, 293–294, 310–311, *312*, 326, 327, 328, 334, 336, 338, 349–352, 353–357, 360, 378–381, 385–388, 391, *403*, 405–407, 429, 432, *433*, *441*, 443–445, 460–461, *462*, 466–468, 469–470, 475, *482*, 482–484, *483*
UK legal system, 427–429
Umbria, 226–227, 229, 305–307, 322
UN agencies for disaster relief, 423–424
UN Disaster Risk Index, 315–316
UN International Decade for National Disaster Reduction, 328
UN Office for Disaster Risk Reduction, 11, 298
UNESCO, 338
uncertainty, 3–4, 84–88, 129–130, 143, 194, *195*, 209, 294, 339, 367–369, 375, 421
    estimated upper and lower bands of landslide information anticipated during stages of site investigation, *85*
    future uncertainty, climate model predictions, 472–474
    future uncertainty, global change, 470–471
    future uncertainty, implications for landslide management, 449, 452
    future uncertainty, valuation of environmental resources, 468–470
    illustration of how degree of assurance might change with intensity of investigation, *86*
    relationship between probability and uncertainty, *131*
    risk assessment, 484–485, *484*
    summary of expected defensibility of different methods for undertaking subjective judgements, *87*
    see also confidence
Undercliff, 246–248, 293–294, 460–461
undercutting, 24, 65, 117, 184, 206, 215, 231, 237, 307, 475
upstream flooding, 68
USA, 3, *27*, 58, 63, 70, 71, 74, 100, 137, 157, 163, 166, 174, 176, 178, 209, 210–211, 225–226, *227*, *228*, 249, 250–252, 268–270, 321, 322, 327, 328, 334, 335, 337, 474
USA disaster relief, 423, 424, 452
USGS, 211

Vaiont disaster, 71, 326, 330

Vancouver Island, British Columbia, 80
Vargas State flooding, 157–158, 325
vegetation, 56, 108, 109, *110*
  see also logging
velocity, 52, 54, 58–60, 63, 64, 68, 70, 74, 76, 77, 78, 79, 119, 206, 184, 185, 291, 299, 300, 302, 307, 308, 310, 330, 352, 422
  classes, *59*
  examples, *75*
  see also intensity; magnitude
Venezuela, 157–158, 325, 368
Ventnor, Isle of White, UK, 246–248, 293–294, 460–461
Ventura County, California, 327
vertical displacement damage, 63
volcanoes, 4, 13, 43, 47, 48, 58, 60, 68, 70–71, 149, 211, 212, 245, 308, 321, 322, 325, 328, 477, 481
vulnerability, 8–12, *12*, 25, 282, 297–316, 322, 323, 343–344, 345, 350, *351,* 352, 353, 354, 358–359, 398, 399, 403, *404*, 405, 406, 407, 412, *413*, 422, 424, 433, *444*, 449, 495
  framework for vulnerability assessment, 313
  matrix, *358*
  meaning of, 297–299
  pipeline vulnerability factor, 347
  theoretical relationship between landslide intensity and vulnerability, *304*
  vulnerability of a range of assets to landslide events, *303*
  vulnerability of buildings according to the type of damage, *303*
vulnerability index, 301

Wales, 47, 216, 217, 218–221, 313, 326, 338, 360
Weibull formula, 166, 218
weight-of-evidence modelling method, 83
West Nile Delta, 17, 49, 139–140, *141*, 173, 208, 218, 221–225, 236, 283–284
Whitby coast, UK, 142, 144, *158*, 267–268
willingness-to-pay/willingness-to-accept, 333, 438, 470, 495
World Health Organisation CRED, 325–326
worst-case scenarios, 29, 323, 368

Yungay, Peru, 321, 335

zoning, 80–81, *106*, 304, 422, 459–461, *462*